# Scaffolds for Tissue Engineering

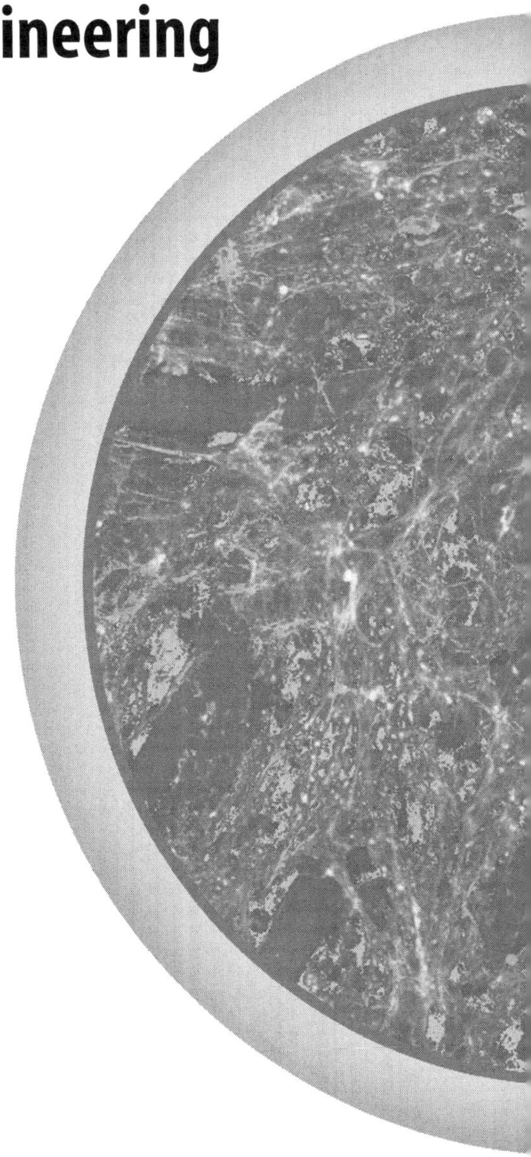

# Scaffolds for Tissue Engineering

## Biological Design, Materials, and Fabrication

edited by

**Claudio Migliaresi**

**Antonella Motta**

PAN STANFORD PUBLISHING

*Published by*

Pan Stanford Publishing Pte. Ltd.
Penthouse Level, Suntec Tower 3
8 Temasek Boulevard
Singapore 038988

Email: editorial@panstanford.com
Web: www.panstanford.com

**British Library Cataloguing-in-Publication Data**
A catalogue record for this book is available from the British Library.

**Scaffolds for Tissue Engineering: Biological Design, Materials, and Fabrication**



ISBN 978-981-4463-20-1 (Hardcover)
ISBN 978-981-4463-21-8 (eBook)

Printed in the USA

# Contents

# Preface

Three years ago, after a conference in Italy, we received an invitation by email from Stanford Chong, Pan Stanford Publishing, to submit a book project. It was to be focused on the functional applications of polymers (the theme of the conference), or another topic of our interest.

After some months of thought, we submitted our idea, which came after long discussions between us, sometimes involving our coworkers, our students, and also scientists who we were visiting then during our travels abroad.

We were well aware of the fact that there were many excellent books on the topic, and that others will have been published before this one. However, we imagined our book as the "negotiated" synthesis between the two different approaches of tissue engineering—that of the biologist (Antonella Motta) and of the engineer (Claudio Migliaresi)—which were much more different at that time, than now.

Our common assumption was that a scaffold should be designed, in terms of materials, architecture, properties, to accomplish not only mechanical but biological requirements, and our book should properly highlight this aspect, in both title and content. This is how the title of the book, *Scaffolds for Tissue Engineering: Biological Design, Materials, and Fabrication*, came into being.

Often scaffolds are made by a random combination of materials, with architecture and properties that evoke unrealistic biological behaviors, ignoring the fact that a scaffold is an *active device* that interacts with cells *in vitro* and with the complex biological system *in vivo*. The chapters of this book, written by leading scientists in the field, many of them also good friends, reflect this idea. And upon reading the book, which we have done several times, we must state that the contributing authors have done a wonderful job.

We are grateful to them for having accepted our invitation and for the valuable time they have spent on the book. We are also grateful to Pan Stanford Publishing, which patiently forgave our delay with respect to the scheduled times, and in particular to Stanford Chong and Shivani Sharma, who were prompt in offering help and assistance.

**Claudio Migliaresi**
**Antonella Motta**
Spring 2014

# Chapter 1

# Overview: An Evolving State of the Art in Tissue Engineering

**Anthony T. DiBenedetto**
*Department of Chemical Engineering, Gant Complex, Unit 3136, University of Connecticut, Storrs, CT 06269, USA*
atdibenedetto@gmail.com

Dr. Anthony Atala of the Wake Forest Institute for Regenerative Medicine in Winston-Salem, North Carolina, encountered a young patient, Luke Masella, with a life-threatening bladder condition that required an organ transplant. Parental permission was obtained to use the patient's own cells in an experimental procedure to create a new bladder. Healthy muscle cells taken from the patient's diseased bladder and urothelial cells from his urinary tract were multiplied by incubation in petri dishes and then applied to a balloon-shaped scaffold made from collagen and other biocompatible organic materials. The cellularized bladder scaffold was incubated *in vitro* at body temperature until the cells generated functioning tissues and then successfully transplanted at Boston Children's Hospital in a procedure that took approximately 14 hours. At this writing, Luke is a healthy, physically active student at the University of Connecticut with

*Scaffolds for Tissue Engineering: Biological Design, Materials, and Fabrication*
Edited by Claudio Migliaresi and Antonella Motta

ISBN 978-981-4463-20-1 (Hardcover), 978-981-4463-21-8 (eBook)
www.panstanford.com

the prospects of living a long and productive life. In 2003, a startup company, Tengion, was founded with an objective of advancing Dr. Atala's basic and clinical research. A successful clinical trial of the use of a tissue-engineered organ in reconstructive surgery on seven patients with end-stage bladder disease was completed after 31 months, during which time the engineered bladders displayed normal behavior.[1]

In 2008, a team of doctors, led by Surgeon Professor Paolo Macchiarini, of the Hospital Clinic of Barcelona, Spain, transplanted a tissue-engineered left bronchus into the windpipe of Claudio Castillo, a 30-year-old mother of two. She needed the transplant to save a lung after her air passage had been damaged by tuberculosis. To make the new bronchus, a donor windpipe was taken from a patient who had recently died. Scientists from the Politecnico di Milano, Italy, the University of Bristol, United Kingdom, and the University of Padua, Italy, assisted in the preparation of a new tissue-engineered bronchus using a portion of the harvested donor windpipe. Strong chemicals and enzymes were used to wash away all of the cells from the donor windpipe, or trachea, leaving only a decellularized tissue scaffold made of the fibrous protein collagen. Cells were taken from Ms. Castillo's bone marrow and windpipe lining to create a tissue scaffold repopulated with Ms. Castillo's own cells. After four days of growth in a rotating bioreactor, the newly formed donor bronchus was transplanted into Ms. Castillo. Four days after transplantation the hybrid windpipe was almost indistinguishable from adjacent normal airways. A biopsy of the site proved that the transplant had developed its own blood supply, with no signs of rejection.

Drs. Harald Ott and Doris Taylor and colleagues at the University of Minnesota harvested donor hearts from adult rats and removed all cells from the hearts by bathing them with a solution of detergent compounds. The decellularized matrix preserved the fibrous morphology and orientation of the myocardial/extracellular matrix (ECM) of the original heart, resulting in an authentic *heart scaffold* with blood vessels, heart valves, and an intact atrial and ventricular geometry that retained the architecture required to pump blood.[2] They seeded these scaffolds with cardiac cells from young healthy rats and maintained them in culture conditions in a bioreactor that simulated cardiac physiology. Four days later, they observed "contractions and, by day eight, their constructs could generate pump function equivalent to about 2% of adult heart function." After a further period of time, both the larger cardiac vessels and the smaller third- and fourth-level branches were capable of supporting the expansions and contractions of a living heart. The techniques were also used to create "breathing rat lungs." The work resulted in a start-up company called Miromatrix Inc. Shortly after that, a team from Massachusetts General Hospital and Harvard University used a mixture of stem cells from rats and humans to recellularize donor rat lungs and transplanted the resulting scaffolds into young rats. The lungs functioned for six hours, albeit imperfectly.[3]

The three above-mentioned accomplishments are examples of a myriad of research and technological advancements that are now occurring in the 21st century. The concept of an interdisciplinary field of "tissue engineering" as a unique field of study was generated during a series of panel meetings and workshops sponsored by the National Science Foundation of the United States.[4] The widespread acceptance of the concept can be traced to Y. C. Fung of the University of California at San Diego, who proposed the idea at a 1988 NSF workshop held at Lake Granlibakken, California, and to Robert Langer and Joseph Vacanti, whose review article in *Science* promoted a rapid increase in collaborative research in the worlds of clinical medicine, developmental biology, biomaterials, biomechanics, and biomedical science and engineering.[5] Studies in developmental biology of cell and tissue coalescence, cell adhesion, and the properties of embryonic tissues provided the biological foundation for modern tissue engineering. The possibility of engineering a tissue was demonstrated as early as 1930s,[6,7] and fundamental studies of stress-dependent morphogenetic effects in tissue remodeling,[8–10] analytical modeling of cell movement, interstitial flow and deformation in an anisotropic ECM,[11] and studies of morphogenetic processes in the growth of vascular networks[12] all contributed to the early development of tissue-engineered scaffolds. Since then, there has been a dramatic increase in the imaging, modeling, and simulation of human anatomy and physiology, promoted by long range strategic plans for basic research in the biological sciences and medicine, for development of computer and engineering technologies related to public health, and by the Human Genome Project.[4,13–15]

*Tissue Engineering* as a field of study was defined as "an interdisciplinary field that applies the principles of engineering and life sciences toward the development of biological substitutes that restore, maintain, or improve tissue function or a whole organ."[5] Research on the development of two-dimensional scaffolds containing cells on sheets of naturally occurring collagen and collagen–glycosaminoglycan composites for the regeneration of new skin was a precursor to the creation of three-dimensional (3D) scaffolds capable of housing the large numbers of cells required for regenerating other organs and functional tissues.[7,16,17] A primary objective in the creation of functional tissue-engineered scaffolds has been to design 3D scaffolds that permit the production of ECM that mimic the ECM of the tissue being replaced or repaired. To accomplish that function, one must provide a mechanically stable environment that hosts the necessary cells, growth factors, and other biological components in a porous structure that allows cell migration, adhesion and proliferation, and vascularization of the growing tissue that mirror that of the original tissue.[18] Critical issues have included enhancing cell survival, maintaining differentiated function, developing a significant cell mass, and achieving vascularization where required. Attempts to engineer functional tissues and organs during the period 1990–1999 included scaffolds to aid nerve regeneration, corneal implants, tissue-engineered liver implants, creation of an artificial pancreas, and the development of a urinary organ using bladder cells encased in a polymer composite matrix. During this period, considerable effort also focused on creating tissue-engineered blood vessels, heart valves, and polymer scaffolds for creating heart muscle tissue.[7]

With regard to the mechanical design of tissues, Roderic Lakes drew attention to the fact that man-made objects, in common with all living objects, including the human body, exhibit structure on more than one length scale and that structural hierarchy in large part determines bulk material properties.[19] One of the earliest examples of the use of hierarchical design principles in architectural engineering has been attributed to Gustave Eiffel's design of the Eiffel tower (exhibited at the International Exhibition of Paris of 1889). At a height of approximately 320 meters and a weight of 7300 metric tons it was the tallest metal structure of its time. It was designed as an open lattice, third-order hierarchical cantilevered structure in which the primary structural shape was strengthened by a lattice network of second-order and third-order girders. His hierarchical design strategy started a revolution in civil, architectural, and mechanical engineering design of skyscrapers, bridges, transportation vehicles, and other loadbearing commercial products. In many respects, the use of computer-aided design and manufacturing (CAD/CAM) techniques that are employed for these purposes have been found to be applicable to the design of tissue-engineered scaffolding. Engineers and material scientists initially explored the application of solid mechanics, fluid mechanics, and CAD/CAM, using "top-down" reverse engineering strategies in the design of orthopedic and orthotic devices manufactured from biocompatible metal alloys, ceramics, and fiber reinforced plastics. Total hip replacements consisting of a femoral ball that fits into the socket of the hip bone and a shaft tapered to fit into a reamed medullary canal of the femur,[20] prosthetic limbs consisting of a socket, shank, and foot for amputees,[21] composite disc prostheses,[22] and tailor-made, and computer-designed cranial implants used to minimize risks of brain surgery[23] are examples of the many biomedical applications described in the literature. In these cases, the primary biological criterion for the materials employed was that they were "biocompatible" in the sense that there was minimal risk of inflammation of surrounding tissues.

The primary objective of creating *biologically* functional tissue-engineered scaffolding required the melding of the mechanical design and manufacturing strategies used in orthopedic and orthotic applications with the biological processes involved in tissue formation. In a series of journal articles, a CAD strategy was proposed that integrates advances in biology, biomedical engineering, information technology, and modern design and manufacturing processes in the design of scaffolds for tissue engineering application.[24–27] The three components of the design strategy, labeled *CATE*, were characterized as "computer-aided tissue and bio-modeling, scaffold informatics and biomimetic design, and biomanufacturing for tissue and organ regeneration."

The first step in the *CATE* process is the use of 2D and 3D voxel-based medical imaging of the tissue or organ to characterize its structure. The voxel-based 3D images are then converted to a vector-based modeling environment needed for a CAD process. A library of CAD-based unit cells are created and used to construct a scaffold that mimics the shape, fibrous morphology, vascular channels, and pore size distributions within tissues and organs. Computer-based modeling is then employed to define an optimized CAD for a specific scaffold, which can then be incorporated into a CAD/CAM process for manufacture of the scaffold.

An *in silico* computational model for tissue-engineered scaffolds that uses finite element analysis (FEA) to incorporate solid and fluid mechanical principles with stochastic network modeling at hierarchical length-scales of tissue and cell and molecular dimensions was developed by Knothe Tate and colleagues.[28–30] The human body was depicted as an ecosystem functioning at multiple hierarchical levels of length. The stresses and fluid flows within the poroelastic structure of the tissue at scale lengths of [$m^{-1}$, $m^{-2}$] were characterized using classical solid and fluid mechanics. Stochastic network modeling was used within the pericellular networks and microporous matrices to characterize the movement of cells and molecules at the tissue-to-cell and cell-to-molecular levels [$m^{-6}$ to $m^{-7}$]. The mechanical and biological requirements were met by designing a porous scaffold with the required structural support and channels of interconnected pores through which cells, signals, and nutrients could be transported. Several versions of stereolithographic rapid prototyped scaffolds for bone replacement were manufactured to test the effectiveness of the design strategy in predicting the permeability of the manufactured scaffolds. Experimental measurements of the permeability of the scaffolds were comparable to values calculated using the *in silico* models, including several of the manufactured scaffolds that were predicted to be impermeable, thereby validating the design strategy as an economic way of determining appropriate designs for *in vitro* and *in vivo* studies. The stochastic modeling also ascertained the most effective morphology for promoting cell mobility and cell adhesion. The limitations of the CATE and similar approaches are associated with the difficulty of producing, from a CAD/CAM designed biodegradable scaffold, a true *biomimetic* reproduction of the complex fractal character of a naturally created ECM. The CAD/CAM-based unit cells used to build the mechanical structure of commercial products are limited by the Euclidean geometric shapes chosen and must be utilized in a manner that more adequately represents the topographical and fractal character of natural materials. Two of the more common CAM techniques utilized in manufacturing processes are rapid prototyping (RP) and electrospinning (ELSP). RP is a generic classification of processes used to fabricate a scaffold by building a sequence of thin layers, one at a time, to form a complete 3D structure directly from CAD/CAM sources. Cells and other biological substances that promote tissue in-growth can be incorporated into the prefabricated layers during the assembly process. ELSP is an alternate route to mimicking the fractal character of natural tissues by spinning fibers from a biomaterial solution using a high voltage electrostatic field to form multiple layers of fibrous 3D nonwoven meshes, the orientations of which are controlled by controlling the path of the delivery nozzle or the direction of the electrical field. Meshes with fiber diameters ranging from micro- to nanometers can be produced, comparable to the dimensions of those in ECM of load-bearing tissues. Comprehensive presentations of the state of the art of these technologies are offered in ***III: Fabrication,*** **11. Fabrication Methods of Tissue Engineering Scaffolds** (by V. M. Correlo, J. M. Oliveira, A. R. C. Duarte, A. Martins, A. L. Oliveira, J. Silva-Correia, R. L. Reis, and N. M. Neves); **12. Engineered Scaffolds: Materials and Microstructure from Nanostructures to Macrostructures for Tissue Engineering** (by V. G. Varanasi, P. S. Shiakolas, and P. B. Aswath); **13. Organ Printing and Cell Encapsulation** (by S. Tasoglu, U. A. Gurkan, S. Guven, and U. Demirci).

The creation of accurate 2D and 3D visual representations of tissue and organ structures for use in RP and other manufacturing techniques are generated by medical scanning devices such as magnetic resonance imaging (MRI) and computed tomography (CT). A combination of CT with positron emission tomography (PET), CT/PET, provides a more refined image that has the potential of measuring tissue functions such as blood flow.[31] Electron beam CT (EBCT) scanners produce images in fractions of a second, even permitting the scanning of a beating heart,[32] and diffraction-enhanced imaging (DEI) delivers more detailed images of soft tissue than is possible with other CT techniques.[33] Imaging techniques are also employed in examination of diseased tissues and organs. Fluorescein angiography, for example, is used for imaging vascular patterns in the human retina for diagnosis of diabetic retinopathy and retino-vascular occlusive disease. Image resolution of the vascular patterns of vessels of various dimensions present a fractal pattern that can be quantified and used to diagnose early-stage vascular disease in the human retina.[34] Development of Fourier domain optical coherence tomography (FD-OCT) has been proposed as an improvement over fluorescein angiography due to improved sensitivity and acquisition time.[35] Synchrotron radiation micro-CT has also been used for both morphological characterization and quantification of cell distributions down to the micrometer level of tissues, thereby extending the range of imaging analysis from the whole tissue down into the cellular range.[36] A comprehensive presentation of microvascular imaging methods for tissue engineering is given in ***IV: Validation and Translation to the Application,* 16. Microvascular Imaging Methods for Tissue Engineering** (by B. W. Thimm, S. Hofmann, and R. Müller).

*In vitro* and *in vivo* testing of the scaffold biomaterials is an indispensable step before acceptance by the medical profession and accrediting agencies of tissue-engineered scaffolding. Dr. Milena Fini and colleagues have pointed out the importance of *in vitro* and *in vivo* testing in determining the utility of biological therapies and biomaterials used for bone remodeling, humoral networks, and smart biomaterial technologies for osteoporosis mitigation.[37] A presentation of the state of the art of *in vitro* and *in vivo* testing is offered in ***IV: Validation and Translation to the Application,* 14. *In vitro* and *in vivo* Testing** (by M. Fini, N. Nicoli Aldini, and P. Torricelli), while bioreactor strategies to develop neo-tissues *in vitro* are presented in ***IV: Validation and Translation to the Application,* 15. Bioreactor Strategies for *in vitro* Conditioning of Engineered Tissues** (by A. S. Goldstein and P. Thayer).

To mimic, more precisely, a true biomimetic reproduction of a naturally created ECM requires one to overcome the difference between an engineering approach and "nature's way":

> While the engineer selects a material to fabricate a part to an exact design, nature goes the opposite way and grows both the material and the whole organism using the principles of self-assembly. This provides control of the material at all levels of hierarchy and is certainly a key to the successful use of polymers and composites as structural materials.[38]

The transplant of a tissue-engineered bladder by Dr. Anthony Atala, mentioned at the beginning of this overview, was accomplished by utilizing the processes of *directed tissue assembly* that occur naturally in an ECM. Prior to the 1990s, the standard procedure for treating patients with neurogenic bladder disease was to remove the bladder and replace it with a section of the patient's bowel, a dangerous surgery with serious limitations. Dr. Atala developed a technique of taking healthy cells from the patients' diseased bladder, multiplying them *in vitro*, and then applying them to a balloon-shaped scaffold, made partly of collagen, with muscle cells on the outside and urothelial cells on the inside to line the urinary tract.

The bladder scaffold was then incubated until a functioning tissue was produced. The manufacture of a tissue-engineered scaffold for the bronchus connecting Claudio Castillo's windpipe to her left lung was accomplished by using a decellularized donor windpipe that was recellularized with cells taken from Ms. Castillo. The creation of a beating heart by Ott, Taylor, and colleagues and the experimental transplantation of rat lungs by the team from Massachusetts General Hospital and Harvard University, also mentioned at the beginning of this overview, were accomplished by using similar decellularization/recellularization processes. Each of these approaches circumvented the problems of a top-down engineering design by inventing technologies that produce a 3D "scaffold" that more closely retains the original architecture of the nature-created ECM of the host tissue. Although the approach is a substantial advance in the field of tissue engineering, it will likely be limited by the availability of suitable donor tissues and complete donor organs. Thus, the search for natural and synthetic biomaterials that meet both the architectural and biological criteria for a functional tissue-engineered biomimetic scaffold has become a major objective of Tissue Engineering. A part of that objective has been the search for new molecular-modified biopolymers that elicit cellular responses at the molecular level by stimulating specific interactions with cell integrins that promote cell proliferation, differentiation, and the production of the host ECM. A thorough review of the materials used for scaffolds is offered in ***II: Materials for Scaffolds,*** **8. Biomaterials for Scaffolds: Synthetic Polymers** (by L. Rojo, B. Vazquez, and J. San Roman); **9. Biomaterials for Scaffolds: Natural Polymers** (by L. S. Wray and D. L. Kaplan); and **10. Hydrogels: Characteristics and Properties** (by R. Barbucci and D. Pasqui).

Recent efforts to generate biocompatible biomimetic scaffolds have been to modify the surfaces in order to promote the appropriate biomechanical and chemical processes that support the tissue-specific cell functions that occur in the host material.[39] Extensive research has focused on the structure and biological influence of the ECM on cell behavior. It has become clear that new technologies should focus on constructing well-defined microenvironments for cells at the nanoscale that match those of the host ECM. A tissue-engineered scaffold is most effective when all chemical and biomechanical processes progress by a *bottom-up* assembly of cell clusters matching that occurring in the ECM of the host tissue. Since their compatibility depends upon the similarity of their morphology and composition, and on the cell adhesion, migration, growth, and differentiation processes encountered in each, the importance of understanding the central role natural ECMs play in the repair and regeneration of tissues and organs is critical. A natural, self-assembled ECM is formed by the secreted products of the cells that reside within tissues and consists of a mixture of structural proteins, hyaluronic acid, proteoglycans, growth factors, enzymes, and other components that are organized in a nature-determined morphology that is specific to a particular tissue. The role of the ECM and the biological processes that occur during the maintenance and repair of living tissues and organs are described in ***I: Biological Design,*** **2. The Functional Role of Extracellular Matrix** (by E. Carletti, M. Stoppato, C. Migliaresi, and A. Motta).

Of primary importance in the choice of biomaterial for a scaffold is the response of the host tissue to the biomaterial implant. When it is implanted in the host tissue, it immediately adsorbs blood and other bodily fluids on its internal surfaces. Macrophages and dendritic cells of the host tissue interact with the protein-coated surfaces of the biomaterial and regulate the inflammatory or tolerogenic response to the scaffold by sensing its presence and providing the intracellular signals that determine acceptance or rejection of the foreign body.[40] The present status of research on the biological processes of dendritic cells and macrophages and their potential for optimizing the host responses to the presence of biomaterial scaffolds is described in ***I: Biological Design,*** **5. Host Response to Biomaterials** (by S. Srinivasan and J. E. Babensee).

In order to generate intracellular signals that indicate biocompatibility and an acceptably minimal inflammatory response to a biomaterial scaffold, the internal surfaces of a porous scaffold must be similar to that of the topography, composition, and molecular character of the ECM of the host tissue. Advances in an understanding of the mechanisms involved in the host response have led to techniques for modification of the surfaces of scaffold biomaterials using natural polymers such as heparin, hyaluronan, and hydrophilic polysaccharides to enhance their intrinsic bioactivity and specific interaction with cell surface receptors.[41] In addition to molecular modification of the internal surfaces of a biomaterial, biocompatibility with the host also requires a similarity in the anisotropy and fractal character of the surfaces. These processes are guided by the molecular composition and nanoscale topography of the scaffold surfaces. ELSP, for example, has been employed to mimic the circumferential and longitudinal mechanical properties of natural vessels to produce tubular, multilayer nanofibrous composites suitable as engineered vascular constructs.[42] The rebuilding of tissues using a biomaterial scaffold requires the creation of an endothelium as a single layer of endothelial cells on the surfaces of the construct. The surface chemistry of the construct and the hemodynamic blood flow through the vascular channels determine the functionality of the resulting vasculature. There has been considerable research on optimizing the use of endothelialized biomaterials for creation of 3D vascularized constructs for tissue engineering applications and regenerative medicine, but many factors causing graft failure related to cell adhesion, matrix interactions that alter the phenotype of the endothelial cells and difficulty of predicting the responses to hemodynamic forces, and other factors have left many issues still to be dealt with.[43] The deployment of biomaterial scaffolds for tissue vascularization is discussed in ***I: Biological Design,* 6. Scaffold Vascularization** (by L. E. Fitzpatrick, A. Lisovsky, E. C. Ciucurel, and M. V. Sefton).

Hydrogels are attractive for tissue-engineered scaffolding since they are biocompatible, hydrophilic and highly permeable, and their mechanical properties are easily manipulated. Thixotropic hydrogels are of special interest since they can be injected directly to a host site as a fluid, return to their original gel-like state *in situ*, and, when crosslinkable, become suitable as scaffolds for cell growth while allowing exchange of oxygen, nutrients, and soluble metabolites. The development of injectable hydrogel scaffolds containing natural polymers, soluble synthetic polymers and interpenetrating polymer networks (IPNs), and their adaptability for *in vivo* tissue formation have been widely employed.[44–46] Aggregates of cells can be readily blended by *direct assembly* into the gel and then reorganized into larger complex patterns. Recognizing the importance of surface topography in promoting a biomimetic self-assembly of preformed cell clusters, a 3D tissue-engineered hydrogel-based scaffold was devised to direct the assembly process by partially restricting reorganization of the cell clusters on the free surface of a disposable form, acting as a template.[47] The ability to also tailor the microenvironment within the gel leads to the possibility of incorporating small-scale chemical and physical gradients that act as signals to cells during morphogenesis and homeostasis.[48] A review of the techniques used for the fabrication of 3D functional tissue-engineered scaffolds and their potential use as biomimetic scaffolds for tissue and organ regeneration is presented in ***I: Biological Design,*** **7. Engineering Biomimetic Scaffolds** (by N. Annabi, N. E. Vrana, P. Zorlutuna, F. Dehghani, and A. Khademhosseini).

A primary characteristic of tissues and organs is their ability to adapt to changes in external stimuli. In choosing a biomimetic material for tissue engineering applications, researchers and practitioners are looking for biologically compatible materials that also can adapt positively to external stimuli by undergoing reversible physical or chemical changes at the host site. This has led to the search for *in situ* and "smart" materials that possess reversible phase transitions at body temperature and have a molecular structure with the necessary proportions of hydrophobicity and hydrophilicity that meet the requirements of a biomimetic response and can be introduced *in vivo* at the host site.[49–50]

On behalf of the National Academy for Engineering, a panel entitled Grand Challenges in Biomaterials was convened at the 2010 Society for Biomaterials Annual Meeting in Seattle, Washington, USA, to consider challenges that, if successfully met, would significantly speed up the transition from commodity off-the-shelf biomaterials to bioengineered, biomaterial devices that both address patient needs and create new market opportunities. The six members of the panel offered six "challenges" that must be overcome to further advance these goals. The six challenges called for (1) "redefining biocompatibility to address new materials that tap into the body's own reconstructive mechanisms and show improved healing and integration;" (2) finding an "appropriate rationale and use of biological experimental design and methodology in biocompatibility studies;" (3) a focus on *in vitro* applications and called for "smart new applications for smart polymers;" (4) a "grand challenge for moving medicine forward is full limb regeneration;" (5) a need to "harness the principles of developmental biology to control collective cell movement and differentiation *in vivo* and *in vitro*;" and (6) an all-encompassing challenge for "the necessity to continue to improve conventional "replacement" medicine by making better use of existing materials, developing better materials, producing better device designs, and improving surgical techniques" and, furthermore, suggested "the need to develop the materials and technologies that optimize the inevitable transition from replacement medicine to regenerative medicine and tissue engineering."[51]

Much of the research, development, and transition to practice in the field of Scaffolds for Tissue Engineering during the 21st century have in fact been directed toward meeting the challenges exhorted by members of the above-mentioned panel. The efficacy of "top-down" designed functional tissue-engineered scaffolds will continue to advance by the increasing sophistication of engineering design principles, advances in manufacturing techniques, and the discovery of new "smart" biomaterials. The ability to build a biomimetic tissue-engineered scaffold will continue to advance using the principles of "bottom-up" self-assembly of the molecules and cells required by the host tissue or organ. Self-assembly at all dimensional scales has been substantially advanced by recent clarification of the intricate schedules of gene expression and proteome activity in cells and developing organisms.[52–54] The same strategies used in the Human Genome project are now being applied to analyze the behavior of proteins and cells in tissues and organs. The spread of computer software designed for the analysis and importation of the data into *in silico* models of cell biochemistry has also enabled the study of numerous biological pathways for creating biomimetic scaffolding. One of the many techniques of building ordered assemblies of nanoparticles has been the directed self-assembly of nanoparticles carrying complementary strands of DNA molecules with flexible linkers of varying size. Tailoring the length of the linkages makes possible the self-assembly of nearly any material at a nanoscale level, and prescription of a specific thermal pathway makes possible the formation of a variety of ordered assemblies, thus opening up potential new avenues for studying biological interactions in a variety of 3D hybrid assemblies ranging from amorphous to highly ordered morphologies.[55,56] The self-assembly of polymers, nanoparticles, and DNA can also be attained by controlling the evaporation of sessile droplets of a solute on a solid substrate to construct self-organized, hierarchically structured patterns at nanoscale levels.[57]

The biomimetic restoration of damaged tissues requires the delivery of cells and biological factors, *in vivo*, to the scaffold surfaces to which they must adhere, organize, and develop in the same way as in the native tissue. Biosignals are incorporated to control the tissue-specific cell responses. One area of recent research has been the incorporation of natural ECM-derived peptides and proteins to mimic natural cell-specific responses. Peptide and protein nanoscale patterning techniques are being studied to advance the technology of biomaterial scaffolding.[58] The incorporation of short synthetic peptide sequences, either by surface modification or by bulk incorporation, to elicit a biomimetic response, and the effects of peptide immobilization on subsequent cell responses are also being pursued.[59] The stabilization, *in vivo*, of relatively unstable fibroblast growth factors, (FGFs), by adsorbing or encapsulating them in a scaffold, is being investigated as a method of securing an acceptable level of biological activity in a targeted delivery system for biomimetic tissue regeneration.[60] Tissue remodeling also requires cell-specific biological pathways that promote the appropriate metabolic pathways for the regeneration process. A morphogenic group of secreted lipid-modified glycoproteins (Wnts) that contribute to metabolic adaptation and regulate tissue maintenance and remodeling, for example, is being studied to map the linkage between Wnt signaling pathways and cellular metabolism.[61] An implanted biocompatible scaffold often incurs a "tolerable," but necessary irritation at the host site. Methods of controlling the extent of irritation by using naturally occurring peptides specifically at the target site have been shown to be effective and are critical in the acceptance of a tissue-engineered scaffold by medical regulatory agencies.[62] A general introduction on aspects of molecular biorecognition will be discussed in ***I: Biological Design,*** **3. The Premise of Tissue Engineering: Molecular Recognition** (by C. J. Pateman and J. W. Haycock). While the use of ECM macromolecules from natural sources in a tissue-engineered scaffold is often advantageous for biological recognition by a host tissue, the technique is limited by variations in materials isolated from mammalian sources. The use of artificial ECM proteins to modulate cellular behavior has been advanced as a means of avoiding such problems. Synthetic biomaterials are being designed that reproducibly mimic the functions of a natural extracellular microenvironment.[63] Biological pathways to tissue regeneration will be discussed in ***I: Biological Design,*** **4. Principles and Biological Pathways To Tissue Regeneration: The Tissue Regenerative Niche** (by R. Cancedda and C. Lo Sicco). The advance of potential medical products from the laboratory to the market place is presented in ***IV: Validation and Translation to the Application,*** **17. Tissue Engineering: Scope, Products, Commercialization Strategies** (by A. J. Coury).

The three accomplishments mentioned at the beginning of this chapter are examples of the judicious use of fundamental research in developmental biology and bioengineering in the translation from the laboratory to medical techniques that will improve the quality of life and extend life expectancy of human beings. The inevitable advancement of these technologies will also create new social challenges for society and governments that will require a broadening of cooperation and participation beyond the bounds of science and engineering. This will be a critical challenge for humanity.

## References


1. Atala, A., Bauer, S., Soker, S., Yoo, J., and Retik, A. (2006). Tissue engineered autologous bladders for patients needing cystoplasty. *Lancet*, **367**, 1241–1246.
2. Ott, H. C., Matthiesen, T. S., Goh, S-K, Lauren D Black, L. D., Kren, S. M., Netoff, T. I., and Taylor, D. A. (2008) Perfusion-decellularized matrix: using nature's platform to engineer a bioartificial heart. *Nature Med.*, **14**, 213–221.
3. Ott, H. C., Conrad, C., Schuetz, C., Pomerantseva, I., Ikonomou, L., Kotton, D., and Vacanti, J. P. (2010) Regeneration and orthotopic transplantation of a bioartificial lung. *Nature Med.*, **16**, 927–933. doi:10.1038/nm.2193.
4. Viola, J., Lal, B., and Grad, O. (2003) *The Emergence of Tissue Engineering as a Research Field*, NSF, Abt Associates, Contract # EEC-9815425, Task Order A-07.
5. Langer, R., and Vacanti, J. P. (1993) Tissue engineering. *Science*, **260**(5110), 920–926.
6. Bisceglie V. (1993) Uber die antineoplastische immunitat; heterologe Einpflanzung von Tumoren in Huhner-embryonen. *Ztschr. Krebsforsch.*, **40**, 122–140.
7. Langer, R. (2000) Tissue engineering, molecular therapy. *Am. Soc. Gene Ther.*, **1**(1), 12–15. doi:10.1006/mthe.1999.0003.
8. Traqui, P., Woodward, D. E., Cruywagen, G. C., and Murray, J. D. (1993) A mechanical model for fibro-blast driven wound healing. *J. Biol. Syst.*, **3**, 1095.
9. Rodriguez, E. K., Hoger, A., and McCulloch, A. D. (1994) Stress-dependent finite growth in soft tissues. *J. Biomech.*, **27**, 455.
10. Taber, L. A. (1995) Biomechanics of growth, remodeling, and morphogenesis. *Appl. Mech. Rev.*, **48**, 487.

11. Barocas, V. H., and Tranquillo, R. T. (1997) An isotropic biophasic theory of tissue-equivalent mechanics: The interplay among cell traction, fibrillar network deformation, fibril alignment, and cell contact guidance. *J. Biomech. Eng.*, **119**, 137.
12. Kiani, M. F., and Hudetz, A. G. (1998) Computer simulation of growth of anastomosing microvascular networks. *J. Theor. Biol.*, **192**, 117.
13. *NIH Roadmap for Medical Research,* October 29, 2008, US Dept. of Health and Human Services. Available at: http://nihroadmap.nih.gov/.
14. DG INFSO & DG JRC (Eds.) (2005) *Towards Virtual Physiological Human: Multilevel Modelling and Simulation of the Human Anatomy and Physiology-White Paper,* Available at: http://ec.europa.eu/information_society/activities/health/docs/events/barcelona2005/ec-vph-white-paper2005nov.pdf
15. Opinion: "The Human Genome" (2001) *Nature,* **409**, 745.
16. Bell, E., Ehrlich, P., Buttle, D. J., and Nakatsuji, T. (1981) Living tissue formed *in vitro* and accepted as skin-equivalent of full thickness. *Science,* **221**, 1052–1054.
17. Burke, J. F., Yannas, I. V., Quimby, W. C., Bondoc, C. C., and Jung, W. K. (1981). Successful use of a physiologically acceptable artificial skin in the treatment of extensive burn injury. *Ann. Surg.*, **194**, 413–448.
18. Lanza, R. P., Langer, R., and Vacanti, J. (Eds.) (2000) *Principles of Tissue Engineering*, Second Edition. San Diego, USA: Academic Press.
19. Lakes, R. (1993) Materials with structural hierarchy. *Nature,* **361**, 511–515. doi:10.1038/361511a0.
20. De Santis, R., Gloria, A., and Ambrosio, L. (2010) Composite Materials for hip joint prostheses, in *Biomedical Composites*, L. Ambrosio (Ed.) Cambridge: CRC Press, Woodhead Publishing, 276–295.
21. Coombs, A. G. A., Greenwood, C. D., and Shorter, J. J. (1996) Plastic materials for external prostheses and orthoses, in *Human Biomaterial Applications*, D. L. Wise, D. J. Trantolo, D. E., Altobelli, M. J. Yaszemski, J. D. Gresser (Eds.), Totowa, NJ: Humana Press Inc., 215–255.
22. Gloria, A., De Santis, R., and Ambrosio, L. (2010) Composite Materials for spinal implants, in *Biomedical Composites*, L. Ambrosio (Ed.). Cambridge: CRC Press, Woodhead Publishing, 178–200.
23. Chacon-Moya, E., Gallegos-Hernández, J. F., Piña-Cabrales, S., Cohn-Zurita, F., and Goné-Fernández, A., (2009) Cranial vault reconstruction using computer-designed polyetheretherketone (PEEK) implant: Case report, *Cir. Ciruj.*, **77**, 437–440.

24. Sun, W., Darling, A., Starly, B., and Nam, J. A. (2004) Computer-aided tissue engineering: Overview, scope and challenges. *J. Biotecnol. Appl. Biochem.*, **39**(1), 29–47.

25. Sun, W., and Lal, P. (2002) Recent development on computer-aided tissue engineering: A review. *Computer Met. Progr. Biomed.*, **67**, 85–103.

26. Sun, W., Starly, B., Darling, A., and Gomez, C. (2004) Computer-aided tissue engineering: Application to biomimetic modeling and design of tissue scaffolds. *J. Biotechnol. Appl. Biochem.*, **39**(1), 49–58.

27. Sun, W., Starly, B., Nam, J., and Darling, A. (2005) Bio-CAD modeling and its applications in computer-aided tissue engineering. *Computer-Aided Design*, **37**, 1097–1114.

28. Knothe Tate, M. L. (2007) Multiscale computational engineering of bones: State-of-the-art insights for the future, in *Engineering of Functional Skeletal Tissues*, F. Bronner, M. C. Farach-Carson, and A. G. Mikos, A.G. (Eds.). London: Springer-Verlag.

29. Anderson, E. J., Savrin, J., Cooke, M., Dean, D., and Knothe Tate, M. L. (2005) Evaluation and optimization of tissue engineering scaffolds using computational fluid dynamics. *Annual Meeting of the Biomedical Engineering Society*, Baltimore, Maryland.

30. Sidler, H., Steck, R., and Knothe Tate, M. L. (2006) Site-specific porosity and its impact on load-induced fluid movement in cortical bone. *52nd Annual Meeting of the Orthopaedic Research Society*, Chicago.

31. Di Carli, M. F., Dorbala, S., Meserv, J., El Fakhri, G., Sitek, A., Moore, S. C. (2007) Clinical myocardial perfusion PET/CT. *The Journal of Nuclear Medicine*, **48**(5), 783–793.

32. Tallisetti, A., Jelnin, V., Ruiz, C., John, E., Benedetti, E., Testa, G., Holterman, A., and Holterman, M. (2004) Electron beam CT scan is valuable and safe imaging tool for the pediatric surgical patient. *J. Pediatric Surg.*, **39**(12), 1859–1862.

33. Zhong, Z. (2004) New medical imaging technique first to use low dose X-rays to reveal soft tissue. *Discover Brookhaven*, **2**(1), 1–3.

34. Avakian, A., Kalina, R. E., Helene Sage, E., Rambhia, A. H., Elliott, K. E., Chuang, E. L., Clark, J. I., Hwang, J.-N, and Parsons-Wingerter, P. (2002) Fractal analysis of region-based vascular change in the normal and non-proliferative diabetic retina. *Curr. Eye Res.*, **24**(4), 274–280.

35. Fingler, J., Zawadzki, R. J., Werner, J. S., Schwartz, D., and Fraser, S. E. (2009) Volumetric microvascular imaging of human retina using optical coherence tomography with a novel motion contrast technique. *Opt. Express.*, **17**(24), 22190–22200.

36. Thurner, P., Muller, R., Raeber, G., Sennhauser, U., and Hubbell, J. A. (2005) 3D Morphology of cell cultures: A quantitative approach using micrometer synchrotron light tomography. *Microsc. Res. Tech.*, **66**, 289–298.

37. Fini, M., Carpi, A., Borsari, V., Tschon, M., Nicolini Aldini, N., Sartori, M., Mechanick, J., and Giardino, R. (2010) Bone remodeling humoral networks, and smart biomaterial technology for osteoporosis. *Front. Biosci.*, **S2**, 468–482.

38. Fratzl, P. (2007) Biomimetic materials research: What can we really learn from nature's structural materials? *J.R. Soc. Interface*, **4**, 637–642.

39. von der Mark, K., Park, J., Bauer, S., and Schmuki, P. (2010) Nanoscale engineering of biomimetic surfaces: Cues from the extracellular matrix. *Cell Tissue Res.*, **339**, 131–153. doi: 10.1007/s00441-009-0896-5.

40. Kou, P. M., and Babensee, J. E. (2011) Macrophage and dendritic cell phenotypic diversity in the context of biomaterials. *J. Biomed. Mater. Res. Part A*, **96A**, 239–260.

41. Morra, M., Cassinelli, C., Cascardo, G., Nagel, M.-D., Della Volpe, C., Siboni, S., Maniglio, D., Brugnara, M., Schol, H. A., and Ulvskov, P. (2004) Effects on interfacial properties and cell adhesion of surface modification by pectic hairy regions. *Biomacromolecules*, **5**, 2094–2104.

42. Bonani, W., Maniglio, D., Motta, A., Tan, W., and Migliaresi, C. (2010) Biohybrid nanofiber constructs with anisotropic biomechanical properties. *J. Biomed. Mater. Res. Part B: Appl. Biomater.*, **96B**, 276–286.

43. Khan, O. F., and Sefton, M. V. (2011) Endothelialized biomaterials for tissue engineering applications *in vivo*. *Trends Biotechnol.*, **29**(8), 379–387.

44. Kachouie, N. N., Du, Y., Bae, H., Khabiry, M., Ahari, A. F., Zamanian, B., Fukuda, J., and Khademhosseini, A. (2010) Directed assembly of cell-laden hydrogels for engineering functional tissues, *Organogenesis*, **6**(4), 234–244.

45. Barbucci, R., Giardino, R., De Cagna, M., Golini, L., and Pasqui, D. (2010) Inter-penetrating hydrogels (IPHs) as a new class of injectable polysaccharidehydrogels with thixotropic nature and interesting mechanical and biological properties. *Soft Matter*, **6**, 3524–3532.

46. Martins, A. M., Alves, C. M., Kasper, F. K., Mikos A. G., and Reis, R. L. (2010) Responsive and *in situ*-forming chitosan scaffolds for bone tissue engineering applications: An overview of the last decade. *J. Mater. Chem.*, **20**, 1638–1645.

47. Fernandez, J. G., and Khademhosseini, A. (2010) Micro-masonry: Construction of 3D structures by microscale self-assembly. *Adv. Mater.*, **22**, 2538–2541.

48. Sant, S., Hancock, M. J., Donnelly, J. P., Iyer, D., and Khademhosseini, A. (2010) Biomimetic gradient hydrogels for tissue engineering. *Canadian J. Chem. Eng.*, **88**, 899–911.

49. Kumar, A., Srivastava, A., Yu Galaev, I., and Mattiasson, B. (2007) Smart polymers: Physical forms and bioengineering applications. *Prog. Polym. Sci.*, **32**, 1205–1237.

50. Aguilar, M. R., Elvira, C., Gallardo, A., Vázquez, B., and Román, J. S. (2007) Smart Polymers and Their Applications as Biomaterials, Chapter 6 in, *Topics in Tissue Engineering, Volume 3*, Ashammakhi N., Reis R., and Chiellini E. (Eds.) Available at: www.oulu.fi/spareparts/ebook_topics_in_t_e_vol3/list_of_contr.html

51. Reichert, W., Ratner, B. D., Anderson, J., Coury, A., Hoffman, A. S., Laurencin, C. T., and Tirrell, D. (2011) Society for biomaterials 2010 panel on the Biomaterials Grand Challenges. *J. Biomed Mater Res Part A*, **96**(2), 275–287. doi: 10.1002/jbm.a.32969.

52. Whitesides, G. M., and Bonchiva, M. (2002) Beyond molecules: Self-assembly of mesoscopic and macroscopic components.*Proc. Natl. Acad. Sci. USA*, **99**(8), 4769–4774. doi:10.1073/pnas.082065899.

53. Whitesides, G. M., and Grzybowski, B. (2002) Self-assembly at all scales. *Science*, **295**(5564), 2418–2421.

54. Jakab, K., Neagu, A., Mironov, V., Markwald, R. R., and Forgacs, G. (2004) Engineering biological structures of prescribed shape using self-assembling multicellular systems. *Proc. Natl. Acad. Sci. USA*, **101**(9), 2864–2869. doi:10.1073/pnas.0400164101.

55. Crocker, J. C. (2008) Golden handshake. *Nature*, **451**(7178), 528–529. doi: 10.1038/451528a.

56. Nykypanchuk, D., Maye, M. M., van der Lelie, D., and Gang, O. (2008) DNA-guided crystallization of colloidal nanoparticles. *Nature*, **451**, 549–552. doi: 10.1038/nature06560.

57. Lin, Z. (2011) *Evaporative self-assembly of ordered complex structures*. New Jersey: World Scientific Publishing Co.

58. Shekaran, A., and Garcia, A. J. (2011) Nanoscale engineering of extracellular matrix-mimetic bioadhesive surfaces and implants for tissue engineering, *Biochim. Biophys. Acta*, **1810**, 350–360.

59. Sreejalekshmi, K. G., and Nair, P. D. (2011) Biomimeticity in tissue engineering scaffolds through synthetic peptide modifications—

Altering chemistry for enhanced biological response. *J. Biomed. Mater. Res. Part A*, **96A**, 477–491. doi: 10.1002/jbm.a.32980.

60. Ye-Rang, Y., Jong Eun, W., Eunyi, J., Sujin, L., Wonmo, K., Hyejin, J., Jun-Hyeog, J., Ueon, S. S., and Hae-Won, K. (2010) Fibroblast growth factors: Biology, function, and application for tissue regeneration. *J. Tissue Eng.*, 1–18. doi:10.4061/2010/218142.
61. Sethi, J. K., Vidal-Puig, A. (2010) Wnt signalling and the control of cellular metabolism. *Biochem. J.*, **427**, 1–17. doi:10.1042/BJ20091866.
62. Charnley, M., Fairfull-Smith, K., Haldar, S., Elliott, R., McArthur, S. L., Williams, N. H., and Haycock, J. W. (2009) Generation of bioactive materials with rapid self-assembling resorcinarene-peptides. *Adv. Mater.*, **21**, 2909–2915. doi: 10.1002/adma.200802731.
63. Gentili, C., and Cancedda, R. (2009) Cartilage and bone extracellular matrix. *Current Pharm. Design*, **15**, 1334–1348.

# Chapter 2

# The Functional Role of Extracellular Matrix

**Eleonora Carletti,**[a, b, c] **Matteo Stoppato,**[a, b, c] **Claudio Migliaresi,**[a, b, c] **and Antonella Motta**[a, b, c]

[a]*Department of Industrial Engineering and BIOtech Research Center, University of Trento, Trento, Italy*

[b]*INSTM Research Unit, Trento, Italy*

[c]*European Institute of Excellence on Tissue Engineering and Regenerative Medicine, Trento, Italy*

antonella.motta@unitn.it

Physical, chemical, and mechanical properties of all body tissues depend on the composition and architecture of the extracellular matrix (ECM).

Besides providing support and mechanical properties and being the reservoir of substances participating to the body metabolism, ECM is the microenvironment of cells, communicates with them with specific signaling molecules, and regulates their behavior and intercellular communications.

ECM participates in cell activities, such as cell proliferation, adhesion, migration, differentiation showing a structure extremely

---

Eleonora Carletti and Matteo Stoppato have contributed equally to this work.

*Scaffolds for Tissue Engineering: Biological Design, Materials, and Fabrication*
Edited by Claudio Migliaresi and Antonella Motta

ISBN 978-981-4463-20-1 (Hardcover), 978-981-4463-21-8 (eBook)
www.panstanford.com

dynamic. Its components continuously undergo a remodeling process comprising deposition, degradation, and modifications, which are fundamental for the formation of the tissue architecture.

The principal components of ECM comprehend a complex mixture of structural proteins, proteoglycans, and multi-adhesive matrix proteins that are produced intracellularly by resident cells, secreted into the ECM via exocytosis and finally assembled.

The chapter reviews ECM composition and main functions, focusing in particular on ECM biochemical and physical properties and on the cell–ECM interactions. The role of ECM on the properties of some body tissues is described, and the mechanisms of ECM scarless healing in fetuses are compared with those occurring in adults.

## 2.1 Introduction

The ECM is the major component of the cellular microenvironment, or niche, and it provides multiple functions in addition to its structural role. The principal constituents of ECM are fibrous proteins with a support function (i.e., collagens, fibronectin, laminins, vitronectin, elastin), other specific proteins [i.e., growth factors (GFs), small matricellular proteins and small integrin-binding glycoproteins], proteoglycans (PGs), and water. Every tissue owes its physical, mechanical, and biochemical properties to a specific concentration and assembling of these constituents.[1]

ECM undergoes to a continuous remodeling driven by cells that cleave, rearrange, and deposit the different components and define the tissue architecture, with reciprocal interactions (Fig. 2.1). All cells are in connection with their ECM, either continuously or at certain stage of their life, such as at stem or progenitor cell phase or during cell migration and proliferation.

ECM can guide and regulate cell activity (i.e., proliferation, adhesion, migration, differentiation, apoptosis, and gene expression) through patterned chemical signals (ligands) properly exposed to cell receptor (transmembrane proteins) as well as mechanical and structural stimuli.[2]

The complex mechanism of ECM remodeling allows regulation of stem cell differentiation and it is central during growth and wound healing. Moreover, anomalous ECM dynamics generate a deregulation of cell activities, resulting in congenital defects and causing pathological tissue growths (fibrosis and cancer). Understanding and regulating the remodeling mechanisms of ECM is strategic for tissue engineering and regenerative medicine perspectives.[3]

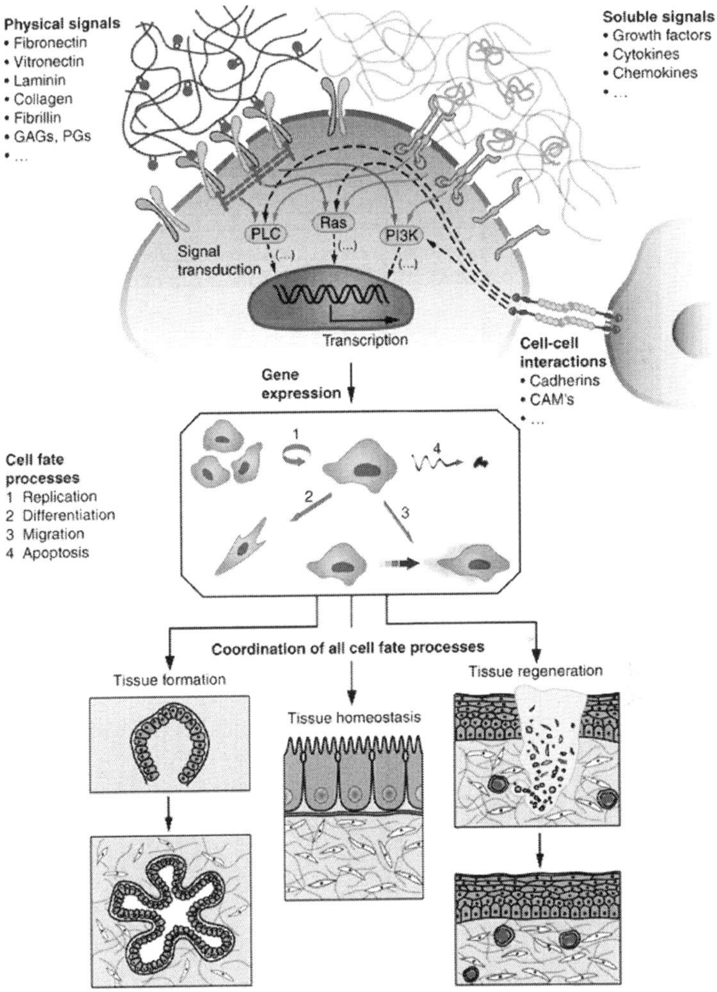


**Figure 2.1** Tissue dynamics, formation-function regeneration, results of temporal and spatial coordination of cell processes, induced by a myriad of signals (included mechanical stimuli) originating from the ECM. Reprinted by permission from Macmillan Publishers Ltd: *Nature Biotechnology*, Ref. 4, Copyright 2005.

## 2.2 ECM Functions

ECM defines the functional organ architecture while acting as a bioactive substrate for cell adhesion, migration, growth, and differentiation. Moreover, ECM serves as storage for molecules and GFs, transmits and distributes the loads preventing mechanical failure, partitions and distributes cells in specific functional units.[5]

The chemical composition, architecture, physical and mechanical properties of ECM vary according to the tissues and their specific developmental stage, age, pathological state, and during the wound healing process (Fig. 2.2).

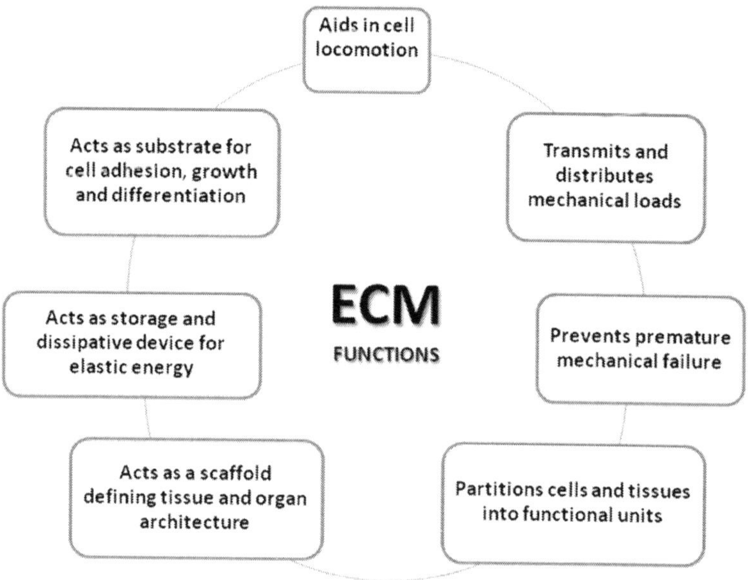


**Figure 2.2** Scheme of the main ECM functions.

### 2.2.1 Biochemical Properties

ECM is typically a highly hydrated macromolecular network composed of various amounts of proteins and PGs.

All macromolecular components of the ECM are synthesized by cells and subsequently secreted in the extracellular space where further physical–chemical modifications can take place.

Nearly all of the proteins are glycoproteins,[6] characterized by oligosaccharide chains covalently attached. Most of them are fibrous structural proteins and include the large family of collagens, laminins, fibronectins, and nidogens. Elastin is also a protein but without chains of carbohydrate residues attached. PGs are other proteins but consist of big clusters of carbohydrate chains, glycosaminoglycans (GAGs), attached to a core protein. Some examples are heparin sulfate (which regulates a wide variety of biological activities, including developmental processes, angiogenesis, and blood coagulation), chondroitin sulfate (which is considered to stabilize brain synapses), and keratin sulfates (which are involved in various tissues development and wound healing). Among the GAGs, hyaluronic acid (HA) does not have a protein component. HA has many functions: it stimulates cytokine and angiogenesis, attracts water, influences damaged or growing tissues, and allows nutrients and metabolites diffusion. Moreover, particularly during embryogenesis, HA creates cell-free space filled with water into which cells can proliferate and migrate.

As a highly charged protein network, rich in polysaccharide modifications, the ECM shows a wide variety of unique functions reflecting buffering, hydration, binding, and force-resistance properties.[7] GAGs can bind to a myriad of GFs controlling the diffusive range and accessibility of ligands to their receptors. Heparan sulfate PG, for example, facilitates interactions between ligands and their receptors and has the capability to bind to various GFs.[8]

ECM acts as a molecular "reservoir" binding to ligands and creating a concentration gradient of many GFs produced by cells, such as bone morphogenetic proteins (BMPs), fibroblast growth factors (FGFs), hedgehogs (HHs), and Wnt proteins (signaling molecules that control cell–cell communication).[9,10]

It is widely recognized that the unique chemical composition of each tissue, due to the mix of the components described, maintains biological and biochemical information, guiding cells behavior and tissue organization.

### 2.2.2 Physical Properties

The physical properties of ECM refer to its morphology (i.e., shape, structure, porosity, pattern), topography (related to the surface), water content, and biomechanical characteristics. ECM has a clear structural function, fundamental in connective tissues, but still significant in all the other tissues where a support for cells is provided.

Some of the ECM proteins, such as collagens and elastin, are organized in fibrils/fibers forming the ECM skeletal structure and providing the tensile strength and viscoelasticity of the tissue. The network solidity is instead guaranteed by linking proteins, working as natural crosslinkers, including collagens, fibronectin, laminin, and nidogen.[1,2,10]

The most evident role of ECM is related to its scaffolding/template support function. More generally, ECM supports tissues structure, ensuring their integrity and favoring cell anchorage and migration or, in the case of basement membrane, creating a barrier for cells and factors migration.[2,9,11-14] The cell–ECM constant interaction is fundamental to maintain the proper ECM composition and assembling, which, in turn, can promote cell movement and orientation. It has been demonstrated the importance of the ECM structure on the formation of tissue during its healing process and, in particular, that the orientation of collagen fibrils can significantly impact on cell arrangement and, consequently, on matrix formation.[15,16]

ECM is a composite material that comprises a fibrous structural net (fibrillar and elastic proteins) embedded in a porous hydrated gels matrix of GAGs. The ratio between the constituents varies according to the tissue.

GAGs and PGs can combine to form huge polymeric complexes, and they are also connected with fibrous matrix proteins (i.e., collagen) and with protein networks (i.e., basal lamina), creating a continuous complex fiber arrangement.[17-19] GAGs tend to arrange in a highly extended conformations that occupy a large volume, forming gels even at low concentrations. The high density of negative charges in the chain attracts cations, $Na^+$, that are osmotically active, thus forcing a large amount of water into the matrix. This creates a swelling pressure that enables the matrix to withstand compressive forces (in contrast to collagen fibrils, which resist the stretching forces). For example, the hyaline cartilage of articular joints can support pressures of hundreds of atmospheres thanks to the swelling of its hydrophilic component (GAGs). As mentioned before, one of the main GAGs is the HA, also important for its resistance to compressive forces due to its capacity to attract water. Chondroitin sulfate is another GAG that contributes to the tensile strength of cartilage, tendons, ligaments, and walls of the blood vessels.

The ECM biomechanical behavior describes the reaction of ECM to loads, such as compressive, tensile, bending, torsion, and shear stresses applied by cells or external forces.[20] These loads have a great impact on the ECM characteristics, influencing its elasticity/rigidity. ECM mechanical properties range from those of soft and compliant structures, as in adipose tissue, brain, or muscles, to those of stiff and rigid tissues, as bones.[21,22] Clinical studies on tissues suggest that ECM elasticity plays a significant role in organ homeostasis and function, recognizing that diseased tissues have markedly altered elasticity that healthy ones and that changes in tissue stiffness may indicate a pathologic condition.[23,24] Indeed, the extent of tissue stiffness is often considered a good prognostic indicator for disease.[7] Atherosclerosis, for instance, is characterized by a premature increased stiffness of the arterial walls[25] and fibrotic tissues are usually stiffer than healthy ones,[26] due to a reduction of the water content and to an increase of the fibrous component.

### 2.2.3 Mechanical Transduction

It is widely accepted that mechanical stimuli have an important role in tissue-organ development and physiological processes. In fact, cells and ECM have a dynamic 3D interrelationship that is kept in balance and influenced by internal and external stimuli, to control cell phenotype and matrix maintenance process.

Local mechanical crosstalk between cells and the surrounding environment is described by Provenzano *et al.*[27] as a unifying principle that connects ECM micro- and nanoscale structure with cell shape, distribution, and differentiation.

Cells have indeed developed sophisticated mechanosensors and systems, such as filaments, integrins, and cytoskeleton, connecting cells with ECM. In this way, external forces can modify nuclear architecture, chromatin organization, and, consequently, gene expression and cell differentiation.[28-30]

This means that external mechanical stimuli are transduced to cells through the ECM that they are connected to; then cells are able to convert sensed mechanical stimuli into biochemical pathways leading to different cell activities (Fig. 2.3).

Part of such bidirectional communication between ECM and cells is a result of a direct mechanical connection between fibrous protein in the ECM and cell cytoskeleton via integrins. Integrin-mediated cell–ECM interactions promote the assembly of cytoskeletal and signaling molecule complexes at focal adhesion sites. These complexes include Src-family members, focal adhesion kinase (FAK), phosphatidylinositol-3′-kinase (PI3K), and phospholipase C (PLC).[31,32] Accordingly, cells do not only exert forces but also respond to the resistance sensed through cytoskeleton organization/tension, triggering several intracellular signaling pathways.[33]

In coordination, it seems that there is a synergy between mechanical stimuli and hormones, GFs, and inflammatory mediators that collectively activate the same signaling pathways (Fig. 2.3).[30,31,34-36]

During development, ECMs are able to adapt to the higher mechanical demands, by increasing the size of the necessary tissue components.[35,37]

Furthermore, mechano-chemical transduction is altered in aging connective tissues. In fact, recent evidence suggests that tension applied by cells to ECM is reduced or lost during aging and diseases such as osteoarthritis.[38] The reduction observed is similar to changes that occur as a result of disuse.[39]

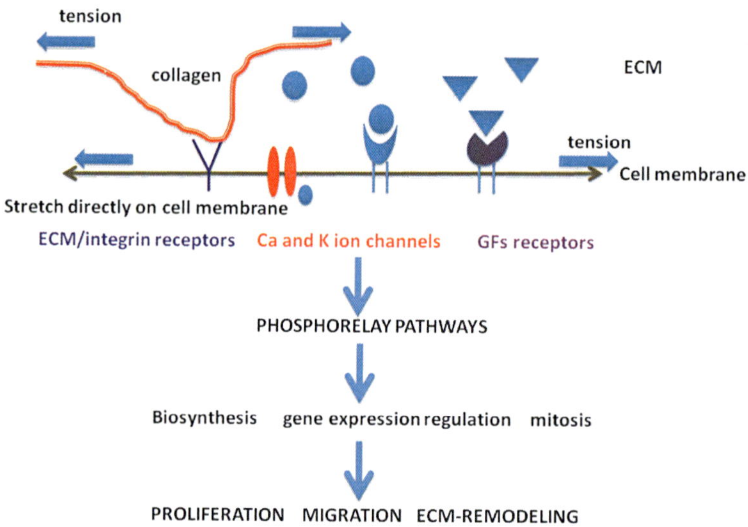


**Figure 2.3** Scheme of relationship between external stimuli transduced through ECM and cell bioactivity and subsequent activation of phosphorelay pathway. The biochemical cell activation is activated by direct stretch of membrane, chemically connected with ECM fibers through receptor-ligand connection. Adapted from Ref. 5.

Mechanical transduction seems to be particularly important even during wound healing. In this case, cells that are moving into a wounded area to synthesize new ECM produce cellular contractile forces, and, simultaneously, in the final phases of the process, the ECM tension transmitted through integrins can modulate cell fate, removing the unneeded cells and organizing the new tissue.

Mechano–transduction is also known to have a role in several diseases, as tumor formation and development.

## 2.3 ECM Main Components

### 2.3.1 Collagens: The Major Fibrous Component

Collagen is the most abundant family of proteins constituting the ECM in animals and it represents one-third of the total proteins content in humans.[40-42] Collagen molecules form fibrils and microfibrils are the major structural elements of all connective tissues, providing stability and structural integrity.[43] If initially only fibroblasts were believed to generate collagen, in the last decade the knowledge about collagen family increased considerably, identifying that also osteoblasts, numerous epithelial cells and other cell types are able to produce certain types of collagens.[44]

Three parallel polypeptide chains arranged in a helical conformation characterize the central feature of collagen. The amino acid composition of collagen is commonly characterized by a repeating X–Y–Gly sequence, where X and Y can be any amino acid (usually proline and hydroxyproline) while glycine is found at almost every third residue.[45]

Collagen, as other secreted proteins, is produced and assembled following a multistep pathway: (1) Collagen chains are synthesized as longer precursors called procollagens, (2) delivered and packaged within the Golgi apparatus into secretory vesicles, (3) released into the extracellular space, and finally (4) assembled and crosslinked. During or following exocytosis, procollagen trimers are processed, depending on the collagen type. The resultant protein consists almost completely of a triple-stranded helix, which has a diameter of 1.5 nm and is 300 nm long.[44,46]

The importance of collagen is attributed to many of its vital characteristics, including the ability to interact with other biomolecules and cells, thermal stability, and mechanical strength. These properties are derived from the triple-helix structure with its interstrands hydrogen bonds, hydroxylation of Pro residues in the Y position, and covalent crosslinking within the triple helices.

The hierarchical nature of collagen organization permits the assessment of its mechanical properties, at different levels of structural complexity, including the tropocollagen monomer, individual collagen fibrils, and collagen fibres.[42]

For example, it was determined that the strength of a tropocollagen monomer is higher than that of a collagen fibril, 11 Gpa versus 0.5 Gpa, probably due to the covalent bonds that must be broken in the first case. Hydration of collagen fibrils reduced their shear modulus significantly, which increases with crosslinking. It is worth noting that a certain level of crosslinking improves the mechanical properties of collagen fibrils, while higher crosslinking results in more brittle collagen fibrils structures,[47] a general symptom of aging. An excessively high crosslinking density does not permit to dissipate energy during deformation and leads to a dehydrated collagen that results strong but brittle.

#### 2.3.1.1 Chemical composition: collagen types

So far, 28 genetically distinct collagen types have been described in vertebrates.[48–51]

The different collagen types are characterized by considerable complexity and diversity in structure, splice variants, the presence of additional non-helical domains, assembly, and function.

Based on their structure and supramolecular organization, they can be grouped in different families.

The most widespread family of collagens with about 90% of the total collagen is represented by the fibril-forming collagens, comprising collagen types I, II, III, V, and XI. These collagens are characterized by their ability to assemble into highly orientated supramolecular aggregates with a characteristic suprastructure, the typical quarter-staggered fibril-array with diameters between 25 and 400 nm. Types I and V collagen fibrils contribute to the structural backbone of bone tissue and types II and XI collagens predominantly contribute to the fibrillar matrix of articular cartilage.[52,53] Type I collagen fibrils are the most abundant and best known and are present in many tissues. Different tissues, in addition to this dominant collagen, contain type III (in skin and reticular fibers)[54,55] or type V collagen (in bone, tendon, cornea).[56] Type II is the most abundant collagen in cartilage and its fibers are usually incorporated with types XI and IX collagens, which are supposed to limit the fibril diameter.[57] The fibril-forming type II collagen is similar in size and biomechanical properties to that of type I but have a higher content of amino acids and residues that mediate the interaction with PGs, another typical component of the highly hydrated matrix of hyaline cartilage.[53,58] Type III collagen is an important component of reticular fibers in the interstitial tissue of the lungs, liver, dermis, spleen, and vessels. Furthermore, it is often found combined with type I collagen in the elastic tissues. Type V and XI seem to have a core structure function with other collagen types, polymerizing around this central axis. In particular, types I and III collagens surround type V, while type XI collagen is supposed to form the core of collagen II heterofibrils.[43,59]

The fibril-associated collagens with interrupted triple helices (FACIT) collagens comprehend collagen types IX, XII, XIV, XVI, XIX, and XX. Collagenous domains interrupted by short non-helical domains characterize the structures of these collagens and the trimeric molecules are associated with the surfaces of various fibrils. The function of these collagens, as well as of collagen types XIX and XX, is still poorly understood and presumably play a role in regulating the diameter of collagen fibrils involved in the linkage of various collagen fibers as well as their interaction with molecules of the ECM.[50,60,61]

Type VI collagen is a microfibrillar collagen with highly disulphide crosslinked, short triple helical domains and rather extended globular termini.[62,63] Type VI collagen fibrils appear on the ultra-structural level as fine filaments, micro fibrils or segments, and form an independent microfibrillar network in virtually all connective tissues, except bone.[64–66]

The hexagonal network-forming collagens comprise collagen types X and VIII. They are short-chain collagens. The function of type X collagen seems to be related to endochondral ossification and matrix calcification.[67,68] Type VIII collagen, produced by endothelial cells and assembled in hexagonal lattices, shows a different distribution and may therefore have different functions.[69]

### 2.3.2 Elastin: The Elastic Component

The ECM flexibility and extensibility are due to a resilient network structure made by elastin. It is a fibrillar protein, which consists of approximately 36 domains, and it is synthesized by fibroblasts, endothelial cells, chondroblasts, and smooth muscle cells.[70]

The elastin polypeptide chain is characterized by the presence of alternating hydrophobic and crosslinks domains (dominated by proline, alanine, valine, leucine, isoleucine, and glycine residues), occurring in repeated sequences and responsible for the long-range deformation of elastin and its high molecular mobility.[71-73]

Elastin is produced and assembled during a highly complex process called elastogenesis. It is synthesized from a monomer, tropoelastin (molecular weight >70 kDa), which is released on the external cell surface, through the cell membrane, in association with a 67 kDa galactolectin, an effective chaperone able to prevent intracellular molecules aggregation.[74] The increasing concentration of the elastin precursor and the alignment allow the crosslinking between monomers, leading to the stabilization of the network.[75] Periodic sequences of elastin have been analyzed, through the synthesis and structure studies of similar polypeptides and through molecular dynamics simulation systems. Therefore, several models have been proposed.[74, 76–78]

In particular, Tamburro[72] suggested a model in which isolated and labile $\beta$ turns result in a flexible elastin polypeptide chain with chaotic motions in the relaxed state. Such protein mobility is due to hydration. Conversely, in the stretched form, the protein is less hydrated and characterized by the presence of $\beta$-sheets structures, causing a more rigid matrix.[79]

Important mechanical properties of connective tissues are established by the interactive combination of elastin and collagen: elastin provides elastic deformability while collagen provides stiffness.

Elastin fibers are crucial in determining the mechanical behavior of many tissues, such as blood vessels, skin, lung, cartilage. For example, in the walls of arteries, elastin is in the form of layers called laminae and it is mostly responsible for elasticity necessary for vessel expansion and recovery during blood flow.[80] In the lung, elastin arranged similarly to a latticework supporting the opening and closing of the alveoli.[81] In skin, elastin fibers confer flexibility and extensibility.[82,83] The Young's modulus for elastic fibers can measure as low as 100 kPa but usually ranges from 300 to 600 kPa.[84,85] Numerous pathological situations (i.e., cardiovascular diseases, aortic aneurysms, aging of connective tissue) involve remodeling and degradation of elastin.[85,86] Studies on elastin show that it does not play only a mechanical role, but it is involved in the regulation of cells in the elastic tissues, mediating cell attachment and proliferation.[87–89] Elastin expresses a specific receptor complex, formed by three subunits, one of them named elastin-binding protein (EBP). The interaction between EBP and elastin-derived peptides can trigger different intracellular pathways. Among them, a rapid and transient increasing in the intracellular $Ca^{2+}$ with a consequent activation of PLC and protein kinase C, can be observed in lymphocytes and fibroblasts.[90]

### 2.3.3 Fibronectin: A Binding Protein

Fibronectin (FN) is a high molecular weight (about 440 kDa) ubiquitous glycoprotein, secreted as a dimer and stabilized by disulphide bonds.[91]

According to its solubility properties, FN can be distinguished in soluble plasma FN, produced by hepatocytes and representing the major protein component of blood plasma, and the less-soluble cellular FN, secreted by fibroblasts and assembled in the matrix in a complex cell-mediated process.

Functionally, FN is the classic example of adhesive glycoprotein, binding and interconnecting ECM components such as collagen, fibrin, heparin sulfate, and PGs.

FN is a ligand for several integrins and many research studies contributed to narrow the regions along the FN molecule destined to cell adhesion to less integrin-recognition sequences. Among these regions, RGD sequence is the best known. The RGD loop is strategically placed to undergo strong conformational changes and constitutes a mechano-sensitive switch for recognition by integrin receptors. The binding property of FN to cell surface integrin receptors is essential for cell adhesion and migration,[92,93] particularly important during tissue remodeling, wound healing, and embryonic development. For example, FN is one of the active components of the blood clot formed in the injury site (FN is covalently crosslinked to the fibrin with the support of factor XIII), forming the first natural regenerative template that is then degraded by proteases secreted by fibroblast and macrophages. In addition, FN plays a role as chemo-attractant for monocytes, fibroblasts, and endothelial cells into the damaged site. Interestingly, the FN fragments produced during the healing step, seem to have a role, promoting wound contraction (important phenomenon in the healing process).[94] FN matrix assembly begins when soluble, compact fibronectin dimers are secreted from cells (in general fibroblasts). The FN soluble dimers bind to $\alpha5\beta1$ integrin receptors on the cellular surface, participating in clustering the integrins. The increase of local concentration of integrin-bound fibronectin allows the bound fibronectin molecules to interact more readily through the fibronectin-binding domains. Short fibronectin fibrils then begin to form between adjacent cells. As matrix assembly proceeds, the soluble fibrils are converted into larger insoluble fibrils that comprise the ECM. It also demonstrated the role of FN to enhance cell cytoskeletal reorganization, gene expression, and collagen deposition in the ECM. The specific 3D organization of FN in the insoluble matrix forms a specific ligand patterning so modulating both cell adhesion and migration into ECM.[92]

A number of pathologies, including cancer and fibrosis, are associated to an altered FN degradation and organization, as well as altered gene expression.

## 2.4 ECM Assembling and Organization

ECM is a complex highly organized three dimensional structure characterized by different compositions and proteins arrangements depending on the tissue and the specific functions required.

The components of ECM play key roles in stimulating cell activity, such as proliferation, migration, and differentiation. The continuous communication between cells and the surrounding ECM, together with the external stimuli to which every tissue is subjected, allows specific assembling and organization.

Matrix assembling and organization is therefore the result of a perfect synergism between proteins, molecules, and cells arranged in a functional hierarchical architecture and acting at different scale levels (Fig. 2.4).

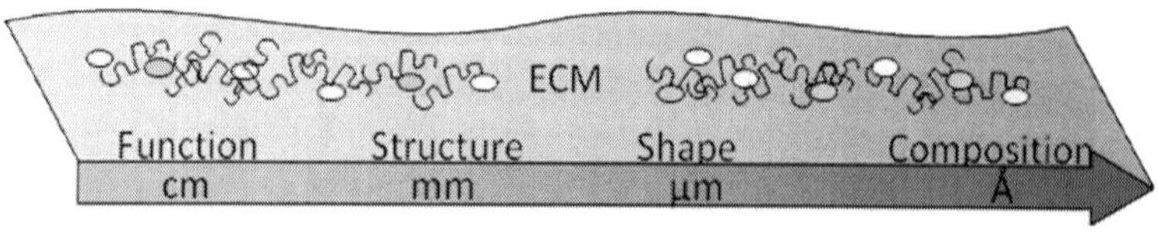


**Figure 2.4** Scheme illustrating the biological interactions, at different size scale levels, controlling tissues organization.

In fact, collagen types and their fibrils/fibers organization, tailor the mechanical properties of specific connective tissues.

In the following paragraphs, different tissues will be then analyzed in terms of their functionality requirements achieved by a specific ECM organization and assembling (Fig. 2.5).

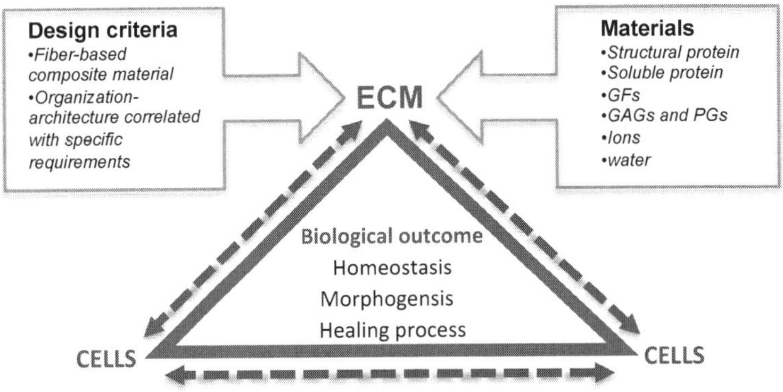


**Figure 2.5** Scheme of cells–ECM–cells dynamic interaction modulated by ECM composition (materials) and structure (design criteria). GF, growth factor; GAGs, glycosaminoglycans; PGs, proteoglycans.

## 2.4.1 ECM Organization Function Dependent: Some Examples

### 2.4.1.1 Aorta wall

*Mechanical requirements:* Specific compliance and strength in both longitudinal and circumferential directions.

*Structural strategy:* Anisotropic and elastic multi-layers laminate, different fibers orientation through the thickness of the aorta wall.

*ECM main components:* 3D fibrous network of collagen combined with elastic fibers (elastin).

Considering the three main layers of aorta wall, the orientation and organization of collagen fibers and their association with elastin component sensibly change, as structural adaptive response to the mechanical requirements.

*In details:* Sub-endothelium presents a distinctive organization of multilayered fabric of collagen. Within the same layer collagen fibers are uniformly oriented. Collagen fibers, in the different layers, are oriented differently, to ensure the proper resistance in each direction. Elastin is also present and it is arranged following a 3D network of elastic fibers.[95,96]

The middle layer of the artery, the media, consists of a complex three-dimensional network of smooth muscle cells, elastin, and bundles of collagen fibrils.[97,98] As described by Rhodin *et al.*,[99] the fenestrated elastic laminae separate the media into a several number of concentrically fiber-reinforced medial layers. The repeating of basic structural and functional units, characterizing the medial thickness, forms alternating layers called musculoelastic fascicles.[98] In this section of aorta wall, the elastin fiber architecture loses its organization near the periphery.[100]

The media is separated from the intima and adventitia by the so-called internal elastic lamina and external elastic lamina, respectively (the latter is absent in cerebral blood vessels). Elastin, combined with bundles of collagen fibers and smooth muscle cells constitute a continuous fibrous helix. The helix within the media is almost circumferentially oriented. Analysis by polarized light microscopy performed by Canham *et al.*[95] showed that the collagen composition was 30% of type I and 70% of type III.[101] This fibrous organization gives the media the ability to resist high loads in the circumferential direction.

In the adventitia, the outermost layer of the artery, collagen fibers are organized in thick bundles. A layer of loose connective tissue surrounds the adventitia and its outer boundary is not well defined. The thickness of the outer section depends strongly on its topographical site and the collagen fibers, primarily of type I form a typical fibrous tissue.[95]

The intima is the more internal layer. It is formed by one layer of endothelial cells, directly in contact with the blood flow, and it is supported by an internal elastic lamina, rich in laminin, fibronectin and type IV collagen.

In conclusion, the collagen fibrous network contributes significantly to the stability and strength of the arterial wall. In the unstressed tissue, collagen fibers are immersed in a soft matrix, which causes the external layer to decrease the stiffness, however, at higher levels of strain, collagen fibers reach their straightened lengths and the mechanical response of the adventitia changes preventing the artery from overstretch and rupture.[100]

### 2.4.1.2 Bone

*Mechanical requirements:* High strength and rigidity combined with good toughness.

*Structural strategy:* In cortical bone, collagen fibrils are densely packed and form concentric lamellae. Fibrils in adjacent lamellae are arranged in perpendicular planes. Trabecular bone has a loosely organized porous matrix.

*ECM main components:* The bone structure is composed of approximately 65–70 %wt mineral, 20–25% protein, 8–9% water, and less than 1% cell amount.[102] Collagen constitutes approximately 95% of the organic matrix (99% is type I collagen); the remaining 5% is composed of PGs and numerous non-collagenous proteins. The major component of bone is, however, mineral. The main bone mineral components are calcium phosphates, especially hydroxyapatite ($Ca_{10}(PO_4)_6(OH)_2$), forming prism crystals, with a length of 20 nm and a thickness of 2 nm.

*In details:* Bone is a dynamic and highly specialized form of connective tissue. The main function of bone is to sustain and protect organs in the body, and act as a storage facility for systemic mineral homeostasis.[103,104]

Mechanically, bone tissue is characterized by a unique composition, which gives a marked rigidity and strength while still maintaining a good degree of elasticity.

The presence of minerals combined with an appropriate distribution of collagen fibers in the ECM provides strong mechanical properties in terms of hardness and compressive, tensile and torsion strengths.[105]

The relatively cell-free bone tissue comprises four different kinds of cells: osteoprogenitor cells, osteoblasts, osteocytes, and osteoclasts. Osteoprogenitor cells, deriving from periosteum, present a high proliferative ability and they are able to differentiate in bone cells.[106] Osteoblasts tend to form sheets on the bone surface during tissue formation and their role is to secrete the type I collagen and the non-collagenous proteins, synthesizing and mineralizing new ECM. After the appropriate bone deposition, osteoblasts decrease their metabolism and finally differentiate into osteocytes. The osteocyte is the most abundant cell type of bone and is responsible for its maintenance and remodelling.[107] These cells have the capacity not only to synthesize, but also to resorb matrix to a limited extent. Each osteocyte occupies a space, or lacunae, within the matrix. Instead, osteoclasts are large (100–200 μm), multinucleated cells, develop from hemopoietic cells of the monocyte–macrophage lineage, which resorb bone.[108]

Morphologically, there are two main types of bone: compact (cortical) and trabecular (cancellous or spongy). In cortical bone, densely packed collagen fibrils form concentric lamellae, and the fibrils in adjacent lamellae run in perpendicular planes like in plywood. Trabecular bone has a loosely organized spongy arrangement. Differences between cortical and cancellous bone are both structural and functional. Differences in the structural arrangements of the two bone types are related to their primary functions: cortical bone provides mechanical and protective functions and cancellous bone provides metabolic functions.[103]

Thanks to cell activity, bone tissue undergoes a continuous renewal in a delicate equilibrium between synthesis and degradation of the ECM. Its metabolism and regeneration are under the influence of a variety of molecules and local factors, such as Insulin-like growth factor (IGF), FGF, and BMP.

#### 2.4.1.3 Hyaline Cartilage

*Mechanical requirements:* Hyaline cartilage is a flexible tissue but, contemporary, strong, and resistant to extremely high pressure.

*Structural strategy:* Cartilage structure is a biphasic poroelastic medium, with a heterogeneous structure both at macro and microscopic level. Many layers are distinguished, with different properties and different orientations of collagen fibers.[109]

*ECM main components:* Cartilage is an aneural and avascular tissue, made by water (60–80 wt%), collagen fibers (mainly collagen type II, 10–20 wt%), and PGs (10–20 wt%). Cartilage cells, chondrocytes (5–10 wt% of the total volume of the tissue) are spread in the matrix, with no cell–cell contacts.

*In detail:* Collagen fibrils architecture in the superficial zone is parallel to the surface, providing the great tensile and shear strength. Cartilage has a gradient structure, with collagen fibers assuming different orientations in the superficial (parallel), transitional (random), and radial (perpendicular) zones. Alterations or disruptions of these zones alter the biomechanics and contribute to osteoarthritis development.

Almost 80% of the tissue is represented by water; mechanical properties are provided by macromolecules, collagen (mainly type II), and PGs, that make up the solid matrix.[110] Collagen fibrils (which constitutes up to 65% of the cartilage dry weight) provide tensile strength and biomechanical resistance; PGs, which aggregate within the collagen network, are saturated with water and provide compression resistance as well as the ability to re-expand after loading (viscoelastic behavior). ECM promotes cells organization (clusters) and distribution and permits very little cells mobility, to avoid migration of other type of cells, that is, macrophages. Chondrocytes have many connections to the matrix in microvilli, to assess the state of the tissue and to monitor the biomechanical behavior. These connections are likely to be crucial for the maintenance of a normal chondrocyte phenotype.

#### 2.4.1.4 Fibrocartilage (the meniscus case)

*Mechanical requirements:* Shock absorption, lubrication, load transmission stability, and load-bearing resistance.

*Structural strategy:* The function of meniscus is reflected in its architecture. The ECM is 3D assembled to optimally endure and redirect compressive forces, shear stresses, and circumferentially directed forces.

*ECM main components:* Water, fibrillar components, PGs, and adhesion proteins compose the meniscus ECM. Fibrous components are collagen (60–70% of the meniscus dry weight) and elastin (<1%).

*In details*: Meniscus is a C-shaped fibro cartilaginous tissue. The fibrous part is mainly formed by the common collagen type II with the addition of the collagen type I. Collagen fibers, in the superficial and median layer, are randomly oriented, while the deep zone consists of circumferentially oriented fibers with some radial fibers, to prevent radial extrusion of the meniscus and to maintain its structural integrity during load bearing.[111,112]

GAGs play an important role in the maintenance of optimal viscoelastic behavior, compressive stiffness, and tissue hydration (78% is water). In addition, GAGs and surface zones proteins are thought to facilitate a smooth frictionless movement of the meniscus over the articular surfaces of the femur.[113]

The degree of vascular penetration is 10–30% of the width of the medial meniscus. However, the majority of the meniscus is avascular and must therefore derive its nutrition through synovial fluid diffusion.

Superficial and deep zone of the meniscus contain cells of different morphologies whose characters resemble those of both fibroblasts and chondrocytes, defined as fibrochondrocytes.

#### 2.4.1.5 Skeletal muscle

*Mechanical requirements:* Muscles are subjected to passive tensions, stretched beyond their slack length. In addition, muscle individual fibers support very rapid and powerful contractions.

*Structural strategy:* Skeletal muscles are composite tissues composed by connective tissue, blood vessels, and nerves as well as contractile materials.

*ECM main components:* Collagen is the major structural protein in skeletal muscle ECM. The fibrillar collagen types I and III are the most abundant in adults. Muscle basement membrane consists primarily of collagen type IV, although types VI, XV, and XVIII are also present. There are different PGs in muscle ECM and many of them belong to the family of small leucine-rich PGs (SLRPs).

*In detail:* In the muscle ECM three main parts are distinguished[114]:

- epimysium (surrounding the muscle)
- perimysium (surrounding muscle fascicles and may be continuous with the tendon)
- endomysium (surrounding muscle fibers, contained within the muscle fascicles).

Endomysium is an ordered network that surrounds individual muscle fibers and deforms nonlinearly with increasing the length of sarcomere (the basic unit of the muscle). Endomysium is a load-bearing structure having mechanical properties dependent on its geometry rather than the constitutive properties of the composite collagen fibers. Due to the morphological features, forces are transmitted by shear via the endomysium.

Perimysial collagen fibers are organized into distinct groups that extend along and across muscle fibers. Basically, fibers having a longitudinal orientation form a dense series of bands and transverse collagen fibers connect muscle fibers at specific sites. Such transverse connection points are co-localized with focal adhesions and intracellular subdomains, and, for this reason, perimysium may be involved in cellular signaling.

The epimysium surfaces are formed by large collagen bundles having the "crimp" pattern typical of tendons.

Regarding the main components of skeletal muscle ECM, collagen is the major structural protein.[114,115]

In particular, collagen types I, III, IV, V, VI, XI, XII, XIV, XV, and XVIII are expressed during skeletal muscle development, while fibrillar types I and III predominate in adult endo-, peri-, and epimysium. Collagen type I is predominant in perimysial, whereas collagen type III is more distributed between endomysium and epimysium. However, it is still not well studied if collagen ratios differ in muscles with different functions. Type V collagen, another fibril-forming collagen, combines with types I and III forming, in perimysium and endomysium, a core for type I collagen fibrils. Types XII and XIV collagen are FACITs mainly found in perimysium.

Muscle basement membrane consists primarily of collagen type IV; collagen types VI, XV, and XVIII are also present.[116] Types XV and XVIII collagen are categorized as multiplexins, that is heparin sulfate PGs. Multiplexins can bind GFs and they are also involved in linking the basement membrane to other basement membrane glycoproteins and endomysium. Although the basement membrane is distinct from endomysium, these two components are closely connected and probably responsible for the transmission of force from the myofiber to the tendon.

Numerous PGs are present in the ECM of muscle tissue and most of them belong to the family of SLRPs (a core protein with attached GAG chains, i.e., decorin, biglycan, fibromodulin, and lumican).[115] The most abundant PGs in skeletal muscle ECM have chondroitin sulphate and dermatan sulphate GAG chains, including decorin and biglycan.

## 2.5 ECM Development and Remodeling

It is important to understand the ECM synthesis, assembling, and remodeling, in order to trigger and guide functional tissues and organs regrowth. The ECM development in the healing process of adult and fetus will be revised. Fetal healing results in full regeneration of the tissue architecture and function, without scarring by processes that involve a large pool of stem/progenitor cells and a reduced immune response. In contrast, adult healing, in specific tissues or in presence of critical size defects, leads to fibrous scar formation through the inflammatory response activation, with a lower number of progenitor cell involvement.[117]

### 2.5.1 ECM in the Healing Process

Wound healing is a dynamic and complex biological event involving interactions between cells, soluble mediators and ECM, resulting in the restoration of tissue anatomic stability and function.

Wounds are usually classified as either acute or chronic,[118] and in both cases, ECM is a key feature in the healing process.

Acute wounds normally proceed through an orderly and efficient manner characterized by four different, but overlapped biologic events: hemostasis, inflammation, proliferation, and remodeling and results in restoration of anatomic and functional integrity. During the whole process, ECM has a crucial role in regulating and integrating the development of the healing.[118–121] Initially, a temporary ECM is formed, rich in fibrin and fibronectin, working as scaffolding support for cells, which are migrating into the injury site. These cells are also stimulated by the provisional template to proliferate, differentiate, and synthesize the new ECM.

Chronic wounds do not follow the normal healing process.

In a first instance, chronic wound evolution is similar to the initial phase of acute wound and a fibrin clot is formed. However, instead of developing throughout the four phases, a healing condition of extended inflammatory phase is created, during which there is an increase of proteases [i.e., matrix metalloproteinases (MMPs), elastase, plasmin, and thrombin], responsible for degradation of ECM components such as GFs and their receptors, necessary for healing. (More details are in Chapter 4 by Cancedda and Lo Sicco, *Principles of biological pathway to tissue regeneration (The tissue regenerative niche)* and Chapter 5 by Srinivasan and Babensee, *Host response in biomaterials*).

For this reason, throughout the healing process, the inflammatory phase results particularly interesting to be deeply investigated.

#### 2.5.1.1 Inflammation

As extensively described in Chapter 5 by Srinivasan and Babensee, *Host response in biomaterials*, the initial phase of wound healing is characterized by an up regulation of inflammatory mediators. Inflammatory cells are recruited in the site of tissue injury and ECM has a significant role in influencing immune cell activity in inflamed tissues.

The *in situ* scenario, when an inflammation condition is present, is extremely complex and not only endothelial and immune cells are involved but also the ECM, which includes the basement membrane beneath of endothelium and the interstitial matrix of the tissue stroma.

The components of ECM, together with its biophysical features and geometrical characteristics send specific signals to cells and regulate important immune functions, such as immune cell migrations into the inflamed tissues, cell activation, proliferation, and differentiation.[122]

If tissue inflammation is chronic, an anomalous expression and remodeling of ECM can affect immune cell activation and viability, influencing the immune response within these sites. In fact, it has been demonstrated how ECM fragments or ECM molecules, upregulated during inflammation, can induce an immune cell activation.

Therefore, ECM is not only a structural support or a barrier for cells involved in the inflammation event, but also its biophysical and biochemical properties are necessary to deliver signals to immune cells.

In particular, such signals are used to guide inflammatory cells from blood to the inflammation site, determining the migration mechanism across the endothelial basement membranes and enhancing their capacity to promote inflammation. There is an increasing interest on the laminin composition of the endothelial basement membrane defining the sites of extravasation for leukocytes, the white blood cells, as well as different mechanisms used by immune cells to enter different basement membranes.[123,124] The biochemical mechanism behind the cells/basement membranes communication has not been evaluated so far and it is not clear if the signal is given by forces, specific signal-transduction mechanisms, or by a synergistic effect of the two. If an inflammation is chronic, ECM is altered by cytokines and proteases. In fact numerous cytokines, and among them TNF and IFNγ, are up regulated at inflammation sites and they can modify the synthesis and dynamics of some ECM molecules of the interstitial matrix and the secretion of proteases (i.e., MMPs).[122]

### 2.5.2 A Lesson from Fetal Healing: Scarless Regeneration

Wound healing in adults results in tissue restoration but fibrosis and scar formation are present in the final phase of such biological event.[125] Conversely, it is interesting to consider the healing phenomenon in fetus, characterized by the ability to heal skin wounds without a scar, organizing a fine and reticular pattern indistinguishable from the surrounding uninjured area.[126] The fetus' scarless repair is independent of the intrauterine environment and, in general, is characterized by a low level of inflammatory signals and it depends on fetal cells properties, ECM, cytokine profile, and gene expression.

Understanding the biology of fetal repair and the related cellular and matrix events may provide insights to help to modulate a fetal-like wound repair in adults.

The scenarios of adult and fetal wound healing are complex and several different dynamics have to be considered to better understand the phenomena. It is known that during inflammation in adults, polymorphonucleate leukocytes (PMNs) infiltrate into the wound site as well as macrophages and lymphocytes. Conversely, fetal wounds lack the PMN and fetal macrophages are fewer.[127,128]

In section 2.5.1, it is extensively described the central role of ECM during wound healing. Fibroblasts are responsible for ECM macromolecules synthesis and ECM remodeling. It has been proved how fetal fibroblasts are able to organize a scarless matrix[129] and, in particular, the higher ability of fetal fibroblasts to synthesize more collagen than adult fibroblasts. Besides, fetal fibroblasts infiltrate faster than adult fibroblasts, affecting collagen deposition and crosslinking. The migration ability of fibroblasts is influenced by the HA content, which, in the case of fetus, is more abundant. HA content inside the matrix regulates the hydration of the ECM and fibroblasts are responsible for its production.

Among the ECM dynamics, the role of collagen is important to understand differences between fetus and adult healing development. In wound healing of fetus collagen is organized in a reticular pattern, which is similar to the surrounding uninjured area, resulting in a scarless behavior. Furthermore, the collagen-type profile is different in fetus and in adult skin. Even if collagen type I is abundant in both fetus and adult skin, type III is prevalent in fetus. Collagen type III fibers are smaller and finer than type I fibers. The higher content of collagen type III guarantees a more reticular pattern of fiber deposition. In general, collagens proportions and their architecture modify during fetus development, affecting the ability to heal with or without scar formation. Collagen crosslinking affects matrix rigidity and a majority of a non-crosslinked collagen in early fetal wounds produces a more flexible matrix, thereby contributing to scarless repair.[130–133]

Moreover, the presence of cell adhesion molecules (i.e., fibronectin), anti-adhesive proteins (i.e., tenascin), which facilitate cell movement and surface receptors (i.e., integrins), occurs earlier in fetus than in adult wounds, leading to a faster fetal fibroblast migration and resulting in a rapid matrix deposition end epithelization.

The remodeling processes of ECM and matrix degradation phenomenon are also different in fetus and adults, and, in particular, there are variations among the enzymes needed for matrix breakdown. MMPs are proteinases involved in ECM remodeling and, among these, collagenases, gelatinases, and stromelysins are the most important. MMPs overexpress in fetal wound healing while in adult skin proteinases are found in similar sites but in less cells.[134]

Finally, even if the knowledge of GF distribution during fetus wound healing is still not well defined, a correlation between GFs and the fetus wound healing characteristics (i.e., absence of inflammatory response and neovascularisation) seems to be evident.[128,135,136]

**Table 2.1** Summary of the main ECM components and cell type/activity involved in the fetal scarless-healing process: main differences with the adult healing [HA, GAG, and TGF-$\beta$ (transforming GF-$\beta$)]

| ECM Composition | Cells Type and activity |
|---|---|
| • More collagen synthesis | • Fetal fibroblasts |
| • High content of collagen type III and V | • High migration rate (affecting collagen deposition and cross-linking) |
| • Major components: HA and non-sulfonated GAGs | • Increased number of HA receptors |
| • More FN and Tenascin | • Earlier activation of myofibroblasts |
| • Low TGF-$\beta$ | • No inflammatory effector cells: Platelets less releasing of active cytokines |

## References


1. Daley, W. P., Peters, S. B. and Larsen, M. (2008) Extracellular matrix dynamics in development and regenerative medicine, *J. Cell. Sci.*, **121**, 255–264.
2. Hynes, R. O. (2009) The extracellular matrix: Not just pretty fibrils, *Science*, **326**, 1216–1219.
3. Lu, P., Takai, K., Weaver, V. M. and Werb, Z. (2011) Extracellular matrix degradation and remodeling in development and disease, *Cold Spring Harb. Perspect. Biol.*, **3**, doi: 10.1101/cshperspect.a005058.
4. Lutolf, M. P. and Hubbell, J. A. (2005) Synthetic biomaterials as instructive extracellular microenvironments for morphogenesis in tissue engineering, *Nat. Biotechnol.*, **23**, 47–55.
5. Silver, F. H., Devore, D. and Siperko, L. M. (2003) Invited review: Role of mechanophysiology in aging of ECM: Effects of changes in mechanochemical transduction, *J. Appl. Physiol.*, **95**, 2134–2141.
6. Hynes, R. O., and Naba, A. (2012) Overview of the matrisome-An inventory of extracellular matrix constituents and functions, *Cold Spring Harb. Perspect. Biol.*, **4.1**, doi: 10.1101/cshperspect.a004903.
7. Frantz, C., Stewart, K. M. and Weaver, V. M. (2010) The extracellular matrix at a glance, *J. Cell. Sci.*, **123**, 4195–4200.
8. Lonai, P. (2003) Epithelial mesenchymal interactions, the ECM and limb development, *J. Anat.*, **202**, 43–50.
9. Rozario, T. and Desimone, D. W. (2010) The extracellular matrix in development and morphogenesis: A dynamic view, *Dev. Biol.*, **341**, 126–140.
10. Vakonakis, I., and Campbell, I. D. (2007) Extracellular matrix: From atomic resolution to ultrastructure, *Curr. Opin. Cell Biol.*, **19**, 578–583.
11. Egeblad, M., Rasch, M. G. and Weaver, V. M. (2010) Dynamic interplay between the collagen scaffold and tumor evolution, *Curr. Opin. Cell Biol.*, **22**, 697–706.
12. Li, L. and Xie, T. (2005) Stem cell niche: Structure and function, *Annu. Rev. Cell. Dev. Biol.*, **21**, 605–631.
13. Tanentzapf, G., Devenport, D., Godt, D. and Brown, N. H. (2007) Integrin-dependent anchoring of a stem-cell niche, *Nat. Cell Biol.*, **9**, 1413–1418.
14. Taddei, I., Deugnier, M.-A., Faraldo, M. M., Petit, V., Bouvard, D., Medina, D., *et al.* (2008) [beta]1 Integrin deletion from the basal compartment of the mammary epithelium affects stem cells, *Nat. Cell Biol.*, **10**, 716–722.

15. Stoppato, M., Carletti, E., Maniglio, D., Migliaresi, C. and Motta, A. (2013) Functional role of scaffold geometries as a template for physiological ECM formation: Evaluation of collagen 3D assembly, *J. Tissue Eng. Regen. M.*, **7**, 161–168.
16. Stoppato, M., Carletti, E., Sidarovich, V., Quattrone, A., Unger, R. E., Kirkpatrick, C. J., *et al.* (2013) Influence of scaffold pore size on collagen I development: A new *in vitro* evaluation perspective, *J. Bioact. Compat. Pol.*, **28**, 16–32.
17. Pellegrini, L., Burke, D., Von Delft, F., Mulloy, B. and Blundell, T. (2000) Crystal structure of fibroblast growth factor receptor ectodomain bound to ligand and heparin, *Nature,* **407**, 1029–1034.
18. Ruoslahti, E. and Yamaguchi, Y. (1991) Proteoglycans as modulators of growth factor activities, *Cell,* **64**, 867–869.
19. Kainulainen, V., Wang, H., Schick, C. and Bernfield, M. (1998) Syndecans, heparan sulfate proteoglycans, maintain the proteolytic balance of acute wound fluids, *J. Biol. Chem.*, **273**, 11563–11569.
20. Yu, H., Mouw, J. K. and Weaver, V. M. (2011) Forcing form and function: Biomechanical regulation of tumor evolution, *Trends Cell Biol.*, **21**, 47–56.
21. Reilly, G. C. and Engler, A. J. (2010) Intrinsic extracellular matrix properties regulate stem cell differentiation, *J. Biomech.,* **43**, 55–62.
22. Lopez, J. I., Mouw, J. K. and Weaver, V. M. (2008) Biomechanical regulation of cell orientation and fate, *Oncogene,* **27**, 6981–6993.
23. Wolfe, J. N. (1976) Risk for breast cancer development determined by mammographic parenchymal pattern, *Cancer,* **37**, 2486–2492.
24. Lieber, M. M. (2006) Towards an understanding of the role of forces in carcinogenesis: A perspective with therapeutic implications, *Riv. Biol.*, **99**, 131–160.
25. Claridge, M. W., Bate, G. R., Hoskins, P. R., Adam, D. J., Bradbury, A. W. and Wilmink, A. B (2009) Measurement of arterial stiffness in subjects with vascular disease: Are vessel wall changes more sensitive than increase in intima-media thickness? *Atherosclerosis,* **205**, 477–480.
26. Discher, D. E., Mooney, D. J. and Zandstra, P. W. (2009) Growth factors, matrices, and forces combine and control stem cells, *Science,* **324**, 1673–1677.
27. Provenzano, P. P. and Keely, P. J. (2011) Mechanical signaling through the cytoskeleton regulates cell proliferation by coordinated focal adhesion and Rho GTPase signalling, *J. Cell Sci.*, **124**, 1195–1205.

28. Butcher, D. T., Alliston, T. and Weaver, V. M. (2009) A tense situation: Forcing tumour progression, *Nat. Rev. Cancer*, **9**, 108–122.

29. Wang, N., Tytell, J. D. and Ingber, D. E. (2009) Mechanotransduction at a distance: Mechanically coupling the extracellular matrix with the nucleus, *Nat. Rev. Mol. Cell. Biol.*, **10**, 75–82.

30. Liu, M., Tanswell, A. K. and Post, M. (1999) Mechanical force-induced signal transduction in lung cells, *Am. J. Physiol-Lung C.*, **277**, L667–L683.

31. Silver, F. H., Siperko, L. M. and Seehra, G. P. (2003) Mechanobiology of force transduction in dermal tissue, *Skin Res. Technol.*, **9**, 3–23.

32. Silver, F. H., and Cowin, F-S. S. (2006) Mechanosensing & mechanochemical transduction in extracellular matrix, in *Biological, Chemical, Engineering and Physiological Aspects*, Springer New York.

33. Miyamoto, S., Teramoto, H., Coso, O. A., Gutkind, J. S., Burbelo, P. D., Akiyama, S. K., *et al.* (1995) Integrin function: Molecular hierarchies of cytoskeletal and signaling molecules, *J. Cell. Biol.*, **131**, 791–805.

34. Plopper, G. E., Mcnamee, H. P., Dike, L. E., Bojanowski, K. and Ingber, D. E. (1995) Convergence of integrin and growth factor receptor signaling pathways within the focal adhesion complex, *Mol. Biol. Cell*, **6**, 1349–1365.

35. Huebsch, N. and Mooney, D. J. (2009) Inspiration and application in the evolution of biomaterials, *Nature*, **462**, 426–432.

36. Guilak, F., Cohen, D. M., Estes, B. T., Gimble, J. M., Liedtke, W. and Chen, C. S. (2009) Control of stem cell fate by physical interactions with the extracellular matrix, *Cell Stem Cell*, **5**, 17–26.

37. Moore, S. W. (2003) Scrambled eggs: Mechanical forces as ecological factors in early development, *Evol. Dev.*, **5**, 61–66.

38. Silver, F. H. and Bradica, G. (2002) Mechanobiology of cartilage: How do internal and external stresses affect mechanochemical transduction and elastic energy storage? *Biomech. Model. Mechan.*, **1**, 219–238.

39. Vandenburgh, H., Chromiak, J., Shansky, J., Del Tatto, M. and Lemaire, J. (1999) Space travel directly induces skeletal muscle atrophy, *FASEB J.*, **13**, 1031–1038.

40. Brinckmann, J.(2005) *Collagens at a Glance*, Springer, Berlin, Heidelberg.

41. Veit, G., Kobbe, B., Keene, D., Paulsson, M., Koch, M. and Wagener, R. (2006) Collagen XXVIII, a novel von Willebrand factor: A domain-containing protein with many imperfections in the collagenous domain, *J. Biol. Chem.*, **281**, 3494–3504.

42. Shoulders, M. D. and Raines, R. T. (2009) Collagen structure and stability, *Annu. Rev. Biochem.*, **78**, 929–958.

43. Gelse, K., Poschl, E. and Aigner, T. (2003) Collagens-structure, function, and biosynthesis, *Adv. Drug. Deliv. Rev.*, **55**, 1531–1546.

44. Lodish, H., Berk, A. and Zipursky, S. L., *et al.* (2000) Collagen: The fibrous proteins of the matrix, in *Molecular Cell Biology,* Fourth edition, (Freeman, W. H., ed), New York, available at http://www.ncbi.nlm.nih.gov/books/NBK21582/.

45. Fratzl, P.(2008) Collagen: Structure and mechanics, an introduction, in *Collagen: Structure and Mechanics* (Wess, T. J., Fratzl, P., eds), Springer, New York.

46. Dee, K. C., Puleo, D. A. and Bizios, R.(2002) *An Introduction to Tissue-Biomaterial Interactions*, Wiley-Liss, p. 26.

47. Buehler, M. J. (2006) Nature designs tough collagen: Explaining the nanostructure of collagen fibrils, *Proc. Natl. Acad. Sci. USA*, **103**, 12285–12290.

48. Berisio, R., Vitagliano, L., Mazzarella, L. and Zagari, A. (2002) Crystal structure of the collagen triple helix model (Pro-Pro-Gly), *Protein Sci.,* **11**, 262–270.

49. Brazel, D., Oberbaumer, I., Dieringer, H., Babel, W., Glanville, R. W., Deutzmann, R., *et al.* (1987) Completion of the amino acid sequence of the alfa1- chain of human basement membrane collagen (type IV) reveals 21 non-triplet interruptions located within the collagenous domain, *Eur. J. Biochem.*, **168**, 529–536.

50. Ramshaw, J. A. M., Shah, N. K. and Brodsky, B. (1998) Gly-X-Y tripeptide frequencies in collagen: A context for host-guest triple-helical peptides, *J. Struct. Biol.,* **122**, 86–91.

51. Astbury, W. T. and Bell, F. O. (1940) Molecular structure of the collagen fibres, *Nature,* **145**, 421–422.

52. Pauling, L. and Corey, R. B. (1951) The structure of fibrous proteins of the collagen-gelatin group, *Proc. Natl. Acad. Sci. USA*, **37**, 272–281.

53. Ramachandran GN, Kartha G (1954) Structure of collagen, *Nature,* **174**, 269–270.

54. Khoshnoodi, J., Cartailler, J., Alvares, K., Veis, A. and Hudson, B. G. (2006) Molecular recognition in the assembly of collagens: Terminal noncollagenous domains are key recognition modules in the formation of triple helical protomers, *J. Biol. Chem.*, **281**, 38117–38121.

55. Sweeney, S. M., Orgel, J. P., Fertala, A., Mcauliffe, J. D., Turner, K. R., Di Lullo, G. A., *et al.* (2008) Candidate cell and matrix interaction domains on the collagen fibril, the predominant protein of vertebrates, *J. Biol. Chem.*, **283**, 21187–21197.

56. Jenkins, C. L., Vasbinder, M. M., Miller, S. J. and Raines, R. T. (2005) Peptide bond isosteres: Ester or (E)-alkene in the backbone of the collagen triple helix, *Org. Lett.*, **7**, 2619–2622.

57. Myllyharju, J. and Kivirikko, K. (2001) Collagens and collagen-related diseases. *Ann. Med.*, **33**, 7–21.

58. Boryskina, O. P., Bolbukh, T. V., Semenov, M. A., Gasan, A. I. and Maleev, V. Y. (2007) Energies of peptide-peptide and peptide-water hydrogen bonds in collagen: Evidences from infrared spectroscopy, quartz piezogravimetry and differential scanning calorimetry, *J. Mol. Struct.*, **827**, 1–10.

59. Comper, W. D.(1996) *Extracellular matrix: Molecular components and interactions,* Harwood Academic publishers, Amsterdam.

60. Holmgren, S. K., Taylor, K. M., Bretscher, L. E. and Raines, R. T. (1998) Code for collagen's stability deciphered, *Nature*, **392**, 666–667.

61. Holmgren, S. K., Bretscher, L. E., Taylor, K. M. and Raines, R. T. (1999) A hyperstable collagen mimic, *Chem. Biol.*, **6**, 63–70.

62. Bretscher, L. E., Jenkins, C. L., Taylor, K. M., Derider, M. L. and Raines, R. T. (2001) Conformational stability of collagen relies on a stereoelectronic effect, *J. Am. Chem. Soc.*, **123**, 777–778.

63. Giacovazzo, C., Monaco, H., Viterbo, D., Scordari, F., Gilli, G., Zanotti, G., *et al.* (1993) Fundamentals of crystallography (IUCr Texts on Crystallography No. 2.), *Acta Cryst.*, **49**, 373–374.

64. Ramachandran, G. N. and Kartha, G. (1955) Structure of collagen, *Nature,* **176**, 593–595.

65. Panasik Jr, N., Eberhardt, E. S., Edison, A. S., Powell, D. R. and Raines, R. T. (1994) Inductive effects on the structure of proline residues, *Int. J. Pept. Prot. Res.*, **44**, 262–269.

66. Improta, R., Benzi, C. and Barone, V. (2001) Understanding the role of stereoelectronic effects in determining collagen stability. 1. A quantum mechanical study of proline, hydroxyproline, and fluoroproline dipeptide analogues in aqueous solution, *J. Am. Chem. Soc.*, **123**, 12568–12577.

67. Doi, M., Nishi, Y., Uchiyama, S., Nishiuchi, Y., Nakazawa, T., Ohkubo, T., *et al.* (2003) Characterization of collagen model peptides containing 4-fluoroproline;(4 (S)-fluoroproline-Pro-Gly) 10 forms a triple helix,

but (4 (R)-fluoroproline-Pro-Gly) 10 does not, *J. Am. Chem. Soc.*, **125**, 9922–9923.

68. Barth, D., Milbradt, A. G., Renner, C. and Moroder, L. (2003) A (4R)- or a (4S)-fluoroproline residue in position Xaa of the (Xaa-Yaa-Gly) collagen repeat severely affects triple helix formation, *Chem. Bio. Chem.*, **5**, 79–86.

69. Schumacher, M. A., Mizuno, K. and Bachinger, H. P. (2006) The crystal structure of a collagen-like polypeptide with 3 (S)-hydroxyproline residues in the Xaa position forms a standard 7/2 collagen triple helix, *J. Biol. Chem.*, **281**, 27566–27574.

70. Nivison-Smith, L. and Weiss, A. (2011) Elastin based constructs, in *Regenerative Medicine and Tissue Engineering- Cells and Biomaterials*, InTech, Croatia, 323–340.

71. Chalmers, G., Gosline, J. and Lillie, M. (1999) The hydrophobicity of vertebrate elastins, *J. Exp. Biol.*, **202**, 301–314.

72. Tamburro, A. (1981) Elastin: Molecular and supramolecular structure, *Progr. Clin. Biol. Res.*, **54**, 45–62.

73. Keeley, F. W., Bellingham, C. M. and Woodhouse, K. A. (2002) Elastin as a self-organizing biomaterial: use of recombinantly expressed human elastin polypeptides as a model for investigations of structure and self-assembly of elastin, *Philos. T. Roy. Soc. B*, **357**, 185–189.

74. Debelle, L. and Tamburro, A. (1999) Elastin: Molecular description and function, *Int. J. Biochem. Cell B*, **31**, 261–272.

75. Vrhovski, B., Jensen, S. and Weiss, A. S. (2004) Coacervation characteristics of recombinant human tropoelastin, *Eur. J. Biochem.*, **250**, 92–98.

76. Urry, D. W. (1984) Protein elasticity based on conformations of sequential polypeptides: The biological elastic fiber, *J. Prot. Chem.*, **3**, 403–436.

77. Mackay, J. P., Muiznieks, L. D., Toonkool, P. and Weiss, A. S. (2005) The hydrophobic domain 26 of human tropoelastin is unstructured in solution, *J. Struct. Biol.*, **150**, 154–162.

78. Tamburro, A. M., Pepe, A. and Bochicchio, B (2006) Localizing helices in human tropoelastin: Assembly of the elastin "puzzle," *Biochemistry*, **45**, 9518–9530.

79. Debelle, L., Alix, A. J., Jacob, M.-P., Huvenne, J.-P., Berjot, M., Sombret, B., *et al.* (1995) Bovine elastin and k-elastin secondary structure determination by optical spectroscopies, *J. Biol. Chem.*, **270**, 26099–26103.

80. Glagov, S., Vito, R., Giddens, D. and Zarins, C. K. (1992) Micro-architecture and composition of artery walls: Relationship to location, diameter and the distribution of mechanical stress, *J. Hypertens.*, **10**, S101–S104.

81. Mariani, T. J., Sandefur, S., and Pierce, R. A. (1997) Elastin in lung development. *Exp. Lung. Res.*, **23**, 131–145.

82. Roten, S. V., Bhat, S. and Bhawan, J. (1996) Elastic fibers in scar tissue, *J. Cutan. Pathol.*, **23**, 37–42.

83. Pasquali-Ronchetti, I. and Baccarani-Contri, M. (1998) Elastic fiber during development and aging, *Micros. Res. Techniq.*, **38**, 428–435.

84. Mithieux, S. M. and Weiss, A. S. (2005) Elastin, *Adv. Protein Chem.*, **70**, 437–461.

85. Zou, Y. and Zhang, Y. (2009) An experimental and theoretical study on the anisotropy of elastin network, *Ann. Biomed. Eng.*, **37**, 1572–1583.

86. Campa, J. Greenhalgh, R. And Powell, J. T. (1987) Elastin degradation in abdominal aortic aneurysms, *Arteriosclerosis*, **65**, 13–21.

87. Ito, S., Ishimaru, S. and Wilson, S. E. (1998) Effect of coacervated alpha-elastin on proliferation of vascular smooth muscle and endothelial cells, *Angiology*, **49**, 289–297.

88. Wilson, B. D., Gibson, C. C., Sorensen, L. K., Guilhermier, M. Y., Clinger, M., Kelley, L. L., *et al.* (2011) Novel approach for endothelializing vascular devices: Understanding and exploiting elastin-endothelial interactions, *Ann. Biomed. Eng.*, **39**, 337–346.

89. Williamson, M. R., Shuttleworth, A., Canfield, A. E., Black, R. A. and Kielty, C. M. (2007) The role of endothelial cell attachment to elastic fibre molecules in the enhancement of monolayer formation and retention, and the inhibition of smooth muscle cell recruitment, *Biomaterials*, **28**, 5307–5318.

90. Mochizuki, S., Brassart, B. and Hinek, A. (2002) Signaling pathways transduced through the elastin receptor facilitate proliferation of arterial smooth muscle cells, *J. Biol. Chem.*, **277**, 44854–44863.

91. Pankov, R. and Yamada, K. M. (2002) Fibronectin at a glance, *J. Cell Sci.*, **115**, 3861–3863.

92. Hirschi, S. D., Gray, S. D. and Thibeault, S. L. (2002) Fibronectin: An interesting vocal fold protein, *J. Voice*, **16**, 310–316.

93. Plow, E. F., Haas, T. A., Zhang, L., Loftus, J. and Smith, J. W. (2000) Ligand binding to integrins, *J. Biol. Chem.*, **275**, 21785–21788.

94. Valenick, L. V., Hsia, H. C. and Schwarzbauer, J. E. (2005) Fibronectin fragmentation promotes alfa4beta1 integrin-mediated contraction of a fibrin-fibronectin provisional matrix, *Exp. Cell Res.*, **309**, 48–55.

95. Canham, P. B., Finlay, H. M., Dixon, J. G., Boughner, D. R. and Chen, A. (1989) Measurements from light and polarised light microscopy of human coronary arteries fixed at distending pressure, *Cardiovasc. Res.*, **23**, 973–982.

96. Finlay, H., Mccullough, L. and Canham, P. (1995) Three-dimensional collagen organization of human brain arteries at different transmural pressures, *J. Vasc. Res.*, **32**, 301–312.

97. Driessen, N. J., Cox, M. A., Bouten, C. V. and Baaijens, F. P. (2008) Remodelling of the angular collagen fiber distribution in cardiovascular tissues, *Biomech. Model. Mechan.*, **7**, 93–103.

98. Clark, J. M. and Glagov, S. (1985) Structural integration of the arterial wall, *Lab Invest.*, **40**, 587–602.

99. Rhodin, J. A.(2011) Architecture of the vessel wall, in *Handbook of Physiology, the Cardiovascular System,* Wiley, J. and Sons, Ltd., Online Library, DOI 10.1002/cphy.cp020201.

100. Gasser, T. C., Ogden, R. W. and Holzapfel, G. A. (2006) Hyperelastic modelling of arterial layers with distributed collagen fibre orientations, *J. Roy. Soc. Interf.*, **3**, 15–35.

101. Shekhonin, B., Domogatsky, S., Muzykantov, V., Idelson, G. and Rukosuev, V. (1985) Distribution of type I, III, IV and V collagen in normal and atherosclerotic human arterial wall: Immunomorphological characteristics, *Collagen Rel. Res.*, **5**, 355–368.

102. Olszta, M. J., Cheng, X., Jee, S. S., Kumar, R., Kim, Y.-Y., Kaufman, M. J., *et al.* (2007) Bone structure and formation: A new perspective, *Mat. Sci. Eng. R.,* **58**, 77–116.

103. Bilezikian, J. P., Raisz, L. G. and Rodan, G. A. (2002) Principles of bone biology, in *Principles of Bone Biology* (Bilezikian JP, Raisz LG, Rodan GA, eds), Academic Press, San Diego, USA, 51–68.

104. Butler, D. L., Goldstein, S. A., and Guilak, F. (2000) Functional tissue engineering: The role of Biomechanics, *J. Biomech. Eng.,* **122**, 570–575.

105. Carter, D. and Hayes, W. (1977) The compressive behavior of bone as a two-phase porous structure, *J. Bone Joint Surg. Am.*, **59**, 954–962.

106. Karageorgiou, V. and Kaplan, D. (2005) Porosity of 3D biomaterial scaffolds and osteogenesis, *Biomaterials,* **26**, 5474–5491.

107. Buckwalter, J., Glimcher, M., Cooper, R. and Recker, R. (1996) Bone biology. I: Structure, blood supply, cells, matrix, and mineralization, *Instr. Course Lect.* **45**, 371–386.

108. Watanabe, H., Yanagisawa, T. and Sasaki, J. (1995) Cytoskeletal architecture of rat calvarial osteoclasts: Microfilaments, intermediate filaments, and nuclear matrix as demonstrated by detergent perfusion, *Anat. Record*, **243**, 165–174.

109. Kafka, V. (2002) Surface fissures in articular cartilage: New concepts, hypotheses and modelling, *Clin. Biomech.*, **17**, 73–80.

110. Muldrew, K. (2002) Osteoarthritis as an inevitable consequence of the structure of articular cartilage, *Med. Hypotheses*, **59**, 389–397.

111. Buma, P., Meel, M. V., Tienen, T. G. V. and Veth, R. P. (2008) Tissue engineering of the meniscus, in *Orthopedic Biology and Medicine: Musculoskeletal Tissue Reg*eneration (Pietrzk WS, ed), Humana Press, Totowa, NJ, 327–346.

112. Aspden, R., Yarker, Y. and Hukins, D. (1985) Collagen orientations in the meniscus of the knee joint, *J. Anat.*, **140**, 371–380.

113. Setton, L. A., Guilak, F., Hsu, E. W. and Vail, T. P. (1999) Biomechanical factors in tissue engineered meniscal repair, *Clin. Orthopaed. Relat. R.*, **367S**, S254–S272.

114. Gillies, A. R. and Lieber, R. L. (2011) Structure and function of the skeletal muscle extracellular matrix, *Muscle Nerve*, **44**, 318–331.

115. Velleman, S. G. (1999) The role of the extracellular matrix in skeletal muscle development, *Poultry Sci.*, **78**, 778–784.

116. Hinds, S., Bian, W., Dennis, R. G. and Bursac, N. (2011) The role of extracellular matrix composition in structure and function of bioengineered skeletal muscle, *Biomaterials*, **32**, 3575–3583.

117. Ferguson, M. W. and O'kane, S. (2004) Scar-free healing: From embryonic mechanisms to adult therapeutic intervention, *Philos. T. Roy. Soc. B*, **359**, 839–850.

118. Schultz, G. S., Ladwig, G. and Wysocki, A. (2005) Extracellular matrix: review of its roles in acute and chronic wounds, *World Wide Wounds*, Published online URL: http://www.worldwidewounds.com/2005/august/Schultz/Extrace-Matric-Acute-Chronic-Wounds.html

119. Lawrence, W. T. (1998) Physiology of the acute wound, *Clin. Plast. Surg.*, **25**, 321–340.

120. Bennett, N. T. and Schultz, G. S. (1993) Growth factors and wound healing: Part II. Role in normal and chronic wound healing, *Am. J. Surg.*, **166**, 74–81.

121. Bennett, N. T. and Schultz, G. S. (1993) Growth factors and wound healing: biochemical properties of growth factors and their receptors, *Am. J. Surg.*, **165**, 728–737.

122. Sorokin, L. (2010) The impact of the extracellular matrix on inflammation, *Nat. Rev. Immunol.*, **10**(10), 712–723.

123. Nourshargh, S., Hordijk, P. L. and Sixt, M. (2010) Breaching multiple barriers: Leukocyte motility through venular walls and the interstitium, *Nat. Rev. Mol. Cell Biol.*, **11**, 366–378.

124. Vestweber, D. (2007) Adhesion and signaling molecules controlling the transmigration of leukocytes through endothelium, *Immunol. Rev.*, **218**, 178–196.

125. Occleston, N. L., Metcalfe, A. D., Boanas, A., Burgoyne, N. J., Nield, K., O'kane, S., *et al.* (2010) Therapeutic improvement of scarring: mechanisms of scarless and scar-forming healing and approaches to the discovery of new treatments, *Dermatol. Res. Prac.*, Article **ID 405262**, doi:10.1155/2010/405262

126. Bullard, K. M., Longaker, M. T. and Lorenz, H. P. (2003) Fetal wound healing: current biology, *World J. Surg.*, **27**, 54–61

127. Scott Adzick, N., Harrison, M. R., Glick, P. L., Beckstead, J. H., Villa, R. L., Scheuenstuhl, H., *et al.* (1985) Comparison of fetal, newborn, and adult wound healing by histologic, enzyme-histochemical, and hydroxyproline determinations, *J. Pediatr. Surg.*, **20**, 315–319.

128. Krummel, T. M., Nelson, J. M., Diegelmann, R. F., Lindblad, W. J., Salzberg, A. M., Greenfield, L. J., *et al.* (1987) Fetal response to injury in the rabbit, *J. Pediatr. Surg.*, **22**, 640–644.

129. Lorenz, H., Lin, R., Longaker, M., Whitby, D. and Adzick, N. (1995) The fetal fibroblast: The effector cell of scarless fetal skin repair, *Plast. Reconstr. Surg.*, **96**, 1251–1261.

130. Lin, R. Y., Sullivan, K. M., Argenta, P. A., Peter Lorenz, H. and Scott Adzick, N. (2002) Scarless human fetal skin repair is intrinsic to the fetal fibroblast and occurs in the absence of an inflammatory response, *Wound Repair Regen.*, **2**, 297–305.

131. Merkel, J. R., Dipaolo, B. R., Hallock, G. G. and Rice, D. C.(1988) Type I and type III collagen content of healing wounds in fetal and adult rats, *Roy. Soc. Med.*, **187**, 493–497.

132. Burd, D. A. R., Longaker, M. T., Scott Adzick, N., Harrison, M. R. and Paul Ehrlich, H. (1990) Fetal wound healing in a large animal model: The deposition of collagen is confirmed, *Brit. J. Plast. Surg.*, **43**, 571–577.

133. Olutoye, O. O. and Cohen, I. K. (1996) Fetal wound healing: An overview, *Wound Repair Regen.*, **4**, 66–74.

134. Bullard, K. M., Cass, D. L., Banda, M. J. and Adzick, N. S. (1997) Transforming growth factor beta-1 decreases interstitial collagenase in healing human fetal skin, *J. Pediatr. Surg.*, **32**, 1023–1027.

135. Whitby, D. J. and Ferguson, M. W. J. (1991) Immunohistochemical localization of growth factors in fetal wound healing, *Dev. Biol.*, **147**, 207–215.

136. Sullivan, K. M., Lorenz, H. P., Meuli, M., Lin, R. Y. and Adzick, N. S. (1995) A model of scarless human fetal wound repair is deficient in transforming growth factor beta; papers presented at the 25th Annual Meeting of the American Pediatric Surgery Association, **30**, 198–203.

# Chapter 3

# The Premise of Tissue Engineering: Molecular Recognition

**Christopher J. Pateman and John W. Haycock**
*Department of Materials Science and Engineering, Kroto Research Institute, University of Sheffield, United Kingdom*
j.w.haycock@sheffield.ac.uk

Tissue engineering and regenerative medicine requires an intimidate understanding of the native extracellular matrix (ECM), together with the complexities of cell and tissue biology. The ECM is known to control many physiological functions including the maintenance of homeostasis, growth during development, and coordinating the stages of wound healing. For therapeutic applications, investigators have attempted to mimic the basic structure of the ECM using a variety of synthetic or naturally derived materials. Both approaches have associated strengths and weakness, but a major drive in modern regenerative medicine research is in the development of three-dimensional (3D) growth environments. This is a departure from the "traditional" use of 2D cell cultures and the understanding that to recapitulate native tissue physiology, or to create an implant with directed function, one can create biomimetic scaffolds. These

*Scaffolds for Tissue Engineering: Biological Design, Materials, and Fabrication*
Edited by Claudio Migliaresi and Antonella Motta

ISBN 978-981-4463-20-1 (Hardcover), 978-981-4463-21-8 (eBook)
www.panstanford.com

approaches combine the basic structural properties of the ECM, and in addition, integrate molecular cues for controlling biological response, by incorporating cell-adhesive ligands, protease sensitive sites for enabling migration, and the local delivery of soluble factors for stimulating tissue remodelling. In this review we consider the development of molecular recognition in combination with the importance of physical and chemical cues for regenerative strategies.

## 3.1 Introduction

Unicellular organisms are the simplest known forms of life and are the most prolific living organisms due to their inherent ability to inhabit a diverse range of environments. However, in order for a more complex multicellular organism to evolve, the communicative processes between discrete cellular units and their environmental interpretation must be such that a constructive unison of purpose can take place. In this situation, individual cells typically perform highly specialized roles within the organism to perform specific tasks. This is clearly evident when considering the number of different cell types present within a large multicellular organism; the diverse range of functions they adopt; and the careful interplay between cells, tissues, and organs. The increased size and greater complexity of external environmental response provides an organism with a powerful evolutionary advantage that has driven this process to take place.

The ability for individual cells to interpret and react to their microenvironment is fundamental in both the development and maintenance of physiological function *in vivo*, but is an equally powerful factor when considering *in vitro* cell culture or the design of materials and constructs for tissue engineering and regenerative medicine.[1,2] Cell–surface interactions have a single unifying dependency, and that is the character of the interfacial region between the cell and material surface. This region in turn is dependent on a number of specific factors, and by understanding the principles behind these interactions surfaces can now be tailored for a wide variety of biological functions, from increased cellular adhesion and migration to the synthesis of new tissue *in situ*.

## 3.2 Biomaterials: From Medical Devices to Tissue Scaffolds

Traditionally, materials for medical devices have been chosen due to their favourable physical bulk properties, availability, and tolerance within the host. The use of medical-grade stainless steels or titanium for orthopedic replacements through to the development of ceramic materials for dental applications is now extremely commonplace. Implantation in several thousands of individuals reveals the many successes of these biomaterials. However, not all materials (or surgical approaches) are without problems. Orthopedic implants suffer from wear debris formation limiting the useful lifespan of a hip device typically to 20 years.[3,4] With improved medical care and extended human lifespan there is an increasing number of prosthetic implants being inserted into increasingly older patients. Due to this and the limited capacity for a patient to accept a multiple number of successive implants, a drive exists to improve the efficacy of prosthetic implants. Many materials while being approved at a regulatory level, and consequently deemed to be safe, are quite often inert. While the body does not generate a particular response, the material arguably has a limited usefulness as a strategy for stimulating tissue repair. Through an array of modern investigative techniques however it has become apparent that many discrete cellular events are controlled by the external environment, and in particular through the interfacial region. This is now regarded as of key importance in order to induce a favorable host response. Generally speaking, this environment falls into two areas, the cellular response induced by manipulating the physical properties of the underlying bulk material and the physicochemical environment at the cell–material surface interface.

Biomaterials play a central role in regenerative medicine and their design can directly influence cellular function, either as a three-dimensional (3D) cell-containing construct, or as a cell-free device. Cell-containing approaches can be aimed at controlling tissue morphogenesis by the delivery of constructs manufactured *ex vivo*, whereas cell-free materials can stimulate cell and tissue growth *in situ*. Cell-free biomaterials have an advantage in usually being available as "off the shelf" for clinical use, but may have limited effectiveness compared to cell-containing devices. Modern types of biomaterials are typically classified as being either natural or synthetic in origin.[1]

Natural materials have a distinct advantage of containing intrinsic recognition sites such as native ligand sequences for stimulating biological recognition. However, difficulties in the translatable use of natural materials such as contamination, purity, and immunogenicity have underpinned the development of synthetic materials for regenerative strategies. Over the last decade rapid advances have been made in the complexity of synthetic biomaterials. Rather than the utilisation of simple polymeric materials as implantable devices, materials are now being fabricated that take in to account the intricate design of extracellular matrix (ECM) motifs for controlling specific biological function.[1]

In order to understand the complexity of how cells behave within a tissue, scientists need to utilise appropriate experimental models. The most widely conducted cell biology technique is *in vitro* cell culture in two dimensions, most often on a "treated" polystyrene "Tissue Culture Plastic" well plate, or Petri dish.[5] Treatment is usually by ozonolysis or ultraviolet irradiation to impart a surface charge to the polystyrene groups, and in turn the increased adsorption of soluble proteins in the culture, necessitating a permissive level of cell adhesion and growth. However, a major limiting factor of 2D cell culture is the relatively poor physiological representation and structural organisation compared to cell physiology *in vivo*. The function of a normal cell from division, through proliferation to migration and apoptosis requires an accurately controlled series of events that inherently relies on the principles of spatial and temporal organisation. The culture of cells in two dimensions is arguably far too simple and overlooks many parameters known to be important for reproducing native cell and tissue physiology. These include mechanical cues, communication between the cell and its matrix, and communication between adjacent cells. On the point of intercellular communication, many 2D culture experiments do not consider the paracrine interplay between different cell types, with many cell culture studies being of a single type.

A number of 3D methods have therefore been developed for a range of tissues in which the culture environment takes in to account the spatial organisation of the cell (reviewed in Ref. 6). A common goal for many of these studies is to bridge the gap between the uses of whole animals at one end of the spectrum, with cellular monolayers at the other. It is therefore necessary to create a growth environment that mimics the native tissue as closely as possible, and a simple starting point is the introduction of cells in to a porous biocompatible scaffold. Three-dimensional culture models can be grouped in to the study of whole animals and organotypic explant cultures (including embryos), cell spheroids, microcarrier cultures, and tissue engineered models.[6] While not all 3D culture models require a scaffold, the development of scaffolds for 3D models has increased rapidly in the past 10 years (reviewed in Ref. 7).

### 3.2.1 Natural ECM materials

ECMs in nature provide a physical scaffold or framework upon which physiologically relevant forces are exerted.[8] Examples of these materials include laminin, elastin, and collagen. Collagen in particular has remarkable tensile strength due to the organisation of a microfibrillar triple helical structure. Likewise, elastin has a physical role in the matrix, but in contrast to collagen it has a disorganised fibrous structure, which facilitates its elastic properties. The mechanical properties of matrix proteins are being increasingly recognised as important for influencing cell function. Cells contained within a matrix are in intimate contact with it through specific binding motifs. Transmembrane integrin receptors in particular enable a continuous connection between intracellular cytoskeletal proteins such as filamentous actin with extracelluar matrix proteins.[9,10] Through these tight formations, mechanisms exist that enable cells to sense external mechanical stimuli, and by conversion to chemical signals, respond with changes in migration and adhesion.[11–13]

## 3.3 Physical Properties of Scaffolds that Influence Molecular Bio-Recognition

An increasing number of reports document the physical property effects of the adherent surface, usually expressed as the modulus of elasticity, having a distinct biological importance to cellular function.[13–15] When considering cells in their natural physiological state, one can easily appreciate the different physical properties imparted on them due to their particular anatomical site, whether a high-modulus bone material or a lower modulus softer tissue such as nerve or cartilage. The character of the ECM, which is synthesised and continually maintained by stromal cells, plays a key role in maintaining tissue homeostasis and structural integrity. The nature of the cell–matrix interaction in many ways is conserved, with integrin receptors binding to the arginine–glycine–aspartic acid (RGD) motif commonly arising in many tissues.[16–19] However, the underlying physical properties that dictate the extent of movement between these points are different. One can make the analogy of pins in a pincushion, where the pinheads are the adhesion points and the cushion is the underlying matrix. If the cushion is soft, the pinheads can readily move with a large range of motion. But if the pins are embedded in a stiffer matrix such as a solidified resin, a much more limited range of movement results. Thus, while the points of interaction are similar, the underlying matrix stiffness can communicate different levels of mechanical stress.

As mentioned above, cells interact with their extracellular environment by specific membrane-bound proteins receptors. These include cadherins, selectins, the immunoglobulin superfamily of receptors, and the transmembrane receptor glycoprotein family of integrins.[9,10] Integrins contain inter- and intra-membrane chains and transduce information between the ECM and the intracellular environment and vice versa. Specifically, they are formed from heterodimers composed of a pair of $\alpha$- and $\beta$-glycoprotein chains. Different $\alpha$ and $\beta$ chain combinations allows for a high degree of specificity between integrin pairs and ECM motifs, and thus a high level of control for the accurate modulation of the cell–ECM response. A detailed review of integrins is available elsewhere, including a comprehensive table of chain types and complementary recognition sites.[9]

Cells not only interact with other cells via cell junctions, but also take cues from the ECM, which is commonly comprised of various collagens. The physical interaction between cells and collagen fibres is relatively low, with an interwoven matrix of fibronectin and laminin predominantly acting as the adhesion mechanism between cells and the collagen-containing ECM. This allows cells to detect physical changes within their host environment and modulate their response. Instances of where the modulation of cellular responses via cell surface integrins takes place includes embryogenesis;[20] growth and development and wound healing, such as in peripheral nerve repair.[21–23] Consequently, integrins play a key role in determining cell metabolic rate, functionality, protein synthesis, motility, and adherence.

Laminin is also a major component of the ECM responsible for influencing cell function.[24,25] These molecules are glycoproteins with a relatively large molecular weight (900 kDa). They are usually found in the basement membrane and form an important component of cell–ECM interactions via integrin receptors. A number of specific sequence motifs have been isolated from laminin that give rise to integrin conjugation and include isoleucine-lysine-valine-alanine-valine (IKVAV)[26] and tyrosine–isoleucine–glycine–serine–arginine (YIGSR).[27,28] It has been found that laminin has wide number of effects on cell function, including adhesion, migration, growth, differentiation, and neurite extension.[26] Because of these features, laminin and its motifs are of great interest for improving biomaterials with advantageous properties by incorporation or modification of these sequences.[29–34] In a comprehensive study of laminin, through motif simulation using synthetic peptides and blocking, it has been shown that laminin can specifically stimulate cell adhesion and neurite extension *in vitro*.[26] Indeed, IKVAV synthetic peptides when coated on a number of materials including expanded polytetrafluoroethylene (ePTFE) and titanium increase cell attachment.[30] With respect to neural regeneration, it has been shown that laminin and more specifically the IKVAV motif is particularly important.[35]

Many cellular responses to short peptide motifs arise through the integrin–RGD interface, which arguably is the most studied of the cell–matrix interfaces. RGD is present in many adhesion-mediating proteins and has been known for its role in integrin binding for a number of years. Significant advancements in the 1970s by Grinnell *et al.*[36–38] highlighted the significance of serum proteins within tissue culture medium for adhesion and proposed two minimum requirements for cell adhesion—the cell-surface adhesive bond itself and the ability of the cell to utilize this bond for attachment and spreading. The requirement of just the RGD three amino acid motif for adhesion, rather than a complete protein structure was elucidated by studies that employed targeted antibody binding.[16] This allowed for synthetic replication of the sequence.[17] Key studies have since identified critical parameters for dictating cellular response. For example, the minimum distance between consecutive RGD motifs on a 2D surface for determining fibroblast adhesion and spreading is 440 nm (via the $\alpha V\beta 3$ integrin receptor).[39] When this distance is decreased to 140 nm, fibroblast focal contact and stress fibre formation is observed.

Such work has led to the inclusion of RGD and analogous motifs in to biomaterials for forming tissue scaffolds with improved cell responsive properties. A number of comprehensive reviews on this topic are available in detail elsewhere.[40–42] However, particular examples include RGD-functionalized nerve guides for improving cellular adhesion and motility and in turn improving injury gap regeneration distances. This has been investigated *in vivo* using rat sciatic nerve injury models[43] and *in vitro* using dendritic cells.[44] Results too show that nerve guides made from collagen functionalized with RGD increase regeneration distance significantly compared to collagen only non-functionalized guides.[45,46] More detail on the role of adhesion molecules for neural development can be found in Ref. 47.

### 3.3.1 Modification at the Interface

The ability to create bioactive materials from otherwise inert biomaterials has seen some interesting developments in the technologies of plasma polymerization.[48,49] Here, a nearly atomically thin highly controllable pinhole-free layer of a specific chemistry can be deposited on a wide range of materials for controlling cell function. Considerable developmental work has been conducted using 2D substrates for defining the plasma chemical groups and surface density, with parallel chemical evaluation by X-ray photoelectron spectroscopy (XPS) and surface ion mass spectrometry (SIMS). The principle has been to "convert" poorly supportive materials in respect of cell adhesion and proliferation in to materials that are active in supporting growth. Acrylic acid deposited layers have been found to enable the growth of a wide range of eukaryotic cells, including keratinocytes,[50–54] osteoblasts,[55,56] and Schwann cells.[57] The principle of applying this deposition method has also been translated through to the surface modification of 3D porous scaffolds of poly (lactic-co-glycolic) acid.[58,59] The authors of these studies conclude that the plasma deposition of a charged surface group improves cellular attachment and growth via the adsorbed protein layer, which either increases the amount of protein binding at the surface or facilitates the optimal organization of a favorable combination or conformation of proteins at the interface. A major advantage of this technology is the ability to scale up and translate plasma surface modification once deposition conditions are defined.[60]

The ability to control particular cellular responses has also been exploited by the development of rapid "dip-and-dry" approaches. For example, the use of highly adhesive asymmetric molecules called resorcinarene peptides[61] can convert inert materials in to anti-inflammatory materials. Here, the authors designed a chiral molecule containing three sections—an adhesive calix, a lining polyethylene glycol (PEG) chain, and a short peptide (GKPV) known to inhibit inflammatory signaling. The chiral molecule could be dissolved in a solvent and the biomaterial surface "activated" simply by immersion in the solution. The adhesive groups are reported to adhere to the biomaterial via non-covalent interactions, with the result of organizing a layer of anti-inflammatory peptides uppermost. XPS and SIMS analysis revealed the presence of amine and GKPV groups, and *in vitro* culture of fibroblast and Schwann cells when grown on active surfaces and challenged with the pro-inflammatory cytokine TNF-$\alpha$ demonstrated a reduced NF-$\kappa$B response compared to the same cells grown on a control surface (containing the calix and PEG sections, but without the GKPV peptide). This approach, as with plasma polymerization, is readily scalable, broadly applicable to a number of different biomaterials, and simple to use, if the desired response of the biomaterial is for inhibiting inflammation.

### 3.3.2 Stem Cell Differentiation: Physical and Topographic Properties

Stem cell research has greatly intensified over the last decade as their potential for therapeutic use has been recognized, which has excited both the scientific community and the wider public. Stem cells are of fundamental importance during embryogenesis and lifelong tissue maintenance, but the potential offered in treating numerous clinical disorders from Parkinson's disease,[62–65] heart tissue,[66] diabetes,[67–71] or spinal cord repair is considerable. While plasticity is of interest, an ability to proliferate stem cells without losing plasticity provides a route for the *in vitro* expansion for producing therapeutic relevant cell numbers. A great deal of research has traditionally been given to the role of soluble signaling agents such as growth factors for inducing stem cell behavior.[72] However, increasing evidence is emerging on the role of the microenvironment known as the stem cell niche[73] and the mechanical role this plays on stem cell phenotype.[74–76] The stem cell niche is influenced by a particular arrangement and interaction of ECM proteins, cell adhesion receptors, and underlying tissue properties, which can vary from being relatively soft and elastic through to stiff and brittle depending upon the tissue type. By quantifying and subsequently adopting similar physical conditions, the potential exists for inducing stem cells to differentiate based on the specific culture surfaces *in vitro*.[77,78] For example, it has been recently described that when mesenchymal stem cells are cultured on surfaces of varying stiffness, harder materials direct cells along an osteoblastic lineage, whereas softer substrates direct cells along chrondrogenic and neural lineages.[14,77]

Topography has an important influence in the natural development of an organism and also in processes such as wound healing following tissue damage. These features are produced naturally by the local ECM. However, studies are emerging on the relationship between substrate topography and cell behaviour, which reveals a critical dependence on length scale, feature size, and order. Cells are entirely dependent upon their particular membrane bound adhesion molecules and their relationship with internal cytoskeletal structures for sensing such features. Consequently, surface features of a similar or greater size than that of a given cell are unlikely to have a discernible effect at this length scale, as most eukaryotic stromal cells are of the order of 10–100 μm. Consequently relevant feature sizes are one to three orders of magnitude below this, predominantly in the range of 1 nm to 1 μm.

More recently, the role of topography in influencing stem cell differentiation has been reported. Nano-topographies produced on poly (methyl methacrylate) surfaces using electron beam lithography (EBL) are reported to direct bone marrow stromal cells to differentiate into osteoblast-like cells.[79] Cell spreading and cytoskeletal organisation was increased on modified surfaces compared to flat controls. This study was followed up by a second report that showed measured increases in osteopontin and osteocalcin production.[80] The effect of nano-feature sized disorder was than studied, and it was found that both fully ordered and completely random patterns produced less osteogenesis, in contrast to features that were slightly displaced from the ordered array.[80–82] The authors suggest this may be due to the length of fibrillar adhesions formed by the stem cells. Slightly displaced features may permit longer structures to be formed, which in turn encourages differentiation. A particularly interesting observation was that the slightly displaced features supported osteogenic differentiation in the absence of dexamethasone, a classical soluble mediator of stem cell osteoblastic differentiation.

The length scale by which the substrate influences cell response impacts upon the design of scaffolds for tissue regeneration. It is quite feasible to produce structures with nanometer-sized features by techniques such as electrospinning, whereby fibre diameters down to 10 nm can be readily formed.[83–91] The use of such nano-scale dimensions for scaffolds is argued to be more relevant as a method for forming biologically inspired materials compared to micro-sized features, as adherent interactions between the cell and the substrate are analogous to the number and 3D position of the native cell-matrix interface.[2] Nano-fibrillar matrixes with biophysical properties can also be formed by self-assembly techniques. For example, the use of self-complementary amphiphilic peptides can be gelled under isotonic and neutral pH conditions to facilitate the growth of cartilage and nerve cells.[92–94] These approaches provide physical scaffolds appropriate for supporting cell growth, without the presence of biologically active motifs. However, the approach can be extended to the self-assembly of oligomeric amphiphiles with the inclusion of molecules such as the IKVAV laminin-binding motif.[95,96] Such materials have been shown to stimulate the differentiation of progenitor cells specifically along a neural lineage.

While the value of stem cells therapeutically is considerable, a number of practical issues must be addressed before their use can be considered as safe in the clinic.[97] Controlling stem cell differentiation is of primary importance and this is relevant when cells divide and leave the stem cell niche.[98] Here, a complex combination of both soluble factors and ECM protein interactions control the cell response, either toward self-renewal or differentiation. To address this, synthetic materials are now being designed that can recapitulate the natural stem cell niche. Recent studies have reported on the creation of a set of synthetic ECM proteins that mimic this environment and in turn control stem cell function. For example, the synthesis of a recombinant elastin core coupled to the Notch receptors, hDelta1 and hJagged1, has been shown to control neural phenotype and function.[99,100] Parallel work focusing on combinatorial polymer surfaces has been described for the selection of stem cells via high-throughput methods.[101] This may be of future value for selecting target cells expressing desired functional groups for inclusion in regenerative scaffolds.

### 3.3.3 Multi-functional Biological Matrices for Directing Cellular Response

The design of synthetic ECMs that recapitulate the natural matrix, but in addition encode within their structure signals for directing specific biological responses are now being developed. This includes motifs integrated within hydrogel scaffolds that stimulate migration, cellular remodeling, and tissue regeneration. This includes the work of Hubbell and colleagues in the design of multi-functional hydrogels, which importantly are created under physiological conditions.[1] They typically employ the use of precursor materials such as self-assembling peptides comprised of adhesion ligands. But for stimulating processes such as 3D migration, cells must combine a series of highly coordinated events. This includes the behavior of polarization, pseudopodia protrusion, firm adhesion of protrusions with the ECM, remodeling the intracellular cytoskeleton, and contraction for forward movement, together with detachment at the rear of the cell. Importantly, these processes must be simultaneously conducted while overcoming the physical resistance of moving through the matrix bulk. Migration through the ECM involves cellular-mediated proteolysis, a process that is both cell and matrix dependent.[102] This is in contrast to non-proteolytic migrations assumed by specialized cells such as leucocytes, which typically "squeeze" through matrix pores. Proteolytic migration enables cells to create a transient path through the matrix by carefully regulating the secretion of matrix metalloproteinase (MMP), serine proteases, and hyaluronidase. Degradation is a very controlled process, as the proteases that are typically membrane bound, restrict the remodeling of proteoglycans solely to the migratory path.

Consequently, these proteolytic pathways have been exploited and utilized in the design of synthetic biomimetic materials, which contain a combination of adhesive ligands and protease sensitive sites to direct a particular cell response. For example, the photo-initiated polymerization of acrylated PEGs containing pendant RGD peptides with substrates for MMPs and fibrin (or fabricated by conjugation) have been made.[103–105] Biological evaluation of these materials has demonstrated a controllable ability to support the 3D migration of fibroblast and endothelial cells within a synthetic ECM. This idea can be applied to materials for stimulating tissue regeneration where primary the requirements include the presence of cell adhesion ligands, a mechanism for transient and controllable local matrix degradation and the delivery of a signal for directing cellular differentiation. Materials have been designed which are comprised of an MMP (or plasmin-sensitive) cleavage site, integrin-binding ligands in the form of RDG peptides, and a growth factor component as BMP-2 for stimulating bone growth.[106,107] When the construct was implanted in to a bone defect the biomimetic material remodelled into bone over time. A similar approach has also been described for blood vessel formation, whereby biomimetic scaffolds have been fabricated containing MMP protease sensitive sites, integrin-binding motifs, and a soluble factor in the form of VEGF for stimulating blood vessel formation.[108] Implantation revealed the formation of healthy vessels, but of particular interest was the relationship between the soluble factor concentration and host response.[109] Low VEGF concentrations in the matrix were associated with a well-formed vasculature, whereas high concentrations stimulated the formation of hyper-permeable structures.

## 3.4 Conclusions and Future Directions

Tremendous progress has been made to date on the development of biomaterials for tissue engineering and regenerative medicine; however, much work remains if biomaterials are to be used routinely for repair and regeneration. In this review, we consider the approaches relevant to tissue engineering and regenerative medicine that combine the basic structural properties of the ECM, with the integration of molecular cues for controlling biological response. The development of "traditional" medical device biomaterials is still very active, but new avenues in the design of implantable materials are seeing increased complexity by incorporating cell-adhesive ligands, protease sensitive sites for enabling migration, and the local delivery of soluble factors for stimulating tissue remodelling. The biochemical environment and the chemical interface between the matrix and the cell is a major determining factor for controlling cell function, but new research on the potential of stem cells and the niche is seeing research being translated in to the fabrication of biomaterials. New areas of development are seeing the physical environment of the matrix as additionally influential, with studies demonstrating the relationship between the modulus of materials and the differentiation of stem cells. Similar work is also emerging on the nano-topography of materials and ordered versus disordered structures having a similar role.

## References

1. Lutolf, M. P., and Hubbell, J. A. (2005) Synthetic biomaterials as instructive extracellular microenvironments for morphogenesis in tissue engineering, *Nat. Biotechnol.*, **23**, 47–55.
2. Steven, M. M., and George, J. H. (2005) Exploring and engineering the cell surface interface, *Science*, **310**, 1135–1138.
3. Hatton, A., Nevelos, J. E., Nevelos, A. A., Banks, R. E., Fisher, J., and Ingham, E. (2002) Alumina-alumina artificial hip joints. Part I: A histological analysis and characterisation of wear debris by laser capture microdissection of tissues retrieved at revision, *Biomaterials*, **23**, 3429–3440.
4. Tipper, J. L., Hatton, A., Nevelos, J. E., Ingham, E., Doyle, C., Streicher, R., *et al.* (2002) Alumina-alumina artificial hip joints. Part II: Characterisation of the wear debris from *in vitro* hip joint simulations, *Biomaterials*, **23**, 3441–3448.
5. Amstein, C. F., and Hartman, P. A. (1975) Adaptation of plastic surfaces for tissue-culture by glow-discharge, *J. Clin. Microbiol.*, **2**, 46–54.
6. Haycock, J. W. (2011) 3D cell culture: A review of current approaches and techniques, *Methods. Mol. Biol.*, **695**, 1–15.
7. Carletti, E., Motta, A., and Migliaresi, C. (2011) Scaffolds for tissue engineering and 3D cell culture. *Methods Mol. Biol.*, **695**, 17–39.
8. Kleinman, H. K., Philp, D., and Hoffman, M. P. (2003) Role of the extracellular matrix in morphogenesis, *Curr. Opin. Biotech.*, **14**, 526–532.
9. Hynes, R. O. (1992) Integrins: Versatility, modulation, and signaling in cell adhesion, *Cell*, **69**, 11–25.
10. Richard, O. H. (1987) Integrins: A family of cell surface receptors, *Cell*, **48**, 549–554.
11. Galbraith, C. G., Sheetz, M. P. (1998) Forces on adhesive contacts affect cell function, *Curr. Opin. Cell. Biol.*, **10**, 566–571.
12. Geiger, B., Bershadsky, A., Pankov, R., and Yamada, K. M. (2001) Transmembrane extracellular matrix-cytoskeleton crosstalk, *Nat. Rev. Mol. Cell. Bio.*, **2**, 793–805.

13. Yang, Y., Magnay, J. L., Cooling, L., and El Haj, A. J. (2002) Development of a 'mechano-active' scaffold for tissue engineering, *Biomaterials*, **23**, 2119–2126.

14. Rehfeldt, F., Engler, A. J., Eckhardt, A., Ahmed, F., and Discher, D. E. (2007) Cell responses to the mechanochemical microenvironment--Implications for regenerative medicine and drug delivery, *Adv. Drug Deliver. Rev.*, **59**, 1329–1339.

15. Ayala, R., Zhang, C., Yang, D., Hwang, Y., Aung, A., Shroff, S. S., *et al.* (2011) Engineering the cell-material interface for controlling stem cell adhesion, migration, and differentiation, *Biomaterials*, **32**, 3700–3711.

16. Pierschbacher, M. D., Hayman, E. G., Ruoslahti E. (1981) Location of the cell-attachment site in fibronectin with monoclonal antibodies and proteolytic fragments of the molecule, *Cell*, **26**, 259–267.

17. Pierschbacher, M. D., and Ruoslahti, E. (1984) Cell attachment activity of fibronectin can be duplicated by small synthetic fragments of the molecule, *Nature*, **309**, 30–33.

18. Ruoslahti, E., Hayman, E. G., Pierschbacher, M., and Engvall, E. (1982) Fibronectin: Purification, immunochemical properties, and biological activities, in *Methods in Enzymology* (Leon, W., Cunningham, D. W. F., eds.), Academic Press, pp. 803–831.

19. Ruoslahti, E., Pierschbacher, M., Hayman, E. G., and Engvall, E. (1982) Fibronectin: A molecule with remarkable structural and functional diversity, *Trends Biochem. Sci*, **7**, 188–190.

20. Giros, A., Grgur, K., Gossler, A., and Costell, M. (2011) Alpha5beta1 Integrin-mediated adhesion to fibronectin is required for axis elongation and somitogenesis in mice, *Plos One*, **6,** e22002.

21. Pham, K., Nassiri, N., and Gupta, R. (2009) c-Jun, krox-20, and integrin beta 4 expression following chronic nerve compression injury, *Neurosci. Lett.*, **465**, 194–198.

22. Wang, A. G., Yen, M. Y., Hsu, W. M., and Fann, M. J. (2006) Induction of vitronectin and integrin alpha v in the retina after optic nerve injury, *Mol. Vis.*, **12**, 76–84.

23. Tsuda, M., Toyomitsu, E., Komatsu, T., Masuda, T., Kunifusa, E., Nasu-Tada, K., *et al.* (2008) Fibronectin/integrin system is involved in P2X(4) receptor upregulation in the spinal cord and neuropathic pain after nerve injury, *Glia*, **56**, 579–585.

24. Robey, P. G., Foidart, J. M., Martin, G. R., and Timpl, R. (1979) Structural components of basement-membrane - type-iv collagen and laminin, *Fed. Proc.*, **38**, 816.

25. Timpl, R., Rohde, H., Robey, P. G., Rennard, S. I., Foidart, J. M., and Martin, G. R. (1979) Laminin—glycoprotein from basement-membranes, *J. Biol. Chem.*, **254**, 9933–9937.

26. Tashiro, K., Sephel, G. C., Weeks, B., Sasaki, M., Martin, G. R., Kleinman, H. K., *et al.* (1989) A synthetic peptide containing the IKVAV sequence from the A chain of laminin mediates cell attachment, migration, and neurite outgrowth, *J. Biol. Chem.*, **264**, 16174–16182.

27. Graf, J., Iwamoto, Y., Sasaki, M., Martin, G. R., Kleinman, H. K., Robey, F. A., *et al.* (1987) Identification of the major epithelial-cell attachment site (yigsr) in the b1-chain of laminin, *J. Invest. Dermatol.*, **88**, 491.

28. Iwamoto, Y., Robey, F. A., Graf, J., Sasaki, M., Kleinman, H. K., Yamada, Y., *et al.* (1987) Yigsr, a synthetic laminin pentapeptide, inhibits experimental metastasis formation, *Science*, **238**, 1132–1134.

29. Armstrong, S. J., Wiberg, M., Terenghi, G., Kingham, P. J. (2008) Laminin activates NF-kappa B in Schwann cells to enhance neurite outgrowth, *Neurosci. Lett.*, **439**, 42–46.

30. Lin, X., Takahashi, K., Liu, Y., and Zamora, P. O. (2006) Enhancement of cell attachment and tissue integration by a IKVAV containing multi-domain peptide, *Biochim. Biophys. Acta (BBA)—General Subjects*, **1760**, 1403–1410.

31. Matsumoto, K., Ohnishi, K., Kiyotani, T., Sekine, T., Ueda, H., Nakamura, T., *et al.* (2000) Peripheral nerve regeneration across an 80-mm gap bridged by a polyglycolic acid (PGA)-collagen tube filled with laminin-coated collagen fibers: A histological and electrophysiological evaluation of regenerated nerves, *Brain Res.*, **868**, 315–328.

32. Matthias, Chiquet L. M-N, and Konrad, B. Attachment toan endogenous laminin-like protein initiates sprouting by leech neurons, *J. Cell. Biol.*, **107**, 1189–1198.

33. Wen, X.J., and Tresco, P.A. (2006) Effect of filament diameter and extracellular matrix molecule precoating on neurite outgrowth and Schwann cell behavior on multifilament entubulation bridging device *in vitro*, *J. Biomed. Mat. Res. Part A.*, **76A**, 626–637.

34. Yu, T. T., and Shoichet, M. S. (2005) Guided cell adhesion and outgrowth in peptide-modified channels for neural tissue engineering, *Biomaterials*, **26**, 1507–1514.

35. Agius, E., Sagot, Y., Duprat, A. M., and Cochard, P. (1996) Antibodies directed against the [beta]1-integrin subunit and peptides containing the IKVAV sequence of laminin perturb neurite outgrowth of peripheral neurons on immature spinal cord substrata, *Neuroscience*, **71**, 773–786.

36. Grinnell, F., Hays, D. G., and Minter, D. (1977) Cell adhesion and spreading factor : Partial purification and properties, *Exp. Cell. Res.*, **110**, 175–190.

37. Grinnell, F., and Minter, D. (1979) Cell adhesion and spreading factor. Chemical modification studies, *Biochim. Biophys. Acta (BBA) - Biomembranes*, **550**, 92–99.

38. Grinnell, F. (1974) Studies on the mechanism of cell attachment to a substratum with serum in the medium: Further evidence supporting a requirement for two biochemically distinct processes, *Arch. Biochem. Biophys.*, **165**, 524–530.

39. Massia, S. P., and Hubbell, J. A. (1991) An Rgd spacing of 440 nm is sufficient for integrin alpha-v-beta-3-mediated fibroblast spreading and 140 nm for focal contact and stress fiber formation, *J. Cell. Biol.*, **114**, 1089–1100.

40. Bellis, S.L. (2011) Advantages of RGD peptides for directing cell association with biomaterials, *Biomaterials*, **32**, 4205–4210.

41. García, A.J. (2005) Get a grip: Integrins in cell-biomaterial interactions, *Biomaterials*, **26**, 7525–7529.

42. Hersel, U., Dahmen, C., and Kessler, H. (2003) RGD modified polymers: Biomaterials for stimulated cell adhesion and beyond, *Biomaterials*, **24**, 4385–4415.

43. Gunn, J. W., Turner, S. D., and Mann, B. K. (2005) Adhesive and mechanical properties of hydrogels influence neurite extension, *J. Biomed. Mat. Res. Part A*, **72A**, 91–97.

44. Acharya, A. P., Dolgova, N. V., Moore, N. M., Xia, C-Q, Clare-Salzler, M. J., Becker, M. L., *et al.* (2010) The modulation of dendritic cell integrin binding and activation by RGD-peptide density gradient substrates, *Biomaterials*, **31**, 7444–7454.

45. Ahmed, M. R., and Jayakumar, R. (2005) Peripheral nerve regeneration in cell adhesive peptide incorporated collagen tubes in rat sciatic nerve - early and better functional regain, *J. Peripher. Nerv. Syst.*, **10**, 390–391.

46. Ahmed, M. R., Venkateshwarlu, U., and Jayakumar, R. (2004) Multilayered peptide incorporated collagen tubules for peripheral nerve repair, *Biomaterials* **25**, 2585–2594.

47. Jessell, T. M. (1988) Adhesion molecules and the hierarchy of neural development, *Neuron*, **1**, 3–13.

48. Beck, A. J., Jones, F. R., and Short, R. D. (1996) Plasma copolymerization as a route to the fabrication of new surfaces with controlled amounts of specific chemical functionality, *Polymer*, **37**, 5537–5539.

49. Beck, A. J., Short, R. D., Jones, F. R., France, R. M., Kettle, A. P., Daw, R., *et al.* (1998) Plasma polymers for suitable surfaces, *Mater. World*, **6**, 70–72.

50. France, R. M., Short, R. D., Dawson, R. A., and MacNeil, S. (1998) Attachment of human keratinocytes to plasma co-polymers of acrylic acid octa-1,7-diene and allyl amine octa-1,7-diene, *J. Mater. Chem.*, **8**, 37–42.

51. France, R. M., Short, R. D., Duval, E., Jones, F. R., Dawson, R. A., and MacNeil, S. (1998) Plasma copolymerization of allyl alcohol 1,7-octadiene: Surface characterization and attachment of human keratinocytes, *Chem. Mater.*, **10**, 1176–1183.

52. Haddow, D. B., France, R. M., Short, R. D., MacNeil, S., and Dawson, R. A. (1999) Comparison of proliferation and growth of human keratinocytes on plasma co-polymers of acrylic acid/1,7-octadiene and self-assembled monolayers, *Mater. Res. Soc. Symp. P.*, **544**, 39–44.

53. Haddow, D. B., France, R. M., Short, R. D., MacNeil, S., Dawson, R. A., Leggett, G. J., *et al.* (1999) Comparison of proliferation and growth of human keratinocytes on plasma copolymers of acrylic acid/1,7-octadiene and self-assembled monolayers, *J. Biomed. Mater. Res.*, **47**, 379–387.

54. Neil, S. M., Beck, A., Haddow, D., France, R., Dawson, R., Phillips, J., *et al.* (2001) Carrier surfaces for culture of keratinocytes and melanocytes for clinical use, *J. Invest. Dermatol.*, **117**, 422-.

55. Daw, R., Candan, S., Beck, A. J., Devlin, A. J., Brook, I. M., MacNeil, S., *et al.* (1998) Plasma copolymer surfaces of acrylic acid/1,7 octadiene: Surface characterisation and the attachment of ROS 17/2.8 osteoblast-like cells, *Biomaterials*, **19**, 1717–1725.

56. Kelly, J. M., Scutt, A. M., Short, R. D., and Brook, I. M. (2000) Osteoblast attachment and spreading an acrylic acid plasma polymer, *J. Dent. Res.*, **79**, 1184-.

57. Murray-Dunning, C., McArthur, S. L., Sun, T., McKean, R., Ryan, A. J., and Haycock, J. W. (2011) Three-dimensional alignment of schwann cells using hydrolysable microfiber scaffolds: Strategies for peripheral nerve repair, *Methods Mol. Biol.*, **695**, 155–166.

58. Barry, J. J. A., Howard, D., Shakesheff, K. M., Howdle, S. M., and Alexander, M. R. (2006) Using a core-sheath distribution of surface chemistry through 3D tissue engineering scaffolds to control cell ingress, *Adv. Mater.*, **18**, 1406-+.

59. Barry, J. J. A., Silva, M. M. C. G., Shakesheff, K. M., Howdle, S. M., and Alexander, M.R. (2005) Using plasma deposits to promote cell

population of the porous interior of three-dimensional poly(D,L-lactic acid) tissue-engineering scaffolds, *Adv. Funct. Mater.,* **15**, 1134–1140.

60. MacNeil, S. (2007) Progress and opportunities for tissue-engineered skin, *Nature,* **445**, 874–880.

61. Charnley, M., Fairfull-Smith, K., Haldar, S., Elliott, R., McArthur, S. L., Williams, N. H., *et al.* (2009) Generation of bioactive materials with rapid self-assembling resorcinarene-peptides, *Adv. Mater.,* **21**, 2909–2915.

62. Ernest, A. (2010) Towards stem cell replacement therapies for Parkinson's disease, *Biochem. Biophys. Res. Commun.,* **396**, 152–156.

63. McCoy, M. K., Martinez, T. N., Ruhn, K. A., Wrage, P. C., Keefer, E. W., Botterman, B. R., *et al.* (2008) Autologous transplants of adipose-derived adult stromal (adas) cells afford dopaminergic neuroprotection in a model of Parkinson's disease, *Exp. Neurol.,* **210**, 14–29.

64. Preynat-Seauve, O., Burkhard, P. R., Villard, J., Zingg, W., Ginovart, N., Feki, A., *et al.* (2009) Pluripotent stem cells as new drugs? The example of Parkinson's disease, *Int. J. Pharm.,* **381**, 113–121.

65. Torres, E. M., Monville, C., Gates, M. A., Bagga, V., and Dunnett, S. B. (2007) Improved survival of young donor age dopamine grafts in a rat model of Parkinson's disease, *Neuroscience,* **146**, 1606–1617.

66. Bolli, R., Chugh, A. R., D'Amario, D., Loughran, J. H., Stoddard, M. F., Ikram, S., *et al.* (2011) Cardiac stem cells in patients with ischaemic cardiomyopathy (SCIPIO): Initial results of a randomised phase 1 trial, *Lancet,* **378**, 1847–1857.

67. Berná, G., León-Quinto, T., Enseñat-Waser, R., Montanya, E., Martín, F., and Soria, B. (2010) Stem cells and diabetes, *Biomed. Pharmacother.,* **55**, 206–212.

68. Gonez, L. J., and Knight, K. R. (2010) Cell therapy for diabetes: Stem cells, progenitors or beta-cell replication?, *Mol. Cell. Endocrinol.,* **323**, 55–61.

69. Limbert, C., Päth, G., Jakob, F., and Seufert, J. (2008) Beta-cell replacement and regeneration: Strategies of cell-based therapy for type 1 diabetes mellitus, *Diabetes Res. Clin. Prac.,* **79**, 389–399.

70. Street, C. N., Sipione, S., Helms, L., Binette, T., Rajotte, R. V., Bleackley, R. C., *et al.* (2004) Stem cell-based approaches to solving the problem of tissue supply for islet transplantation in type 1 diabetes, *Int. J. Biochem.,* **36**, 667–683.

71. Lumelsky, N., Blondel, O., Laeng, P., Velasco, I., Ravin, R., and McKay, R. (2001) Differentiation of embryonic stem cells to insulin-secreting structures similar to pancreatic islets, *Science,* **292**, 1389–1394.

72. Thomson, M., Liu Siyuan, J., Zou, L-N, Smith, Z., Meissner, A., and Ramanathan, S. (2011) Pluripotency factors in embryonic stem cells regulate differentiation into germ layers, *Cell*, **145**, 875–889.

73. Fuchs, E., Tumbar, T., and Guasch, G. (2004) Socializing with the neighbors: Stem cells and their niche, *Cell*, **116**, 769–778.

74. Morrison, S. J., and Spradling, A. C. (2008) Stem cells and niches: Mechanisms that promote stem cell maintenance throughout life, *Cell*, **132**, 598–611.

75. Wojcik-Stanaszek, L., Gregor, A., and Zalewska, T. (2011) Regulation of neurogenesis by extracellular matrix and integrins, *Acta Neurobiol. Exp. (Wars)*, **71**, 103–112.

76. Wang, N., Tolić-Nørrelykke, I. M., Chen, J., Mijailovich, S. M., Butler, J. P., Fredberg, J. J., *et al.* (2002) Cell prestress. I. Stiffness and prestress are closely associated in adherent contractile cells, *Am. J. Physiol. Cell Physiol.*, **282**, C606–C616.

77. Engler, A. J., Sen, S., Sweeney, H. L., and Discher, D. E. (2006) Matrix elasticity directs stem cell lineage specification, *Cell*, **126**, 677–689.

78. Reilly, G. C., and Engler, A. J. Intrinsic extracellular matrix properties regulate stem cell differentiation, *J. Biomech.*, **43**, 55–62.

79. Dalby, M. J., McCloy, D., Robertson, M., Agheli, H., Sutherland, D., Affrossman, S., *et al.* (2006) Osteoprogenitor response to semi-ordered and random nanotopographies, *Biomaterials*, **27**, 2980–2987.

80. Dalby, M. J., Gadegaard, N., Tare, R., Andar, A., Riehle, M. O., Herzyk, P., *et al.* (2007) The control of human mesenchymal cell differentiation using nanoscale symmetry and disorder, *Nat. Mater.*, **6**, 997–1003.

81. Biggs, M. J. P., Richards, R. G., Gadegaard, N., Wilkinson, C. D. W., Oreffo, R. O. C., and Dalby, M. J. (2009) The use of nanoscale topography to modulate the dynamics of adhesion formation in primary osteoblasts and ERK/MAPK signalling in STRO-1+ enriched skeletal stem cells, *Biomaterials*, **30**, 5094–5103.

82. Dalby, M. J., Gadegaard, N., Herzyk, P., Sutherland, D., Agheli, H., Wilkinson, C. D. W., *et al.* (2007) Nanomechanotransduction and interphase nuclear organization influence on genomic control, *J. Cell. Biochem.*, **102**, 1234–1244.

83. Badami, A. S., Kreke, M. R., Thompson, M. S., Riffle, J. S., and Goldstein, A. S. (2006) Effect of fiber diameter on spreading, proliferation, and differentiation of osteoblastic cells on electrospun poly(lactic acid) substrates, *Biomaterials*, **27**, 596–606.

84. Baker, S. C., Atkin, N., Gunning, P. A., Granville, N., Wilson, K., Wilson, D., *et al.* (2006) Characterisation of electrospun polystyrene scaffolds for

three-dimensional *in vitro* biological studies, *Biomaterials*, **27**, 3136–3146.

85. Bhardwaj, N., and Kundu, S.C. (2010) Electrospinning: A fascinating fiber fabrication technique, *Biotechnol. Adv.*, **28**, 325–347.
86. Huang, C. B., Chen, S. L., Lai C. L., Reneker, D. H., Qiu, H., Ye, Y., *et al.* (2006) Electrospun polymer nanofibres with small diameters, *Nanotechnology*, **17**, 1558–1563.
87. Li, M., Mondrinos, M. J., Gandhi, M. R., Ko, F. K., Weiss, A. S., and Lelkes, P. I. (2005) Electrospun protein fibers as matrices for tissue engineering, *Biomaterials*, **26**, 5999–6008.
88. Megelski, S., Stephens, J. S., Chase, D. B., and Rabolt, J. F. (2002) Micro- and nanostructured surface morphology on electrospun polymer fibers, *Macromolecules*, **35**, 8456–8466.
89. Rutledge, G. C., and Fridrikh, S. V. (2007) Formation of fibers by electrospinning, *Adv. Drug Del. Rev.*, **59**, 1384–1391.
90. Sun, T., Mai, S. M., Norton, D., Haycock, J. W., Ryan, A. J., and MacNeil, S. (2005) Self-organization of skin cells in three-dimensional electrospun polystyrene scaffolds, *Tissue Eng.*, **11**, 1023–1033.
91. Yang, F., Murugan, R., Wang, S., and Ramakrishna, S. (2005) Electrospinning of nano/micro scale poly(L-lactic acid) aligned fibers and their potential in neural tissue engineering, *Biomaterials*, **26**, 2603–2610.
92. Zhang, S. G. (2003) Fabrication of novel biomaterials through molecular self-assembly, *Nat. Biotechnol.*, **21**, 1171–1178.
93. Holmes, T. C., de Lacalle, S., Su, X., Liu, G. S., Rich, A., and Zhang, S. G. (2000) Extensive neurite outgrowth and active synapse formation on self-assembling peptide scaffolds, *Proc. Natl. Acad. Sci.*, **97**, 6728–6733.
94. Kisiday, J., Jin, M., Kurz, B., Hung, H., Semino, C., Zhang, S., *et al.* (2002) Self-assembling peptide hydrogel fosters chondrocyte extracellular matrix production and cell division: Implications for cartilage tissue repair, *Proc. Natl. Acad. Sci.*, **99**, 9996–10001.
95. Niece, K. L., Hartgerink, J. D., Donners, J. J. J. M., and Stupp, S. I. (2003) Self-assembly combining two bioactive peptide-amphiphile molecules into nanofibers by electrostatic attraction, *J. Am. Chem. Soc.*, **125**, 7146–7147.
96. Silva, G. A., Czeisler, C., Niece, K. L., Beniash, E., Harrington, D. A., Kessler, J. A., *et al.* (2004) Selective differentiation of neural progenitor cells by high-epitope density nanofibers, *Science*, **303**, 1352–1355.

97. Weissman, I. L. (2000) Translating stem and progenitor cell biology to the clinic: Barriers and opportunities, *Science*, **287**, 1442–1446.

98. Watt, F. M., and Hogan, B. L. M. (2000) Out of eden: Stem cells and their niches, *Science*, **287**, 1427–1430.

99. Liu, C. Y., Apuzzo, M. L. J., and Tirrell, D. A. (2003) Engineering of the extracellular matrix: Working toward neural stem cell programming and neurorestoration - Concept and progress report, *Neurosurgery*, **52**, 1154–1165.

100. Liu CY, Westerlund U, Svensson M, Moe MC, Varghese M, Berg-Johnsen J, *et al.* Artificial niches for human adult neural stem cells: Possibility for autologous transplantation therapy. *J. Hematoth. Stem Cell.*, 2003, **12**, 689–699.

101. Anderson, D. G., Levenberg, S., and Langer, R. (2004) Nanoliter-scale synthesis of arrayed biomaterials and application to human embryonic stem cells, *Nat. Biotechnol.*, **22**, 863–866.

102. Friedl, P. (2004) Prespecification and plasticity: Shifting mechanisms of cell migration, *Curr. Opin. Cell. Biol.*, **16**, 14–23.

103. Gobin, A. S., and West, J. L. (2002) Cell migration through defined, synthetic extracellular matrix analogues, *FASEB J.*, **16,** 751–753.

104. West, J. L., and Hubbell, J. A. (1999) Polymeric biomaterials with degradation sites for proteases involved in cell migration, *Macromolecules*, **32**, 241–244.

105. Mann, B. K., Tsai, A. T., Scott-Burden, T., and West, J. L. (1999) Modification of surfaces with cell adhesion peptides alters extracellular matrix deposition, *Biomaterials*, **20**, 2281–2286.

106. Pratt, A. B., Weber, F. E., Schmoekel, H. G., Muller, R., and Hubbell, J. A. (2004) Synthetic extracellular matrices for *in situ* tissue engineering, *Biotechnol. Bioeng.*, **86**, 27–36.

107. Lutolf, M. R., Weber, F. E., Schmoekel, H. G., Schense, J. C., Kohler, T., Muller, R., *et al.* (2003) Repair of bone defects using synthetic mimetics of collagenous extracellular matrices, *Nat. Biotechnol.*, **21**, 513–518.

108. Zisch, A. H., Lutolf, M. P., Ehrbar, M., Raeber, G. P., Rizzi, S. C., Davies, N., *et al.* (2003) Cell-demanded release of VEGF from synthetic, biointeractive cell-ingrowth matrices for vascularized tissue growth, *FASEB J.*, **17**, 2260–2262.

109. Ozawa, C. R., Banfi, A., Glazer, N. L., Thurston, G., Springer, M. L., Kraft, P. E., *et al.* (2004) Microenvironmental VEGF concentration, not total dose, determines a threshold between normal and aberrant angiogenesis, *J. Clin. Invest.*, **113**, 516–527.

# Chapter 4

# Principles and Biological Pathways to Tissue Regeneration: The Tissue Regenerative Niche

**Ranieri Cancedda and Claudia Lo Sicco**
*Department of Experimental Medicine, University of Genova & IRCCS AOU San Martino-IST, Genova, Italy*
ranieri.cancedda@unige.it

According to the traditional tissue engineering approach, stem cells are explanted from a patient, expanded *in vitro* and re-administered to the same patient as a cell suspension or in association with a suitable biomaterial. Nevertheless, the use of these *ex vivo* tissue engineered constructs is highly limited because of economic and regulation-related issues and should be confined to extreme life- or organ-saving situations. Indeed the patient organism itself could provide the key elements necessary for tissue repair in less dramatic clinical conditions. A major goal of a modern regenerative medicine approach to tissue repair/regeneration should be the definition of strategies and the development of standardized products capable of reactivating endogenous repair/regeneration mechanisms.

*Scaffolds for Tissue Engineering: Biological Design, Materials, and Fabrication*
Edited by Claudio Migliaresi and Antonella Motta

ISBN 978-981-4463-20-1 (Hardcover), 978-981-4463-21-8 (eBook)
www.panstanford.com

The first response of a tissue to a wound is the platelet activation and the concomitant inflammatory cascade. During this phase, soluble mediators released by platelets, the stimulated resident cells, and the first recruited inflammatory cells trigger the whole healing process. Most of these activated pathways after a wound are common to all tissues. For the purpose of this chapter, we will especially consider the skeletal tissue with particular regard to platelet-derived growth factors, inflammation process, hypoxic environment of the wound, angiogenic and the vasculogenic response, locally resident pluripotent cells and circulating stem/progenitor cells, and other mechanisms involved in the bone responses to healing as a paradigm for other tissues.

## 4.1 Introduction

Regenerative medicine is a rapidly developing field that constitutes one of the most challenging novel therapeutic approaches and represents a shift of paradigms with respect to present medical treatment for tissue reconstruction and repair of damages caused by disease, trauma, or resection of a tumor.

Up to now, two major tissue engineering/regenerative medicine schemes have been used: tissue replacement and endogenous regeneration. Tissue replacement uses *ex vivo* construction and transplantation of new tissue, based on the triad of cells, growth factors (GFs), and biomaterial scaffolds. During the last two decades, remarkable progress has been made with respect to *in vitro* fabrication of substitutes for tissues; however, this approach has not yet reached the stage of routine clinical applications. The use of *ex vivo* tissue engineered constructs is highly limited because of economic and regulation-related issues, which confine this approach only to extreme life- or organ-saving situations. Differently, endogenous regeneration aims at the stimulation of the intrinsic potential of a tissue to heal or regenerate by using "of-the-shelf" standardized products without cells, which mimic the microenvironment of endogenous tissues. The use of minimally invasive procedures qualifies the endogenous regeneration strategy for a broader application.

Since its beginning, tissue engineering focused on the identification of exogenous sources of cells with increased differentiation capacity without considering that the recipient organism could provide key elements necessary for tissue repair. According to the traditional tissue engineering approach, stem cells are explanted from a patient, expanded *in vitro* and re-administered to the patient as a cell suspension or in association with a suitable biomaterial. The rapidly developing knowledge about the developmental pathways occurring during the healing process adds fascinating information about the natural response of the body to injury, suggesting a novel approach based on the activation of the endogenous regenerative capacity of the tissue itself. The proper functioning of all tissues of the body depends upon innate regenerative processes that maintain proper cell numbers (homeostasis) and replace damaged cells after injury (repair). This regenerative potential correlates with the presence of a dedicated population of stem/progenitor cells in the tissues, which respond to exogenous signal to produce replacement cells when needed for a physiological turnover or for repair/regeneration of tissues damaged by diseases or injuries. The balance between tissue homeostasis and appropriate responsiveness for cell replacement and repair is accomplished by maintaining stem cells in a specialized microenvironment, or niche, which provides spatial and temporal cues to support and coordinate stem cell activities [Wagers, 2012]. The stem cell niche is originally recognized as a specific anatomic location that regulates stem cell fate decisions, including choices between quiescence or proliferation, self-renewal or differentiation, migration or retention, and cell death or survival [Doetsch, 2003].

Although stem cells have been predominantly investigated in the bone marrow, adult stem cells have been identified and isolated from an increasing array of tissues of the adult body, including skeletal muscle, adipose tissue, tooth pulp, hair follicle, peripheral circulation, intestinal epithelium, myocardium, lung, breast, kidney, and brain [Wagers and Weissman, 2004]. Adult stem cell maintenance involves a fine balance between genetic and molecular cell mechanisms, external factors in the local niches and systemic environment in the body, and multiple signaling pathways that regulate adult stem cell homeostasis.

Nowadays, the main goal of alternative regenerative medicine and tissue engineering strategies is to stimulate the mobilization and recruitment of endogenous stem/progenitor cells to the site of injury in order to induce the body to heal itself. Novel tissue engineering and regenerative medicine schemes have been developed to design biomaterials that, by mimicking the tissue microenvironment, promote tissue development and regeneration *in vivo*. Although every tissue has a specific response to healing, there is a general pattern of resolution shared by most tissues. The initial stage after damage is characterized by hematoma formation and subsequent inflammation. Growing evidence indicates that immune cells and their secreted factors are crucial for the physiologic response following injury. The interaction between the immune cells, their secreted factors, and the hematoma creates, at the site of the wound, an inflammatory microenvironment, which then brings about the recruitment of regenerative progenitor cells. The inflammatory microenvironment elicits both a local proliferative response, through stimulation of locally resident stem cells [Barker *et al.*, 2010] as well as migration of stem cells present surrounding the tissues and mobilization of circulating stem/progenitor cells derived from the bone marrow. The locally resident stem cells guarantee a continuous self-renewal of the tissue and hence play a major role in tissues continuously exposed to environmental assault, such us epithelia of the intestine, stomach, and skin, or exposed to intense exercise or minor trauma, that is, the skeletal muscle [Cassano *et al.*, 2009] (Fig. 4.1). Among resident stem cells, bone marrow mesenchymal stem cells (MSCs) are multi-potent and present abilities of self-renewal and differentiation into several types of cells, especially in the case of connective tissues. In severe trauma, endogenous bone marrow MSCs or exogenously delivered MSCs migrate to the injury site and participate in the repair process [Li and Jiang, 2011]. These cells are considered effective therapeutic agents promoting the repair of injured tissues not only for their differentiation potential, but also for their role as cellular modulators [Tasso *et al.*, 2009; Caplan and Correa, 2011]. Indeed, in some cases, transplanted bone marrow (BM) it-derived MSC directly contribute to form a new tissue; whereas in others, they activate endogenous mechanisms leading to new tissue formation by the host cells [Tasso *et al.*, 2009]. This could be explained taking into account experimental evidences that indicate how different MSC-cultured conditions can promote the selection and expansion of specific cell clones or sub-populations. Tasso and co-workers reported that MSC's capacity to activate host endogenous regenerative mechanisms is critically dependent on its commitment level, and, in particular, it was associated with the presence of fibroblast growth factor-2 (FGF-2) in the culture medium during the *ex vivo* cell expansion [Tasso *et al.*, 2012].

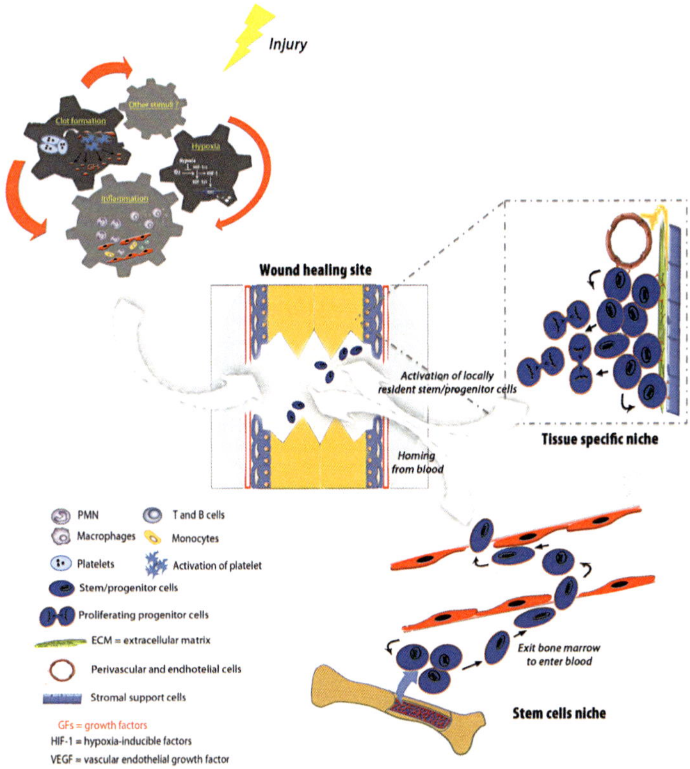


**Figure 4.1** Schematics of the activation of endogenous regenerative responses.

In mammals, the healing process consists of three consecutive and overlapping stages: inflammation, new tissue formation, and remodeling [Gurtner *et al.*, 2008]. The transition from one stage to another depends on the maturation and differentiation of the cell populations involved and is controlled and regulated by cell-released mediators, which are common to most tissues during the inflammatory, reparative, and remodeling phases. Wound healing is primarily triggered by the initial platelet activation, lysis, and an inflammatory cascade leading to new tissue formation. During the initial phase, namely inflammatory phase, soluble mediators released by platelets and by resident and first recruited inflammatory cells play a crucial role for the healing of the damaged tissues [Kasemkijwattana *et al.*, 2000; Molloy *et al.*, 2003; Nagumo *et al.*, 2005; Pujol *et al.*, 2008].

The existence of common pathways has enabled research conducted on a specific tissue to be broadly applied to our understanding of healing in the other tissues. For the purpose of this chapter, we will especially consider the skeletal tissue with particular regard to the mechanisms involved in the bone responses to wound healing as a paradigm for other tissues.

This schematic illustration depicts the critical stimuli, clot formation, inflammation, and hypoxia activated during wound-healing process. In response to injury, the interaction between immune cells and the platelet-released factors under hypoxic condition creates a microenvironment, which brings about the recruitment of regenerative stem/progenitor cells from either the blood or the tissue-specific niche to the wound healing site.

## 4.2 Platelet-Derived Growth Factors

At the injury site of all tissues, the first regenerative/inflammatory/healing events are the clot formation and the platelet degranulation that results in the release of an orchestra of factors involved in the wound-healing process. Upon vessel injury, the platelets stick to the sub-endothelial extracellular matrix, forming the platelets plug that stabilizes the clot within the fibrin meshwork. Once activated by thrombin, platelets release their polypeptides content, which is crucial for signaling the local resident cells and activating a cascade of events eventually leading to mesenchymal and epidermal stem/progenitor cells migration, division, and extracellular matrix synthesis (collagen, glycosaminoglycans, and other macromolecules) [Tsukamoto *et al.*, 1991; Sanchez *et al.*, 2003; Eppley *et al.*, 2004; Grageda, 2004; Kilian *et al.*, 2004].

The platelets' alpha granules are storage of many GFs including, among the others, platelet-derived growth factor (PDGF), transforming growth factor beta (TGF-beta), platelet-derived epidermal GF, vascular endothelial growth factor (VEGF), insulin-like growth factor 1 (IGF-1), fibroblast growth factor-2 (FGF-2), and epidermal GF [Weyrich *et al.*, 2003; Anitua *et al.*, 2004, 2005; Yeaman, 2010]. The alpha granules are also the source of several cytokines/chemokines and other active molecules that possess a strong chemotactic activity toward leukocytes (granulocytes, monocytes, macrophages) and are differently involved in the control of cell proliferation, differentiation, and maturation [Weyrich *et al.*, 2003].

In 1998, Marx and co-workers [Marx *et al.*, 1998] introduced a platelet concentrate, the platelets rich plasma (PRP), as therapeutic agent for healing. The PRP offers the advantage of being the concentrate of the "right" factors in the "right" proportions required for the healing process [Anitua *et al.*, 2004]. In an inflammatory microenvironment and/or in the presence of an infection, PRP exerts both an enhancing regulatory function and an antimicrobial activity. Today, PRP is used in the clinical practice for the treatment of difficult to heal wounds such as chronic vascular and diabetic ulcers [Mazzucco *et al.*, 2004; Pietramaggiori *et al.*, 2008; Scimeca *et al.*, 2010; Villela and Santos, 2010], traumatic injuries, and other soft tissue pathologies [Chandra *et al.*, 2007; Rozman and Bolta, 2007; DeRossi *et al.*, 2009; Lopez-Jornet *et al.*, 2009; Nakamura *et al.*, 2010] as well as corneal lesions [Geremicca *et al.*, 2010; Lopez-Plandolit *et al.*, 2010; Tanidir *et al.*, 2010]. In addition, PRP is currently applied in orthopaedics and maxillofacial surgery [Marx *et al.*, 1998]. Platelet concentrates and their derivatives have also been suggested for bone augmentation and other regeneration therapies as well as cell culture supplement substitute [Zaky *et al.*, 2008]. The PRP can be used as supplement substitute also in an activated form as platelet lysate (PL) that contains all platelet-growth factors without platelet membrane residues and shows properties similar to PRP.

The PRP not only plays a major role in tissue regeneration by enhancing recruitment and proliferation of locally resident and circulating stem/progenitor cells of mesenchymal and/or epithelial nature, but also favors vascularization of the new forming tissues by promoting endothelial cells recruitment, migration, and organization [Kitaguchi *et al.*, 1992] either directly or via the induction of an increased expression of factors such as VEGF and PDGF in cells present in the microenvironment [Hu *et al.*, 2009].

Recent *in vitro* studies on the mechanism of action of the PRP treatment in wound healing showed an increased rate in the closure of scratches made in confluent monolayer cultures of cells of different tissue origin such as keratinocytic cell lines, c2c12 myoblast cell line, dermal fibroblasts, endothelial cells [Ranzato *et al.*, 2008, 2009 (a, b), 2010; El Backly *et al.*, 2011]. This effect is strongly increased in inflammatory conditions, that is, in cells stimulated with IL-1, a factor physiologically present in the wound microenvironment.

These data suggest that PL enhances the immediate physiological inflammatory response of the tissue to the wound, induces a local defense from bacterial infection and leads to a more rapid tissue repair/regeneration [El Backly *et al.*, 2011].

Several studies reported the positive effect of a PL cell culture medium supplement on the proliferation and differentiation of human mesenchymal stem/progenitor cells of different origin [Johansson *et al.*, 2003; Doucet *et al.*, 2005; Muller *et al.*, 2009]. In particular, using the human PL as a medium supplement to support the culture of human MSC provides two advantages: avoiding the use of animal sera and enhancing the cell proliferation and differentiation potentials [Zaky *et al.*, 2008].

The complexity of the issue is clear: each GF is involved in several pathways and the activation of a single pathway may require several GFs [Kassolis *et al.*, 2000; van den Dolder *et al.*, 2006]. The use of a single recombinant GF would, theoretically, miss the natural interaction of this factor with other GFs involved in the pathways eventually leading to transcription activation of specific genes and synthesis of the specific proteins required for the healing process. As a result, adopting the PRP as source of GFs for healing appears the best simulation of natural healing.

## 4.3 Inflammation

As mentioned above, the wound-healing process occurs through the following overlapping phases: hematoma formation, inflammation, angiogenesis, and tissue construction/maturation. These events involve numerous cell types, including tissue resident cells and cells recruited to the site of tissue damage as well as the expression of specific genes during these stages [Cho *et al.*, 2002]. The cell interactions taking place during the healing processes are of particular interest and create a microenvironment supporting the entire cascade of events. It is becoming clearer that the inflammatory microenvironment resulting from the initial cell interactions dictates how the healing process will proceed.

Innate immunity characteristic cells lead the inflammatory cascade reaction guiding revascularization and repair/regeneration of tissue at injury sites [Anghelina *et al.*, 2006; Lolmede *et al.*, 2009; Chen *et al.*, 2012]. These cells exert the function by migrating to the wound site and by secreting inductive cytokines that promote progenitor cells migration. During the early inflammation phase, the local immune innate system is activated with an initial involvement of polymorphonuclear (PMN) leukocytes followed by invasion of blood monocytes that differentiate in macrophages. PMN leukocyte and neutrophil cells entrapped in the fibrin cloth release factors that amplify the aggregation response, the coagulation cascade, and favor the chemo-attraction of cells involved in the inflammatory phase. Monocytes extravase from the blood vessels become activated and mature into macrophages. Macrophages play an integral role in a successful healing response through release of factors such as TGF-beta, TGF-alpha, b-FGF, and VEGF. These molecules promote cell proliferation and cell synthesis of extracellular matrix molecules. Two distinct states of activation of macrophages define different functions in repair [Mantovani *et al.*, 2004]. Classical activated macrophages, also called M1, whose prototypical activating stimulus is interferon-gamma alone or in combination with microbial products such as lipopolysaccharide or cytokines such as tumor necrosis factor (TNF) [Hamilton *et al.*, 2002] exhibit potent antimicrobial properties, high capacity to present antigen, high interleukin-12 (IL-12) and IL-23 production [Verreck *et al.*, 2004], and consequent activation of Th1 responses. Cytokines such as IL-4 and IL-13 inhibit macrophage classical activation to the M1 phenotype and induce a distinct activation program, referred to as "alternative activation" [Verreck *et al.*, 2004]. Alternatively activated macrophages called M2 show a pro-resolving profile as they consist of high levels of scavenger, mannose, and galactose receptors; support Th2-associated effector functions; and possess the capacity to facilitate tissue repair and regeneration [Sica and Mantovani, 2012]. M2 cells play a critical role in promoting the angiogenic process by releasing a specific spectrum of cytokines and GFs.

Inflammatory cytokines have a negative effect on bone, joints, and implanted material when prolonged or chronic expression occur; however, a brief and highly regulated secretion of pro-inflammatory cytokines at the time of the acute injury is crucial for the healing process and the tissue regeneration as it orchestrates a microenvironment suitable for wound healing and repair.

IL-1, IL-6, and TNF-alpha, secreted not only by macrophages and inflammatory cells but also by locally resident cells of mesenchymal origin that have been stimulated by the macrophage secretome, are playing a role in the initiation of the repair cascade [Einhorn *et al.*, 1995; Kon *et al.*, 2001]. These cytokines induce the downstream responses to injury and have a chemotactic effect on other inflammatory cells. Furthermore, they enhance extracellular matrix synthesis and stimulate angiogenesis [Kon *et al.*, 2001] In the early phase of the inflammatory response, TNF-alpha induce secondary inflammatory signals and act as a chemotactic agent to recruit necessary cells [Kon *et al.*, 2001; Lehmann *et al.*, 2005]. The macrophage produced IL-1 induces synthesis and release of IL-6 by osteoblasts, promotes the formation of the primary cartilaginous callus, and stimulates angiogenesis at the injured site by activating either of its two receptors, IL-1RI or IL-1RII [Kon *et al.*, 2001]. Subsequently, the IL-6, produced during the acute phase, stimulates angiogenesis, VEGF production, along with the differentiation of osteoblasts and osteoclasts [Yang *et al.*, 2007].

The switch from a pro-inflammatory M1 macrophage stage to a pro-resolving M2 stage is crucial for the healing and the regeneration processes. Locally resident cells stimulated by platelet and macrophage-derived factors could have a "feedback like" regulatory activity of the M1–M2 switch. Tasso *et al.* [2012] tested the hypothesis that in an ectopic bone formation murine model, MSC treated with FGF-2 (a factor abundantly present also in the PL) could cause an increased percentage of alternatively activated pro-resolving (CD206+CD51+) M2 macrophages infiltrating the ectopic implants and a decrease in the percentage of infiltrating pro-inflammatory (CD86+CD40+) M1 macrophages. It is known that the secretory pattern of MSC is highly influenced by their microenvironment, and that these cells are very sensitive to inflammatory stimuli [Trento and Dazzi, 2010; Caplan and Correa, 2011; Singer and Caplan, 2011]. Tasso and co-workers [2012] mimicked *in vitro* the inflammatory environment surrounding the implanted MSC and tested whether this change in the microenvironment could activate specific intracellular pathways and the release of factors possibly involved in the macrophage functional switch. MSC from the bone marrow express high levels of IL-1R-alpha [Ortiz *et al.*, 2007] which activates the nuclear factor kappa-light-chain-enhancer of activated B cells (NF-κB)-dependent inflammatory pathways leading to enhanced expression of Cyclo Oxygenase 2 (COX-2) and acute phase proteins, such as KC/CXCL1, involved in the control of the angiogenesis process [Scapini *et al.*, 2004], and Lcn-2, controlling the expression of the stromal cell-derived factor-1 (SDF-1), a chemokine with a strong chemotactic activity for the mobilisation of progenitor cells [Costa *et al.*, 2010].

The prostaglandin $PGE_2$, a COX-2 downstream product, plays an important role in modulating the macrophage cytokine profile [Maggini *et al.*, 2010]. The MSC-mediated secretion of $PGE_2$ is responsible for the macrophage switch to the pro-resolving phenotype and suggests that the enhanced percentage of alternatively activated macrophages in the MSC-seeded implants is related to the MSC-mediated secretion of $PGE_2$ [Tasso *et al.* 2012]. Furthermore, $PGE_2$ controls angiogenesis (formation of new blood vessels by sprouting from pre-existing vessels) and vasculogenesis (development of blood vessels through mobilization and recruitment of circulating endothelial progenitor cells, EPCs) by binding to the EP4 receptor present on the surface of endothelial cells [Hristov *et al.*, 2007; Aso *et al.*, 2012] and possibly by secreting keratinocyte chemoattractant (KC) an important inducer of neutrophils leading to the release of VEGFA, which sustains new vessel formation [Scapini *et al.*, 2004].

The polarization of macrophages, from M1 to M2, distorts the secretion of damage associated molecular patterns (DAMPS), such as high mobility group box 1 (HMGB1), TNF-alpha, VEGF, and metalloproteinase-9 (MMP-9), all molecules involved in the control of cell migration. The secreted DAMPS (HMGB1 and others), interacting with specific pattern recognition receptors, including RAGE and members of the Toll-like receptor (TLR) family, amplifies the inflammatory response, and enhances vasculogenesis and migration of specific progenitor cells [Pistoia and Raffaghello, 2011].

## 4.4 Hypoxia

Cells' functions require an adequate supply of oxygen that is consumed in aerobic metabolism to turn, carbohydrates into the energy required to power all main cellular processes. Therefore, it is critical that cells sense and respond adequately to hypoxia or low oxygen levels. In particular, the oxygen tension plays an important role in the stem cell niches, whereas low oxygen level is involved in maintaining the quiescence of niche [Das *et al.*, 2010]. Even in physiological condition, the bone marrow is low in oxygen and limited in nutrient supply. Investigations on the partial oxygen pressure in the bone marrow revealed hypoxic values compared to ambient air, but higher than for the threshold initiating a frank hypoxic molecular response. It is reasonable to assume that gradients of $O_2$ from below 1% in the hypoxic niche to 6% in the vascular niche exist in the body. Hypoxic conditions easily occur at the wound site because of the concurring vasculature damages. In mammals, a family of transcription factors, the hypoxia-inducible factors (HIFs) is responsible for orchestrating cellular responses to hypoxia.

HIF-1 is composed by two subunits, where the subunit beta is constitutively expressed, whereas the subunit alpha is an oxygen-sensitive inducible protein. Under sufficient oxygen concentration (normoxic condition), prolyl-hydroxylase enzymes add hydroxyl groups to two proline residues of the HIF-1-alpha. The prolyl-hydroxilated HIF-1-alpha is recognized by the von Hippel–Lindau protein, which adds a destruction molecular tag (polyubiquitination) to the HIF-1-alpha [Giaccia *et al.*, 2004]. Under limited oxygen conditions the degradation rate decreases, and HIF-1-alpha protein accumulates and associates with HIF-1-beta to form a functional transcription complex [Hewitson and Schofield, 2004]. The active HIF-1 transcription factor induces the expression of several survival genes counteracting hypoxia, including gene coding for angiogenic factors [Potier *et al.*, 2007].

Hypoxia conditions favor angiogenesis not only via HIF-induced expression of angiogenic factors, but also by promoting concomitant inflammation and release of pro-angiogenic factor by immune cells. For instance, leukocytes produce a myriad of angiogenic factors such as ILs, TGF-beta-1, monocyte chemoattractant protein 1 (MCP-1), and specific proteinases [Vacca *et al.*, 1998; Norrby, 2002]. Hypoxia has also been shown to affect angiogenic and osteogenic properties of MSCs by modulating their paracrine activity. MSCs subjected to hypoxia show increased expression of various secretable factors, among which are important angiogenic and osteogenic factors such as VEGF, IL-8, and osteopontin [Potier *et al.*, 2007]. This increased secretion of osteopontin-enhancing macrophage infiltration, at the wound site, which also interacts with bone formation. Finally, hypoxia plays an important role in mobilization and homing of MSCs, primarily by its ability to induce SDF-1 expression along with its receptor CXCR4 [Das *et al.*, 2010].

## 4.5 Angiogenesis and Vasculogenesis

Among the angiogenic factors, VEGF is probably the most remarkable, and it stimulates angiogenesis in a strictly dose-dependent manner

[Ferrara *et al.*, 2003]. The HIF-1 transcription complex can induce a 30-fold increase of VEGF expression. Different members of the VEGF family and different spliced form of the same VEGF protein have been identified (VEGF-A, VEGF-B, VEGF-C, placenta-derived growth factor). These variants interact at variable efficiency with the endothelial tyrosine kinase VEGF receptors (VEGFR-1, VEGFR-2, VEGFR-3) resulting in different final biological effects. VEGF family members promote angiogenesis by activating both cell protein kinases and the endothelial nitric oxide synthase [Silvestre *et al.*, 2003]. Other angiogenic factors include angiopoietin-2/ angiopoietin-1 [Phelan *et al.*, 1998; Kietzmann *et al.*, 1999], Tie2 [Kuwabara *et al.*, 1995], PDGF [Negus *et al.*, 1998; Wykoff *et al.*, 2000], basic fibroblast growth factor (bFGF) [Sakuda *et al.*, 1992], and MCP-1 [Fuentes *et al.*, 1995]. These factors not only increase vascular permeability and endothelial sprouting, but also promote endothelial cell recruitment, proliferation, and migration as well as the enhancement of endothelial lumen assembly. Factors that promote endothelial cell recruitment, proliferation, migration are named direct angiogenic factors. Differently, indirect angiogenic factors function upstream the direct angiogenic factors, promoting their secretion by cells recruited into the wound site.

In a regenerative microenvironment, both angiogenesis and vasculogenesis occur [Madeddu, 2005]. Many of the angiogenic factors that stimulate proliferation and sprouting of endothelial cells are also involved in the recruitment of circulating, bone marrow derived endothelial cell progenitors. Factors such as the stem cell factor (SCF or sKitL) and enzymatic activities such as the matrix metalloproteinase-9 (MMP-9) certainly play an important role in the vasculogenesis process. Unfortunately, molecular and cellular mechanisms that control vasculogenesis are only partially known.

Perivascular cells, often know as vascular mural cells, are the cells that surround tunica intima of the blood vessels that constitute a major component of the vascular wall. Pericytes, though similar but distinct from vascular smooth muscle cells, closely encircle capillaries and micro-vessels in most connective tissues. Pericytes have been postulated as the *in situ* counterpart of the bone marrow-derived MSCs [Sacchetti *et al.*, 2007]. The CD146+CD105+ pericytes developed around the walls of capillaries regarded as structural elements of blood vessels that regulate vascular contractility and are important for the stabilization of the nascent capillaries and in the prevention of vascular regression. During the angiogenesis process, pericytes are recruited to growing micro-vessels, where they directly inhibit endothelial cell proliferation and migration through contact and secretion of growth inhibitory factors [Hirschi and D'Amore, 1996]. These cells communicate with endothelial cells via gap junctions, causing endothelial cells to proliferate or be selectively inhibited. Endothelial cell and pericytes are interdependent: if cell communication is impaired, vessel hyperplasia and abnormal morphogenesis occur.

Inhibiting the PDGF pathway leads to a pericyte deficiency, as well as upregulation of VEGF expression, resulting in a vascular leakage and haemorrhage [Hellstrom *et al.*, 2001]. Similarly to the inhibition of the PDGF pathway, increased expression of angiopoietin 2 reduces the number of pericytes [Hammes *et al.*, 2004].

Pericytes maintain cell plasticity and, depending upon the specific tissue microenvironment, can differentiate into several cell types such as smooth muscle cells, dermal fibroblasts, osteoblasts, and MSCs.

## 4.6 Locally Resident Pluripotent Cells and Circulating Stem/Progenitors Cells

The existence of pluripotent mesoderm derived cells, often named MSCs in adult tissues and organs, is widely accepted. MSCs should contribute to the cell turnover in the tissue and also to the tissue repair/regeneration after damage; however, their exact location is often unclear and they are present in small numbers. MSCs can be isolated as plastic adherent fibroblast-like cells from a variety of fetal and adult tissues and display strong immunomodulatory activity; self-renewal capacity; as well as ability to differentiate in osteoblasts, chondrocytes, and adipocytes in appropriate culture conditions [Glennie *et al.*, 2005; Spaggiari *et al.*, 2006]. MSCs express CD73, CD90, CD105, CD146 and do not express CD3, CD34, CD45, and CD80. The archetype MSC population has been identified in the bone marrow [Friedenstein *et al.*, 1974; Caplan, 1991; Pittenger *et al.*, 1999], but cells with similar properties have been identified in a wide range of connective tissues, including adipose [Zuk *et al.*, 2002], dermis [Toma *et al.*, 2001], and teeth [Gronthos *et al.*, 2000, 2002; Miura *et al.*, 2003; Sonoyama *et al.*, 2008]. Despite the common properties, MSCs maintain the memory of the tissue of origin with regard to their default differentiation potential [Kern *et al.*, 2006; Covas *et al.*, 2008; Huang *et al.*, 2009; Hegyi *et al.*, 2010]. It has been suggested that these pluripotent mesenchymal cells coincide with the pericytes, which are broadly distributed in all the organs [Nehls and Drenckhahn, 1993; Satokata *et al.*, 2000; Alliot-Licht *et al.*, 2001; Shi and Gronthos, 2003; Crisan *et al.*, 2008; da Silva Meirelles *et al.*, 2008; Lin *et al.*, 2008]. Indeed, perivascular cells expressing CD146, NG2, and PDGF-R-beta and with a potential to differentiate in different cell phenotype *in vitro* have been isolated from multiple tissues and organs, such as skeletal muscle, pancreas, adipose tissue, and placenta [Satokata *et al.*, 2000; Crisan *et al.*, 2008; da Silva Meirelles *et al.*, 2008].

Although pericytes have been identified as ancestors of MSCs, it has been proposed that not all pericytes can give rise to MSCs and that same MSCs could derive from other perivascular cells. The pericyte ability to act as MSCs remains controversial. Nevertheless, the potential for pericytes, not only to contribute *in vivo* to the angiogenesis/vasculogenesis process, but also to participate in the formation of different types of tissues has been shown for osteoblasts, chondroblasts, fibroblasts, adipocytes [Nehls and Drenckhahn, 1993; Lin *et al.*, 2008], myogenic cells [Crisan *et al.*, 2008], and odontoblasts [Alliot-Licht *et al.*, 2001; Shi and Gronthos, 2003]. Although pericytes are capable of acting as a source of MSC and of differentiating into cells of many tissues of mesodermal origin, the *in vivo* healing process also involves MSC of a non-pericyte origin.

In this regard, the reactivation of endogenous stem cell potential and their "homing" to the injury site might help the repair/regeneration of damaged tissues. The term "homing" refers to the stem/progenitor cell capacity of being recruited to a specific anatomic site from neighboring healthy tissue or to find its way to a specific anatomic destination, even at a great distance, via the blood circulation. The recruitment and homing of stem/progenitor cells at the injury site may be facilitated by external stimuli, such as the release by the first wave of cells migrated to the wound of cell homing factors, including the SDF-1, the hepatocyte growth factor (HGF), and the VEGF.

Pluripotent/multipotent stem/progenitor cells can be isolated, although with major technical difficulties, from the peripheral blood. It is assumed that these circulating cells are derived principally from the bone marrow. In 2010, the group of Méndez-Ferrer identified in the bone marrow a stromal nestin-expressing MSC population. The Nestin + MSCs were strictly perivascular, located in more central areas of the BM, but they were also present in the immediate vicinity of the endosteum, although with a lower frequency [Ehninger and Trumpp, 2011]. The circulating cells are described in the literature as EPCs [Asahara *et al.*, 1997; Shi *et al.*, 1998], MSCs [Prockop, 1997; Peister *et al.*, 2004], multi-potent adult progenitor cells [Jiang *et al.*, 2002], marrow-isolated adult multi-lineage inducible cells [D'Ippolito *et al.*, 2004], multipotent adult [Beltrami *et al.*, 2007], and very small embryonic-like [Kucia *et al.*, 2006; Ratajczak *et al.*, 2008]. It is likely that different names were assigned to the same population of primitive stem cells isolated using different experimental strategies.

An increasing body of evidences suggest that all these circulating bone marrow-derived stem/progenitor cells could possibly play a role also in the homeostasis of peripheral tissues; however, their major contribution occur within few days from the tissue injury (i.e., after a fracture) when an increase in the number of these circulating cells was observed [Iwakura *et al.*, 2012].

Human circulating stem/progenitor cells have been reported to be quite different from mice and rabbits' ones. The biggest difference is probably their number, as these cells are rare in adult humans compared to laboratory animals, even during fracture healing [Alm *et al.*, 2010; Kuznetsov *et al.*, 2007].

It has been suggested that the relative contribution of pericyte-derived and non–pericyte-derived MSCs to cell differentiation in any given tissue depends on the extent of the vascularity and the kinetics of growth and/or repair. In tissues with low vascularity, such as articular cartilage, the pericyte contribute less to MSCs as compared to tissues with more extensive blood supplies. [Feng *et al.*, 2011].

## 4.7 Fracture Healing Cascade: A Paradigm of Tissue Repair/Regeneration

Fracture healing shares many similarities with soft-tissue healing with the exception of its ability to complete without the formation of a scar. Fracture healing is a complex regenerative process initiated in response to injury and involving intracellular and extracellular molecular signaling for bone induction and conduction to restore broken bones to their original shape and function. The body itself responds to the injury to protect the tissue integrity. There are four components at injury site taking part in the healing process: the cortex of the bone, the periosteum, the bone marrow, and the external soft tissue. The extent to which each component is involved depends on specific conditions at the injured tissue, such as the level of GFs, nutrients, oxygen tension, the electrical environment, and the mechanical stability of the fracture.

Many of the molecular mechanisms and tissue differentiation profiles observed during embryonic development are repeated during the bone fracture repair. Similar to bone development, bone repair occurs through a combination of intra-membraneous and endochondral ossification. Intra-membraneous ossification involves direct formation of the bone without previous formation of the cartilage by committed osteoprogenitors and undifferentiated mesenchymal cells that reside in the periosteum. This process results in the formation of a callus within a few millimeters from the fracture site, described histologically as "hard callus." Endochondral ossification consists of recruitment, proliferation, and differentiation of undifferentiated mesenchymal cells into cartilage, which becomes calcified and eventually replaced by bone [Tsiridis *et al.*, 2007].

The pattern of fracture healing based on endochondral ossification develops through hematoma formation, inflammation, angiogenesis, tissue formation, and tissue maturation. Immediately after the occurrence of the fracture, the repair process begins with the formation of a blood clot, or hematoma, and the degranulation of platelets. The clot serves as a reservoir for cytokines and GFs that begin the healing cascade. The hematoma also establishes a fibrin network that provides a pathway for further cell migration. Due to the disruption of skeletal integrity, normal vascular structures, and nutrient flow at the fracture site are affected, causing a reduction in oxygen tension and a disruption of the bone marrow architecture. During this process, a variety of GFs regulating cell migration and differentiation, which are normally trapped in the bone matrix, are released into the local environment [Shapiro, 2008], leading to the infiltration of inflammatory cells, granulocytes, and macrophages. This initial (within the first 24 hours) inflammatory cascade, necessary for a proper fracture healing, is controlled by major signaling molecules: pro-inflammatory cytokines such as IL-1, IL-6 and tumour TNF-alpha secreted by inflammatory cells that have a chemotactic effect on other inflammatory cells and on the recruitment of mesenchymal cells [Kon *et al.*, 2001]. Other secreted growth and differentiation factors, such as TGF-beta superfamily including bone morphogenetic protein (BMP), as well as platelet-derived growth factor (PDGF), FGFs, and IGF, orchestrate crucial events for chondro-osteogenesis, including mesenchymal and osteoprogenitor cell chemotaxis, proliferation and differentiation, and extracellular matrix ossification. Finally, other important signaling molecules are the angiogenic factors such as VEGFs and Angiopoietin 1 and 2. VEGFs are essential mediators of neoangiogenesis and endothelial-cell specific mitogens. Angiopoietin 1 and 2 are regulatory vascular morphogenetic molecules related to the formation of larger vessels and development of collateral branches from existent vessels.

Simultaneously BMP are released from the bone matrix and by primary recruited MSC, inducing a sequential cascade of events for chondro-osteogenesis. In fracture healing of mice and rats, these proteins have shown a variety of osteogenic effects, temporal expressions, and mitogenic capacities [Cho *et al.*, 2002] suggesting an osteogenic hierarchical model of BMPs. BMP-2, BMP-6, and BMP-9 are the most potent inducers of osteoblast differentiation of mesenchymal progenitor cells; however, most BMPs (except BMP-3 and BMP-13) promote the terminal differentiation of committed osteoblastic precursors and osteoblasts. BMPs may also stimulate the synthesis and secretion of other bone and angiogenic GFs such as IGF and VEGF as well as bone formation by directly activating endothelial cells to stimulate angiogenesis [Valdimarsdottir *et al.*, 2002].

The inflammation phase is followed by osteoprogenitor cell migration and differentiation to develop the fracture callus, occurring over the first week after injury. These cells potentially come from three sources: the periosteum, the surrounding soft tissues, and the marrow space within the bone. The periosteum is commonly considered the primary source of osteo/chondroprogenitors, but the potential of intramuscular implantation of demineralized bone to induce ectopic bone formation and the ability of bone marrow stromal cells to undergo osteogenic and chondrogenic differentiation [Dupont *et al.*, 2010] suggest that the surrounding tissues and the endosteal bone marrow also play important roles in the repair process [Gerstenfeld *et al.*, 2003]. During these processes,

MSC differentiate into chondroblasts forming a soft callus. First, chondroblasts synthesize cartilage specific matrix such as type II collagen and proteoglycans; then the cartilage undergoes into hypertrophy and mineralization in a spatial organized manner. During this phase, new vessels begin to invade calcified hypertrophic chondrocyte and woven bone formation occurs with a recruitment of new MSCs. Recruited MSCs play a key role in the mobilization of macrophages and in the induction of their functional switch from a pro-inflammatory to a pro-resolving phenotype, as well as the subsequent formation of a bone regenerative niche through recruitment of specific bone marrow-derived circulating endothelial and pluripotent mesenchymal progenitors with vasculogenic and osteogenic properties, respectively.

The cartilagineous avascular callus formation serves to stabilize the fracture gap and to reduce interfragmentary motion. Its maturation into hypertrophic cartilage, occurring at approximately two weeks after injury, promotes growth of vasculature into the callus [Shapiro, 2008], introducing a blood supply at the fracture site. GFs that stimulate angiogenesis, such as FGF and VEGF, are present in the fracture site and potentially enhance the healing process by affecting chondrocytes and osteoblasts. Vascularization is necessary during this stage and is thought to be necessary for recruitment of osteoblastic progenitors necessary for bone deposition. As osteoblastic progenitors are brought into the area with the new blood vessels, osteoclasts begin to resorb the calcified tissue. Chondrocytes in the area are removed by a combination of apoptotic cell death and cell necrosis. The osteoblastic progenitors differentiate into osteoblasts that produce the woven bone to unite the fracture ends, facilitating rapid woven bone formation, chondrocyte apoptosis, and osteo/chondroclast recruitment to completely replace the cartilage template with disorganized woven bone. Finally, starting from about the third week after injury, the external callus is resorbed by osteoclasts, the disorganized woven bone is replaced by mature lamellar bone and the marrow is re-established [Gerstenfeld *et al.*, 2003].

Previously published data [Otsuru *et al.*, 2008; Pignolo and Kassem, 2011] suggest that the number of pluripotent mesenchymal cells in the peripheral blood increases after a fracture and that homing of osteogenic cells from systemic circulation participate in fracture repair. Homing pathways might be modulated to enhance the contribution of circulating progenitors at the site of skeletal injury. Little is known about the signals that trigger the mobilization and recruitment of pluripotent mesenchymal cells into the bloodstream after fracture. One possible signalling pathway to recruit progenitors from the systemic circulation involves SDF-1, a CXC chemokine, which binds to the CXC receptor 4 (CXCR4) on the cell surface of responsive cells [Nagasawa, 2000]. Otsuru *et al.* [2008] demonstrated that circulating bone marrow-derived osteoblast progenitor cells migrated to the bone-forming site by SDF-1 chemoattraction in a mouse model of boBMP-2 induced ectopic bone formation. In addition, Kitaori *et al.* [2009] demonstrated that SDF-1 recruits multi-potent mesenchymal cells to bone repair sites during the early phase of bone repair. These data suggest that a SDF-1 chemoattraction may play an important role in the recruitment of circulating pluri/multipotent mesenchymal cells to the fracture site. Local expression of SDF-1 has been shown to recruit hematopoietic/EPCs to ischemic sites [Kucia *et al.*, 2006(a, b)], suggesting that SDF-1 may enhance fracture repair through a second possible mechanism. SDF-1 could recruit EPCs, contributing to increased angiogenesis [Kucia *et al.*, 2006], which is a key event in the early phase of fracture repair. Monocyte chemotactic protein 3 (MCP-3), a member of the cysteine–cysteine chemokine family, is another possible signaling pathway to recruit osteoprogenitors from the systemic circulation [Mezey *et al.*, 2000].

The scaffold used for the repair of critical size defects such as large bone defects, remains a crucial part of the therapeutic strategy. New generations of scaffolds aim at mimicking the morphology and function of the extracellular matrix, by incorporating properties such as "bioactivity" and ability of inducing host-cell recruitment and homing to the graft site. In this context, the true knowledge of the mechanisms of healing, as well as the GFs and cytokines involved is important for its exploitation in the clinical practices.

### 4.7.1 NF-κB: A Main Biological Pathway Activated during Tissue Regeneration

Taking advantage of the closure scratch test performed with confluent monolayer cultures of cells of different tissue origin such as keratinocytic cell lines, C2C12 myoblast cell line, dermal fibroblasts, endothelial cells, it has been reported that the effect of PRP in speeding up the closure rate depends upon the activation of the P38 and ERK pathways. In the human skin epithelial-like cells (NCTC strain 2544) PRP induces a strong pro-inflammatory response involving P38 and NF-κB and leading to the secretion of IL-8 and the antimicrobial lipocalin NGAL.

This enhancing effect of PRP is strongly increased under inflammatory conditions, that is, in cells stimulated by treatment with IL-1, a condition that is physiological in a wound site. These findings suggest that platelet released factors increases the physiological response of the tissue to the wound leading to more rapid tissue repair and, at the same time, establish a defense to bacterial infection [El Backly *et al.*, 2011].

The activation of common pathways has been described during cell differentiation and during the response of cells to inflammation. In the musculoskeletal tissue, NF-κB controls differentiation or activity of the major skeletal cell types, indicating that NF-κB exerts a regulatory role in bone growth and development. In chick embryo, the overexpression of IκB-alpha, which prevents NF-κB activation, results in abnormal limb development [Kanegae *et al.*, 1998]. The activation of the NF-κB pathway is required for the osteoclast and chondrocyte differentiation. Indeed, mice deficient in both the NF-κB subunits p50 and p52 could not form osteoclast with consequent retarded growth and shortened long bones [Franzoso *et al.*, 1997; Iotsova *et al.*, 1997]. The resulting phenotype was osteopetrosis, in which dense, misshapen bones are highly susceptible to fracture and teeth fail to erupt. Both canonical and non-canonical pathways are involved in osteoclast differentiation. The canonical pathway is activated by virtually all stimuli that affect NF-κB expression, including receptor activator of nuclear factor kappa-B ligand (RANKL), the master osteoclastogenic cytokine, TNF-alpha, and other inflammatory mediators [Vaira *et al.*, 2008]. In the alternative NF-κB pathway, the protein NIK (NF-κB-inducing kinase) is the regulatory switch [Soysa *et al.*, 2010].

Studies in chicken chondrocytes supported involvement of inflammatory pathways in chondrocyte differentiation [Cermelli *et al.*, 2000; Zerega *et al.*, 2000]. Similarly, studies in murine chondrocytes, suggested the involvement of NF-κB during differentiation and inflammation [Ulivi *et al.*, 2006, 2008]. In addition, NF-κB has been shown to promote longitudinal bone growth by stimulating growth plate chondrogenesis and preventing chondrocyte apoptosis [Wu *et al.*, 2007, 2011]. On the contrary, activation of NF-κB promotes proliferation and inhibits differentiation of osteoblast, myoblast, and smooth muscle cells via activation of cyclin D1 [Guttridge *et al.*, 1999; Chang *et al.*, 2009].

Lastly, the p38 mitogen-activated protein kinases (MAPK) subfamilies that are activated by most environmental stresses, and by pro-inflammatory cytokines, such as IL-1 and TNF-alpha, are involved in a number of cell functions apparently unrelated to stress response such as chondrogenesis, osteogenesis, epidermal stem cells differentiation, myoblast differentiation [Nakamura *et al.*, 1999; Noth *et al.*, 2003; Greenblatt *et al.*, 2010; Lovett *et al.*, 2010; Connelly *et al.*, 2011].

In summary, a variety of evidences suggest that mammalians adult tissues exposed to insults leading to wound healing, retrace the same events occurred during embryogenesis. Many of the molecular mechanisms that control cellular differentiation and growth during embryogenesis recur during healing; however, these processes take place in a post-natal environment that is unique and distinct from the embryonic one.

## References

Alliot-Licht, B., Hurtrel, D. and Gregoire, M. (2001) Characterization of alpha-smooth muscle actin positive cells in mineralized human dental pulp cultures, *Arch. Oral Biol.*, **46**(3), 221–228.

Alm, J. J., Koivu, H. M., Heino, T. J., Hentunen, T. A., Laitinen, S. and Aro, H. T. (2010) Circulating plastic adherent mesenchymal stem cells in aged hip fracture patients, *J. Orthop. Res.*, **28**(12), 1634–1642.

Anghelina, M., Krishnan, P., Moldovan, L. and Moldovan, N. I. (2006) Monocytes/macrophages cooperate with progenitor cells during neovascularization and tissue repair: Conversion of cell columns into fibrovascular bundles, *Am. J. Pathol.*, **168**(2), 529–541.

Anitua, E., Andia, I., Ardanza, B., Nurden, P. and Nurden, A. T. (2004) Autologous platelets as a source of proteins for healing and tissue regeneration, *Thromb. Haemost.*, **91**(1), 4–15.

Anitua, E., Andia, I., Sanchez, M., Azofra, J., del Mar Zalduendo, M., de la Fuente, M., Nurdenet Al. (2005) Autologous preparations rich in growth factors promote proliferation and induce VEGF and HGF production by human tendon cells in culture, *J. Orthop. Res.*, **23**(2), 281–286.

Asahara, T., Murohara, T., Sullivan, A., Silver, M., van der Zee, R., Li, T., *et al.* (1997) Isolation of putative progenitor endothelial cells for angiogenesis, *Science*, **275**(5302), 964–967.

Aso, H., Ito, S., Mori, A., Morioka, M., Suganuma, N., Kondo, M., *et al.* (2012) Prostaglandin E2 enhances interleukin-8 production via EP4 receptor in human pulmonary microvascular endothelial cells, *Am. J. Physiol.*, **302**(2), L266–L273.

Barker N, Bartfeld S, Clevers H (2010) Tissue-resident adult stem cell populations of rapidly self-renewing organs, *Cell Stem Cell*, **7**(6), 656–670.

Beltrami, A. P., Cesselli, D., Bergamin, N., Marcon, P., Rigo, S., *et al.* (2007) Multipotent cells can be generated *in vitro* from several adult human organs (heart, liver, and bone marrow), *Blood*, **110**(9), 3438–3446.

Caplan, A. I. (1991) Mesenchymal stem cells, *J. Orthop. Res.*, **9**(5), 641–650.

Caplan, A. I. and Correa, D. (2011) The MSC: an injury drugstore, *Cell Stem Cell*, **9**(1), 11–15.

Cassano, M., Quattrocelli, M., Crippa, S., Perini, I., Ronzoni, F. and Sampaolesi, M. (2009) Cellular mechanisms and local progenitor activation to regulate skeletal muscle mass, *J. Muscle Res. Cell. Motil*, **30**(7–8), 243–253.

Cermelli, S., Zerega, B., Carlevaro, M., Gentili, C., Thorp, B., Farquharson, C., *et al.* (2000) Extracellular fatty acid binding protein (Ex-FABP) modulation by inflammatory agents: "Physiological" acute phase response in endochondral bone formation, *Eur. J. Cell. Biol.*, **79**(3), 155–164.

Chandra, R. K., Handorf, C., West, M., Kruger, E. A. and Jackson, S. (2007) Histologic effects of autologous platelet gel in skin flap healing, *Arch. Facial. Plast. Surg.*, **9**(4), 260–263.

Chang, J., Wang, Z., Tang, E., Fan, Z., McCauley, L., Franceschi, R., *et al.* (2009) Inhibition of osteoblastic bone formation by nuclear factor-kappaB, *Nat. Med.*, **15**(6), 682–689.

Chen, F. M., Sun, H. H., Lu, H. and Yu, Q. (2012) Stem cell-delivery therapeutics for periodontal tissue regeneration, *Biomaterials*, **33**(27), 6320–6344.

Cho, T. J., Gerstenfeld, L. C. and Einhorn, T. A. (2002) Differential temporal expression of members of the transforming growth factor beta superfamily during murine fracture healing, *J. Bone Miner Res.*, **17**(3), 513–520.

Connelly, J. T., Mishra, A., Gautrot, J. E. and Watt, F. M. (2011) Shape-induced terminal differentiation of human epidermal stem cells requires p38 and is regulated by histone acetylation, *PloS One*, **6**(11), e27259.

Costa, D., Biticchi, R., Negrini, S., Tasso, R., Cancedda, R., Descalzi, F., *et al.* (2010) Lipocalin-2 controls the expression of SDF-1 and the number of responsive cells in bone, *Cytokine*, **51**(1), 47–52.

Covas, D. T., Panepucci, R. A., Fontes, A. M., Silva, W. A., Jr., Orellana, M. D., Freitas, M. C., *et al.* (2008) Multipotent mesenchymal stromal cells obtained from diverse human tissues share functional properties and gene-expression profile with CD146+ perivascular cells and fibroblasts, *Exp. Hematol.*, **36**(5), 642–654.

Crisan, M., Yap, S., Casteilla, L., Chen, C. W., Corselli, M., Park, T. S., *et al.* (2008) A perivascular origin for mesenchymal stem cells in multiple human organs, *Cell Stem Cell*, **3**(3), 301–313.

D'Ippolito, G., Diabira, S., Howard, G. A., Menei, P., Roos, B. A. and Schiller, P. C. (2004) Marrow-isolated adult multilineage inducible (MIAMI) cells, a unique population of postnatal young and old human cells with extensive expansion and differentiation potential, *J. Cell Sci.*, **117**(Pt 14), 2971–2981.

Da Silva Meirelles, L., Caplan, A. I. and Nardi, N. B. (2008) In search of the *in vivo* identity of mesenchymal stem cells, *Stem Cells*, **26**(9), 2287–2299.

Das, R., Jahr, H., van Osch, G. J. and Farrell, E. (2010) The role of hypoxia in bone marrow-derived mesenchymal stem cells: Considerations for regenerative medicine approaches, *Tissue Eng. Part B Rev.*, **16**(2), 159–168.

DeRossi, R., Coelho, A. C., Mello, G. S., Frazilio, F. O., Leal, C. R., Facco, G. G., *et al.* (2009) Effects of platelet-rich plasma gel on skin healing in surgical wound in horses, *Acta Cir. Bras.*, **24**(4), 276–281.

Doetsch, F. (2003) A niche for adult neural stem cells, *Curr. Opin. Genet. Dev.*, **13**(5), 543–550.

Doucet, C., Ernou, I., Zhang, Y., Llense, J. R., Begot, L., Holy, X., *et al.* (2005) Platelet lysates promote mesenchymal stem cell expansion: A safety substitute for animal serum in cell-based therapy applications, *J. Cell. Physiol.*, **205**(2), 228–236.

Dupont, K. M., Sharma, K., Stevens, H. Y., Boerckel, J. D., Garcia, A. J. and Guldberg, R. E. (2010) Human stem cell delivery for treatment of large segmental bone defects, *Proc. Natl. Acad. Sci. USA*, **107**(8), 3305–3310.

Ehninger A, Trumpp A (2011) The bone marrow stem cell niche grows up: Mesenchymal stem cells and macrophages move in, *J. Exp. Med.*, **208**(3), 421–428.

Einhorn, T. A., Majeska, R. J., Rush, E. B., Levine, P. M. and Horowitz, M. C. (1995) The expression of cytokine activity by fracture callus, *J. Bone Miner. Res.*, **10**(8), 1272–1281.

El Backly, R., Ulivi, V., Tonachini, L., Cancedda, R., Descalzi, F. and Mastrogiacomo, M. (2011) Platelet lysate induces *in vitro* wound healing of human keratinocytes associated with a strong proinflammatory response, *Tissue Eng. Part A*, **17**(13–14), 1787–1800.

Eppley, B. L., Woodell, J. E. and Higgins, J. (2004) Platelet quantification and growth factor analysis from platelet-rich plasma: Implications for wound healing, *Plast. Reconstr. Surg.*, **114**(6), 1502–1508.

Feng, J., Mantesso, A., De Bari, C., Nishiyama, A. and Sharpe, P. T. (2011) Dual origin of mesenchymal stem cells contributing to organ growth and repair, *Proc. Nat. Acad. Sci. USA*, **108**(16), 6503–6508.

Ferrara, N., Gerber, H. P. and LeCouter, J. (2003) The biology of VEGF and its receptors, *Nat. Med.*, **9**(6), 669–676.

Franzoso, G., Carlson, L., Xing, L., Poljak, L., Shores, E. W., Brown, K. D., *et al.* (1997) Requirement for NF-kappaB in osteoclast and B-cell development, *Genes Dev.*, **11**(24), 3482–3496.

Friedenstein, A. J., Chailakhyan, R. K., Latsinik, N. V., Panasyuk, A. F. and Keiliss-Borok, I. V. (1974) Stromal cells responsible for transferring the microenvironment of the hemopoietic tissues. Cloning *in vitro* and retransplantation *in vivo*, *Transplantation*, **17**(4), 331–340.

Fuentes, M. E., Durham, S. K., Swerdel, M. R., Lewin, A. C., Barton, D. S., Megill, J. R., *et al.* (1995) Controlled recruitment of monocytes and macrophages to specific organs through transgenic expression of monocyte chemoattractant protein-1, *J. Immunol.*, **155**(12), 5769–5776.

Geremicca, W., Fonte, C. and Vecchio, S. (2010) Blood components for topical use in tissue regeneration: Evaluation of corneal lesions treated with platelet lysate and considerations on repair mechanisms, *Blood Transfus.*, **8**(2), 107–112.

Gerstenfeld, L. C., Cullinane, D. M., Barnes, G. L., Graves, D. T. and Einhorn, T. A. (2003) Fracture healing as a post-natal developmental process:

Molecular, spatial, and temporal aspects of its regulation, *J. Cell. Biochem.*, **88**(5), 873–884.

Giaccia, A. J., Simon, M. C. and Johnson, R. (2004) The biology of hypoxia: The role of oxygen sensing in development, normal function, and disease, *Genes Dev.*, **18**(18), 2183–2194.

Glennie, S., Soeiro, I., Dyson, P. J., Lam, E. W. and Dazzi, F. (2005) Bone marrow mesenchymal stem cells induce division arrest anergy of activated T cells, *Blood*, **105**(7), 2821–2827.

Grageda, E. (2004) Platelet-rich plasma and bone graft materials: A review and a standardized research protocol, *Implant. Dent.*, **13**(4), 301–309.

Greenblatt, M. B., Shim, J. H., Zou, W., Sitara, D., Schweitzer, M., Hu, D., *et al.* (2010) The p38 MAPK pathway is essential for skeletogenesis and bone homeostasis in mice, *J. Clinical Investig.*, **120**(7), 2457–2473.

Gronthos, S., Brahim, J., Li, W., Fisher, L. W., Cherman, N., Boyde, A., *et al.* (2002) Stem cell properties of human dental pulp stem cells, *J. Dent. Res.*, **81**(8), 531–535.

Gronthos, S., Mankani, M., Brahim, J., Robey, P. G. and Shi, S. (2000) Postnatal human dental pulp stem cells (DPSCs) *in vitro* and *in vivo*, *Proc. Natl. Acad. Sci. USA*, **97**(25), 13625–13630.

Gurtner, G. C., Werner, S., Barrandon, Y. and Longaker, M. T. (2008) Wound repair and regeneration, *Nature*, **453**(7193), 314–321.

Guttridge, D. C., Albanese, C., Reuther, J. Y., Pestell, R. G. and Baldwin, A. S., Jr. (1999) NF-κB controls cell growth and differentiation through transcriptional regulation of cyclin D1, *Mol. Cell. Biol.*, **19**(8), 5785–5799.

Hamilton, T. A., Ohmori, Y. and Tebo, J. (2002) Regulation of chemokine expression by antiinflammatory cytokines, *Immunol. Res.*, **25**(3), 229–245.

Hammes, H. P., Lin, J., Wagner, P., Feng, Y., Vom Hagen, F., Krzizok, T., *et al.* (2004) Angiopoietin-2 causes pericyte dropout in the normal retina: Evidence for involvement in diabetic retinopathy, *Diabetes*, **53**(4), 1104–1110.

Hegyi, B., Sagi, B., Kovacs, J., Kiss, J., Urban, V. S., Meszaros, G., *et al.* (2010) Identical, similar or different? Learning about immunomodulatory function of mesenchymal stem cells isolated from various mouse tissues: Bone marrow, spleen, thymus and aorta wall, *Int. Immunol.*, **22**(7), 551–559.

Hellstrom, M., Gerhardt, H., Kalen, M., Li, X., Eriksson, U., Wolb urg, H., *et al.* (2001) Lack of pericytes leads to endothelial hyperplasia and abnormal vascular morphogenesis, *J. Cell. Biol.*, **153**(3), 543–553.

Hewitson, K. S. and Schofield, C. J. (2004) The HIF pathway as a therapeutic target, *Drug. Discov. Today*, **9**(16), 704–711.

Hirschi, K. K. and D'Amore, P. A. (1996) Pericytes in the microvasculature, *Cardiovasc. Res.*, **32**(4), 687–698.

Hristov, M., Zernecke, A., Liehn, E. A. and Weber, C. (2007) Regulation of endothelial progenitor cell homing after arterial injury, *Thromb. Haemost.*, **98**(2), 274–277.

Hu, Z., Peel, S. A., Ho, S. K., Sandor, G. K. and Clokie, C. M. (2009) Platelet-rich plasma induces mRNA expression of VEGF and PDGF in rat bone marrow stromal cell differentiation, *Oral Surg. Oral Med. Oral Pathol. Oral Radiol. Endod.*, **107**(1), 43–48.

Huang, G. T., Gronthos, S. and Shi, S. (2009) Mesenchymal stem cells derived from dental tissues vs. those from other sources: Their biology and role in regenerative medicine, *J. Dent. Res.*, **88**(9), 792–806.

Iotsova, V., Caamano, J., Loy, J., Yang, Y., Lewin, A. and Bravo, R. (1997) Osteopetrosis in mice lacking NF-kappaB1 and NF-kappaB2, *Nat. Med.*, **3**(11), 1285–1289.

Iwakura, T., Lee, S. Y., Miwa, M., Sakai, Y., Niikura, T., Oe, K., *et al.* (2012) Analysis of circulating mesenchymal progenitor cells in arterial and venous blood after fracture, *J. Tissue. Eng. Regen. Med.*, **7**(6), 501–504.

Jiang, Y., Vaessen, B., Lenvik, T., Blackstad, M., Reyes, M. and Verfaillie, C. M. (2002) Multipotent progenitor cells can be isolated from postnatal murine bone marrow, muscle, and brain, *Exp. Hematol.*, **30**(8), 896–904.

Johansson, L., Klinth, J., Holmqvist, O. and Ohlson, S. (2003) Platelet lysate: A replacement for fetal bovine serum in animal cell culture? *Cytotechnology*, **42**(2), 67–74.

Kanegae, Y., Tavares, A. T., Izpisua Belmonte, J. C. and Verma, I. M. (1998) Role of Rel/NF-kappaB transcription factors during the outgrowth of the vertebrate limb, *Nature*, **392**(6676), 611–614.

Kasemkijwattana, C., Menetrey, J., Bosch, P., Somogyi, G., Moreland, M. S., Fu, F. H., *et al.* (2000) Use of growth factors to improve muscle healing after strain injury, *Clin. Orthop. Relat. Res.*, **370**, 272–285.

Kassolis, J. D., Rosen, P. S. and Reynolds, M. A. (2000) Alveolar ridge and sinus augmentation utilizing platelet-rich plasma in combination with freeze-dried bone allograft: case series, *J. Periodontol.*, **71**(10), 1654–1661.

Kern, S., Eichler, H., Stoeve, J., Kluter, H. and Bieback, K. (2006) Comparative analysis of mesenchymal stem cells from bone marrow, umbilical cord blood, or adipose tissue, *Stem Cells*, **24**(5), 1294–1301.

Kietzmann, T., Roth, U. and Jungermann, K. (1999) Induction of the plasminogen activator inhibitor-1 gene expression by mild hypoxia via a hypoxia response element binding the hypoxia-inducible factor-1 in rat hepatocytes, *Blood*, **94**(12), 4177–4185.

Kilian, O., Flesch, I., Wenisch, S., Taborski, B., Jork, A., Schnettler, R., *et al.* (2004) Effects of platelet growth factors on human mesenchymal stem cells and human endothelial cells *in vitro*, *Eur. J. Med. Res.*, **9**(7), 337–344.

Kitaguchi, H., Funahara, Y., Butthep, P. and Bunyaratvej, A. (1992) Red blood cell and platelet induced differentiation of endothelial cells into capillary-like structure, *Southeast Asian J. Trop. Med. Public Health*, **23**(Suppl), 295–100.

Kitaori, T., Ito, H., Schwarz, E. M., Tsutsumi, R., Yoshitomi, H., Oishi, S., *et al.* (2009) Stromal cell-derived factor 1/CXCR4 signaling is critical for the recruitment of mesenchymal stem cells to the fracture site during skeletal repair in a mouse model, *Arthritis Rheum.*, **60**(3), 813–823.

Kon, T., Cho, T. J., Aizawa, T., Yamazaki, M., Nooh, N., Graves, D., *et al.* (2001) Expression of osteoprotegerin, receptor activator of NF-kappaB ligand (osteoprotegerin ligand) and related proinflammatory cytokines during fracture healing, *J. Bone Miner. Res.*, **16**(6), 1004–1014.

Kucia, M., Reca, R., Campbell, F. R., Zuba-Surma, E., Majka, M., Ratajczak, J., *et al.* (2006a) A population of very small embryonic-like (VSEL) CXCR4(+)SSEA-1(+)Oct-4+ stem cells identified in adult bone marrow, *Leukemia*, **20**(5), 857–869.

Kucia, M., Wojakowski, W., Reca, R., Machalinski, B., Gozdzik, J., Majka, M., *et al.* (2006b) The migration of bone marrow-derived non-hematopoietic tissue-committed stem cells is regulated in an SDF-1-, HGF-, and LIF-dependent manner, *Arch. Immunol. Ther. Exp.*, **54**(2), 121–135.

Kuwabara, K., Ogawa, S., Matsumoto, M., Koga, S., Clauss, M., Pinsky, D. J., *et al.* (1995) Hypoxia-mediated induction of acidic/basic fibroblast growth factor and platelet-derived growth factor in mononuclear phagocytes stimulates growth of hypoxic endothelial cells, *Proc. Natl. Acad. Sci. USA*, **92**(10), 4606–4610.

Kuznetsov, S. A., Mankani, M. H., Leet, A. I., Ziran, N., Gronthos, S. and Robey, P. G. (2007) Circulating connective tissue precursors: extreme rarity in humans and chondrogenic potential in guinea pigs, *Stem Cells*, **25**(7), 1830–1839.

Lehmann, W., Edgar, C. M., Wang, K., Cho, T. J., Barnes, G. L., Kakar, S., *et al.* (2005) Tumor necrosis factor alpha (TNF-alpha) coordinately regulates the expression of specific matrix metalloproteinases (MMPS) and angiogenic factors during fracture healing, *Bone*, **36**(2), 300–310.

Li, L. and Jiang, J. (2011) Regulatory factors of mesenchymal stem cell migration into injured tissues and their signal transduction mechanisms, *Front. Med.*, **5**(1), 33–39.

Lin, G., Garcia, M., Ning, H., Banie, L., Guo, Y. L., Lue, T. F., *et al.* (2008) Defining stem and progenitor cells within adipose tissue, *Stem Cells Dev.*, **17**(6), 1053–1063.

Lolmede, K., Campana, L., Vezzoli, M., Bosurgi, L., Tonlorenzi, R., Clementi, *et al.* (2009) Inflammatory and alternatively activated human macrophages attract vessel-associated stem cells, relying on separate HMGB1- and MMP-9-dependent pathways, *J. Leukoc. Biol.*, **85**(5), 779–787.

Lopez-Jornet, P., Camacho-Alonso, F., Molina-Minano, F. and Vicente-Ortega, V. (2009) Effects of plasma rich in growth factors on wound healing of the tongue. Experimental study on rabbits, *Med. Oral. Patol. Oral. Cir. Bucal.*, **14**(9), e425–e428.

Lopez-Plandolit, S., Morales, M. C., Freire, V., Etxebarria, J. and Duran, J. A. (2010) Plasma rich in growth factors as a therapeutic agent for persistent corneal epithelial defects, *Cornea*, **29**(8), 843–848.

Lovett, F. A., Cosgrove, R. A., Gonzalez, I. and Pell, J. M. (2010) Essential role for p38alpha MAPK but not p38gamma MAPK in Igf2 expression and myoblast differentiation, *Endocrinology*, **151**(9), 4368–4380.

Madeddu, P. (2005) Therapeutic angiogenesis and vasculogenesis for tissue regeneration, *Exp. Physiol.*, **90**(3), 315–326.

Maggini, J., Mirkin, G., Bognanni, I., Holmberg, J., Piazzon, I. M., Nepomnaschy, I., *et al.* (2010) Mouse bone marrow-derived mesenchymal stromal cells turn activated macrophages into a regulatory-like profile, *PloS One*, **5**(2), e9252.

Mantovani, A., Sica, A., Sozzani, S., Allavena, P., Vecchi, A. and Locati, M. (2004) The chemokine system in diverse forms of macrophage activation and polarization, *Trends Immunol.*, **25**(12), 677–686.

Marx, R. E., Carlson, E. R., Eichstaedt, R. M., Schimmele, S. R., Strauss, J. E. and Georgeff, K. R. (1998) Platelet-rich plasma: Growth factor enhancement for bone grafts, *Oral Surg. Oral Med. Oral Pathol. Oral Radiol. Endod.*, **85**(6), 638–646.

Mazzucco, L., Medici, D., Serra, M., Panizza, R., Rivara, G., Orecchia, S., *et al.* (2004) The use of autologous platelet gel to treat difficult-to-heal wounds: A pilot study, *Transfusion*, **44**(7), 1013–1018.

Mezey, E., Chandross, K. J., Harta, G., Maki, R. A. and McKercher, S. R. (2000) Turning blood into brain: Cells bearing neuronal antigens generated *in vivo* from bone marrow, *Science*, **290**(5497), 1779–1782.

Miura, M., Gronthos, S., Zhao, M., Lu, B., Fisher, L. W., Robey, P. G., *et al.* (2003) SHED: Stem cells from human exfoliated deciduous teeth, *Proc. Natl. Acad. Sci. USA*, **100**(10), 5807–5812.

Molloy, T., Wang, Y. and Murrell, G. (2003) The roles of growth factors in tendon and ligament healing, *Sports Med.*, **33**(5), 381–394.

Muller, A. M., Davenport, M., Verrier, S., Droeser, R., Alini, M., Bocelli-Tyndall, C., *et al.* (2009) Platelet lysate as a serum substitute for 2D static and 3D perfusion culture of stromal vascular fraction cells from human adipose tissue, *Tissue Eng. Part A*, **15**(4), 869–875.

Nagasawa, T. (2000) A chemokine, SDF-1/PBSF, and its receptor, CXC chemokine receptor 4, as mediators of hematopoiesis, *Int. J. Hematol.*, **72**(4), 408–411.

Nagumo, A., Yasuda, K., Numazaki, H., Azuma, H., Tanabe, Y., Kikuchi, S., *et al.* (2005) Effects of separate application of three growth factors (TGF-beta1, EGF, and PDGF-BB) on mechanical properties of the *in situ* frozen-thawed anterior cruciate ligament, *Clin. Biomech.*, **20**(3), 283–290.

Nakamura, K., Shirai, T., Morishita, S., Uchida, S., Saeki-Miura, K. and Makishima, F. (1999) p38 mitogen-activated protein kinase functionally contributes to chondrogenesis induced by growth/differentiation factor-5 in ATDC5 cells, *Exp. Cell Res.*, **250**(2), 351–363.

Nakamura, S., Ishihara, M., Takikawa, M., Murakami, K., Kishimoto, S., Nakamura, S., *et al.* (2010) Platelet-rich plasma (PRP) promotes survival of fat-grafts in rats, *Ann. Surg.*, **65**(1), 101–106.

Negus, R. P., Turner, L., Burke, F. and Balkwill, F. R. (1998) Hypoxia down-regulates MCP-1 expression: Implications for macrophage distribution in tumors, *J. Leukoc. Biol.*, **63**(6), 758–765.

Nehls, V. and Drenckhahn, D. (1993) The versatility of microvascular pericytes: From mesenchyme to smooth muscle? *Histochemistry*, **99**(1), 1–12.

Norrby, K. (2002) Mast cells and angiogenesis, *APMIS*, **110**(5), 355–371.

Noth, U., Tuli, R., Seghatoleslami, R., Howard, M., Shah, A., Hall, D. J., *et al.* (2003) Activation of p38 and Smads mediates BMP-2 effects on human trabecular bone-derived osteoblasts, *Exp. Cell Res.*, **291**(1), 201–211.

Ortiz, L. A., Dutreil, M., Fattman, C., Pandey, A. C., Torres, G., Go, K., *et al.* (2007) Interleukin 1 receptor antagonist mediates the antiinflammatory and antifibrotic effect of mesenchymal stem cells during lung injury, *Proc. Natl. Acad. Sci. USA*, **104**(26), 11002–11007.

Otsuru, S., Tamai, K., Yamazaki, T., Yoshikawa, H. and Kaneda, Y. (2008) Circulating bone marrow-derived osteoblast progenitor cells are recruited to the bone-forming site by the CXCR4/stromal cell-derived factor-1 pathway, *Stem Cells*, **26**(1), 223–234.

Peister, A., Mellad, J. A., Larson, B. L., Hall, B. M., Gibson, L. F. and Prockop, D. J. (2004) Adult stem cells from bone marrow (MSCs) isolated from different strains of inbred mice vary in surface epitopes, rates of proliferation, and differentiation potential, *Blood*, **103**(5), 1662–1668.

Phelan, M. W., Forman, L. W., Perrine, S. P. and Faller, D. V. (1998) Hypoxia increases thrombospondin-1 transcript and protein in cultured endothelial cells, *J. Lab. Clin. Med.*, **132**(6), 519–529.

Pietramaggiori, G., Scherer, S. S., Mathews, J. C., Alperovich, M., Yang, H. J., Neuwalder, J., *et al.* (2008) Healing modulation induced by freeze-dried platelet-rich plasma and micronized allogenic dermis in a diabetic wound model, *Wound Repair Regen.*, **16**(2), 218–225.

Pignolo, R. J. and Kassem, M. (2011) Circulating osteogenic cells: Implications for injury, repair, and regeneration, *J. Bone Miner. Res.*, **26**(8), 1685–1693.

Pistoia, V. and Raffaghello, L. (2011) Damage-associated molecular patterns (DAMPs) and mesenchymal stem cells: A matter of attraction and excitement, *Eur. J. Immunol.*, **41**(7), 1828–1831.

Pittenger, M. F., Mackay, A. M., Beck, S. C., Jaiswal, R. K., Douglas, R., Mosca, J. D., *et al.* (1999) Multilineage potential of adult human mesenchymal stem cells, *Science*, **284**(5411), 143–147.

Potier, E., Ferreira, E., Andriamanalijaona, R., Pujol, J. P., Oudina, K., Logeart-Avramoglou, D., *et al.* (2007) Hypoxia affects mesenchymal stromal cell osteogenic differentiation and angiogenic factor expression, *Bone*, **40**(4), 1078–1087.

Prockop, D. J. (1997) Marrow stromal cells as stem cells for nonhematopoietic tissues, *Science*, **276**(5309), 71–74.

Pujol, J. P., Chadjichristos, C., Legendre, F., Bauge, C., Beauchef, G., Andriamanalijaona, R., *et al.* (2008) Interleukin-1 and transforming growth factor-beta 1 as crucial factors in osteoarthritic cartilage metabolism, *Connect. Tissue Res.*, **49**(3), 293–297.

Ranzato, E., Balbo, V., Boccafoschi, F., Mazzucco, L. and Burlando, B. (2009a) Scratch wound closure of C2C12 mouse myoblasts is enhanced by human platelet lysate, *Cell Biol. Int.*, **33**(9), 911–917.

Ranzato, E., Boccafoschi, F., Mazzucco, L., Patrone, M. and Burlando, B. (2010) Role of ERK1/2 in platelet lysate-driven endothelial cell repair, *J. Cell. Biochem.*, **110**(3), 783–793.

Ranzato, E., Mazzucco, L., Patrone, M. and Burlando, B. (2009b) Platelet lysate promotes *in vitro* wound scratch closure of human dermal fibroblasts: different roles of cell calcium, P38, ERK and PI3K/AKT, *J. Cell. Mol. Med.*, **13**(8B), 2030–2038.

Ranzato, E., Patrone, M., Mazzucco, L. and Burlando, B. (2008) Platelet lysate stimulates wound repair of HaCaT keratinocytes, *Br. J. Dermatol.*, **159**(3), 537–545.

Ratajczak, M. Z., Zuba-Surma, E. K., Machalinski, B., Ratajczak, J. and Kucia, M. (2008) Very small embryonic-like (VSEL) stem cells: Purification from adult organs, characterization, and biological significance, *Stem Cell Rev.*, **4**(2), 89–99.

Rozman, P. and Bolta, Z. (2007) Use of platelet growth factors in treating wounds and soft-tissue injuries, *Acta Dermatovenerol. Alp. Panonica. Adriat.*, **16**(4), 156–165.

Sacchetti, B., Funari, A., Michienzi, S., Di Cesare, S., Piersanti, S., Saggio, I., *et al.* (2007) Self-renewing osteoprogenitors in bone marrow sinusoids can organize a hematopoietic microenvironment, *Cell*, **131**(2), 324–336.

Sakuda, H., Nakashima, Y., Kuriyama, S. and Sueishi, K. (1992) Media conditioned by smooth muscle cells cultured in a variety of hypoxic environments stimulates *in vitro* angiogenesis. A relationship to transforming growth factor-beta 1, *Am. J. Pathol.*, **141**(6), 1507–1516.

Sanchez, A. R., Sheridan, P. J. and Kupp, L. I. (2003) Is platelet-rich plasma the perfect enhancement factor? A current review, *Int. J. Oral. Maxillofac. Implants*, **18**(1), 93–103.

Satokata, I., Ma, L., Ohshima, H., Bei, M., Woo, I., Nishizawa, K., *et al.* (2000) Msx2 deficiency in mice causes pleiotropic defects in bone growth and ectodermal organ formation, *Nat. Genet.*, **24**(4), 391–395.

Scapini, P., Morini, M., Tecchio, C., Minghelli, S., Di Carlo, E., Tanghetti, E., *et al.* (2004) CXCL1/macrophage inflammatory protein-2-induced angiogenesis *in vivo* is mediated by neutrophil-derived vascular endothelial growth factor-A, *J. Immunol.*, **172**(8), 5034–5040.

Scimeca, C. L., Bharara, M., Fisher, T. K., Kimbriel, H. and Armstrong, D. G. (2010) Novel use of platelet-rich plasma to augment curative diabetic foot surgery, *J. Diabetes Sci. Technol.*, **4**(5), 1121–1126.

Shapiro, F. (2008) Bone development and its relation to fracture repair. The role of mesenchymal osteoblasts and surface osteoblasts, *Eur. Cells Mater.*, **15,** 53–76.

Shi, Q., Rafii, S., Wu, M. H., Wijelath, E. S., Yu, C., Ishida, A., *et al.* (1998) Evidence for circulating bone marrow-derived endothelial cells, *Blood*, **92**(2), 362–367.

Shi, S. and Gronthos, S. (2003) Perivascular niche of postnatal mesenchymal stem cells in human bone marrow and dental pulp, *J. Bone Miner. Res.*, **18**(4), 696–704.

Sica, A. and Mantovani, A. (2012) Macrophage plasticity and polarization: *In vivo* veritas, *J. Clin. Investig.*, **122**(3), 787–795.

Silvestre, J. S., Tamarat, R., Ebrahimian, T. G., Le-Roux, A., Clergue, M., Emmanuel, F., *et al.* (2003) Vascular endothelial growth factor-B promotes *in vivo* angiogenesis, *Circulat. Res.*, **93**(2), 114–123.

Singer, N. G. and Caplan, A. I. (2011) Mesenchymal stem cells: mechanisms of inflam mation, *Ann. Rev. Pathol.*, **6,** 457–478.

Sonoyama, W., Liu, Y., Yamaza, T., Tuan, R. S., Wang, S., Shi, S., *et al.* (2008) Characterization of the apical papilla and its residing stem cells from human immature permanent teeth: A pilot study, *J. Endod.*, **34**(2), 166–171.

Soysa, N. S., Alles, N., Weih, D., Lovas, A., Mian, A. H., Shimokawa, H., *et al.* (2010) The pivotal role of the alternative NF-κB pathway in maintenance of basal bone homeostasis and osteoclastogenesis, *J. Bone Miner. Res.*, **25**(4), 809–818.

Spaggiari, G. M., Capobianco, A., Becchetti, S., Mingari, M. C. and Moretta, L. (2006) Mesenchymal stem cell-natural killer cell interactions: evidence that activated NK cells are capable of killing MSCs, whereas MSCs can inhibit IL-2-induced NK-cell proliferation, *Blood*, **107**(4), 1484–1490.

Tanidir, S. T., Yuksel, N., Altintas, O., Yildiz, D. K., Sener, E. and Caglar, Y. (2010) The effect of subconjunctival platelet-rich plasma on corneal epithelial wound healing, *Cornea*, **29**(6), 664–669.

Tasso, R., Augello, A., Boccardo, S., Salvi, S., Carida, M., Postiglione, F., *et al.* (2009) Recruitment of a host's osteoprogenitor cells using exogenous mesenchymal stem cells seeded on porous ceramic, *Tissue Eng. Part A*, **15**(8), 2203–2212.

Tasso, R., Gaetani, M., Molino, E., Cattaneo, A., Monticone, M., Bachi, A., *et al.* (2012a) The role of bFGF on the ability of MSC to activate endogenous regenerative mechanisms in an ectopic bone formation model, *Biomaterials*, **33**(7), 2086–2096.

Tasso, R., Ulivi, V., Reverberi, D., Lo Sicco, C., Descalzi, F. and Cancedda, R. (2012b) Mesenchymal stem cells (MSC) induce homing of endogenous stem/progenitor cells through the activation of alternatively activated macrophages in an ectopic bone formation model, Paper Presented at the 3rd TERMIS Word Congress, Vienna, Austria.

Toma, J. G., Akhavan, M., Fernandes, K. J., Barnabe-Heider, F., Sadikot, A., Kaplan, D. R., *et al.* (2001) Isolation of multipotent adult stem cells from the dermis of mammalian skin, *Nat. Cell Biol.*, **3**(9), 778–784.

Trento, C. and Dazzi, F. (2010) Mesenchymal stem cells and innate tolerance: Biology and clinical applications, *Swiss Med. Wkly*, **140,** w13121.

Tsiridis, E., Upadhyay, N. and Giannoudis, P. (2007) Molecular aspects of fracture healing: Which are the important molecules? *Injury*, **38**(Suppl. 1), S11–S25.

Tsukamoto, T., Matsui, T., Nakata, H., Ito, M., Natazuka, T., Fukase, M., *et al.* (1991) Interleukin-1 enhances the response of osteoblasts to platelet-derived growth factor through the alpha receptor-specific up-regulation, *J. Biol. Chem.*, **266**(16), 10143–10147.

Ulivi, V., Giannoni, P., Gentili, C., Cancedda, R. and Descalzi, F. (2008) p38/NF-κB-dependent expression of COX-2 during differentiation and inflammatory response of chondrocytes, *J. Cell. Biochem.*, **104**(4), 1393–1406.

Ulivi, V., Tutolo, G., Mallein-Gerin, F., Daga, A., Cancedda, R. and Cancedda, F. D. (2006) A common pathway in differentiation and inflammation: p38 mediates expression of the acute phase SIP24 iron binding lipocalin in chondrocytes, *J. Cell. Physiol.*, **206**(3), 728–737.

Vacca, A., Ribatti, D., Iurlaro, M., Albini, A., Minischetti, M., Bussolino, F., *et al.* (1998) Human lymphoblastoid cells produce extracellular matrix-degrading enzymes and induce endothelial cell proliferation, migration, morphogenesis, and angiogenesis, *Int. J. Clin. Lab. Res.*, **28**(1), 55–68.

Vaira, S., Alhawagri, M., Anwisye, I., Kitaura, H., Faccio, R. and Novack, D. V. (2008) RelA/p65 promotes osteoclast differentiation by blocking a RANKL-induced apoptotic JNK pathway in mice, *J. Clin. Investig.*, **118**(6), 2088–2097.

Valdimarsdottir, G., Goumans, M. J., Rosendahl, A., Brugman, M., Itoh, S., Lebrin, F., *et al.* (2002) Stimulation of Id1 expression by bone morphogenetic protein is sufficient and necessary for bone morphogenetic protein-induced activation of endothelial cells, *Circulation*, **106**(17), 2263–2270.

Van den Dolder, J., Mooren, R., Vloon, A. P., Stoelinga, P. J. and Jansen, J. A. (2006) Platelet-rich plasma: Quantification of growth factor levels and the effect on growth and differentiation of rat bone marrow cells, *Tissue Eng.*, **12**(11), 3067–3073.

Verreck, F. A., de Boer, T., Langenberg, D. M., Hoeve, M. A., Kramer, M., Vaisberg, E., *et al.* (2004) Human IL-23-producing type 1 macrophages promote but IL-10-producing type 2 macrophages subvert immunity to (myco) bacteria, *Proc. Nat. Acad. Sci. USA*, **101**(13), 4560–4565.

Villela, D. L. and Santos, V. L. (2010) Evidence on the use of platelet-rich plasma for diabetic ulcer: A systematic review, *Growth Factors*, **28**(2), 111–116.

Wagers, A. J. (2012) The stem cell niche in regenerative medicine, *Cell Stem Cell*, **10**(4), 362–369.

Wagers, A. J. and Weissman, I. L. (2004) Plasticity of adult stem cells, *Cell*, **116**(5), 639–648.

Weyrich, A. S., Lindemann, S. and Zimmerman, G. A. (2003) The evolving role of platelets in inflammation, *J. Thromb. Haemost.*, **1**(9), 1897–1905.

Wu, S., Flint, J. K., Rezvani, G. and De Luca, F. (2007) Nuclear factor-kappaB p65 facilitates longitudinal bone growth by inducing growth plate chondrocyte proliferation and differentiation and by preventing apoptosis, *J. Biol. Chem.*, **282**(46), 33698–33706.

Wu, S., Morrison, A., Sun, H. and De Luca, F. (2011) Nuclear factor-kappaB (NF-kappaB) p65 interacts with Stat5b in growth plate chondrocytes and mediates the effects of growth hormone on chondrogenesis and on the expression of insulin-like growth factor-1 and bone morphogenetic protein-2, *J. Biol. Chem.*, **286**(28), 24726–24734.

Wykoff, C. C., Beasley, N. J., Watson, P. H., Turner, K. J., Pastorek, J., Sibtain, A., *et al.* (2000) Hypoxia-inducible expression of tumor-associated carbonic anhydrases, *Can. Res.*, **60**(24), 7075–7083.

Yang, X., Ricciardi, B. F., Hernandez-Soria, A., Shi, Y., Pleshko Camacho, N. and Bostrom, M. P. (2007) Callus mineralization and maturation are delayed during fracture healing in interleukin-6 knockout mice, *Bone*, **41**(6), 928–936.

Yeaman, M. R. (2010) Platelets in defense against bacterial pathogens, *Cell. Mol. Life Sci.*, **67**(4), 525–544.

Zaky, S. H., Ottonello, A., Strada, P., Cancedda, R. and Mastrogiacomo, M. (2008) Platelet lysate favours *in vitro* expansion of human bone marrow stromal cells for bone and cartilage engineering, *J. Tissue Eng. Regen. Med.*, **2**(8), 472–481.

Zerega, B., Cermelli, S., Michelis, B., Cancedda, R. and Cancedda, F. D. (2000) Expression of NRL/NGAL (neu-related lipocalin/neutrophil gelatinase-associated lipocalin) during mammalian embryonic development and in inflammation, *Eur. J. Cell Biol.*, **79**(3), 165–172.

Zuk, P. A., Zhu, M., Ashjian, P., De Ugarte, D. A., Huang, J. I., Mizuno, H., *et al.* (2002) Human adipose tissue is a source of multipotent stem cells, *Mol. Biol. Cell*, **13**(12), 4279–4295.

# Chapter 5

# Host Response to Biomaterials

**Sangeetha Srinivasan and Julia E. Babensee**
*Wallace H. Coulter Department of Biomedical Engineering,*
*Georgia Institute of Technology and Emory University, 313 Ferst Drive,*
*Atlanta, GA 30332, USA*
julia.babensee@bme.gatech.edu

The host response to biomaterials can include aspects of the inflammatory response, wound healing, neovascularization and regeneration, and if in the context of combination products, an adaptive immune response. The cascade of events associated with the inflammatory response and foreign body reaction to an implanted biomaterial is presented. To put the biomaterial-induced inflammatory response in context, key components of the more generalized innate immune response are presented. Adaptive immunity is an antigen-specific immune response, the quality and magnitude of which is affected by the earlier innate immune response, when antigen is taken up by antigen-presenting cells such as dendritic cells. Thus, interconnections between innate and adaptive immunity are important when in a biomaterial context, there is an inflammatory response toward the biomaterial component and the potential for an antigen-specific immune response to a biologi-

*Scaffolds for Tissue Engineering: Biological Design, Materials, and Fabrication*
Edited by Claudio Migliaresi and Antonella Motta

ISBN 978-981-4463-20-1 (Hardcover), 978-981-4463-21-8 (eBook)
www.panstanford.com

cal component in a combination product (e.g., tissue-engineered construct). Throughout the chapter, biomaterial strategies for controlling aspects of the inflammatory response and its deleterious consequents are presented. Furthermore, biomaterial strategies to control immune responses in contexts such as vaccines and anti-tumor scenarios as well as to direct regeneration, vascularization, and appropriate remodeling of biomaterials, with capabilities to constructively interact with the host, are discussed.

## 5.1 Introduction

The host response toward implanted biomaterials determines the success of the biomedical implant. The host response, upon biomaterial implantation, includes protein adsorption, blood-material interactions with the formation of provisional fibrin matrix, acute inflammation, chronic inflammation, development of granulation tissue that is integral to wound healing, and finally the foreign body reaction (FBR). A typical FBR to a biomaterial implant is the formation of a dense, avascular fibrous capsule that is comprised of fibroblasts with associated collagen. The fibrous capsule walls of the implant from its surroundings, and within it, some macrophages (MΦs) or foreign body giant cells (FBGCs) are retained at the surface of the material. Such a reaction often has deleterious consequences such as mass transfer limitations, contraction, and degradation of the material surface due to aggressive agents released by the FBGCs.[1–3] Alternatively, more appropriate host responses to implanted biomaterials can be observed that facilitate device integration such as vascularization. Of all the cell types involved in the host response to a biomaterial, MΦs are particularly important in the inflammatory response but also in directing repair and healing, depending on their phenotype. Thus, a shift in thinking about the host response from one of a stereotypical FBR to one that can be controlled or "designed for" through the biomaterial, is emerging. Similarly, a shift of goals from an "inert" material that will inhibit all protein adsorption and in turn avoid cellular responses,[4–7] to one that can interact with the surrounding tissue to direct events that will help and guide the tissue to heal itself, is developing. Hence, controlled tissue remodeling and regeneration is the new goal for biomaterial-induced host responses.

## 5.2 A Typical FBR

### 5.2.1 Cellular and Molecular Aspects

Biomaterial implantation is associated with surgically induced injury that results in the damage of surrounding blood vessels. This permits blood contact with the material leading to activation of the coagulation cascade; thrombus formation accompanied with activation of platelets; and activation of the complement-, fibrinolytic-, and kinin-generating systems.[8,9]

Implanted biomaterials are instantaneously coated with adsorbed proteins from the interstitial fluid or blood. In the adsorbed protein layer, key proteins such as fibrinogen provide molecular signals for host cell interaction. In addition, they facilitate leukocyte adhesion to the material surface, while the chemotactic effects of soluble complement activation fragments such as C5a, bring about leukocyte recruitment. The cellular response is hence directed by the adsorbed protein layer. Endogenously secreted extracellular matrix (ECM) proteins along with the fibrin clot act as a provisional matrix to support cell migration.[1,2]

Inflammation is a protective mechanism of the body in response to cellular or tissue injury or infection and is an important first step of wound healing as will be discussed later.

Acute inflammation is the first stage of inflammation, and it is characterized by the early and rapid tissue infiltration of polymorphonuclear (PMN) cells or neutrophils.[9] Tissue-resident MΦs and dendritic cells (DCs) at the injured site, initiate the early events of acute inflammation.[3] These cells recognize pathogens and danger signals via pattern recognition receptors (PRRs). They also secrete molecular mediators specifically chemokines, which provide chemotactic signals for the migration of PMNs from flowing blood to the site of inflammatory insult and cytokines, which activate the endothelium to support this cell recruitment.[10] Though recruited in large numbers, PMNs are fairly short-lived and generally undergo apoptosis within 24 hours of the inflammatory response. Their main role is to cleanse the injury site, phagocytosing cellular and tissue debris, dead cells, and any pathogens. PMNs also release cytokines/chemokines that control the intensity of the acute inflammation and recruit monocytes from circulation that differentiate to MΦs in the tissue. Acute inflammation normally resolves within 1–4 days.

Persistence of inflammatory stimulus or intracellular microorganisms, repeated bouts of acute inflammation, prolonged exposure to non-degradable substances, immune reactions, or micromotion of an implant can lead to a state of chronic inflammation. MΦs are the predominant cell type that control the mediation of events during this phase of inflammation. Consequently, MΦ functions are diverse, ranging from phagocytosis, antigen uptake, and presentation for initiation of an immune response, to angiogenesis and tissue remodeling.[11,12]

Activation of MΦs is characteristic of the causative agent namely (1) direct Toll-like receptor (TLR) binding to pathogens or danger signals, (2) antibody (Ab) binding, (3) pro-inflammatory cytokines such as interferon-$\gamma$(IFN-$\gamma$) or tumor necrosis factor-alpha (TNF-$\alpha$), via the classical pathway, and (4) cytokines such as interleukin-4 (IL-4), IL-13, IL-10, transforming growth factor-beta (TGF-$\beta$) via the alternative pathway.[12] Classically activated MΦs of M1 phenotype destroy pathogens by phagocytosis and release of reactive nitrogen and oxygen species. Classically activated MΦs secrete pro-inflammatory cytokines, possess anti-proliferative functions, and induce Th1 responses. Furthermore, classically activated MΦs are crucial in matrix destruction and tissue reorganization at injured tissues via the production of proteolytic enzymes such as matrix metalloproteinases (MMPs). This allows the MΦs to quickly migrate through the injured tissues to clear pathogens and debris. However, prolonged activation of classically activated MΦs can lead to tissue destruction. Therefore, mechanisms have evolved to replace the inflammatory MΦs with reparative or alternatively activated MΦs at later stages of inflammation to facilitate tissue remodeling. Alternatively activated M2 MΦs (composed of three specific subtypes M2a, M2b, M2c) are phenotypically distinct from M1 MΦs[12] with an IL-12$^{low}$, IL-23$^{low}$, IL-10$^{high}$ phenotype, and can produce signal-dependent inflammatory cytokines. Specifically, artificially induced M2a MΦs are responsible for generating FBGCs and a Th2 response while retaining the ability to kill pathogens.[13] M2b MΦs can also generate a Th2 response whereas M2c MΦs primarily focus on supporting immunosuppression, ECM remodeling, and inducing fibrosis.[13] A spectrum of MΦ functions with characteristics of M1 and M2 phenotypes are expected during the host response based on the temporal environmental signals that the MΦs receive.

The FBR to implanted biomaterials, is characterized by the occurrence of (1) mature MΦs, (2) FBGCs apposed to the biomaterial surface, and (3) a commonly avascular fibrous capsule comprised of fibroblasts and their associated collagen.[2] Generally, it is considered that phagocytosis of particulates by MΦs occurs in a size-dependent manner. When MΦs encounter a surface that is too large for them to phagocytose (e.g., >10 μm) substances, they fuse with one another to form FBGCs.[14] In other words, phenotypically, M2 MΦs involved in FBGC formation are different from M1 MΦs involved in phagocytosis. Whether this change from phagocytic to fusogenic states is due to complete physiological transformation or is of graded nature is yet to be fully understood. To gain a better understanding of this phenomenon, a study showed that when IL-4-induced fusogenic MΦs of a murine cell line accompanied with a phagocytic stimulus of 3 μm polystyrene microspheres *in vitro*, the formation of FBGCs is not abrogated.[15] Further, *in vivo*, intra-peritoneal implantation of cellulose ester filters in mice followed by an injection of the microspheres into the same site did not prevent fusion of MΦs.[15] These results, in turn, suggest that tailoring biomaterial design to increase phagocytic nature of MΦs may not be sufficient in avoiding FBGC formation and such inhibition would require material approaches that can endogenously reduce IL-4-induced fusogenic MΦs. Other key components that influence FBGC formation such as MΦ chemoattractant protein-1 (MCP-1)[16] and MMP-9[17] may also be targeted artificially through biomaterials to abate the intensity of the FBR. Resolution of PMN presence is presumably needed before a chronic inflammatory response or FBR can occur, as PMNs have been shown to inhibit monocyte adhesion, MΦ maturation, and FBGC formation.[18]

Studies of spatial and temporal components surrounding an implant *in vivo* in FBR have determined that locally, a predominantly immunosuppressive environment prevailed, mediated mainly by Th2 cytokines (IL-4, IL-13, IL-10, and TGF-$\beta$); moreover, there was localized loss of IL-6, loss of other pro-inflammatory cytokines, and lysozyme expression in FBGCs and MΦs.[19] This profile likely indicates the resolution of an inflammatory response as these immune-suppressive cytokines are released to balance pro-inflammatory cytokines.

The foreign body response, tissue healing, and matrix remodeling are associated and co-existing processes, of which the former is deleterious while the latter two, are necessary for effective biomaterial–tissue integration. Importantly, MΦs exhibit plasticity and a spectrum of phenotype, in that they are able to modulate their functions from pro-inflammatory to reparative. Presumably, relative distributions of MΦ phenotypic subtypes or aggregate functions vary temporally in response to environmental stimuli. Biomaterial-based control of their phenotype and functions is expected to facilitate appropriate host responses.

## 5.2.2 Deleterious Outcomes

### 5.2.2.1 Fibrosis, FBGC, and material degradation

MΦs at the site of an implanted biomaterial participate in and respond to the molecular signals in the microenvironment. Such signals include inflammatory cytokines and chemokines as well as anti-inflammatory molecules such as IL-10. An excessive inflammatory response can result in chronic inflammation with persistence of inflammatory cells and associated inflammatory mediators.[2,20] MΦs of an M1 phenotype would be expected to present a pro-inflammatory, phagocytic phenotype while MΦs in contact with a non-phagocytosable surface can fuse and undergo frustrated phacocytosis. This MΦs fusion occurs under the influence of cytokines such as IL-4 and IL-13 associated with supporting an M2 MΦ phenotype. It is important to note that all MΦs do not undergo this transformation and there are both phagocytosing (M1) MΦs and FBGCs present concurrently at the material surface surrounding the biomaterial surface microenvironement. Material-adherent MΦs and FBGCs unleash a destructive environment of reactive oxygen species (ROS), and degrading proteolytic enzymes.[2,3,20] Evidence of this destructive environment has been observed as implant failure such as stress cracking of polyurethane pacemaker leads.[21] Biomaterial-adherent MΦs that have undergone frustrated phagocytosis can be left exhausted and incapable of fighting a bacterial infection. The dependence of FBGC formation on the presence of various extracellular and matricellular proteins and proteolytic enzymes has been demonstrated. For example, vitronectin-coated polystyrene, strongly supported FBGC formation *in vitro*, relative to other plasma proteins.[22] Similarly, in another study, fibronectin knockout mice were shown to have significantly higher biomaterial-adherent FBGCs and fibrous capsules twice as thick compared to wild type controls.[23] Furthermore, MMP-9 has also been shown to be important in MΦs fusion as MMP-9 knockout mice implanted with mixed cellulose acetate or polyvinyl alcohol (PVA) sponges displayed lower extents of FBGC formation.[17] In addition, there is evidence that the FBR is compromised in the absence of supplementary wound microenvironment components, specifically, matricellular proteins such as thrombospondin-2 (TSP-2)[24] and secreted protein, acidic and rich in cysteine,[25] implying their key roles in the progression of the response.

### 5.2.3 Biomaterial Strategies to Affect Aspects of the Foreign Body Response

Several biomaterial strategies have been developed to direct the tissue reaction. Of these, modification of material-surface microarchitecture has been particularly successful in guiding a vascularized tissue reaction. The microarchitecture of the implant, especially at the surface, is an important parameter that influences the host response.[26] Highly microporous materials have been developed by rapidly stretching polytetrafluoroethylene (PTFE) fibers to vary the internodal distance.[27] As such, implantation into mice of discs of expanded PTFE (ePTFE) materials of varying internodal distance, and thus varying topographic pore size, resulted in the formation of fibrous capsules of differential thickness.[27] Specifically, ePTFE discs of 4.4 μm internodal distance (abbreviated 4.4-ePTFE) resulted in a less organized and significantly thinner tissue reaction as compared to the non-porous PFTE control. Mechanistically, *in vitro* culture of MΦs on these discs resulted in higher levels of pro-inflammatory cytokines namely, IL-1$\beta$, IL-6, TNF-$\alpha$, and chemokines such as MCP-1, MΦ inflammatory protein-1$\beta$ (MIP-1$\beta$) for cells cultured on 4.4-ePTFE.[27] Thus, increasing intranodal distance between ePTFE fibers (and thus larger pore sizes on the disc surface) directly correlated with the increasing pro-inflammatory state of MΦs *in vitro* and decreasing fibrotic capsule thickness around the implant *in vivo*.[27] Similarly, uniform microporous pHEMA scaffolds (SCs) have been used as test bed to determine the more precise correlation between micropore size and extent of vascularization.[28]

Biomaterial surface modification techniques have been widely employed in the context of protein adsorption studies. Self-assembly of monolayers and surface modifying agents can characteristically alter the surface properties such as charge and hydrophobicity visible to endogenous proteins. Since the adsorbed protein layer on devices is thought to mediate successive events in FBR, many researchers have utilized surface modification in regulating FBR. In this regard, surface modified poly(hydroxyethyl methacrylate) (polyHEMA) carrying positive charge has displayed the potential to adsorb specifically oriented osteopontin *in vitro* and mediate cell adhesion relative to corresponding polyHEMA negative surfaces.[29] Further, the former, when implanted in mice, showed a decreased fibrous capsule thickness and a diminished FBR.[29] A different strategy demonstrated that manipulating polyurethane surface with Tween 80 (surfactant) rather than PLL-RGD (cell adhesive) greatly diminished the foreign body response viewed in drastically reduced FBGC numbers and a thin fibrous capsule of (2–3 μm).[30] Surface modification with polyethylene oxide (PEO) to form tetraglyme surfaces are discussed later in section 5.4.1.1.1.4 of this chapter.

## 5.3 Instant Blood-Mediated Inflammatory Reaction

The clinical transplantation of cells, such as islets or hepatocytes, into sites such as the liver via intraportal vein injection, exposes the cells to the blood of the recipient leading to their destruction. This damage is mainly due to a far-reaching series of inflammatory events, collectively known as instant blood-mediated inflammatory reaction (IBMIR).[31,32] IBMIR is characterized by (1) macroscopic coagulation occurring within minutes after transplantation; (2) platelet activation, aggregation, adhesion, and high consumption; (3) activation of the complement system; (4) fibrin formation; and (5) activation of PMNs (Fig. 5.1). These events bring about insulin dumping and rapid islet cell-death post blood exposure.[32]

Interestingly, expression of tissue factor and MCP-1 on the islet cell surface have been recognized as key players in triggering these complex cascades.[33] In a related manner, IBMIR is also likely to amplify the immune response in allotransplantation.[34] Due to transplanted islet cell-death induced by IBMIR, cells from multiple donors may be required to meet sufficient graft size for one recipient. Efficient abrogation of IBMIR could lead to engraftment of a higher proportion of islets, leading to a reduction in the number of donors

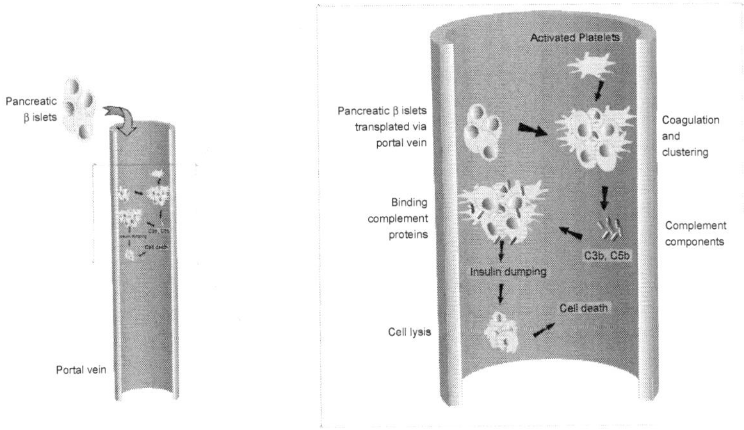


**Figure 5.1** IBMIR. Islet transplantation into the portal vein of the liver

required for each patient; preferably a 1:1 donor to recipient ratio would be an optimal condition.

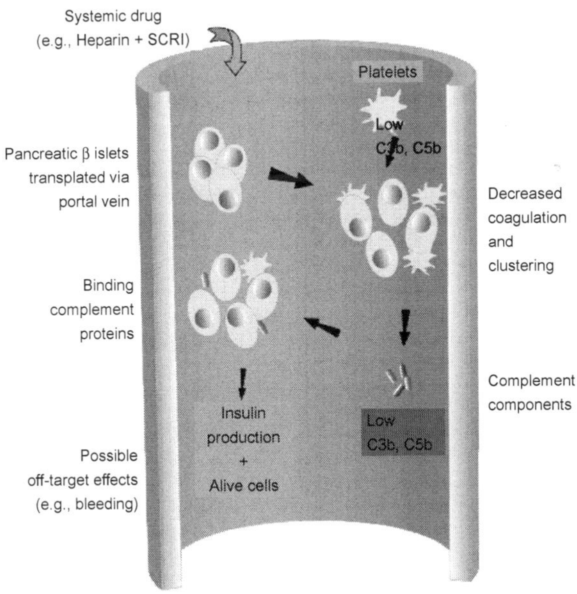


**Figure 5.2** Strategies to avoid IBMIR in islet transplant patients include systemic administration of the minimal doses of immunosuppressive medications.

### 5.3.1 Biomaterial Intervention for IBMIR

Avoidance of IBMIR in patients undergoing transplantation has the potential to reduce associated islet cell damage and allorejection. However, two such strategies described below necessitate systemic administration of the minimal doses of immunosuppressive medications with potentially harmful systemic side effects (e.g., bleeding); instead, a localized, targeted approach in treating IBMIR may be more effective, as described subsequently. Systemic strategies are illustrated schematically in Fig. 5.2. One such systemic strategy involves delivery of substances that can decrease thrombin production, coagulation, platelet, and complement activation. In this regard, when islets were delivered in a combination of heparin and complement inhibitor (soluble complement receptor inhibitor—sCRI) in an *in vitro* human allotransplatation model, significant suppression of IBMIR and islet cell-death were observed.[31] Low molecular weight–dextran sulfate (LWM-DS) has also been employed in *in vitro* and *in vivo* xenotranplantation models of pig islets. LWM-DS was able to inhibit coagulation, platelet and leukocyte activation, and consumption at low concentrations and at higher concentrations was able to prevent IBMIR.[35,36] When administered in higher non-human primate models, LWM-DS showed similar results with dose dependency, suggesting its potential use in a human clinical trial setup.[37,38] Another substance, a thrombin inhibitor called melgatran, was co-delivered with human islet exposed *in vitro* to ABO-compatible blood (where, A, B, and O are human Rh blood types) to eliminate blood coagulation and complement activation.[39] Simultaneously, it was able to decrease platelet consumption without altering islet cell surface morphology,[39] proposing a role for other thrombin inhibitors to be used in clinical treatments.

Localized, targeted approaches for avoiding IBMIR include *ex vivo* manipulation of donor tissue surface as demonstrated schematically in Fig. 5.3 with strategy details and outcomes for different approaches presented in Table 5.1.

This masks the inherent properties of the cells that support IBMIR or inhibit propagation of the coagulation cascade locally without presenting any systemic side effects. Surface modification of tissue transplant surface offers simple advantages such as its specificity for the transplant tissue without risking safety or immunogenicity as would be concerns with delivery of virally encoded products through gene therapy. Surface-modification approaches to abrogate

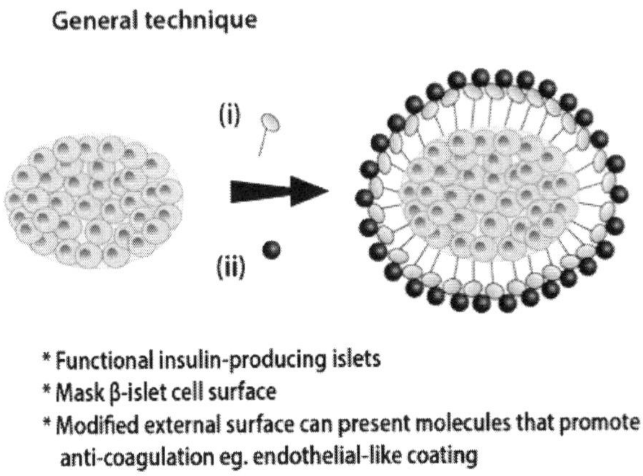


**Figure 5.3** Biomaterial surface modification of $\beta$-islet cell.

**Table 5.1** Biomaterial surface modification of $\beta$-islet cell experimental details and outcomes for Fig. 5.3 schematic

| Components | (i) | (ii) | Outcome |
|---|---|---|---|
| 1 | Targeted biotinylation of surface amines[40–42] | Avidin-modified recombinant thrombomodulin[40–42] and aPC with liposomes[43] | Islet cell viability,[40–43] engraftment efficiency and induction of euglycemia in diabetic mice[43] |
| 2 | Targeted biotinylation of surface amines[33] | Avidin–heparin | Protection against IBMIR, insulin-producing functional islets |
| 3 | Biotin-PEG-lipids[44] further protected by streptavidin and biotin-BSA Mal-PEG-lipid[45] instead of biotin | Urokinase or heparin | Islet insulin production and fibrinolytic activity[44] |
| 4 | (islet cell surface – no group added)[46] | Heparin conjugated dopamine (heparin-DOPA)and MR-1 (immunosuppressive drug)[46] | Islet survival in xenogeneic model and euglycemia maintained in 64% of recipients[46] |

IBMIR and enhance cell engraftment include heparinization of the islet cell surface via a biotin–avidin approach.[33] This technique was used to conjugate preformed heparin complexes to islet cell surface to achieve an anti-coagulant, endothelial-like coating.[33] When analyzed in both an *in vitro* loop model and in an *in vivo* allogeneic porcine model of clinical transplantation, the heparin coating provided protection against the IBMIR. In addition, the insulin-producing functionality was also retained and these cells were able to reverse the diabetic state in mice in comparison to untreated islet controls.[33] Another approach is the immobilization of the anti-thrombotic molecule, thrombomodulin (TM), to the islet surface.[40–42] TM abrogates inflammation through its capacity to support the degradation of thrombin toward generation of activated protein C (aPC—not be confused with antigen presenting cells [APCs]), a potent anti-coagulant and anti-inflammatory agent.[40] Briefly, their novel biosynthetic method involved using recombinant human TM (rTM) bearing a C-terminal azide group, which permits site-specific biotinylation of rTM through Staudinger ligation.[40] The rTM was immobilized on islet surfaces through streptavidin-biotin interactions, resulting in a nearly three-fold increase in the catalytic capacity of islets to generate aPC.[41] Donor islets were biotinylated by targeting surface amines and aldehydes. Islet viability and the TM surface density were maximized through optimizing biotinylation conditions.[40–42] An improvement on this strategy involved TM functionalized with liposome (lipo-TM) due to its tendency for preferential uptake in the liver thereby improving engraftment efficacy of islets.[43] In diabetic C57BL/6 mice that underwent intraportal transplantation with B10. BR murine islets, a single infusion of lipo-TM led to euglycemia in 83% of recipients along with reduction in fibrin accumulation, neutrophil infiltration, and TNF-$\alpha$ and IL-$\beta$ expression. In untreated controls, euglycemic recovery was observed in only 29% of mice receiving allogeneic islets.[43] Teramura and colleagues present a different, multi-faceted surface functionalization technique by modifying islet surface with a polyethylene glycol (PEG)–phospholipid conjugate containing biotin (biotin-PEG-lipids).[44] These biotin-PEG-lipids were positioned on the cell membrane, and the PEG-lipid layer on the islets was subsequently protected by streptavidin and biotin–bovine serum albumin conjugate using a layer-by-layer approach. Furthermore, the surface was activated with oxidized dextran. The fibrinolytic enzyme, urokinase, or anti-coagulant, heparin, was anchored to the islets through Schiff's base or polyion complex formations, respectively.[44] Islet insulin production was maintained and fibrinolytic activity observed for these modified islets.[44] A subsequent development included incorporation of a PEG-lipid conjugate bearing a maleimide group (Mal–PEG–lipid; instead of biotin) to allow for the co-immobilization of TM and urokinase, which enhanced islet cell engraftment while retaining islet functionality.[45] A more recent scheme by Jung and colleagues, proposed a combination method of using heparin conjugated with 3,4-dihydroxy-L-phenylalanine (heparin-DOPA) and a low dose of a T-cell targeting immunosuppressive drug, anti-CD154 monoclonal Ab (MR-1) treatment to improve the *in vivo* survival of islet xenograft.[46] Here, heparin-DOPA was directly grafted on to the islet surface. When MR-1 was administered every other day after transplantation of heparin-DOPA islets to mice, euglycemia was maintained in 64% of the recipients monitored for a period of two months. On the other hand, in MR-1 untreated controls, with either unmodified islets or heparin-DOPA islets, the grafts were completely rejected within two weeks.[46]

It is, therefore, evident that cell surface modification methodologies present a powerful approach to improve islet engraftment. The key here is to modify or mask the islet cell surface and present a combination of thrombosis and fibrin inhibitors. These strategies may soon realize clinical applicability in treating diabetes with islet cell transplantation.

## 5.4 Immune Response in the Context of Biomaterials

The immune response is the body's protective mechanisms against foreign entities for their elimination. Two arms of the immune system work to eliminate foreign entities, the rapid, non-specific innate immune response and the later acting, antigen-specific, adaptive immune response.[47]

### 5.4.1 Types of Immune Responses

#### 5.4.1.1 Innate immunity

Innate immunity is the first line of host defense against foreign entities in a non-specific, rapid manner. It has the ability to respond to microbes, foreign cells, and/or products of injured cells. These responses are non-specific and initiated molecular structures that are common to pathogens and not found on eukaryotic cells. They actively clear foreign antigen, materials, or apoptotic cells and influence repair of damaged host tissue.[48]

The innate immune system, first, provides physical and chemical barriers in the form of epithelia and epithelial surface-bound antimicrobials, respectively. Second, phagocytic cells such as MΦs resident in the sub-epithelial tissue offer protection. Recognition of pathogens by these phagocytes results in the secretion of cytokines and chemokines that initiate the recruitment and activation of leukocytes such as neutrophils and monocytes/MΦs from circulation at the site of infected tissue. These leukocytes subsequently recognize and destroy pathogens.[48]

Conserved effector mechanisms specifically, the complement system, serve to enhance the innate immune response. The complement system comprises a set of plasma proteins that can bind to pathogen surfaces resulting in their activation to generate proteolytic degradation fragments.[49] These cleavage products coat foreign antigen surfaces (opsonin) for leukocyte recognition and clearance, release of soluble chemotactic mediators for leukocyte recruitment and activation with the terminal product of the complement activation cascade being the membrane attack complex (MAC) that lyses the targeted cell.[49]

#### 5.4.1.1.1 Key components of innate immunity

##### 5.4.1.1.1.1 *C-type lectin receptors*

The C-type lectin receptors (CLRs) are expressed on a variety of cells including MΦs, DCs, B-cells, T-cells, PMNs, endothelial cells (ECs), and fibroblasts.[50–52] They can identify carbohydrate structures expressed on pathogen surfaces mediated by carbohydrate recognition domains (CRDs) through calcium-dependent binding,[53,54] thereby mediating cellular processes, involving cellular migration,[55] crucial steps in adaptive immunity, and internalization of pathogens for antigen processing and presentation.[56] These receptors have been classified based on the number of CRDs that they possess[54] or by their affinity to specific carbohydrates, particularly lectins-binding mannose-type or galactose-type carbohydrates.[50,57] CLRs can recognize endogenously and exogenously presented carbohydrate ligands. Also, CLRs that function as PRRs are capable of distinguishing self- and non-self-ligands.[50,56] Ligation of certain CLRs such as dectin-1 and mannose receptor induces phagocytosis.[58] Dectin-1 ($\beta$-glucan receptor), in collaboration with TLRs, specifically TLR2[59] and TLR6,[58] supports an inflammatory response to pathogens.[56,60] Signaling through dectin-1 and TLR2 stimulates the production of ROS and pro-inflammatory cytokines such as TNF-$\alpha$ and IL-12.[61] Important immune tolerance processes occur through CLRs including differential expression of MΦ galactose-like lectin on tolerogenic DCs[62] and participation of dendritic cell-specific intercellular adhesion molecule-3-grabbing non-integrin (DC-SIGN) ligation in the production of regulatory T-cells by DCs.[63]

#### 5.4.1.1.1.2 *Toll-like receptors*

TLRs are expressed on various cell types including DCs, MΦs, T-cells, B-cells, fibroblasts, and ECs.[56] TLRs have the ability to specifically recognize pathogen components or pathogen-associated molecular patterns (PAMPs) such as lipopolysaccharide (LPS), teichoic acids, lipoprotein, "double stranded" RNA (dsRNA), bacterial DNA, and flagellin.[11,56,64] TLRs 1, 2, 4, 5, and 6 are expressed on the cell surface, while TLRs 3, 7, 8, and 9 are expressed on intracellular compartments, these receptors facilitate the internalization of antigen and trigger subsequent cellular responses.[56] Crucial endogenous ligands of TLRs known as damage-associated molecular patterns (DAMPs) or "danger signals" initiate inflammatory responses upon cell death or tissue damage and act as natural adjuvants. Important DAMPs are endogenous molecules that are normally "hidden self" and released upon cell damage or tissue death such as high-mobility group box protein-1 (HMGB1), heat-shock proteins, and mRNA; or upon tissue damage such as heparan sulfate and fibrinogen.[49] The ligation of intracellular TLRs leads to the downstream signaling of nuclear factor kappa-light-chain-enhancer of activated B cells (NF-κB) and mitogen-activated protein kinase activation via a MyD88-dependent pathway.[56,65] TLRs 3 and 4 can activate the transcription factor IFN regulating factor 3 in a MyD88-independent manner. This results in the generation of IFN-$\beta$ and IFN-inducible gene products.[66] NF-κB, also, participates in gene regulation in the MyD88-independent pathway of TLR signaling.[11]

#### 5.4.1.1.1.3 *Complement system*

The complement system, a fundamental arm of the innate immune response, is composed of nearly 30 plasma and membrane-bound proteins. It has the ability to distinguish between self- and non-self-foreign entities. The principal function of this system is the eradication of non–self-entities from the body in particular, microbes, foreign cells, or substances (e.g., biomaterial) and apoptotic cellular debris.[67–69] This is done by labeling the entity for recognition by inflammatory cells for phagocytosis and clearance or direct cell lysis.[47]

Progression of events in the complement cascade can occur via three pathways: (1) the classical pathway triggered by antigen– Ab complexes, (2) the alternative pathway activated by the binding of complement proteins to foreign surfaces, and (3) the lectin pathway mediated by conjugation of mannan-binding lectin or ficolins to particular carbohydrate structures, notably those expressed on pathogens.[11] Biomaterials have been shown to involve mainly the alternative pathway and to some extent, the classical pathway.[70] For biomaterials that contain carbohydrate groups, there could be a contribution from the lectin pathway; however, this has not been assessed to date.

In the classical pathway, the complement fragment C1q attaches to pathogens primarily by the binding to the Fc portion of opsonizing antibodies. Also, the C1q component can directly bind to pathogens by diverse mechanisms such as binding to C-reactive protein and polysaccharides expressed by pathogens.[71] In the alternate pathway, spontaneous cleavage of C3 results in opsonization of pathogens, by covalent binding of C3b specifically to hydroxyl groups on surfaces.[72] Endogenously secreted C3b can also bind to host cells; however, cell surface-bound complement regulatory proteins such as CD46, CD55, and CD59 block the host cell death.[73,74] Further, the alternative pathway offers an amplification loop to strengthen the initially weak signals.[75,76] Activation of the classical or lectin pathways results in the breakdown of C4 into C4a and C4b. Complement component C4b is capable of opsonizing pathogens by covalently binding to cells through hydroxyl, amides, or carbohydrate moieties.[77]

Regardless of the mode of initiation of complement effector function, a common induction of C5 cleavage by C5-convertases into C5a and C5b occurs.[49] Complement component C5b participates in the formation of the MAC, along with of C5b, C6, C7, C8, and polymeric C9. MAC is the cytolytic end product of the complement cascade; it forms a transmembrane channel, which causes pore formation and osmotic lysis of the target cell. Kupffer cells and other MΦ cell types function in removal of circulating pathogens opsonized by C3b and iC3b. Soluble C5b-9 can interact with and activate platelets.[75] Additionally, sC5b-9 is commonly assessed experimentally to determine the extent of complement activation.[78]

Blood coagulation, complement activation, and inflammation are co-existing processes with ample instances of crosstalk between each other for synergistic amplification.[79]

#### 5.4.1.1.1.4 *Biomaterial approaches for inhibiting complement activation effects*

Biomaterial approaches are devised commonly to address coagulation and complement activation. Traditionally, heparin coatings have been used to improve blood compatibility, in the light of reducing coagulation and complement activation. The well-known and accepted hypothesis for using heparin is that the anticoagulation is achieved by sequestration of regulators such as antithrombin and possibly factor H.[80] However, subsequent studies have determined that much lower surface concentration of heparin is required for inhibiting the coagulation than the complement system. Increasing heparin concentration to combat the complement cascade has still proven to be an insufficient mechanism;[81] interestingly, high heparin concentration does not necessary result in an increase of factor H attachment, suggesting that factor H may not be the most significant regulator of complement.[82] Importantly, heparin is capable of interacting with a myriad of plasma proteins, possibly saturating available bound heparin for factor H as well as potential undesirable side effects.

Commonly employed biomaterial techniques involve surface modification of implanted materials with relatively inert PEG-based materials while simultaneously delivering a surface attached regulator of complement activation such as factor H. Polystyrene (PS) surfaces with human factor H bound in its native conformation with a triblock copolymer of polypropylene oxide flanked by two PEO groups on each side (PEO-PPO-PEO), were capable of attenuating complement activation with 55% lower platelet consumption and nearly 60% decrease in thrombin-antithrombin complex formation when compared to untreated controls in an *in vitro* whole blood model.[83] Incidentally, PS, polyethylene (PE), PVA hydrogels, PEG, PEO-based plasma deposited tetraglyme surfaces have been studied in the context of blood compatibility, complement, and leukocyte activation.[84–88] Of these, PEO and PEO-like surfaces have shown to elicit minimal protein adsorption. Specifically, plasma-deposited PEO tetraglyme coatings are likely able to act as steric barriers to proteins by stimulating tight assembly of water in "ice-like" forms.[89] Furthermore, plasma deposition of short ethylene oxide (EO) chains allows for high surface density of the same. PEG and poly(2-methoxyethylacrylate) (PMEA) have similar ability to structure water by possibly minimizing the driving force for protein adsorption.[89]

In an *in vitro*, disturbed blood flow catheter model, tetraglyme coatings on glass, biospan, or PE were studied against untreated controls.[89] Coated surfaces displayed significant reduction in platelet adhesion with concomitant lower levels of thrombin generation and adsorption of plasma proteins such as von Willebrand factor and fibrinogen. Collectively, these results suggested considerable attenuation in contact activation and procoagulant activity of tertraglyme-coated materials. However, the formation of potential thromboemboli or platelet consumption, the more accurate measures of blood compatibility,[90] were not studied.[89]

Although the tetraglyme coatings evidently decreased procoagulant processes *in vitro*, their effect *in vivo* was far less satisfying. *In vivo* results showed MΦ and leukocyte infiltration in response to implanted tetraglyme-coated surfaces and with associated FBR.[91] To test the hypothesis that these surfaces were still able to trigger complement activation *in vivo*, tetraglyme surfaces were studied by Szott and colleagues, as far as extent of complement activation.[92] Even though the coatings they used were found to have low OH group content, (OH groups are known to play an important in complement activation), and low amounts of C3 adsorption, there were very high levels of C3a and SC5b-9 detected in serum. PVA hydrogels, silastic sheeting, and PEG-based self-assembled monolayers with different end groups mainly served as control surfaces.[92] The *in vivo* occurrence of complement activation provided an explanation of MΦ activation and FBR in response to these implanted tetraglyme surfaces.[78,92]

Nanostructuring of surfaces can be used to alter surface pore size. It is conceivable that nanoporous surfaces may not allow for the assembly of multiple, large complement proteins to assemble as complexes to participate in enzymatic cleavage reactions, such as formation of the C3 convertase. In a study by Ferraz and colleagues, pore size-dependent protein adsorption was observed on nanoporous aluminium surfaces.[93] In a whole blood *in vitro* model, these surfaces were analyzed by determining the amount of TSP-1 adsorbed for platelet activation, and C3a and sC5b-9 for complement activation, in a time-dependent manner. Platelet and complement activation was significantly lower on surfaces with a pore diameter of 20 nm as compared to on surfaces with a pore diameter of 200 nm pores, even though the former displayed a larger total surface area.[93] Similarly, using methacrylate-based polymers, it was shown that hydrophobic, uncharged polymers such as pHEMA or styrene-based polymer with ethylene glycol dimethacrylate linker with smaller pores structures of approximately 12 nm sufficiently lowered the levels of complement activation as compared to surfaces of larger pore structures and of less hydrophobicity and charged nature.[94]

Complement activation and typically concomitant blood coagulation promote leukocyte infiltration and activation. From the above-mentioned examples, it is clear that newer biomaterial strategies need to be developed to combinatorially target complement and procoagulant entities in order to bring about a synergistic reduction in innate host response to implanted biomaterials. Additionally, the biomaterial surface properties, such as charge, hydrophobicity, and micro- or nanostructure, can affect extent of coagulation and complement activation.

#### 5.4.1.2 Adaptive immunity

An adaptive immune response is the later arm of the immune response, following the innate immune response. It is more effective than the innate immune response in eliminating antigens as it is antigen specific. An adaptive immune response is achieved via three main strategies:[49]

(1) B cells take up antigen that is processed with their subsequent differentiation into antigen-specific Ab-secreting plasma cells. Secreted Abs recognize and bind to extracellular antigens, marking (or opsonizing) them for clearance by phagocytes. Antigen-Ab complexes also initiate the complement cascade via the classical pathway.[49]

(2) Phagocytosis of antigen is achieved by APCs such as DCs and MΦs. APCs process antigen, load it, and present it in the context of major histocompatibility complex (MHC) molecules on their surface. Activated DCs are unique in their ability to present MHC-bound antigen to naïve T-cells to stimulate their clonal expansion and elicit appropriate effector functions. Engagement of MHC-II molecules on APCs are CD4 restricted presenting their bound antigenic peptides to CD4+ cells such as Helper T ($T_H$) cells to stimulate cytokine release that provides help for activation of APCs and lymphocyte function.

(3) Exogenous antigens can be cross-presented for loading into MHC-I molecules on APCs that are CD8 restricted presenting their bound antigenic peptides to CD8+ cytotoxic T cells (CTLs) for targeted cell cytotoxicity.[95] Foreign cells or cells expressing foreign proteins (e.g., upon viral infection) present these antigenic peptides loaded on the MHC-I molecules to effect CTL responses.

### 5.4.1.2.1 Key components of adaptive immunity

#### 5.4.1.2.1.1 *Dendritic cells (DCs)*

DCs are the most potent APCs of the body. They have not only been elucidated to play a vital role in the instigation and regulation of the host's protective immune response but also in maintaining tolerance to self or achieving tolerance to non–self-antigen, for example, upon organ/cell transplantation.[96]

Originally discovered by Dr. Ralph Steinman in 1973,[97] the importance of DCs has been recognized by the medical community and are officially established as the professional APCs of hematopoietic origin that are present quantitatively sparsely yet in spatially extensive distributions within the body.[96,98,99] Mature DCs (mDCs) and immature DCs (iDCs) are defined by key phenotypic features and by the presence or absence of expression of cell surface markers. The primary morphological characteristic of mDCs is the presence of numerous membrane processes or dendrites that extend out from the main cell body. DCs may be derived from myeloid or plasmacytoid origin.[96]

Myeloid DCs (MDCs), typically express class I and class II MHC molecules that aid in the recognition of MHC-bound processed antigen by T-cells. DCs express a variety of adhesion molecules including

CD11a (LFA-1), CD50 (ICAM-2), CD54 (ICAM-1), CD58 (LFA-3), and CD102 (ICAM-3); and costimulatory molecules namely, CD80 (B7.1) and CD86 (B7.2), expressions of which are upregulated during DC maturation.[96,100] CD86 tends to be a marker of early DC maturation, while CD80 expression increases in fully mDCs.[96] Three distinct subsets of MDCs based on their surface molecule expression exist in the human blood. CD1+ DCs (MDC1) that are BDCA-$1^+$ $Lin^-$ HLA-$DR^+$ $CD11c^+$ MDCs, constitute the chief subset while BDCA-$3^+$ $Lin^-$ HLA-$DR^+$ $CD11c^+$ or MDC2 comprise the minor subset.[101] MDC2 expresses CLEC9A, an endocytosis-mediating C-type lectin with a cytoplasmic immune receptor tyrosine-based activation-like motif.[102] MDC1 was elucidated to possess superior allostimulatory ability than MDC2.[103] The third and final subset was more recently recognized to present a carbohydrate variation P-selectin glycoprotein ligand-1, called 6-sulfo-LacNac. These DCs have been shown to produce TNF-$\alpha$ more strongly than MDC1 in response to bacterial cell membrane LPS stimulation.[104]

The state of maturation of these potent APCs plays a key role in activating naïve T-cells and in stimulating the growth and differentiation of B cells.[96]

##### 5.4.1.2.1.2 *Antigen processing and presentation*

iDCs or tissue-associated DCs continuously sample their extracellular environment for foreign entities such as pathogens.[96] These cells are characterized by high endocytic activity and low T-cell stimulation potential.[100] They trap antigen by receptor-mediated endocytosis or pinocytosis. Captured antigen is proteolytically cleaved in endocytic vesicles at acidic pH. Pathogen identification is done through PRRs such as the TLRs and CLRs.[47,96]

MHC molecules are expressed on most cell surfaces. They are mainly classified into MHC class I and MHC class II molecules. Both MHC classes I and II molecules play pivotal roles in adaptive immunity. MHC I molecules are displayed on most nucleated cells, while MHC II are displayed on APCs, such as DCs, MΦs, and B-cells. MHC I molecules, usually, bind to processed self-peptides or foreign antigens from infected cells or foreign cells. Cross-presentation of exogenous proteins that are reduced by the ubiquitin–proteasome pathway are also displayed on the cell surface of MHC I molecules.[105] The peptides thus produced are transported to the endoplasmic reticulum (ER) by ATP-binding transporters associated with antigen processing (TAP).[106] These peptides are bound to MHC class I molecules in the ER, which are then translocated to the cell membrane.

The MHC class II molecules on APCs display exogenous foreign antigens, such as peptides derived from pathogens or released from a controlled release vaccine delivery system. The MHC class II molecules are produced in the ER, where the invariant chain (Ii) occupies the peptide-binding groove of MHC class II molecules. Upon leaving the ER and the Golgi apparatus, Ii is reduced in the endosome, causing class II-associated peptides (CLIP) to be retained in the peptide-binding groove.[107] Following endocytosis of exogenous antigen, CLIP is removed from the peptide-binding groove for loading with other foreign antigens. The MHC class II molecules with bound antigens are then transported to the APC cell membrane. Cross-presentation is a unique mechanism primarily facilitated by DCs. Antigen processing for cross-presentation occurs via two pathways. The first pathway takes place in a TAP- and proteasome-dependent manner. In this pathway, endocytosed antigens are transported to the cytosol and hydrolyzed by proteasomes.[105] The peptides generated in the cytosol may be reinserted into phagosomes for binding to MHC class I molecules, or may be imported to the ER for binding to MHC class I moleculess.[105] The peptides are then transferred to MHC class I molecules. In contrast, the second pathway is TAP and proteosome independent. In this pathway, peptides are produced from foreign antigens degraded by endosomal proteases and may bind to MHC class I molecules in the endosome.[105]

###### 5.4.1.2.1.3 *T-cell receptor and control pathway of CD4+ T-cell activation*

A T-cell, on an average, expresses about 30,000 T-cell receptors (TCRs) on its surface.[108] Antigen recognition by most T-cells is mediated through the TCR. A TCR is composed of two chains, TCR-$\alpha$ and TCR-$\beta$, bridged by a disulfide bond. The *N*-terminal variable region shares homology with immunoglobulin V, including a constant region similar to immunoglobulin C domains and a hinge region with a disulfide bond.[95] It also has a hydrophobic transmembrane domain and a short C terminal cytoplasmic tail. TCR couples with CD3 complex on the T-cell surface to function as a single unit. CD3 consists of three constant chains and one homodimeric $\zeta$ chain that has a very short extracellular domain. CD3 molecules possess a negative charge that aids in the formation of the complex since the

TCR is positively charged. Signaling processes are controlled by the ten immunoreceptor tyrosine-based activation motifs present in the invariant chains of the TCR complex (TCR $\zeta$ and CD3-$\gamma$, -$\delta$, and -$\varepsilon$).[109,110] Presented peptide-induced stimulation of T-cells is also modulated through co-ligation of CD4 or CD8 surface molecules to along with the TCR-MHC-CD3 complex.[95]

TCRs recognize peptide fragments of antigens bound to MHC molecules. Naïve $T_H$ cells require a second signal for generation of protective effector functions post TCR stimulation.[95,109] This second signal is provided by ligation of costimulatory molecules with B7 molecules on APCs. Upon APC maturation, CD80 and CD86 expressions are upregulated, thus supporting their immunostimulatory ability.[111] The third signal of the cytokine environment polarizes T-cells responses. Absence of the second signal leads to T-cell anergy. T-cell polarization, for example, toward $T_H1$ or $T_H2$ phenotypes is achieved in response to secreted IL-12 or IL-4. Regulatory T-cells or $T_{reg}$s with $CD4^+$ $CD25^+$ $CTLA4^+$ or TGF-$\beta$-induced $CD4^+$ $CD25^-$ $FoxP3^+$[112] have the ability of resolving immunity and promoting tolerance with specificity to antigen. The T-lymphocyte antigen 4 (CTLA4), a molecule homologous to CD28, binds to costimulatory B7 molecules.[113,114]

### 5.4.2 Biomaterial Strategies to Affect Adaptive Immune Responses

With a growing understanding of intrinsic biological pathways of immune responses, biomaterial scientists aim to develop strategies to engineer appropriate interactions with the immune system using biomimetic strategies, controlled release or biomaterial design as required by the context—vaccines for infectious disease or cancer, tissue-engineered constructs, cell/tissue transplants, and autoimmune disorders. The adaptive immune response in any given scenario is vastly influenced by the phenotypic state of DCs, which direct immune responses.[96] Hence, an attempt to successfully modulate DC phenotype through biomaterial strategies can be a way that potentially leads to need-based tailoring of the body's immune response. Biomaterial-based approaches to influence immune outcomes through effects on DC phenotype are described below and summarized in Table 5.2.

**Table 5.2** Targeted DC-based biomaterial therapies

| Application | Adjuvants, vaccines | Antigen delivery | Anti-tumor immunity | TEDs, allograft survival via tolerance induction |
|---|---|---|---|---|
| **Targeting modalities** | CLRs, NLRs and more commonly TLRs | Particle size | TLR ligand and tumor antigen | Localized delivery |
| **Delivery vehicle** | PLGA MPs with antigen poly I:C[131], PLGA NPs with anti-DEC205 Ab and antigen [132] | NPs >500 nm and MP of 3μm with antigen deliver to neutral phagosomes in DCs—ideal for cross-presentation[134] | PLGA NPs with tumor antigen,[138] PLG matrix with GM-CSF, CpG-ODN (vaccination nodes)[141] | Immunosuppressive cytokine cocktail,[144,145] with TLR ligand,[146] PLGA NPs with IL-10 gene-inducing plasmid[147] |
| **Induction strategy** | Injection[131,132] and delivery in LNs[131] | *In vitro* culture | *In vitro* culture and *in vivo* delivery | *In vitro* culture[144–147] |
| **Associated DC phenotypes** | Mature myeloid DCs | DC2.4 cell line | Myeloid[138-143] and plasmacytoid DCs[141] | Tolerogenic DCs[145] and aaDCs[144,146,147] |
| **Outcome** | Enhanced humoral response and expansion of antigen-specific T-cells[131,] differential production of IL-5, IL-12, IL-2, IFN-$\gamma$[132] | Enhanced CD8+ T-cell expansion for larger particles. | Tumor-specific CD8+ T-cells, reduced tumor size | FoxP3+ Treg expansion |

#### 5.4.2.1 DC-mediated biomaterial adjuvant effect

iDCs, typically receive a maturation signal from microbial by-products such as LPSs, peptidoglycans, or cellular damage-associated "danger signals," or DAMPs, via TLR ligation and signaling.[96] In mDCs, there is an upregulation of expression of costimulatory molecules, which are necessary for effective activation of T-cells for an antigen-specific immune response and chemokine receptors such as CCR7 that direct migration of DCs toward T-cell areas of the lymph nodes.[96] Biomaterials such as 75:25 poly (lactic-co-glycolic acid) (PLGA) films or microparticle (MP) were able to affect the phenotype of DCs upon their treatment as demonstrated by their increased expression of costimulatory molecules and MHC class II molecules, as compared to untreated iDC controls. However, biomaterials are a moderate modulator of DC phenotype, much less stimulating than pathogen-derived maturation factors such as LPS, as would be expected.[115] In addition, these PLGA-treated DCs displayed enhanced allostimulatory capacity in a mixed lymphocyte reaction (MLR) supporting a functional phenotype as far as modulating antigen-induced T lymphocyte stimulation.[116] Similarly, PLGA-treated murine bone marrow–derived DCs showed moderate maturation; however, agarose-treated DCs were not significantly different from iDC controls.[117]

Cytokine profiling of microenvironment showed that PLGA-treated DCs secreted higher amounts of pro-inflammatory cytokines than agarose-treated DCs. As part of examining the cytokine profile released by biomaterial treated DCs, it is interesting to note that PLGA-treated DCs also secreted large amounts of IL-10 (anti-inflammatory cytokine); though this may seem contradictory, it in fact strengthens the basis for believing that PLGA influences maturation of DCs since cytokines, specifically pro-inflammatory and anti-inflammatory, are usually secreted in balanced manner. Hence, a heightened inflammatory response is moderated by a corresponding anti-inflammatory response offering a mode of control.[117] DCs treated with PLGA or agarose in the form of porous SCs induced similar effects on DC phenotype with maturation upon PLGA treatment but treatment of DCs with agarose SCs maintaining an iDC phenotype.[117,118]

A differential biomaterial effect on induced DC phenotype has been observed depending on the biomaterial used for treatment.

In particular, PLGA or chitosan films induced DC maturation, while agarose or alginate films did not; hyaluronic acid film, in fact, suppressed DC maturation.[118] Similarly, in MLR assays, allostimulatory ability of DCs treated with chitosan or PLGA revealed higher levels of induced T-cell proliferation as compared to iDC controls, while hyaluronic acid-treated DCs supported lower levels of T-cell proliferation; agarose- or alginate-treated DCs were not significantly different from iDC controls as far as allostimulatory ability.[116]

The immunogenicity of biomaterial-associated antigens can be modulated by the biomaterial component as expected based on the effect of the biomaterial on DC phenotype. As such, co-delivery of the model antigen, ovalbumin (OVA) with PLGA MPs or SCs supported a moderate level of anti-OVA humoral immune response that was primarily a Th2 response as directed by the principal IgG1 isotype during the 18-week period of the experiment.[119] To correlate the *in vitro* differential effects of biomaterials on DC maturation with the resultant adjuvant effect *in vivo*, we hypothesized that PLGA SCs would support a stronger murine adaptive immune response to controlled released co-delivered antigen than release from agarose SCs. At 12 weeks, mice receiving porous PLGA SCs delivering OVA were found to have significantly higher anti-OVA Ab concentrations than mice receiving OVA with PBS (negative control).[120] In contrast, mice receiving porous agarose SCs delivering OVA were not shown to be significantly different the anti-OVA Ab concentrations from mice receiving OVA with PBS. Therefore, porous PLGA SCs were shown to support a biomaterial adjuvant effect to enhance specific Ab production, which was not observed with porous agarose SCs. Furthermore, this validated the *in vitro* assessment of differential biomaterial effects on DC maturation in predicting the *in vivo* biomaterial adjuvant effect to co-delivered antigen.[120]

Evidence of biomaterial-implant site associated "danger signals" or DAMPs was observed as far as priming the implant site for a biomaterial adjuvant effect. To assess the differential level of enhancement of the immune response depending on the form of carrier vehicle (MPs vs. SC) and the extent of tissue injury due to its delivery, the total amounts of polymer and OVA delivered were kept constant as well as the release rate of OVA for both carrier vehicles.[119] The level of the humoral immune response was higher and sustained for OVA released from PLGA SCs, which were implanted with associated tissue damage, and lower and transient when the same amount of polymer and OVA were delivered from PLGA MPs, which were minimally invasively delivered by injection.[119] Tissue injury results in generation of "danger signals" such as damaged heparan sulfate proteoglycan, release of heat-shock proteins, and alarmins.[121] Since damaged heparan sulfate proteoglycan is chemically analogous to bacterial peptidoglycan and is detected by similar TLR mechanisms, it acts as an internal natural adjuvant in the presence of an antigen. Additionally, an alarmin, HMGB-1,[122] released from necrotic but not apoptotic cells, is shown to be present at elevated levels in the surrounding microenvironment, subcutaneously implanted PLGA SCs when compared to sham controls.[115] These results suggest that non-invasively delivering a tissue-engineering construct may serve to mask it from the immune system by minimizing implantation-associated "danger signals."

#### 5.4.2.2 Biomaterials-based adjuvants and vaccines

Adjuvants are employed to improve vaccine immunogenicity and also to potentially direct the immune response toward developing a cell-mediated or humoral response. Such directed control may be greatly valuable in tailoring the immune response toward distinct antigens. This distinction is of principal importance since different mechanisms may be required for the immunity of distinct pathogens.[123] Experimentally used adjuvants such as Freund's adjuvant are extremely toxic for human use. Most commercially used vaccine adjuvants usually contain alum-based compositions. Expectedly, these alum containing adjuvants have the ability to directly activate DCs, enhance the ability of DCs to present antigen to T-cells, and induce different effector cells.[124,125] Biomaterials in particulate forms including liposomes have been shown to mimic adjuvant properties of alum coupled with their ability to induce DC maturation.[126,127]

Certain polymers, especially in particulate form, have been shown to function as potent adjuvants that is mediated by their effects on DC maturation status.[117,118,120,128] Biomaterial nanoparticles (NPs) and MPs have been extensively applied in vaccine delivery approaches. Particulate adjuvants including PLGA and PS caused the dramatic increase is secretion of pro-inflammatory IL-1$\beta$ by LPS-treated DCs. Upon particle uptake by DCs, the subsequent adjuvant effect occurs via activation of the inflammasome similar to alum adjuvant stimulation of DCs.[129] The amplification of Ag-specific Ab production by PLGA MPs was independent of the inflammasome component, NALP3. However, the ability of PLGA MPs to promote antigen-specific IL-6 production by T-cells required NALP3.

Targeted antigen delivery to immune cells, specifically DCs, is currently being explored by several researchers. Target receptors are those constitutively expressed in DCs, namely TLRs, CLRs, and Nod-like receptors. TLRs are the most commonly targeted receptors for their effect on DC maturation.[123] Biomaterial particulates that are surface modified to represent external PAMPs offer an important scheme to control the subsequent induction of immunity.[130] Recently, TLR-targeting PLGA MPs carrying model antigen OVA were studied.[131] TLR-3 ligand poly(inosinic:cytidylic acid) (polyIC) was encapsulated in biodegradable PLGA MPs; these MPs were designed to remain extracellular and release polyIC in the LN over several days. Intra lymph node (iLN) injection of OVA mixed with polyIC-releasing MPs enhanced the humoral response and expanded OVA-specific T-cells following a single injection in comparison to antigen mixed with polyIC-loaded NPs or a 10-fold greater dose of soluble polyIC controls. Thus, slow release-formulated adjuvants from iLN delivery may be potentially applicable to improve vaccine therapeutic efficiency.[131] Bandyopadhyay and colleagues demonstrated the importance of target ligand density on NP surface.[132] In this study, the surfaces of PLGA NPs loaded with antigen were modified with varying densities of the monoclonal Ab to the DC lectin DEC-205 receptor. DEC-205-targeted NPs induced differential production of IL-5, IL-12, IL-2, and IFN-$\gamma$. Surprisingly, anti-inflammatory cytokine, IL-10 production by DCs and T-cells increased in a surface density-dependent manner. Boosting mice with DEC-205 targeted OVA-NPs after immunization with OVA in CFA induced a similar pattern of IL-10 response.[132]

Cross-presentation refers to the ability of antigen-presenting cells to take up, process, and present extracellular antigens with MHC class I molecules to CD8 T-cells and is consequently important in mediating CTL response. OVA carrying PLGA MPs have been shown to induce cytotoxic CD8 response toward an exogenous antigen at an OVA concentration in MP that was 1000-fold lower than its soluble counterpart and 10-fold lower than OVA-coated latex beads.[133]

Tran and Shen examined the importance of PS particle size of model antigen carriers on the induced immune response as a critical factor of vaccine design.[134] Model antigen OVA adsorbed onto 50 nm beads immediately transferred to an acidic environment within 30 minutes after exposure to DC cell line DC2.4, that engendered its rapid and unregulated degradation and hence, inefficient cross-presentation. In contrast, antigen bounded to 500 nm and 3 μm beads remained in a more neutral environment, which preserved the majority of antigens, leaving it available for the generation of peptides to be loaded onto MHC class I molecules. Increase in OVA-specific CD8+ T-cell activation was noticed for the larger particles. Soluble OVA controls did not give to CD8+ T-cell activation. This suggests a possible mechanism of MP and not NPs to induce higher *in vivo* cross-presentation.[134]

Another approach for directing antigen-specific immune responses using polymeric carriers is in controlling intracellular trafficking of materials and antigen load. Specifically, polymers sensitive to pH can intracellularly break down within endosomes or destabilize the endosomal membrane and release antigen into the cytoplasm for efficient antigen loading to MHCs and avoidance of endosomal degradation.[135] Another particle size-dependent strategy to engineer immune responses had been demonstrated using PEG-stabilized poly(propylene sulfide) (PPS). NPs that are of a size that is effectively transported in the lymphatic drainage to lymph nodes then interact with resident LN DCs. Mainly, smaller particle size directly influenced the improved uptake into lymphatics and also prolonged retention in the lymph node. Mechanism of MP and NP uptake by DCs may result in different functions due to the inherent distinction in their respective processing pathways and can offer specific targeting to destination cells.[136] These studies suggest a potential optimization strategy for design parameters of particulate vaccine delivery vehicles for each step of particle interaction with antigen presenting to direct desired immune responses.

#### 5.4.2.3 Biomaterial-based engineering anti-tumor immunity

DCs have been found to play a critical role in anti-tumor immunity. Biomaterial-based strategies have been applied to engineer anti-tumor immunity. For example, iron-oxide zinc oxide core-shell NPs developed can deliver carcinoembryonic antigen to DCs and additionally act as an imaging modality.[137] NP-uptake by DCs occurred shortly after delivery without significant difference in expression of maturation markers based on NP load. An *in vivo* model utilized this NP–antigen complex to immunize mice, consequently showing enhanced tumor antigen specific T-cell responses, delayed tumor growth, and improved survival compared to controls. PLGA NPs carrying the murine B16 melanoma antigenic peptides, hgp100, and TRP2 were capable of eliciting antigen-specific tumor immunity by priming CTL responses *in vitro* and *in vivo*.[138] The response was stronger than that induced by CFA-mixed peptide controls. Simultaneously tumor escape blockade by IFN-$\gamma$ augmented the effect of these NPs. Another strategy used a poly beta amino ester MP system delivering non-viral gene vectors to DCs and MΦs that showed improved transfection of these cells *in vitro* and reduced tumor size *in vivo* than PLGA MPs carrying DNA vaccine.[139]

Localized vaccination nodes involve the use of injectable self-gelling alginate gel matrices containing stimulated DCs that were able to recruit host DCs and T-cells to prolong the interaction of DCs at a local site and concentrate DC-produced factors promoting an inflammatory microenvironment *in situ*.[140] DCs loaded with SIY peptide encapsulated in the hydrogel were capable of generating Ag-specific T-cell responses.[140] In another approach, a PLGA SC was loaded with granulocyte-macrophage colony-stimulating factor (GM-CSF), CpG oligodeoxynucleotides (CpG ODN), and tumor lysate, and implanted subcutaneously in mice to serve as a vaccination node.[141] The controlled release of immunomodulators was achieved using PLGA wherein GM-CSF was used as a chemattractant for host DCs, which were stimulated to undergo maturation with the TLR9 ligand, CpG-ODN, while picking up the tumor lysate antigen to induce an antigen-specific anti-tumor response. The resultant immunogical response demonstrated enhanced recruitment of conventional DCs and plasmacytoid DCs, which are particularly necessary for an anti-tumor response, and generation of tumor-specific CD8+ T-cells. Furthermore, interestingly, there was also an abrogation of immunosuppressive agents such as regulatory T-cells and immunosuppressive cytokines such as IL-10. The effect on tumor load was a reduction in tumor size with improved murine survival. Recently, a new system of artificially generated APCs have shown the ability to prime and "educate" CD8+ T-cells against melanoma tumors to acquire both central and effector memory phenotypes along with the capacity to stay viable in culture for extended durations of time.[142] These *in vitro* tuned CTLs necessarily homed in on tumor sites and elicited their cellular response while retaining anti-tumor memory. Furthermore, other engineered delivery vehicles with a PEGylated liposome carrying anti-CD40 and CpG, were able to compromise tumor formation *in vivo* while preventing systemic leaks and reducing the associated systemic side effects, possibly by the anchoring property offered by the liposome.[143]

#### 5.4.2.4 Biomaterials and tolerance toward tissue-engineered devices (TEDs)/transplants

Tissue engineering and regenerative medicine are futuristic medical approaches of treating disease or trauma. Tissue-engineered devices (TEDs) consist of biologic cellular or tissue and natural or synthetic non-biologic polymer components.[115] The immune response toward such a medical device has significant implications on its ultimate function, performance, and survival in the host. There is an increasing need for developing methods of suppressing immune responses or establishing ultimately tolerance toward transplants and TEDs. As mentioned earlier, DCs are pivotal in establishing and maintaining a tolerogenic response toward self- and non-self-antigen. The DC phenotype involved in such a response is widely disparate from that required to generate protective immunity. Such DCs are termed alternatively activated DCs (aaDCs).[144] They are stimulated with immunosuppressive cytokine cocktail including IL-10 and TGF-$\beta$[145] or with glucocorticoids such as dexamethasone, and activated vitamin D3.[144] Concommitant exposure to soluble antigen in the form of cell lysates, derived from the same cell source as that of the donor transplant/TED, with a maturation stimulus such as a TLR ligand (e.g., LPS) will induce the aaDCs to provide antigen-specific immunosuppression.[144,146] Generally, tolerance is supported by to two main mechanisms: (1) clonal expansion of antigen-specific T-cells and their subsequent deletion or anergy, (2) induction of antigen-specific regulatory T-cells. Few biomaterial-based strategies for inducing a tolerogenic response have been studied. However, one example is the *in vitro* induction of aaDC phenotype in murine bone marrow derived DCs through the delivery of IL-10 gene-inducing plasmid in peptide-modified PLGA NPs.[147] These aaDCs were functionally viable and able to induce expansion of allo-FoxP3+ Treg cells.

## 5.5 Constructive Remodeling and Effective Host Integration of Implants

Cells live in an interactive environment whereby they continuously adapt by modifying their phenotype in response to physical or chemical microenvironmental cues. Cells aim to maintain homeostasis; as they encounter mechanical or pathological stress signals from their environment, they are capable of adjusting themselves by achieving a new steady state. The possible acquired accommodative responses in response to stress signals are hypertrophy, hyperplasia, atrophy, metaplasia, or neoplasia.[3] There may also be a number of changes in gene expression within the cell that may be functionally favorable. Characteristically if the extracellular stresses withdraw, the cells will return to their original state. However, if the stress stimuli persist in the cell environment and if the cell's adaptive ability is surpassed, then cell injury occurs. Within a certain limit, cell injury is reversible, on the contrary if the stress is persistent and extreme, the cells faces irreversible injury and death.[3,148]

### 5.5.1 Cell Injury and Repair

Cell injury can be of several types and may be caused by different agents. Some common injurious agents are chemical contact, mechanical force, extreme temperature, surgery-associated trauma, hypoxia and ischemia, biological agents, and genetic disorders. Reparative capacity of a cell depends on multiple factors such as

(1) Nature, duration, and intensity stress stimuli,
(2) Type and extent of cells damaged,
(3) Differentiating and/or proliferative tendency of the cells,
(4) Supply of nutrition, blood to the cell,
(5) Maintenance of ECM support structure.[3,149]

Repairing tissue involves restoring cells, their protein matrix, and ultimately, tissue function. The protein matrix defines tissue structure and supports cellular adhesion, migration, and stores/present important chemical cues (e.g., immobilized growth factors). Cells impact the activity and composition of the matrix and vice versa. Reparation of injured tissue can occur by distinct altered homoeostatic mechanisms.

(1) Regeneration of tissue-specific parenchymal cells and restoration of the normal tissue structure;
(2) Reorganization and replacement of the injured tissue characterized by the formation of fibrovascular connective tissue that leaves a scar.[150]

Both regeneration and fibrosis are determined by essentially similar cellular processes, involving cell migration, proliferation, and differentiation as well as cell-matrix interactions.

The common goal of several researchers today is to develop TEDs that can enable the restitution of tissue structure and function accompanied by controlled remodeling and minimal scarring. Some degree of reorganization of the tissue would be expected since the wound healing response to device implantation often results in fibrovascular tissue generation or fibrous encapsulation. The functional success of the device may be influenced to varying degrees by the host response.

## 5.5.2 Stages of Wound Healing

Wound healing, caused by trauma, pathogen, or foreign materials, proceeds by means of an overlapping sequence of events including (1) platelet activation and blood coagulation, (2) inflammation, (3) epithelialization (in the case of dermal wounding), (4) formation of granulation tissue, and (5)matrix/tissue remodeling. Molecular mediators such as cytokines initiate and coordinate the multitude of cellular activities that are predominant in inflammation and healing. Complex interplay between various cytokines, cells and ECM is fundamental to the initiation, progression, and resolution of wounds.[151]

### 5.5.2.1 Activation of platelets and blood coagulation

Immediately after injury, coagulation factors of the blood may undergo conformational changes and subsequently become activated enzymes that lead to the eventual formation of thrombin and strands of fibrin that stabilize the clot (coagulation cascade). In addition, platelets undergo activation followed by structural changes that lead to high circulatory platelet consumption and the formation of platelet aggregates that may adhere to the surface. Thrombin and activated platelets are both crucial to the initial coagulation events, with each of them acting as a positive catalyst in the generation of the other.

Injury to vascularized tissue in biomaterial implantation surgery causes rapid development of a provisional matrix at the implant site. The fibrin provisional matrix supports leukocyte migration into the wound site to initiate the resolution, reorganisation, and repair processes. This matrix is believed to provide structural and biochemical (embedded adhesive proteins, cytokines, and growth factors) cues that regulate wound healing.[1,3]

#### 5.5.2.2 Inflammation

Inflammation, long seen as an enemy of transplants and artificial biomaterial implants, but is a necessary step to initiate wound repair and tissue regeneration.[152] The inflammatory response is indispensable in delivering cytokines and growth factors that orchestrate events required for healing. Injury-associated occurrences bring about the recruitment of inflammatory cells. Leukocytes migrate across the endothelium into the provisional matrix, releasing proteases, and function to ingest microbes, cellular debris, and matrix degradation products. Commonly, neutrophils are the majority cell type that migrate within minutes after the injury and last for a few days thereafter. They are soon replaced by monocyte/MΦs as the predominant cell type. Neutrophils and monocytes are responsible for initiating inflammation whereas monocyte/MΦs are capable of regulating the course of inflammation until the tissue is effectively repaired. Clearance of injury-related debris signifies an end to the acute phase making way for the healing phase. The physical and chemical properties of the biomaterial may be responsible for distinctions in the intensity and length of the inflammatory and/or wound-healing response.[3]

The initial phases in wound healing represent an upregulation of inflammatory mediators. However, progression to wound healing requires resolution of inflammatory response, which in turn is directed by the downregulation of pro-inflammatory mediators and reconstitution of normal microvascular permeability.[152,153] This results in the termination of local chemoattractants and synthesis of anti-inflammatory mediators.

Neither an excessive/prolonged nor a diminished initial inflammatory response is ideal for wound healing, thereby suggesting that there needs to be a suitable balance between the sequential pro- and anti-inflammatory events to engender appropriate healing of tissue. MΦs residing in the provisional matrix during the course wound healing may undergo variation in the relative balance of their phenotypic extremes of M1 (pro-inflammatory) to M2 (anti-inflammatory) corresponding the state of inflammatory events; and may accordingly play a crucial role in the mediation of the same. Eventually, a shift in the overall MΦ phenotype toward more reparative or regulatory may be preferable to facilitate tissue remodeling and healing.[97]

#### 5.5.2.3 Re-epithelialization, granulation tissue, and angiogenesis

The repair response encompasses overlapping events involved in granulation tissue, angiogenesis, and re-epithelialization. Within hours, epithelial cells begin to proliferate. Immature keratinocytes secrete MMPs and plasmins. MMPs break away basal attachment and clear fibrillar collagen while plasmin resolves the fibrin clot. Free from the provisional matrix, the cells migrate across the wound bed. TGF-$\beta$ is important for wound healing and induces keratinocyte proliferation.[154,155]

Granulation tissue forms below the epithelium and is composed of a balanced mix of inflammatory cells, fibroblasts, EC, and newly forming blood vessels. Angiogenesis is the process in which pre-existing cells of nearby capillaries send out capillary sprouts to produce new vessels; or may occur by the longitudinal division of existing vessels with peri-ECs, which may then split and branch into pre-capillary arterioles and capillaries.[156] Angiogenic cues include hypoxia, nitric oxide, vascular endothelial cell growth factor (VEGF), fibroblast growth factor-2 (FGF-2) and MCP-1, MIP-1$\alpha$, and is mediated by a complex array of pro- and anti-angiogenic factors and vessel maturation factors (e.g., platelet-derived growth factor, PDGF) coupled with multiple cell and ECM components, all in a tightly regulated manner.[157–159]

#### 5.5.2.4 Matrix, scar formation, and remodeling

Matrix-generating cells move into the granulation tissue post the assembly of new vasculature. These cells degrade the provisional matrix via MMPs and respond to cytokine/growth factors by proliferating and synthesizing new ECM to replace injured tissue with a connective tissue scar. TGF-$\beta$ regulates the fibrotic process by recruiting fibroblasts and stimulating them to produce of collagens

I, III, and V, proteoglycans, fibronectin, and other ECM components. TGF-$\beta$ in parallel impedes protease inhibitors, favoring ECM accumulation. PDGF secreted by activated platelets also contributes to the reparative phase either directly by influencing endothelial and fibroblast functions or indirectly by inducing additional cytokines and growth factors (e.g., IL-6). Fibroblasts can display increased levels of expression of adhesion molecules. Further, they can adopt a contractile myofibroblast phenotype, facilitated mainly by TGF-$\beta$, PDGF to cause wound contraction.[154,155,160] The remodeling phase is tightly cytokine-regulated and implicates continuous synthesis, break-down, and re-organization of collagen and other ECM molecules that is promoted by serine proteases and MMPs.

Scar-free healing is a desirable wound healing outcome. Interestingly, embryonic/fetal but not adult wounds have been shown to heal with minimal or no scarring. This is primarily due to the difference in physicochemical composition of the wound that offer distinct physiological cues to concomitant cell types.[161,162] In a study comparing fetal and post-natal wounds, it has been seen that Wnt signaling, TGF-$\beta$ isoforms, hyaluronan contents, and expression of hyaluronan regulatory enzymes played crucial roles in scar-free healing.[163] Embryonic/fetal wounds heal without scarring demonstrated reduced inflammatory response hence altered cytokine profile, reduced wound angiogenesis, absence of fetal wound fibrin clots, lots of hyaluronic acid, presence of high levels of morphogenetic factors involved in skin growth, remodeling, and morphogenesis in embryonic skin because constantly/rapidly growing, high levels of FGF in embryonic wounds while PDGF (from platelets) is high in adult wounds. Interestingly, in adults, TGF-$\beta$1 and TGF-$\beta$2 are present at high levels while TGF-$\beta$3 is present at low levels.[164] The opposite is true in embryonic would that heals scar-free with high levels of TGF-$\beta$3 and low levels of the other two isoforms.

Wound healing research as a field has mainly been focused on the associated growth factors and migration of related cells, in the context of external or dermal wounds. Comparatively fewer studies have been done on internal wounds such as those occurring in myocardial infarction. Moreover, it would be interesting to know which cellular and molecular agents and events perpetuate healing internally; and whether studies on external wounds are translatable to internal ones.

### 5.5.3 Biomaterial Strategies Employed to Direct Repair and Regeneration

Biomaterial based strategies to affect wound healing and tissue regeneration have included strategies such as unique intrinsic polymer chemistry to drive vascularization and wound healing, microarchitecture-based induction of a vascularized tissue response, delivery of key biological signals such as angiogenic and stabilizing growth factors or ECM components to drive a desirable host response and also delivery of cells, such as mesenchymal stem cells (MSC) with immunomodulatory and accessory cell functions for appropriate tissue remodeling. Research examples of these approaches will be discussed below and are summarized in Table 5.3.

#### 5.5.3.1 Wound closure and vascularization

Use of biomaterials as wound closures has been common practice for a number of years to curtail blood flow and hemorrhaging in case of traumatic injury and surgery related trauma. Here, rapid and effective clot formation is desired. Wound dressings aim to cover the wound, acting as a temporary barrier for the missing skin and ultimately to enhance wound healing. As an example, *in situ* ultraviolet-polymerizable azide-chitosan-lactobionic acid hydrogel as a wound healing dressing was developed and tested in a mouse skin wound model.[165] In this study, the chitosan hydrogel matrix provided directional cues for increased permeation/migration of inflammatory cells and fibroblasts. It was expected that such an environment would enhance wound contraction and accelerate wound closure and healing.

For functional integration of engineered tissues or biomaterial-tissue composites and survival of associated cells, it is critical that these implants are efficiently vascularized in a timely manner upon implantation. Vascularization, which may be particularly complex for large three dimensional tissues, is crucial in transporting nutrients to and secreted products from the cells within the SC.

Patterned biomaterials possessing intricate microarchitecture such as pillared structures[166] or pores[167] can support an angiogenic tissue response. As such, porous membranes wherein pore sizes are greater than 0.8 μm were associated with the formation of closed vascular structures with enhanced cell penetration into the pores of the membrane. Moreover, vascular re-growths were

**Table 5.3** Key biomaterial strategies in wound healing

**Primary Goals**: Wound closure, Vascularization, Scar-free healing and Matrix remodeling

| **Strategy** | **Chemistry-based** | **Microarchitecture-based** | **Biological components** | | **Tissue Composite** |
|---|---|---|---|---|---|
| | | | **Soluble signals** | **Acellular scaffold** | |
| | Negatively charged groups, e.g., surfaces coated with methacrylic acid groups[169–174] | Micropatterned pillared structures[166] and pores[167] (>0.8 μm) | VEGF and PDGF,[176–178] controlled Wnt-signaling, FGF high, PDGF low,[163] TGF-isoform[164] | Porcine Small intestinal submucosa (SIS)[191,192] | Stem cells—in biodegradable scaffold[179,180] or in co-culture[181–184] |
| **Result** | Angiogenesis,[169–174] increase microvessel density,[170] wound closure in diabetic mice[171] | Close vascular structures, high cell penetrations,[166–168] accompanied by inflammatory cells[168] | Vascularization and associated inflammatory response,[176–178] reduced scarring in adult wounds[163,164] | Matrix remodeling, localized immunosuppressive Th2-restricted response | Complex tissue remodeling, regeneration, vascularization and MΦ recruitment |
| **Potential drawback** | Lower specificity without biological agents | Lower specificity without biological agents | Requires a supportive natural or synthetic matrix for delivery | Immune reaction upon implantation | Tumor formation |

often accompanied by inflammatory cells suggesting that some inflammation is perhaps necessary for effective angiogenesis.[168]

An interesting, biomaterial chemistry-based angiogenesis approach has been described wherein porous bulk polymerised SCs prepared from n-butyl methacrylic acid–methyl methacrylate (MMA) co-polymer demonstrated an ability to promote angiogenesis within a 30-day period when implanted in rats.[169] This approach is interesting since the angigogenic observation is dependent solely on a unique material-chemistry (methacrylic acid [MAA] containing) without the additional of any exogenous factors such as angiogenic growth factors. Beads made of the same material were used to treat skin grafts in rodents with an observed benefit on graft engraftment.[170] This was associated with a significantly increased microvessel density and thickness of panniculus carnosus muscle 11 days post implantation. In a similar study, poly(MAA-co-MMA) (45 mol% MAA) beads enhanced wound closure and vascularization in diabetic mice within a 21-day period.[171] Another study compared the effect of surface charge properties on angiogenesis. Surface-modified polymer fibers were plasma coated with either negatively charged MAA, positively charged *N,N*-dimethyl-aminoethyl methacrylate, or neutral hexafluoropropylene.[172–174] *In vivo* implantation showed that MAA-coated constructs promoted angiogenesis better than the other groups. This suggests a possible dependence of angiogenic behavior on the negative charge carried by MAA. Furthermore, other material properties such as polymer fiber diameter and fiber spacing within SC mesh were shown to directly impact the development of neovascular structures.

Porous SCs with the ability to support vascularization have been applied to the transplantation of hES cell-derived cardiomyocytes were able to direct the proliferation and survival of these cells *in vitro*. These constructs contained parallel channels to organize cardiomyocyte bundles, supported by micrometer-sized spherical, interconnected, uniform pores that possibly enhance angiogenesis and reduce scarring. Additionally, an acellular version of this SC with pore diameters of 30–40 μm, was capable of supporting angiogenesis, a reduced fibrotic response and corresponding MΦ M2 phenotype shift when implanted *in vivo* in cardiac tissue of rats.[175]

Angiogenesis not only depends on EC infiltration and proliferation: but also requires pericyte coverage of vascular sprouts for vessel equilibrium and maturation of nascent vessels. These processes are mediated VEGF and PDGF through their associated receptors on ECs and vascular smooth muscle cells (VSMCs), respectively. PDGF supports neovascularization by priming VSMCs/pericytes to release pro-angiogenic mediators but also recruiting and maturing perivascular smooth muscle cells. VEGF on the other hand plays a dual role a promoter of EC function and a negative regulator of VSMCs and vessel maturation.[176] Since VEGF possesses a dichotomous role and PDGF acts in VEGF-related manner, spatially and temporally controlled release of a combination of these factors may offer to enhance endogenous angiogenic capabilities.[177] Implanted fibers with VEGF releasing hydrogel coatings have been shown to increase vascularization.[178] This study clearly demonstrated the co-occurrence of inflammation with angiogenesis upon delivering VEGF. When the local inflammation was suppressed with dexamethasone, the corresponding angiogenesis was also reduced.

#### 5.5.3.2 Stem cell-seeded SCs to direct regeneration

Regeneration of local/damaged tissue, mainly in the context of developing angiogenesis endogenously, has been extensively explored. EC spheroids encapsulated in gelatin modules presumably underwent remodeling *in vivo* and promoted development of blood vessels when implanted in nude rats.[179,180] Micropatterned and multilayer 2D and 3D cell co-cultured ECs and dermal fibroblasts have been fabricated using thermoresponsive polymers such as pNIPAAM.[181] These structures underwent remodeling within six days to form a modified organization of ECs and fibroblasts. A different study shows that an engineered blood vessel with human umbilical vein endothelial cells (HUVECs) in a collagen was stabilized with human bone-marrow derived MSC.[182] Similarly, adipose progenitor cells co-delivered in collagen/fibronectin SC with ECs *in vivo* were capable of generating functional vascular networks when implanted in mice. Successful delivery of multiple cell types in the aforementioned examples suggests that ECs alone are not sufficient in developing neovasculature.[182–184] Interestingly, a collaborative influence of cells, namely MSCs that are perivascular to ECs, has been described to stabilize growing blood vessels. Furthermore, a recent study offers insight as to how the remodeling of BMC-seeded biodegradable SC occurs *in vivo* when implanted into a mouse—briefly, seeded cells secret ample amounts of MCP-1 that results in the recruitment of monocyte/MΦs in the implant. Subsequently, seeded donor cells are completed eliminated from the SC within a 6-month period, instead ECs and SMCs from the host populate the SC that in due course of time developed into functional living blood vessels.[185–187]

Interestingly, recent advances in tissue-engineered development of blood vessel have shown evidence contrary to popular belief that host cells from adjacent blood vessels and not implanted donor cells, entirely make up the neovascular tissue.[185,186] Microfabricated SC with "smart" interconnected microporous architecture and microfluidic network have been shown to promote tissue regeneration. These SCs were seeded with muscle myoblast cells, neuroblastomas, and ES cells.[188] *In vitro*, myoblast, neuroblastomas, and ES attached, proliferated, and differentiated. ES proliferation occurred in a controlled manner. *In vivo* vascular structures were formed indicating the SC–host tissue integration.

#### 5.5.3.3 Remodeling of the ECM

The extracellular protein matrix renders structure and mechanical integrity to the tissue. It also actively supports the adhesion and migration of cells. Aberrant ECM composition post wound healing in adults contributes to scarring and other pathological issues such as fibrotic disease, tumor progression, and metastasis.[189,190] Cells and ECM directly influence the function and activity of each other. Recruitment of inflammatory cells to the site of tissue injury has been established as a key event in host response. If an implanted SC with ECM components is able to provide improved homing of inflammatory cells, suitable tissue remodeling can be programmed. Such strategies have been utilized in developing immune-tolerant allo- and xenografts. One such example is that of using small intestinal submucosa (SIS) for regenerative studies. SIS is a naturally occurring, acellular biomaterial derived from porcine jejunum, which promotes constructive tissue remodeling when applied as a xenogeneic graft material. In addition, implanting SIS SC in mice resulted in its endogenous remodeling accompanied with immunosuppressive Th2-restricted response that is localized.[191,192] Interestingly, Th2-milieu is conventionally consistent with remodeling rather than rejection.

#### 5.5.3.4 Biodegradable and bioresorbable polymers

The number of contributions aimed at introducing degradable new polymeric systems is increasing in the literature in relation to the interest of therapeutic devices for time-limited applications. Subtle semantic distinctions indicate whether a material is degradable (when the cleavage mechanism is controlled by chemistry such as hydrolysis) or biodegradable (when the cleavage is cell mediated). Commonly, a material is not classified as biodegradable if it is cleaved by only by isolated enzymes; that is, lack of these specific isolated enzymes will render the material fairly stable. Additionally, it is extremely relevant to consider the fate of degradation products in terms of storage or elimination and possible induction of local inflammation. Hence it would be highly suitable to exploit the use of biodegradable polymers that are comprised of pro-metabolites constitutive repeating monomers that can be inserted in biochemical pathways or eliminated by the kidney.[193] The epitome of such artificial biopolymers is poly(glycolic acid), poly (lactic acid), and various formulations of copolymer of the two, namely PLGA. Several new systems can be found in the literature but the number is still limited. Such porous SCs allow spatial and temporal exposure of SC components on biodegradation of the polymer. PLA, PLGA continue to be the only FDA-approved synthetic biodegradable biomaterials for human implants.[194] Biodegradable SCs are commonly used in tissue remodeling studies mainly because they present an artificial matrix for controlled delivery or exposure of embedded proteins, cytokines, growth factors, and cells. Additionally, these materials permit flexibility of subsequent matrix compensation with endogenously secreted ECM to replace the temporary polymer SC.[195,196]

## 5.6 Summary

In summary, the molecular and cellular aspects of the host response to biomaterials are being elucidated through several recent investigations. It is becoming recognized that the typical host response to a biomedical implant, can be modulated to an alternative host response through intelligent biomaterial properties such as chemistry or microarchitecture. Furthermore, strategies to direct MΦ phenotype offer opportunities to direct appropriate functional integration of biomaterials into the host. There exist great potential to be able to modulate adaptive immune responses through smart biomaterial design considering parameters such as size, shape, and chemistry to affect key cells that determine immune outcomes, namely DCs. In summary, controlled tissue remodeling and regeneration is the new goal for biomaterial-induced host responses. Furthermore, enhanced control and direction of immune outcomes in vaccines, anti-tumor immunity, and tissue engineering are in view with application of immunological understanding of these processes and innovative biomaterial applications.

## References


1. Anderson, J. M. (2001) Biologic responses to materials, *Annu. Rev. Mater. Res.*, **31**, 81–110.
2. Anderson, J. M., Rodriguez, A. and Chang, D. T. (2008) Foreign body reaction to biomaterials, *Semin. Immunol.*, **20**, 86–100.
3. Ratner, B. D., Hoffman, A.S., Scheon F.S., Lemons, J.E., (2004) *Biomaterials Science: An Introduction to Materials in Medicine,* Academic Press, San Diego, CA.
4. Basmadjian, D., Sefton, M. V. and Baldwin, S. A. (1997) Coagulation on biomaterials in flowing blood: Some theoretical considerations, *Biomaterials*, **18**, 1511–1522.
5. Hench, L. L. and Polak, J. M. (2002) Third-generation biomedical materials, *Science*, **295**, 1014 –1017.
6. Tang, L., Lucas, A. H. and Eaton, J. W (1993) Inflammatory responses to implanted polymeric biomaterials: Role of surface-adsorbed immunoglobulin G, *J. Lab. Clin. Med.*, **122**, 292–300.
7. Hench, L. L. (1998) Biomaterials: A forecast for the future, *Biomaterials*, **19**, 1419–1423.
8. Renné, T. (2012) The procoagulant and proinflammatory plasma contact system, *Semin. Immunopathol.*, **34**, 31–41.
9. Levi, M., Keller, T. T., Van Gorp, E. and Ten Cate, H. (2002) Inflammation and the coagulation system, **60**, 26–39.
10. Weiss, S. J. (1989) Tissue destruction by neutrophils. *N Engl J Med* **320**, 365–376.
11. Janeway, C. A. and Medzhitov, R. (2002) Innate immune recognition, *Ann. Rev. Immunol.*, **20**, 197–216.
12. Mosser, D. M. (2003) The many faces of macrophage activation, *J. Leukoc. Biol.*, **73**, 209–212.

13. Mantovani, Sica and Locati, M. (2005) Macrophage polarization comes of age, *Immunity*, **23**, 344–346.

14. Temenoff J. S. and Mikos, A. G. (2008) Biomaterials: The intersection of biology and materials science, Saddle River, NJ.

15. Jay, S. M., Skokos, E. A., Zeng, J., Knox, K. and Kyriakides, T. R. (2010) Macrophage fusion leading to foreign body giant cell formation persists under phagocytic stimulation by microspheres *in vitro* and *in vivo* in mouse models, *J. Biomed. Mater. Res. A*, **93**, 189–199.

16. Kyriakides, T. R. *et al.* (2004) The CC chemokine ligand, CCL2/MCP1, participates in macrophage fusion and foreign body giant cell formation, *Am. J. Pathol.*, **165**, 2157–2166.

17. MacLauchlan, S. *et al.* (2009) Macrophage fusion, giant cell formation, and the foreign body response require matrix metalloproteinase, **9**, *J. Leukocy. Biol.*, **85**, 617–626.

18. Kirk, J. T., McNally, A. K. and Anderson, J. M. (2010) Polymorphonuclear leukocyte inhibition of monocytes/macrophages in the foreign body reaction, *J. Biomed. Mater. Res. A*, **94**, 683–687.

19. Higgins, D. M. *et al.* (2009) Localized immunosuppressive environment in the foreign body response to implanted biomaterials, *Am. J. Pathol.*, **175**, 161–170.

20. Xia, Z. and Triffitt, J. T. (2006) A review on macrophage responses to biomaterials, *Biomed. Mater.*, **1**, R1–R9.

21. Stokes, K. (2006) Polyurethane pacemaker leads the contribution of clinical expertise to the elucidation of failure modes and biodegradation mechanisms in, *Clinical Evaluation of Medical Devices Principles and Case Studies, 2nd edition,* Becker, K. M., Whyte, J. J. eds., Humana Press, Totowa, NJ.

22. McNally, A. K., Jones, J. A., MacEwan, S. R., Colton, E. and Anderson, J. M. (2008) Vitronectin is a critical protein adhesion substrate for IL-4-induced foreign body giant cell formation, *J. Biomed. Mater. Res. A*, **86A**, 535–543.

23. Keselowsky, B. G. *et al.* (2007) Role of plasma fibronectin in the foreign body response to biomaterials, *Biomaterials*, **28**, 3626–3631.

24. Kyriakides, T. R., Leach, K. J., Hoffman, A. S., Ratner, B. D. and Bornstein, P. (1999) Mice that lack the angiogenesis inhibitor, thrombospondin 2, mount an altered foreign body reaction characterized by increased vascularity, *Proc. Natl. Acad. Sci.*, **96**, 4449 –4454.

25. Puolakkainen, P. *et al.* (2003) Compromised production of extracellular matrix in mice lacking secreted protein, acidic and rich in cysteine

(SPARC) leads to a reduced foreign body reaction to implanted biomaterials, *Am. J. Pathol.*, **162**, 627–635.

26. Ratner, B. D. (1998) Molecular design strategies for biomaterials that heal, *Macromol. Symp.*, **130**, 327–335.
27. Bota, P. C. S. *et al.* (2010) Biomaterial topography alters healing *in vivo* and monocyte/macrophage activation *in vitro*, *J. Biomed. Mater. Res. A*, **95A**, 649–657.
28. Ratner, B. D. and Bryant, S. J. (2004) Biomaterials: Where we have been and where we are going, *Annu. Rev. Biomed. Eng.*, **6**, 41–75.
29. Liu, L. *et al.* (2008) Reduced foreign body reaction to implanted biomaterials by surface treatment with oriented osteopontin, *J. Biomater. Sci. Polym. Ed.*, **19**, 821–835.
30. Khandwekar, A. P., Patil, D. P., Hardikar, A. A., Shouche, Y. S. and Doble, M. (2010) *In vivo* modulation of foreign body response on polyurethane by surface entrapment technique. *J. Biomed. Mater. Res. A* **95A**, 413–423.
31. Bennet, W., Groth, C. G., Larsson, R., Nilsson, B. and Korsgren, O (2000) Isolated human islets trigger an instant blood mediated inflammatory reaction: Implications for intraportal islet transplantation as a treatment for patients with type 1 diabetes, *Ups. J. Med. Sci.*, **105**, 125–133.
32. Inverardi, L. and Ricordi, C. (2006) Therapeutic approaches to counteract immediate blood-mediated inflammatory reaction in islet transplantation, *Transplantation*, **82**, 312–313.
33. Cabric, S. *et al.* (2007) Islet surface heparinization prevents the instant blood-mediated inflammatory reaction in islet transplantation, *Diabetes*, **56**, 2008 –2015.
34. Azzi, J., Geara, A. S., El-Sayegh, S. and Abdi, R. (2010) Immunological aspects of pancreatic islet cell transplantation, *Expert Rev. Clin. Immunol.*, **6**, 111–124.
35. Goto M, *et al.* (2004) Low molecular weight dextran sulfate prevents the instant blood-mediated inflammatory reaction induced by adult porcine islets, *Transplantation*, **77**, 741–747.
36. Goto, M., Groth, C. G., Nilsson, B. and Korsgren, O. (2004) Intraportal pig islet xenotransplantation into athymic mice as an *in vivo* model for the study of the instant blood-mediated inflammatory reaction, *Xenotransplantation*, **11**, 195–202.
37. Goto, M., *et al.* (2008) Dissecting the instant blood-mediated inflammatory reaction in islet xenotransplantation, *Xenotransplantation*, **15**, 225–234.

38. Johansson, H., *et al.* (2006) Low molecular weight dextran sulfate: A strong candidate drug to block IBMIR in clinical islet transplantation, *Am. J. Transpl.*, **6**, 305–312.

39. Ozmen, L., *et al.* (2002) Inhibition of thrombin abrogates the instant blood-mediated inflammatory reaction triggered by isolated human islets: Possible application of the thrombin inhibitor melagatran in clinical islet transplantation, *Diabetes*, **51**, 1779–1784.

40. Stabler, C. L., *et al.* (2007) Surface re-engineering of pancreatic islets with recombinant azido-thrombomodulin, *Bioconjugate Chem.*, **18**, 1713–1715.

41. Wilson, J. T., *et al.* (2010) Biomolecular surface engineering of pancreatic islets with thrombomodulin, *Acta Biomater.*, **6**, 1895–1903.

42. Krishnamurthy, V. R., *et al.* (2010) Chemoselective Immobilization of peptides on abiotic and cell surfaces at controlled densities, *Langmuir*, **26**, 7675–7678.

43. Cui W *et al.* (2009) Thrombomodulin improves early outcomes after intraportal islet transplantation, *Am. J. Transplant.*, **9**, 1308–1316.

44. Teramura, Y. and Iwata, H. (2008) Islets surface modification prevents blood-mediated inflammatory responses, *Bioconjugate Chem.*, **19**, 1389–1395.

45. Chen, H., Teramura, Y. and Iwata, H. (2011) Co-immobilization of urokinase and thrombomodulin on islet surfaces by poly(ethylene glycol)-conjugated phospholipid. *J. Control. Release*, **150**, 229–234.

46. Jung, Y. S., *et al.* (2012) Surface modification of pancreatic islets using heparin-DOPA conjugate and anti-CD154 mAb for the prolonged survival of intrahepatic transplanted islets in a xenograft model, *Biomaterials*, **33**, 295–303.

47. Medzhitov, R. and Janeway Jr., C. A. Jr. (1998) Innate immune recognition and control of adaptive immune responses, *Semin. Immunol.*, **10**, 351–353.

48. Janeway, C. A. and Medzhitov, R. (2002) Innate immune recognition, *Ann. Rev. Immunol.*, **20**, 197–216.

49. Ishii, K. J., Coban, C., Akira, S. (2005) Manifold mechanisms of toll-like receptor-ligand recognition, *J. Clin. Immunol.*, **25**, 511–521.

50. McGreal, E. P., Martinez-Pomares, L. and Gordon, S. (2004) Divergent roles for C-type lectins expressed by cells of the innate immune system, *Mol. Immunol.*, **41**, 1109–1121.

51. Gordon, J. (1994) B-cell signalling via the C-type lectins CD23 and CD72, *Immunol. Today*, **15**, 411–417.

52. He, B. *et al.* (2006) HIV-1 envelope triggers polyclonal Ig class switch recombination through a CD40-independent mechanism involving BAFF and C-type lectin receptors, *J. Immunol.*, **176**, 3931–3941.

53. McGreal, E. P., Miller, J. L. and Gordon, S (2005) Ligand recognition by antigen-presenting cell C-type lectin receptors, *Curr. Opin. Immunol.*, **17**, 18–24.

54. Figdor, C. G., van Kooyk, Y. and Adema, G. J. (2002) C-type lectin receptors on dendritic cells and Langerhans cells, *Nat. Rev. Immunol.*, **2**, 77–84.

55. Geijtenbeek, T. B. H. *et al.* (2000) DC-SIGN–ICAM-2 interaction mediates dendritic cell trafficking, *Nat. Immunol.*, **1**, 353–357.

56. Engering, A., Geijtenbeek, T. B. H. and van Kooyk, Y. (2002) Immune escape through C-type lectins on dendritic cells, *Trends Immunol.*, **23**, 480–485.

57. Weis, W. I., Taylor, M. E. and Drickamer, K. (2006) The C-type lectin superfamily in the immune system, *Immunol. Rev*, **163**, 19–34.

58. Brown, G. D. (2005) Dectin-1: A signalling non-TLR pattern-recognition receptor, *Nat. Rev. Immunol.*, **6**, 33–43.

59. Gantner, B. N., Simmons, R. M., Canavera, S. J., Akira, S. and Underhill, D. M. (2003) Collaborative induction of inflammatory responses by dectin-1 and toll-like receptor 2, *J. Exp. Med.*, **197**, 1107–1117.

60. Akira, S., Uematsu, S. and Takeuchi, O. (2006) Pathogen recognition and innate immunity, *Cell*, **124**, 783–801.

61. Brown, G. D. *et al.* (2003) Dectin-1 mediates the biological effects of beta-glucans, *J. Exp. Med.*, **197**, 1119–1124.

62. van Vliet, S. J., van Liempt, E., Geijtenbeek, T. B. H. and van Kooyk, Y. (2006) Differential regulation of C-type lectin expression on tolerogenic dendritic cell subsets, *Immunobiology*, **211**, 577–585.

63. Smits, H. H. *et al.* (2005) Selective probiotic bacteria induce IL-10-producing regulatory T cells *in vitro* by modulating dendritic cell function through dendritic cell-specific intercellular adhesion molecule 3-grabbing nonintegrin, *J. Allergy Clin. Immunol.*, **115**, 1260–1267.

64. Aderem, A. and Ulevitch, R. J. (2000) Toll-like receptors in the induction of the innate immune response, *Nature*, **406**, 782–787.

65. Helmy, K. Y. *et al.* (2006) CRIg: A macrophage complement receptor required for phagocytosis of circulating pathogens, *Cell*, **124**, 915–927.

66. Akira, S. and Takeda, K. (2004) Toll-like receptor signalling, *Nat. Rev. Immunol.*, **4**, 499–511.

67. Morikis, D. and Lambris, J. D. (2002) Structural aspects and design of low-molecular-mass complement inhibitors, *Biochem. Soc. Trans.*, **30**, 1026–1036.

68. Sahu, A. and Lambris, J. D. (2000) Complement inhibitors: A resurgent concept in anti-inflammatory therapeutics, *Immunopharmacology*, **49**, 133–148.

69. Atkinson, J. P. and Farries, T. (1987) Separation of self from non-self in the complement system, *Immunol. Today*, **8**, 212–215.

70. Nilsson, B. *et al.* (1998) Compstatin inhibits complement and cellular activation in whole blood in two models of extracorporeal circulation, *Blood*, **92**, 1661 –1667.

71. Gasque, P. (2004) Complement: A unique innate immune sensor for danger signals, *Mol. Immunol.*, **41**, 1089–1098.

72. Ajees, A. A. *et al.* (2006) The structure of complement C3b provides insights into complement activation and regulation, *Nature*, **444**, 221–225.

73. Russell, S. (2004) CD46: A complement regulator and pathogen receptor that mediates links between innate and acquired immune function, *Tissue Antigens*, **64**, 111–118.

74. Gelderman, K. A., Blok, V. T., Fleuren, G. J. and Gorter, A. (2002) The inhibitory effect of CD46, CD55, and CD59 on complement activation after immunotherapeutic treatment of cervical carcinoma cells with monoclonal antibodies or bispecific monoclonal antibodies, *Lab. Invest*, **82**, 483–493.

75. Nilsson, B., Ekdahl, K. N., Mollnes, T. E. and Lambris, J. D (2007) The role of complement in biomaterial-induced inflammation, *Mol. Immunol.*, **44**, 82–94.

76. Harboe, M., Ulvund, G., Vien, L., Fung, M. and Mollnes, T. E. (2004) The quantitative role of alternative pathway amplification in classical pathway induced terminal complement activation, *Clin. Exp. Immunol.*, **138**, 439–446.

77. Haas, P.-J. and Van Strijp, J. (2007) Anaphylatoxins: Their role in bacterial infection and inflammation, *Immunol. Res.*, **37**, 161–175.

78. Szott, L. M. and Horbett, T. A. (2010) The role of complement C3 and fibrinogen in monocyte adhesion to PEO-like plasma deposited tetraglyme, *J. Biomed. Mater. Res. A*, **95A**, 1252–1260.

79. Gorbet, M. B. and Sefton, M. V. (2004) Biomaterial-associated thrombosis: Roles of coagulation factors, complement, platelets and leukocytes, *Biomaterials*, **25**, 5681–5703.

80. Videm, V. *et al.* (1999) Heparin-coated cardiopulmonary bypass equipment. II. Mechanisms for reduced complement activation *in vivo*, *J. Thorac. Cardiovasc. Surg.*, **117**, 803–809.

81. Gong, J. *et al.* (1996) Tubing loops as a model for cardiopulmonary bypass circuits: Both the biomaterial and the blood-gas phase interfaces induce complement activation in an *in vitro* model, *J. Clin. Immunol.*, **16**, 222–229.

82. Andersson, J. *et al.* (2003) Optimal heparin surface concentration and antithrombin binding capacity as evaluated with human non-anticoagulated blood *in vitro*, *J. Biomed. Mater. Res. A*, **67A**, 458–466.

83. Andersson, J., Bexborn, F., Klinth, J., Nilsson, B. and Ekdahl, K. N. (2006) Surface-attached PEO in the form of activated pluronic with immobilized factor H reduces both coagulation and complement activation in a whole-blood model, *J. Biomed. Mater. Res. A*, **76**, 25–34.

84. Kim, Y. H., Han, D. K., Park, K. D. and Kim, S. H. (2003) Enhanced blood compatibility of polymers grafted by sulfonated PEO via a negative cilia concept, *Biomaterials*, **24**, 2213–2223.

85. Kim, S. C (2002) PEO-grafting on PU/PS IPNs for enhanced blood compatibility--effect of pendant length and grafting density, *Biomaterials*, **23**, 2015–2025.

86. Amiji, M. and Park, K. (1993) Surface modification of polymeric biomaterials with poly(ethylene oxide), albumin, and heparin for reduced thrombogenicity. *J. Biomater. Sci. Polym. Ed.*, **4**, 217–234.

87. Amiji, M. and Park, K. (1992) Prevention of protein adsorption and platelet adhesion on surfaces by PEO/PPO/PEO triblock copolymers, *Biomaterials*, **13**, 682–692.

88. Mansoor M., A. (1995) Permeability and blood compatibility properties of chitosan-poly(ethylene oxide) blend membranes for haemodialysis, *Biomaterials*, **16**, 593–599.

89. Cao, L. *et al.* (2007) Plasma-deposited tetraglyme surfaces greatly reduce total blood protein adsorption, contact activation, platelet adhesion, platelet procoagulant activity, and *in vitro* thrombus deposition, *J. Biomed. Mater. Res. A*, **81A**, 827–837.

90. Sefton, M. V., Gemmell, C. H. and Gorbet, M. B. (2000) What really is blood compatibility? *J. Biomater. Sci. Polym. Ed.* **11**, 1165–1182.

91. Shen, M. *et al.* (2002) PEO-like plasma polymerized tetraglyme surface interactions with leukocytes and proteins: *In vitro* and *in vivo* studies, *J. Biomater. Sci. Polym. Ed.*, **13**, 367–390.

92. Szott, L. M., Stein, M. J., Ratner, B. D. and Horbett, T. A. (2011) Complement activation on poly(ethylene oxide)-like radiofrequency glow discharge–deposited surfaces, *J. Biomed. Mater. Res. A*, **96A**, 150–161.
93. Ferraz, N., Ott, M. K. and Hong, J. (2010) Time sequence of blood activation by nanoporous alumina: Studies on platelets and complement system, *Microsc. Res. Tech.*, **73**, 1101–1109.
94. Engberg, A. E. *et al.* (2011) Blood protein-polymer adsorption: Implications for understanding complement-mediated hemoincompatibility, *J. Biomed. Mater. Res. A*, **97A**, 74–84.
95. Abbas, A. K. and Lichtman, A. H. (2003) *Cellular and Molecular Immunology*, Saunders, Philadelphia, PA.
96. Banchereau, J. and Steinman, R. M. (1998) Dendritic cells and the control of immunity, *Nature*, **392**, 245–252.
97. Steinman, R. M. and Cohn, Z. A. (1973) Identification of a novel cell type in peripheral lymphoid organs of mice I. Morphology, quantitation, tissue distribution, *J. Exp. Med.*, **137**, 1142–1162.
98. Shortman, K. and Liu, Y.-J. (2002) Mouse and human dendritic cell subtypes, *Nat. Rev. Immunol.*, **2**, 151–161.
99. Shortman, K. and Naik, S. H. (2007) Steady-state and inflammatory dendritic-cell development, *Nat. Rev. Immunol.*, **7**, 19–30.
100. Lipscomb, M. F. and Masten, B. J. (2002) Dendritic cells: Immune regulators in health and disease, *Physiol. Rev.*, **82**, 97–130.
101. Dzionek, A. *et al.* (2000) BDCA-2, BDCA-3, and BDCA-4: Three markers for distinct subsets of dendritic cells in human peripheral blood, *J. Immunol.*, **165**, 6037–6046.
102. Huysamen, C., Willment, J. A., Dennehy, K. M. and Brown, G. D. (2008) CLEC9A is a novel activation C-type lectin-like receptor expressed on BDCA3+ dendritic cells and a subset of monocytes. *J. Biol. Chem.*, **283**, 16693–16701.
103. MacDonald, K. P. A. *et al.* (2002) Characterization of human blood dendritic cell subsets, *Blood*, **100**, 4512–4520.
104. Schäkel, K. *et al.* (2002) 6-Sulfo LacNAc, a novel carbohydrate modification of PSGL-1, defines an inflammatory type of human dendritic cells, *Immunity*, **17**, 289–301.
105. Rock, K. L. and Shen, L. (2005) Cross-presentation: Underlying mechanisms and role in immune surveillance. *Immunol. Rev.*, **207**, 166–183.

106. Koch, J., Guntrum, R., Heintke, S., Kyritsis, C. and Tampé, R. (2004) Functional dissection of the transmembrane domains of the transporter associated with antigen processing (TAP), *J. Biol. Chem.*, **279**, 10142–10147.

107. Busch, R. *et al.* (2005) Achieving stability through editing and chaperoning: Regulation of MHC class II peptide binding and expression, *Immunol. Rev.*, **207**, 242–260.

108. Schrum, A. G. and Turka, L. A. (2002) The proliferative capacity of individual naive CD4+T cells is amplified by prolonged T cell antigen receptor triggering, *J. Exp. Med.*, **196**, 793–803.

109. Pitcher, L. A. and van Oers, N. S. C. (2003) T-cell receptor signal transmission: Who gives an ITAM? *Trends Immunol.*, **24**, 554–560.

110. Ardouin, L. *et al.* (1999) Crippling of CD3-ζ ITAMs does not impair T cell receptor signaling, *Immunity*, **10**, 409–420.

111. Collins, M., Ling, V. and Carreno, B. M. (2005) The B7 family of immune-regulatory ligands, *Genome Biol.*, **6**, 223.1–223.7.

112. Fu, S. *et al.* (2004) TGF-$\beta$ induces Foxp3 + T-regulatory cells from CD4 + CD25 – Precursors, *Am. J. Transplant.*, **4**, 1614–1627.

113. Trivedi, H. L. (2007) Immunobiology of rejection and adaptation, *Transplant. Proc.*, **39**, 647–652.

114. Beissert, S., Schwarz, A. and Schwarz, T. (2006) Regulatory T cells, *J. Investig. Dermatol.*, **126**, 15–24.

115. Babensee, J. E. (2008) Interaction of dendritic cells with biomaterials, *Semin. Immunol.*, **20**, 101–108.

116. Park, J. and Babensee, J. E., (2012) Differential functional effects of biomaterials on dendritic cell maturation, *Acta Biomaterialia*, **8**, 3606–3617.

117. Yoshida, M. and Babensee, J. E. (2004) Poly(lactic-co-glycolic acid) enhances maturation of human monocyte-derived dendritic cells, *J. Biomed. Mater. Res. A*, **71A**, 45–54.

118. Babensee, J. E. and Paranjpe, A. (2005) Differential levels of dendritic cell maturation on different biomaterials used in combination products, *J. Biomed. Mater. Res. A*, **74A**, 503–510.

119. Bennewitz, N. L. and Babensee, J. E. (2005) The effect of the physical form of poly(lactic-co-glycolic acid) carriers on the humoral immune response to co-delivered antigen, *Biomaterials*, **26**, 2991–2999.

120. Norton, L. W., Park, J. and Babensee, J. E. (2010) Biomaterial adjuvant effect is attenuated by anti-inflammatory drug delivery or material selection, *J. Control. Release*, **146**, 341–348.

121. Gallucci, S. and Matzinger, P. (2001) Danger signals: SOS to the immune system, *Curr. Opin. Immunol.,* **13**, 114–119.

122. Bianchi, M. E. (2007) DAMPs, PAMPs and alarmins: All we need to know about danger, *J. Leukoc. Biol.,* **81**, 1 –5.

123. Pulendran B (2004) Modulating vaccine responses with dendritic cells and Toll-like receptors, *Immunol. Rev.,* **199**, 227–250.

124. Sokolovska, A., Hem, S. L. and HogenEsch, H. (2007) Activation of dendritic cells and induction of CD4+ T cell differentiation by aluminum-containing adjuvants, *Vaccine,* **25**, 4575–4585.

125. McKee, A. S., Munks, M. W. and Marrack, P. (2007) How do adjuvants work? Important considerations for new generation adjuvants, *Immunity,* **27**, 687–690.

126. Kwissa, M., Pai Kasturi, S. and Pulendran, B. (2007) The science of adjuvants. *Expert Rev. Vaccines,* **6**, 673–684.

127. Reed, S. G., Bertholet, S., Coler, R. N. and Friede, M. (2009) New horizons in adjuvants for vaccine development, *Trends Immunol.,* **30**, 23–32.

128. Wang, X., Uto, T., Akagi, T., Akashi, M. and Baba, M (2008) Poly(gamma-glutamic acid) nanoparticles as an efficient antigen delivery and adjuvant system: Potential for an AIDS vaccine. *J. Med. Virol.,* **80**, 11–19.

129. Sharp, F. A. *et al.* (2009) Uptake of particulate vaccine adjuvants by dendritic cells activates the NALP3 inflammasome, *Proc. Natl. Acad. Sci. USA,* **106**, 870 –875.

130. Demento, S. L., Siefert, A. L., Bandyopadhyay, A., Sharp, F. A. and Fahmy, T. M. (2011) Pathogen-associated molecular patterns on biomaterials: A paradigm for engineering new vaccines. *Trends Biotechnol.,* **29**, 294–306.

131. Jewell, C. M., Bustamante López, S. C. and Irvine, D. J. (2011) *In situ* engineering of the lymph node microenvironment via intranodal injection of adjuvant-releasing polymer particles. *Proc. Natl. Acad. Sci. USA,* **108**, 15745 –15750.

132. Bandyopadhyay, A., Fine, R. L., Demento, S., Bockenstedt, L. K. and Fahmy, T. M (2011) The impact of nanoparticle ligand density on dendritic-cell targeted vaccines, *Biomaterials,* **32**, 3094–3105.

133. Shen, H. *et al.* (2006) Enhanced and prolonged cross-presentation following endosomal escape of exogenous antigens encapsulated in biodegradable nanoparticles, *Immunology,* **117**, 78–88.

134. Tran, K. K. and Shen, H. (2009) The role of phagosomal pH on the size-dependent efficiency of cross-presentation by dendritic cells, *Biomaterials,* **30**, 1356–1362.

135. Reddy, S. T., Swartz, M. A. and Hubbell, J. A. (2006) Targeting dendritic cells with biomaterials: Developing the next generation of vaccines. *Trends Immunol.* **27**, 573–579.

136. Reddy, S. T. *et al.* (2007) Exploiting lymphatic transport and complement activation in nanoparticle vaccines, *Nat. Biotech,* **25**, 1159–1164.

137. Cho, N.-H. *et al.* (2011) A multifunctional core-shell nanoparticle for dendritic cell-based cancer immunotherapy, *Nat. Nanotechnol,* **6**, 675–682.

138. Zhang, Z. *et al.* (2011) Induction of anti-tumor cytotoxic T cell responses through PLGA-nanoparticle mediated antigen delivery, *Biomaterials,* **32**, 3666–3678.

139. Little, S. R. *et al.* (2004) Poly-$\beta$ amino ester-containing microparticles enhance the activity of nonviral genetic vaccines, *Proc. Natl. Acad. Sci. USA,* **101**, 9534 -9539.

140. Hori, Y., Winans, A. M., Huang, C. C., Horrigan, E. M. and Irvine, D. J (2008) Injectable dendritic cell-carrying alginate gels for immunization and immunotherapy, *Biomaterials,* **29**, 3671–3682.

141. Ali, O. A., Emerich, D., Dranoff, G. and Mooney, D. J. (2009) *In situ* regulation of DC subsets and T cells mediates tumor regression in mice, *Sci. Transl. Med.,* **1**, 8ra19–8ra19.

142. Butler, M. O. *et al.* (2011) Establishment of antitumor memory in humans using *in vitro*-educated CD8+ T cells, *Sci. Transl. Med.,* **3**, 80ra34.

143. Kwong, B., Liu, H. and Irvine, D. J. (2011) Induction of potent anti-tumor responses while eliminating systemic side effects via liposome-anchored combinatorial immunotherapy, *Biomaterials,* **32**, 5134–5147.

144. Steinman, R. M., Hawiger, D. and Nussenzweig, M. C. (2003) Tolerogenic dendritic cells, *Annu Rev Immunol,* **21**, 685–711.

145. Lan, Y. Y. *et al.* (2006) 'Alternatively activated' dendritic cells preferentially secrete IL-10, expand Foxp3+CD4+ T cells, and induce long-term organ allograft survival in combination with CTLA4-Ig, *J. Immunol.,* **177**, 5868–5877.

146. Heath, W. R. and Carbone, F. R. (2001) Cross-presentation, dendritic cells, tolerance and immunity, *Annu. Rev. Immunol.,* **19**, 47–64.

147. Jia, L. *et al.* (2008) Expansion of Foxp3-expressing regulatory T cells *in vitro* by dendritic cells modified with polymeric particles carrying a plasmid encoding interleukin-10, *Biomaterials,* **29**, 1250–1261.

148. Rubin, R. and Strayer, D. S. (2008) *Rubin's Pathology: Clinicopathologic Foundations Of Medicine,* Lippincott Williams & Wilkins, Philadelphia, PA.

149. Reichert, W. M. (2007) *Indwelling Neural Implants: Strategies for Contending with the in vivo Environment,* CRC Press, Boca Raton, FL.

150. Babensee J. E., Anderson J. M., McIntire L. V., Mikos A. G. (1998) Host response to tissue engineered devices, *Adv. Drug Deliv. Rev.,* **33**, 111–139.

151. Witte, M. B. and Barbul, A. (1997) General principles of wound healing, *Surg. Clin. North Am.,* **77**, 509–528.

152. Eming, S. A., Krieg, T. and Davidson, J. M. (2007) Inflammation in wound repair: Molecular and cellular mechanisms, *J. Invest. Dermatol.,* **127**, 514–525.

153. Eming, S. A., Brachvogel, B., Odorisio, T. and Koch, M. (2007) Regulation of angiogenesis: Wound healing as a model. *Prog. Histochem. Cytochem.,* **42**, 115–170.

154. Raghow, R. (1994) The role of extracellular matrix in postinflammatory wound healing and fibrosis, *FASEB J.,* **8**, 823 –831.

155. Werner, S. and Grose, R. (2003) Regulation of wound healing by growth factors and cytokines, *Physiol. Rev.,* **83**, 835–870.

156. Conway, E. M., Collen, D. and Carmeliet, P. (2001) Molecular mechanisms of blood vessel growth, *Cardiovasc. Res.,* **49**, 507–521.

157. Risau, W. (1998) Angiogenesis is coming of age, *Circ. Res.,* **82**, 926–928.

158. Risau, W. (1997) Mechanisms of angiogenesis, *Nature,* **386**, 671–674.

159. Griffioen, A. W. and Molema, G. (2000) Angiogenesis: Potentials for pharmacologic intervention in the treatment of cancer, cardiovascular diseases, and chronic inflammation, *Pharmacol. Rev.,* **52**, 237–268.

160. Moulin, V. *et al.* (1996) *In vitro* models to study wound healing fibroblasts, *Burns,* **22**, 359–362.

161. Redd, M. J., Cooper, L., Wood, W., Stramer, B. and Martin, P. (2004) Wound healing and inflammation: Embryos reveal the way to perfect repair, *Philos. Trans. R Soc. Lond. B Biol. Sci.,* **359**, 777–784.

162. Schwartzfarb, E. and Kirsner, R. S. (2012) Understanding scarring: Scarless fetal wound healing as a model, *J. Invest. Dermatol.,* **132**, 260.

163. Carre, A. L. *et al.* (2010) Interaction of wingless protein (Wnt), transforming growth factor-$\beta$1, and hyaluronan production in fetal and postnatal fibroblasts, *Plast. Reconstr. Surg.,* **125**, 74–88.

164. Ferguson, M. W. J. and O'Kane, S. (2004) Scar-free healing: From embryonic mechanisms to adult therapeutic intervention, *Philos. Trans. R Soc. Lond. B Biol. Sci.*, **359**, 839–850.

165. Ishihara, M. *et al.* (2002) Photocrosslinkable chitosan as a dressing for wound occlusion and accelerator in healing process, *Biomaterials*, **23**, 833–840.

166. Picha, G. J. and Drake, R. F. (1996) Pillared-surface microstructure and soft-tissue implants: Effect of implant site and fixation, *J. Biomed. Mater. Res.*, **30**, 305–312.

167. Brauker, J. H. *et al.* (1995) Neovascularization of synthetic membranes directed by membrane microarchitecture, *J. Biomed. Mater. Res.*, **29**, 1517–1524.

168. Padera, R. F. and Colton, C. K. (1996) Time course of membrane microarchitecture-driven neovascularization, *Biomaterials*, **17**, 277–284.

169. Butler, M. J. and Sefton, M. V. (2007) Poly(butyl methacrylate-co-methacrylic acid) tissue engineering scaffold with pro-angiogenic potential *in vivo*, *J. Biomed. Mater. Res. A*, **82A**, 265–273.

170. Eckhaus, A. A., Fish, J. S., Skarja, G., Semple, J. L. and Sefton, M. V.2008) A preliminary study of the effect of poly(methacrylic acid-co-methyl methacrylate) beads on angiogenesis in rodent skin grafts and the quality of the Panniculus carnosus, *Plast. Reconstr. Surg.*, **122**, 1361–1370.

171. Martin, D. C., Semple, J. L. and Sefton, M. V. (2010) Poly(methacrylic acid-co-methyl methacrylate) beads promote vascularization and wound repair in diabetic mice, *J. Biomed. Mater. Res. A*, **93A**, 484–492.

172. Neumann, T., Nicholson, B. S. and Sanders, J. E. (2003) Tissue engineering of perfused microvessels, *Microvasc Res*, **66**, 59–67.

173. Sanders, J. E., Lamont, S. E., Karchin, A., Golledge, S. L. and Ratner, B. D. (2005) Fibro-porous meshes made from polyurethane micro-fibers: Effects of surface charge on tissue response, *Biomaterials*, **26**, 813–818.

174. Sanders, J. E., Lamont, S. E., Mitchell, S. B. and Malcolm, S. G. (2005) Small fiber diameter fibro-porous meshes: Tissue response sensitivity to fiber spacing, *J. Biomed. Mater. Res. A*, **72A**, 335–342.

175. Madden, L. R. *et al.* (2010) Proangiogenic scaffolds as functional templates for cardiac tissue engineering, *Proc. Natl. Acad. Sci. USA*, **107**, 15211 –15216.

176. Greenberg, J. I. *et al.* (2008) A role for VEGF as a negative regulator of pericyte function and vessel maturation, *Nature,* **456**, 809–813.

177. Richardson, T. P., Peters, M. C., Ennett, A. B. and Mooney, D. J. (2001) Polymeric system for dual growth factor delivery, *Nat. Biotech.,* **19**, 1029–1034.

178. Norton, L. W., Koschwanez, H. E., Wisniewski, N. A., Klitzman, B. and Reichert, W. M. (2007) Vascular endothelial growth factor and dexamethasone release from nonfouling sensor coatings affect the foreign body response, *J. Biomed. Mater. Res. A,* **81A**, 858–869.

179. Chamberlain, M. D., Gupta, R. and Sefton, M. V. (2011) Chimeric vessel tissue engineering driven by endothelialized modules in immunosuppressed Sprague-Dawley rats, *Tissue Eng. Part A,* **17**, 151–160.

180. Gupta, R. and Sefton, M. V. (2011) Application of an endothelialized modular construct for islet transplantation in syngeneic and allogeneic immunosuppressed rat models, *Tissue Eng. Part A,* **17**, 2005–2015.

181. Tsuda, Y. *et al.* (2007) Cellular control of tissue architectures using a three-dimensional tissue fabrication technique, *Biomaterials,* **28**, 4939–4946.

182. Au, P., Tam, J., Fukumura, D. and Jain, R. K. (2008) Bone marrow-derived mesenchymal stem cells facilitate engineering of long-lasting functional vasculature, *Blood,* **111**, 4551 –4558.

183. Traktuev, D. O. *et al.* (2009) Robust functional vascular network formation *in vivo* by cooperation of adipose progenitor and endothelial cells, *Circ. Res.* **104**, 1410–1420.

184. Au, P. *et al.* (2008) Differential *in vivo* potential of endothelial progenitor cells from human umbilical cord blood and adult peripheral blood to form functional long-lasting vessels, *Blood,* **111**, 1302–1305.

185. Hibino, N., Breuer, C. and Shinoka, T. (2011) Restoring blood vessels, in *Tissue Engineering Regenerative Medicine, Harold S. Bernstein ed.* 211–220, Humana Press, Totowa, NJ.

186. Hibino, N. *et al.* (2011) Tissue-engineered vascular grafts form neovessels that arise from regeneration of the adjacent blood vessel, *FASEB J.,* **25**, 2731–2739.

187. Roh, J. D. *et al.* (2010) Tissue-engineered vascular grafts transform into mature blood vessels via an inflammation-mediated process of vascular remodeling, *Proc. Natl. Acad. Sci. USA,* **107**, 4669–4674.

188. Gerald James Bakeine Thesis for PhD Nanotechnology .pdf. at http://www.openstarts.units.it/dspace/bitstream/10077/3268/1/

Gerald%20James%20Bakeine%20Thesis%20for%20PhD%20Nanotechnology%20.pdf

189. Midwood, K. S., Williams, L. V. and Schwarzbauer, J. E. (2004) Tissue repair and the dynamics of the extracellular matrix, *Int. J. Biochem. Cell. Biol.*, **36**, 1031–1037.

190. Cox, T. R. and Erler, J. T. (2011) Remodeling and homeostasis of the extracellular matrix: Implications for fibrotic diseases and cancer, *Dis. Model. Mech.*, **4**, 165–178.

191. Allman AJ, *et al.* (2001) Xenogeneic extracellular matrix grafts elicit a TH2-restricted immune response, *Transplantation,* **71**, 1631–1640.

192. Allman, A. J., McPherson, T. B., Merrill, L. C., Badylak, S. F. and Metzger, D. W. (2002) The Th2-restricted immune response to xenogeneic small intestinal submucosa does not influence systemic protective immunity to viral and bacterial pathogens, *Tissue Eng.*, **8**, 53–62.

193. Vert, M. (2007) Polymeric biomaterials: Strategies of the past vs. strategies of the future, *Prog. Polym. Sci.*, **32**, 755–761.

194. Lucke, A., Kiermaier, J. and Göpferich, A. (2002) Peptide acylation by poly(alpha-hydroxy esters), *Pharm. Res.*, **19**, 175–81.

195. Ravi, S. and Chaikof, E. L. (2010) Biomaterials for vascular tissue engineering, *Regen. Med.*, **5**, 107.

196. Lin, A. S. P., Barrows, T. H., Cartmell, S. H. and Guldberg, R. E. (2003) Microarchitectural and mechanical characterization of oriented porous polymer scaffolds, *Biomaterials,* **24**, 481–489.

# Chapter 6

# Scaffold Vascularization

**Lindsay E. Fitzpatrick,[a] Alexandra Lisovsky,[a] Ema C. Ciucurel,[a,b] and Michael V. Sefton[a,b]**
[a]*Institute of Biomaterials and Biomedical Engineering, University of Toronto, Toronto, Canada*
[b]*Department of Chemical Engineering and Applied Chemistry, University of Toronto, Toronto, Canada*
michael.sefton@utoronto.ca

Scaffold vascularization is one of the main obstacles in engineering complex tissues at clinically relevant scales. Despite the considerable effort that has been devoted to addressing this issue, many challenges still exist in creating a mature and functional vasculature and doing so quickly enough to avoid hypoxia and cell death. This chapter reviews contemporary strategies for generating vascularized tissues; including growth factor delivery, cell transplantation, scaffold prevascularization, decellularized scaffolds, bio-functionalized materials, and microfabrication techniques. These strategies are considered for their effectiveness in rapidly forming mature and functional vascular networks that persist over time. While significant progress has been made in the effort to create vascularized tissue

Lindsay E. Fitzpatrick and Alexandra Lisovsky contributed equally to this work.

*Scaffolds for Tissue Engineering: Biological Design, Materials, and Fabrication*
Edited by Claudio Migliaresi and Antonella Motta

ISBN 978-981-4463-20-1 (Hardcover), 978-981-4463-21-8 (eBook)
www.panstanford.com

engineering constructs, many challenges remain. A major concern is that the time required to form functional and perfusable blood vessel networks using current strategies is on the order of weeks, which is too long as most cells cannot survive without a vascular supply for such extended periods of time. Future efforts must focus on accelerating the formation of functional (i.e., perfused) vasculature and controlling how the vasculature of engineered tissues are remodeled and integrated with host vasculature *in vivo*.

## 6.1 Introduction

Evolution of multicellular organisms required the development of a circulatory system capable of distributing gases, nutrients, and signaling molecules. Due to the poor diffusional distance of oxygen and nutrients, all cells within the human body are within 100–200 μm of a blood vessel with few exceptions (e.g., cells in the cornea, cartilage, ligaments).[1] Like the evolution of larger organisms, engineered tissues of clinically relevant sizes require the incorporation of a vascular network capable of supplying seeded or recruited cells deep within the scaffold a means of exchanging gases, nutrients, signaling molecules, and waste. This chapter focuses on the methods currently being developed to vascularize biomaterial scaffolds.

## 6.2 The Native Vascular Network

The vascular network in mammals is a hierarchical network of vessels that circulate the blood to and from the heart through the tissues. For most vessels, there are three layers (Fig. 6.1). The intima is the innermost layer and is a monolayer of endothelial cells (EC) supported by a basement membrane. EC interface with blood and have a number of functions, which include the regulation of capillary permeability, hemostasis, and leukocyte recruitment and translocation.[2,3] The middle layer is the media, which is composed of layers of contractile vascular smooth muscle cells (SMC) and elastin. The adventitia is the outermost layer and contains fibroblasts, mast cells, nerve endings, and, in large vessels such as the aorta, small capillaries.[4]

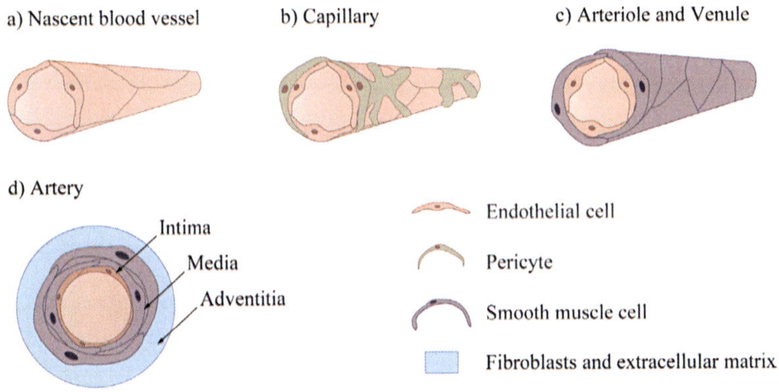


**Figure 6.1** Structure of nascent and mature blood vessels. (a) EC alone assembled into tube-like structures (nascent blood vessel). (b) Capillaries are the smallest type of blood vessels (only 5–10 μm in diameter), and are composed of a single layer of EC supported by a basement membrane of matrix proteins, and an incomplete covering of pericytes that stabilize the vessel. (c) Compared to capillaries, arterioles, and venules are invested with a higher number of mural cells. Arterioles have a thinner media layer than arteries, and regulate the blood flow between the arteries and the capillaries. Venules (and veins) have thin walls and fewer SMC, relative to arteries. (d) Larger blood vessels have three specialized layers: intima (containing EC), media (containing SMC), and adventitia (containing fibroblasts). Arteries have a thick media composed mainly of SMC, which provide mechanical stability to the vessel wall, and allow the vessels to vasodilate and vasoconstrict in order to regulate blood flow and pressure.

In the embryo, the formation of the vascular network starts early on during development, to match the oxygen, nutrients, and waste removal requirements of the growing organism. During vasculogenesis, EC differentiate from mesodermal precursor cells, with vascular endothelial growth factor receptor 2 (VEGFR2) being one of the earliest markers of the EC lineage. The cells proliferate, differentiate, aggregate, and form blood islands. The blood islands then fuse together and ultimately form the primary vascular plexus, a honeycomb-like structure of primitive blood vessels.[5–7] Further remodeling of this primitive vascular structure occurs through either sprouting angiogenesis or intussusception. The arterial–venous differentiation is genetically predetermined, and contrary to what was initially assumed, it does not rely on the start of blood flow. The bidirectional signaling between ephrinB2 (expressed by arteries) and its receptor EphB4 (expressed by veins) plays an important role in defining the boundaries between the arterial and venous sides. A number of genes upstream of ephrinB2/EphB4 are involved in the signaling that ultimately results in the specification of arteries and veins. Sonic hedgehog (Shh), VEGF, neuropilin-1, notch-1, and its ligand delta-like-4 (DLL-4) are among the genes upstream of ephrinB2 and are known to be involved in establishing the arterial cell fate, while neuropilin-2 and COUP-TFII (chicken ovalbumin upstream promoter-transcription factor II) are upstream of EphB4 and are involved in determining the venous identity.[8,9]

### 6.2.1 Angiogenesis: The Formation of New Blood Vessels

In adults, new blood vessels are typically formed through a process called angiogenesis, in which new microvessels sprout from pre-existing vessels. While angiogenesis is not the only mechanism of vascularization, it is the most extensively studied. Two other mechanisms have been described: post-natal vasculogenesis, in which vessels are formed de novo by recruited bone marrow-derived progenitor cells, and intussusception, in which the lumen of an existing blood vessel divides to form two separate vessels.[10]

The process of forming new blood vessels follows the same general pattern in all tissue types. However, the final density of the vascular network depends on the specific requirements of the tissue undergoing vascularization. Tissues with high metabolic requirements, such as muscle, require a rich vascular network to provide sufficient amounts of oxygen, nutrients, and waste removal to match the metabolic demand of the active muscle tissue. In addition, the structure of the blood vessels also reflects the function of the organ or tissue that these vessels supply. While most tissues are supplied by continuous, non-fenestrated capillaries, continuous fenestrated capillaries are present in organs and tissues involved in filtration or secretion, such as the endocrine glands, the gastrointestinal mucosa, or the glomerulus.[11,12] Fenestrations are transcellular pores that extend through EC and facilitate the transport of water and small molecules across the endothelium. Other organs, such as the liver, contain discontinuous capillaries, which have even larger fenestrations (100–200 nm).[11,12]

Numerous conditions can induce angiogenesis including tissue hypoxia, low pH, hypoglycemia and inflammation.[13] The angiogenic process is divided into three phases: (1) vasodilation and increased vessel permeability, (2) EC proliferation and migration, (3) vessel maturation. The first stage of angiogenesis is marked by nitric oxide (NO)-mediated vasodilation, and increased permeability of the blood vessels, which are stimulated by increased levels of VEGF.[14] The increased permeability allows plasma proteins to extravasate into the surrounding tissue to form a provisional matrix for EC migration. This process is tightly regulated by the angiopoietin (Ang)/Tie system. Ang-1 and Ang-2 both bind the Tie2 receptor; however, Ang-1 acts to stabilize blood vessels, while Ang-2 antagonizes Ang-1 and is associated with vessel destabilization and either vessel growth (in the presence of VEGF) or vessel regression.[15–17] Prior to migration, EC loosen their inter-endothelial contacts and release from the supporting mural cells, further destabilizing the blood vessel.[18] Proteinases, such as matrix metalloproteinases (MMPs) and plasminogen activators, degrade the extracellular matrix (ECM) allowing EC migration and liberate matrix-bound pro-angiogenic growth factors (GF), including VEGF and basic fibroblast growth factor (bFGF).[18]

To prevent en masse migration of EC toward an angiogenic signal, and potential dissolution of a pre-existing vessel, EC compete to lead the migrating sprout along the gradient of angiogenic factors.[19] This leading cell, called a tip cell, is selected through a spatial cell patterning mechanism involving VEGF and notch pathway ligand DLL-4. Adjacent to the tip cell are stalk cells, which proliferate and elongate the stalk. An in-depth look at tip and stalk cell selection and function is beyond the scope of this chapter; however, it is reviewed in detail elsewhere.[20–22] As the tip cell migrates and the stalk elongates, the stalk cells form a lumen and stabilize the sprouting vessel by reestablishing tight inter-endothelial junctions, laying down a basement membrane, and recruiting pericytes (mediated by platelet-derived growth factor B [PDGF-B], Ang-1, transforming growth factor beta [TGF-$\beta$], and other factors).[19,21] To become a functional, perfused vessel, the tip cell of one sprout must fuse with a neighboring sprout, a process called anastomosis, that is thought to be facilitated by macrophages.[23] As the new vessel matures, the EC resume their quiescent phalanx phenotype.[21]

Angiogenesis is tightly regulated *in vivo* by maintaining a balance between pro-angiogenic and angiostatic (also called anti-angiogenic) stimuli.[18] Angiogenesis inhibitors (e.g., angiostatin, endostatin, anti-thrombin III, interferon-$\beta$, leukemia inhibitory factor, platelet factor 4, thrombospondin-1) can suppress EC proliferation and migration, or lumen formation.[18,24] Other molecules (TGF-$\beta$1, tumor necrosis factor alpha [TNF-$\alpha$]) can both stimulate or inhibit angiogenesis.[18] In order to form and maintain mature, functional vessels, a balance must exist between factors that induce and inhibit angiogenesis. For example, in cancer the balance is typically shifted too far in favor of angiogenesis, which results in the formation of tortuous, leaky, immature, vessels.[10,21,25] Often times, the vascular network formed within tissue-engineered constructs is also leaky and lacks functionality, due to the absence of an adequate balance or spatio-temporal coordination between signals initiating angiogenesis and signals promoting vessel maturation in the implant area.

An alternative mechanism for forming new blood vessels is vasculogenesis. Previously, it was thought that vasculogenesis occurred only during development; however, recent studies have demonstrated the role of vasculogenesis in adult vascularization (although the extent of contribution is unclear).[13,26–28] During vasculogenesis, blood vessels are formed de novo through the recruitment and differentiation of a heterogeneous (and controversial) family of bone marrow-derived cells called endothelial progenitor cells (EPC). Once recruited to a tissue, EPC proliferate, differentiate, and incorporate themselves into immature vessels or form new vessels.[13,29]

## 6.3 Vascularization of Biomaterial Scaffolds

Vascularization of biomaterial scaffolds is currently a major barrier to creating viable, complex-engineered tissues, such as liver, heart, bone, and adipose, at a clinically relevant scale. There are a few key challenges in creating a vascularized tissue. The first is creating the vascular network within the scaffold. The second is connecting (anastomosing) the scaffold vasculature with the host vasculature so that the scaffold is perfused *in vivo*. Finally, the newly formed vascular network has to mature, be functional, and persist over time. Researchers are currently using a variety of approaches to achieve scaffold vascularization (Fig. 6.2). GF delivery (Section 6.3.1) from biomaterial scaffolds is one of the first methods used to induce vessel formation *in vivo*. Single or multiple GF can be released from the scaffold in a controlled and sustained manner to encourage the ingrowth of host vessels within the scaffold.

Another approach is vascular cell transplantation (Section 6.3.2), where EC alone or in combination with support cells are delivered in a bulk scaffold or in small microtissues (modular tissue engineering). Gene delivery to overexpress angiogenic factors in transplanted cells has also been used. Cells can be cultured and transplanted using scaffold-free approaches or, conversely, scaffolds alone can be used to mobilize and recruit endogenous cells to vascularize the scaffold *in situ* (Section 6.3.3).

A common issue with both GF and cell transplantation approaches is the lack of immediate connection to the host vasculature and perfusion. Scaffold prevascularisation (Section 6.3.4) and decellularization strategies (Section 6.3.5) aim to provide a solution to this problem. In some cases, biomaterials themselves (without the addition of GF or cells) were also shown to induce angiogenesis (Section 6.3.6) and microfabrication methods can be used to create pre-defined, hierarchical vascular networks *in vitro* (Section 6.3.7).

## 6.3.1 GF Delivery

### 6.3.1.1 Single GF delivery

One of the primary approaches to vascularize tissue engineering scaffolds is delivering angiogenic GF, such as VEGF, bFGF, and PDGF, to stimulate the ingrowth of host vasculature into the scaffold. While initial attempts to induce scaffold vascularization used single GF delivery systems, recent efforts have focused on dual or multiple GF delivery to form and maintain a vascular network.

GF have narrow therapeutic windows and short half-times *in vivo*, and in order to induce and maintain vascularization, the supply of exogenous GF must be sustained over several weeks to prevent vessel regression and to allow vessel maturation.[31] Bolus injection of soluble GF into the systemic circulation or directly into an ischemic zone have generally failed to achieve the prolonged, therapeutically relevant concentrations necessary to produce a lasting vascular network.[32,33] In addition, systemic exposure may lead to adverse side

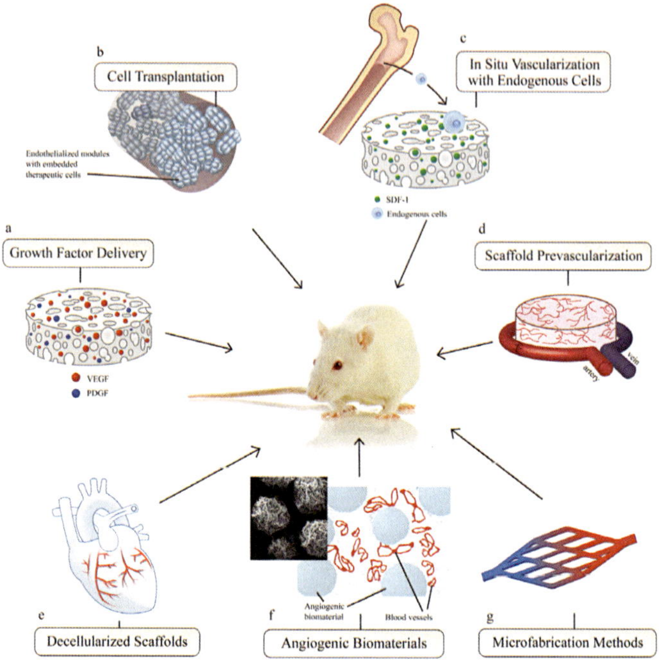


**Figure 6.2** Scaffold vascularization strategies. (a) Delivery of single or multiple angiogenic GF to stimulate the ingrowth of host vasculature into the scaffold. (b) Transplantation of EC (alone or in combination with support cells). The EC are expected to re-assemble *in vivo*, participate in blood vessel formation, and connect to the host vasculature. Support cells stabilize the nascent capillary-like structures. (c) Bio-functionalized materials are used to mobilize and capture endogenous cells at the implant site and stimulate de novo blood vessel formation. (d) *In vivo* prevascularization strategies use the host body as a bioreactor to vascularize the scaffold, prior to relocating the construct to its target implant site. In the case of arteriovenous loops, the prevascularization step takes place in a chamber enclosing an arteriovenous graft. (e) Decellularized scaffolds provide a natural template to recreate the architecture of the vasculature network. After decellularization, organs are reseeded with EC to repopulate the vascular tree and with functional cells to create a functional tissue engineered construct. (f) Upon implantation, some biomaterials induce angiogenesis by themselves, without the addition of any cells or GF. (g) Using microfabrication techniques, predefined, hierarchical pseudo-vasculatures are created (although mostly for *in vitro* applications thus far).

effects such as hemorrhage, hypotension, and the vascularization of undesired sites.[33,34] Consequently, to ensure patient safety and efficacy of treatment, the dose, location, and duration of exogenously administered GF must be tightly controlled. The incorporation of GF into slow-releasing biomaterial scaffolds provides the ability to have sustained local delivery of GF to the site of interest. The pharmacokinetics of protein release can be controlled through the polymer properties (e.g., polymer composition, porosity, degradation kinetics) and the method by which the GF is incorporated into the scaffold (e.g., absorbed, encapsulated, or tethered).

One of the simplest methods to incorporate a GF into a polymer scaffold is to impregnate the scaffold with an aqueous solution containing the desired protein. Gelatin hydrogels loaded with bFGF were used to promote vascularization of the thigh muscle following the excision of the femoral artery in a rabbit model of hind limb ischemia.[35] Angiographic assessment with laser doppler perfusion imaging demonstrated that single intramuscular injection of the bFGF-loaded hydrogel was sufficient to improve tissue blood flow four weeks post-surgery, suggesting the therapeutic potential of the bFGF-containing gelatin scaffold.

To prolong the release of a GF of interest, the protein can be encapsulated into microspheres that are then embedded into the bulk of the scaffold, and this method has been used to deliver a variety of GF.[36,37] For example, Ennet and colleagues used this method to deliver VEGF by pre-encapsulating it in poly(lactide-co-glycolide) (PLG) microspheres prior to incorporation into PLG scaffolds.[36] The encapsulation method resulted in a smaller burst and a slower, prolonged protein release, and the release kinetics were further adjusted by varying the polymer composition and microsphere size. The concentration of VEGF within the tissue infiltrating and surrounding the VEGF-releasing scaffold was present at physiologically relevant concentrations (more than 10 ng/mL) for up to 21 days, with negligible release into the systemic circulation.[36] Similarly, when implanted *in vivo*, composite alginate scaffolds containing encapsulated bFGF increased the von Willebrand factor (vWF)-positive microvessel density, compared to control scaffolds.[37]

Alternatively, GF can be immobilized onto the scaffold to promote desired cell-material interactions.[38–40] GF can be modified to facilitate covalent binding to the scaffold, thereby prolonging their biological activity in the tissue. Ehrbar and collaborators created a variant of VEGF (TG-$VEGF_{121}$) containing a transglutaminase (TG) substrate sequence that allowed it to be directly bound to a fibrin scaffold by spontaneously crosslinking with fibrinogen during factor XIII-mediated fibrin polymerization.[31,38] The TG-$VEGF_{121}$ was released with the gradual degradation of the fibrin gel by enzymes (e.g., plasmin or MMPs) secreted by local cells; providing a sustained, local, low-level release of VEGF. In comparison, the passive release of freely diffusible wild-type VEGF from fibrin resulted in an initial burst release and a shorter sustained delivery (the fibrin gel was almost completely depleted two weeks post-implant). The active release of TG-$VEGF_{121}$ from the fibrin gels induced the formation of blood vessels that were stabilized by SMC at three weeks post-implant. However, the newly formed vessels had regressed by six weeks despite the presence of the SMC at three weeks.[38] The regression was attributed to the absence of a physiological demand, although it could also indicate that administration of a single GF was insufficient for stable vessel development. This is a common issue when exogenously promoting vessel formation.

Other ways to immobilize GF include using carbodiimide chemistry, which covalently links the protein to the scaffold through an amide bond,[40] or physical entrapment of the GF within the crosslinked scaffold.[41] Regardless of the method, immobilization is expected to provide persistent guidance cues for cell behavior and migration,[41] similar to what is observed in the ECM during angiogenesis. However, it also prevents the internalization of the GF upon binding to its receptor, which can eventually lead to the desensitization of the cell to the GF signal.[40] Moreover, certain GF/receptor complex signaling cascades are dependent on internalization[42] and consequently are not suitable for this method of delivery.

More complex scaffold designs exploit synthetic components to direct cell invasion, matrix degradation, and controlled GF release to induce vasculature growth *in vivo*. Phelps *et al.* created a polyethylene glycol (PEG) hydrogel using photo-crosslinkable PEG diacrylate (PEGDA) monomers conjugated to MMP-sensitive peptides, integrin-binding peptides (arginine–glycine–aspartic acid; RGD) and VEGF.[32] This elegant synthetic system encouraged vascularized tissue invasion by allowing cells to bind and remodel the scaffold, while presenting pro-angiogenic signals throughout the matrix. There were higher microvessel densities at two and four weeks, compared to hydrogels that did not contain all three components (i.e., MMP-cleavage sites, RGD, and VEGF), by microCT (indicating also that vessels were connected to the host). Furthermore, in a mouse hind limb ischemia model, injection of the hydrogel in ischemic muscle improved the perfusion of the leg and foot of the ischemic limb seven days after vessel ligation, compared to the negative control.[32]

#### 6.3.1.2 Multiple GF delivery

While single factor delivery typically results in the formation of a branched network of vessels, the vessels often fail to mature and stabilize. A reason for this outcome is that a single factor is likely incapable of inducing the full cascade of events that occur in angiogenesis from sprouting to maturation of the newly formed vessel. Hence, a strategy for promoting mature vessels is to initiate angiogenesis with a pro-angiogenic factor such as VEGF, then promote vessel stabilization by PDGF-mediated SMC and pericyte recruitment.[43–45]

In one study, VEGF and PDGF were included in a calcium phosphate-crosslinked alginate gel to induce mature vessel formation. The alginate scaffold released PDGF more slowly than VEGF, likely due to the difference in affinity of the two GF to the alginate. VEGF had a high release rate during the first 12 days after which it diminished, and PDGF had a slower but sustained release up to 30 days.[44] *In vivo*, the dual delivery system did not increase the vessel density in a myocardial infarct model, compared to VEGF alone. However, it did increase the density of vessels staining positive for smooth muscle actin (SMA; a marker for SMC and pericytes), indicative of greater vessel remodeling and maturation. Cardiac function was improved compared to single GF therapy.[44]

A more controlled, spatially compartmentalized system for sequential VEGF and PDGF delivery was developed in the Mooney laboratory to pattern blood vessel formation and maturation.[43] A bilayered, porous PLG scaffold was designed to deliver the GF with a spatio-temporal gradient. The outer layer (first to be degraded) contained only VEGF, while the second, inner layer contained both VEGF (at a lower concentration) and microencapsulated PDGF. The VEGF was released quickly and created a concentration gradient in the surrounding tissue and within the scaffold that induced the infiltration of sprouting vessels. Encapsulation of the PDGF within PLG microspheres in the inner layer delayed its release, and created a localized gradient of PDGF within the scaffold, concentrated mainly within the inner layer. The porous structure of the scaffold allowed for tissue invasion *in vivo*. When implanted into mouse ischemic hind limbs, the bilayered, dual delivery system yielded a slightly lower blood vessel density, but a significantly more mature vascular network, characterized by large vessels and association of SMA-positive SMC (as compared to single GF systems or empty scaffolds).[43]

One caveat that is relevant to these studies is that there is an antagonistic relationship between VEGF and PDGF; when both VEGF and PDGF are used VEGF activation of VEGFR2 suppresses PDGF-R$\beta$ signaling in SMC, disrupting SMC/pericyte function during neovascularization.[46] How these observations relate to the vascularization results reported above is not clear.

## 6.3.2 Cell Delivery and Transplantation

### 6.3.2.1 Scaffolds seeded with cells

Transplantation of vascular cells (EC alone or in combination with support cells) to facilitate the vascularization of the scaffold is another common approach. Typically, EC are seeded onto the scaffold, which can be either a natural or synthetic biomaterial (or a decellularized scaffold, see Section 6.3.5), and these endothelialized constructs are then used for transplantation. The EC are expected to re-assemble *in vivo*, participate in blood vessel formation (i.e., vasculogenesis), and connect to the host vasculature. Other cell types can be co-cultured and transplanted along with EC to support EC survival and/or to stabilize the nascent capillary-like structures.

A quiescent EC layer is needed to create a non-thrombogenic surface.[47,48] EC produce antithrombotic factors such as thrombomodulin, heparan sulfate, NO, and prostacyclin.[47] Moreover, EC express tissue-type plasminogen activator, urokinase-type plasminogen activator, and plasminogen activator inhibitor 1, and play a key role in balancing fibrinolytic versus coagulation events. An intact, non-activated endothelium also provides a non-adherent surface for the platelets and leukocytes present in the blood, whereas activated EC upregulate the expression of adhesion molecules such as E-selectin, intercellular adhesion molecule 1 and vascular cell adhesion molecule 1 (VCAM-1), and facilitate the binding of circulating leukocytes to the endothelialized surface.[47]

Initial implant studies revealed that EC rapidly undergo apoptosis when transplanted without a supportive scaffold. In a study by Kraehenbuehl and collaborators, only 3% of the EC injected directly into the site of ischemia engrafted, due to extensive cell death.[49] Seeding the EC onto a biomaterial substrate (or embedding the EC within the biomaterial scaffold) can help prevent anoikis (programmed cell death induced by lack of adhesion sites), but the choice of the biomaterial is a factor. For example, seeding human umbilical vein EC (HUVEC) on collagen eliminated anoikis, but the cells still had limited survival *in vivo* after subcutaneous implantation in an immunodeficient mouse model. On the other hand, when the EC were seeded onto fibronectin (another ECM protein)-coated collagen instead of collagen alone, HUVEC survival was significantly improved (30–45% over the first three weeks) and the number of HUVEC-lined blood vessels increased by 100% at days 7 and 14 post-implantation.[50]

As an alternative to the use of ECM-based substrates, synthetic polymers have also been designed to promote EC adhesion, survival, migration, and organization into tube-like structures. MMP-responsive PEG hydrogels with thymosin $\beta 4$ (a small bioactive peptide with pro-angiogenic properties) and RGD sequences, created a 3D environment conducive for HUVEC attachment and survival, and induced vascular-like network formation *in vitro*, although the potential of this system to promote EC survival and increased vascularization remains to be demonstrated *in vivo*.[49] Apotosis in the absence of supporting cells is an issue, regardless of the scaffold.

Angiogenesis does not only involve EC, but is the result of a well-coordinated sequence of interactions among EC, supporting cells, and the surrounding ECM. Consequently, co-culture systems with one or multiple support cell types in addition to EC are a closer mimic of the *in vivo* environment. Several groups reported the development of a functional microcirculation *in vivo* when using this approach, as reviewed elsewhere.[51,52] Seminal work by the Jain group demonstrated the formation of blood vessels and their persistence over at least 1 year *in vivo* when HUVEC and 10T1/2 cells were embedded together in a fibronectin-collagen gel and implanted in a transparent window, immune-compromised mouse model. The constructs prepared with HUVEC alone showed minimal perfusion and no cells were present beyond 60 days. In contrast, when HUVEC and mesenchymal precursor cells were implanted together, the 10T1/2 cells provided an abundant source of mural cells to stabilize the HUVEC-lined blood vessels that were connected to the host vasculature.[53]

The West group also exploited the cellular interactions between EC and 10T1/2 cells in a synthetic PEG hydrogel containing cell attachment (RGD) and protease (MMP)-sensitive substrates. The co-cultured cells actively remodeled the hydrogel, secreted their own ECM proteins, and formed tube-like structures that were stable for at least 28 days *in vitro*.[54]

Levenberg and collaborators used a tri-culture system to create a vascularized muscle tissue, with EC, myoblasts and embryonic fibroblasts seeded on porous biodegradable polymer scaffolds composed of 50% poly(L-lactic acid) (PLLA) and 50% polylactic-glycolic acid (PLGA).[55] *In vitro*, this tri-culture system spontaneously formed tubular structures within the scaffold, with some of the embryonic fibroblasts becoming SMA-positive over time (suggestive of their differentiation into SMC), and co-localizing with the EC to stabilize the newly formed vascular structures. The skeletal muscle constructs were cultured *in vitro* for two weeks and then implanted either subcutaneously or intramuscularly in immunocompromised mouse and rat animal models. Evaluation of the implants two weeks after surgery showed the formation of blood vessels within the construct, with 41% of the blood vessels of human origin being perfused with lectin following lectin injection through the tail vein of the animals, thus confirming their connection to the host vasculature. Control injections with EC alone did not result in vessel formation, again presumably due to apoptosis. Furthermore, the implanted muscle construct continued to differentiate and mature over time, with implanted myotubes elongating, becoming multinucleated and expressing myogenin, a muscle-specific marker.[55]

### 6.3.2.2 Modular tissue engineering

Modular tissue engineering was pioneered by the Sefton group, and provides an alternative platform to creating endothelialized bulk scaffolds. Rather than seeding EC within one large scaffold, modular tissue engineering uses small, EC-coated constructs (modules, ~1 mm long and 0.5 mm in diameter, typically made out of collagen) that randomly pack together within a larger chamber to form a 3D construct (Fig. 6.3).[56] Vascular support cells or therapeutic cells of interest (e.g., hepatocytes) can be embedded during fabrication, prior to EC seeding.[56] The network of EC-lined void spaces formed in between the modules is amenable to perfusion, forming a pseudo-vasculature *in vitro*.[57] Furthermore, the small dimension of the modules (less than 400 μm following contraction by EC) is within the diffusional distance of oxygen and consequently, in individual modules even high densities of embedded cells did not experience the hypoxia that is often observed in thick cell constructs.[58] The 3D tissues assembled using this method have an intrinsic vasculature, are scalable, have uniform cell density throughout the construct, and allow for the mixing of different cell types within the modules or by using modules with different types of embedded cells.

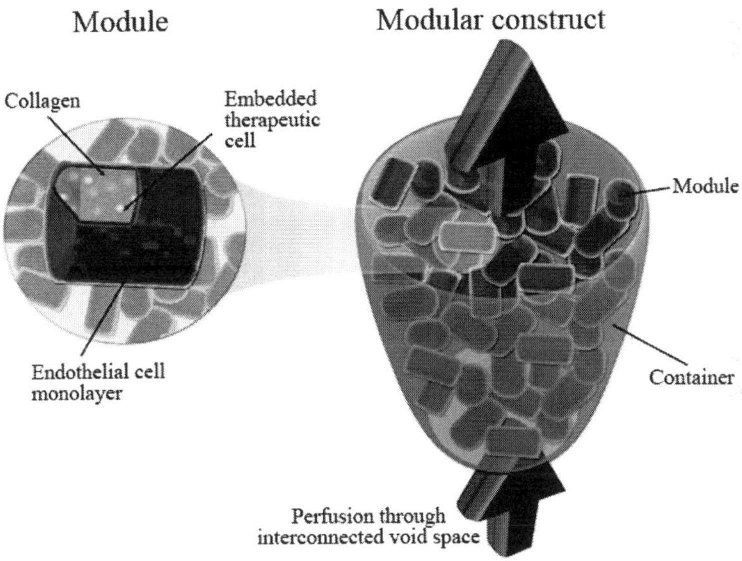


**Figure 6.3** Modular tissue engineering. Submillimeter-sized collagen gels containing therapeutic cells are coated with EC and packed together into a larger structure. The network of void spaces formed in between the endothelialized modules is amenable to perfusion.

When collagen modules coated with rat aortic EC (RAEC) were transplanted into the omental pouch of allogeneic rats with immunosuppressive drug treatment, the RAEC started to form blood vessels within the first seven days.[59] Over time, the blood vessels matured, were supported by endogenous SMC, and anastomosed with the host vasculature as shown by the accumulation of erythrocytes. Although these vessels persisted for at least 60 days, some of them were leaky, as shown by microCT perfusion studies. Moreover, a robust inflammatory response was noted around the implant.[59] When bone marrow-derived MSC were added as support cells to the endothelialized modules, the number of blood vessels did not change, but resulting blood vessels were more stable (i.e., less leaky) compared to vessels formed by RAEC alone.[60] Encapsulated MSC became SMA positive and lined the EC layer suggesting their differentiation into pericyte-like cells. In addition, the presence of MSC decreased the total number of macrophages and shifted macrophages to what was perhaps a more tissue repair phenotype, which further improved RAEC survival and consequently, vascularization.[60]

Modular tissue engineering was also tested in islet transplantation and in cardiac tissue engineering.[61,62] In the case of islet transplantation, vessel density in the implant area was significantly increased, and a trend toward increased islet viability was observed for islets implanted in modules coated with EC as compared to islets implanted in collagen modules without EC at 21 days. However, the endothelialization of the modules did not bring any significant benefit in terms of islet function.[61] In the case of the modular cardiac constructs, RAEC-lined collagen modules supplemented with Matrigel™ were embedded with a neonatal rat heart cell population enriched in cardiomyocytes, and the modules were gathered to form a sheet-like, porous structure. The constructs were electrically responsive and presented cardiac markers at inter-module junctions, suggesting the potential of this approach for cardiac tissue engineering.[62]

Materials other than collagen have been used to fabricate the modules. For example, the surface composition of the modules was altered by coating them with other ECM proteins, such as fibronectin,[50] or the modules were made mechanically stronger by replacing collagen with a photo-crosslinkable, cell adhesive synthetic polymer, such as poloxamine-polylysine acrylate.[63] In addition, other methods such as micromolding were developed to produce similar modular constructs, demonstrating the ease of module manufacturing and the potential of this technology for scale-up and construction of larger 3D tissues.[64]

Although fabrication of modular tissues from different cell types is easily feasible, fabricating modular tissues with specific micro-architectures preserved into larger tissue constructs remains a challenge. The Khademhosseini group introduced a sequential assembly method that allows for better control over the relative spatial arrangement of the different building blocks.[65] The group used PEG micro-gels fabricated by sequential photo-crosslinking through two overlaying masks to generate a concentric design that emulated the EC and SMA vascular layers, with HUVEC encapsulated in the internal layer and SMC encapsulated in the external layer of the micro-gels. Several micro-gels were then reassembled into a tube-like structure upon immersion in mineral oil.[65] However, the cells were immobilized in a non-degradable PEG gel within which cells cannot proliferate or migrate. To make this approach applicable to tissue engineering, other scaffolds with chemical and mechanical properties that better mimic the natural ECM, need to be considered. The remodeling of the initial tissue architecture *in vivo* also needs to be explored for this system.

### 6.3.2.3 Genetically modified GF-producing cells

Transplanted cells (EC or other) can be genetically modified to overexpress angiogenic factors. This approach obviates some limitations associated with the classical scaffold-GF delivery strategy, such as the short half-life of GF *in vivo* and GF degradation during scaffold manufacturing, while, ideally, the genetically modified cells can continuously produce angiogenic factors at constant rates. However, obtaining pharmacologically useful rates is problematic.[66,67]

The Laurencin group combined EC transplantation with *ex vivo* gene transfer for bone tissue engineering applications.[68] Human adipose-derived mesenchymal stromal cells (adMSC) were isolated and transduced with an adenovirus to express VEGF. The co-transplantation of EC and adMSC overexpressing VEGF resulted in significant vascularization of the 3D PLGA sintered microsphere scaffolds implanted subcutaneously in severe combined immunodeficient (SCID) mice. The scaffolds seeded with VEGF-transduced adMSC and EC resulted in the highest number of blood vessels two and three weeks post-implantation, compared to the blank scaffolds, as well as the scaffolds seeded with either EC or adMSC alone, or with co-cultured EC and non-transduced adMSC.

The group speculated that VEGF-transduced adMSC increased migration of endogenous EC into the scaffold and proliferation of transplanted EC, and enhanced the differentiation of EC progenitors present within the adMSC population.[68]

Soker and colleagues suspended VEGF-transfected myoblasts (using Lipofectamine) in collagen and injected them subcutaneously into nude mice.[69] While VEGF expression was noted at one week but was absent at three weeks, increased microvessel density (by vWF staining) was observed. The resulting muscle tissue progressively increased in volume up to four weeks post-implant, a volume that was preserved for the duration of the study (eight weeks). This tissue expressed typical muscle markers (desmin and sarcomeric tropomyosin) and lacked connective tissue formation.[69]

Yang and colleagues used a biodegradable polymer-DNA nanoparticle system based on poly($\beta$-amino esters) to deliver VEGF cDNA into human MSC and hESC-derived cells (hESdC).[70] Scaffolds seeded with VEGF-transfected MSC and implanted subcutaneously into athymic mice led to 2–4 fold increase in blood vessel density at two and three weeks after the implant compared to controls (acellular scaffolds, scaffolds seeded with luciferase transfected cells, and scaffolds seeded with cells transfected with VEGF using Lipofectamine as transfection reagent). A similar trend was noted for the transfected hESdC group. The same study reported enhanced angiogenesis four weeks after intramuscular injection of the VEGF-transfected MSC in a mouse ischemic hind limb model, suggesting that transient, non-viral gene delivery of angiogenic factors has great potential for therapeutic angiogenesis applications (at least in hind limb muscle), while providing a safer alternative to viral gene delivery.[70]

#### 6.3.2.4 Scaffold-free cell delivery approaches

Scaffold-based tissue engineering strategies involve seeding of cells onto scaffolds designed to act as ECM substitutes. An alternative strategy uses the cells' ability to create their own ECM and implant the intact cell-ECM construct.[71] The cell-generated ECM may represent the optimal scaffold, because it recapitulates the native instructive signals to promote tissue growth and maturation. The Okano group cultured cells on temperature-responsive surfaces (cell culture dishes grafted with poly(*N*-isopropylacrylamide), a temperature-responsive polymer) that allowed cell attachment and proliferation at 37°C, then promoted spontaneous cell detachment without enzymatic digestion when the temperature was lowered below 32°C.[72] Once confluent, the cell sheets were detached from the culture surface and stacked together to build a 3D multilayered tissue. When a sparse layer of HUVEC was inserted in between two myoblast cell sheets, EC sprouted and formed capillary-like structures (Fig. 6.4). In a five-sheet construct, EC capillary-like structures connected through all five layers of the construct within four days. The five-layer myoblast sheets, with or without HUVEC, were then placed on a fibrin gel and implanted subcutaneously in nude rats. Histological examination showed that HUVEC-free myoblast sheets had no vessels visible within the graft after seven days. In contrast,

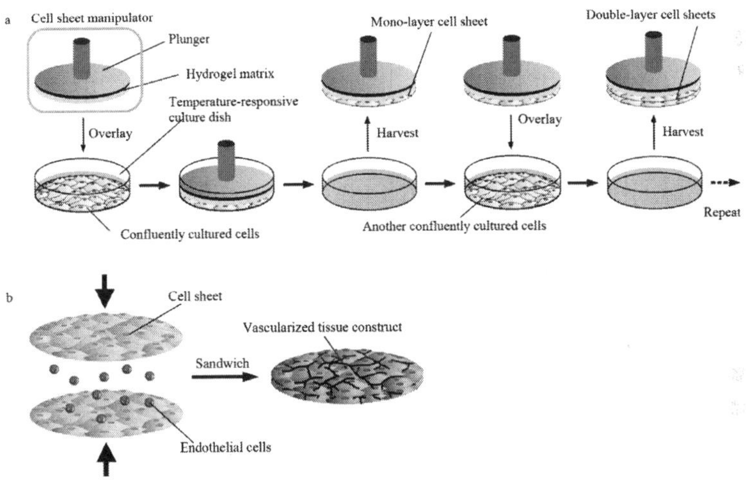


**Figure 6.4** Cell sheet stacking method. (a) Cells (myoblasts) are cultured on a temperature-responsive culture dish. Once confluent, they are harvested as a cell sheet using a hydrogel-coated plunger and overlaid on top of a second cell sheet. After allowing the two cell sheets to attach, the plunger is lifted and the double-layered construct is placed on top of a third cell sheet. The procedure is repeated several times until a multi-layered construct is formed. (b) To create a vascularized construct, layers of sparsely seeded EC are inserted in between the myoblast cell sheets, using a similar stacking method. Reprinted from Biomaterials, Vol 31(7), Sasagawa T., Shimizu T., Sekiya S., Haraguchi Y., Yamato M., Sawa Y., Okano T. Design of prevascularized three-dimensional cell-dense tissues using a cell sheet stacking manipulation technology, 1646–1654, Copyright (2010), with permission from Elsevier.

the presence of HUVEC resulted in the formation of blood vessels containing endogenous erythrocytes, suggesting these blood vessels were connected to the host vasculature. One can envision that more complex tissues could be made by incorporating micro-fabrication techniques to integrate patterning and micro-texture, as discussed in Section 6.3.7.

Another scaffold-free approach was introduced by Kelm and collaborators.[73] They used a microtissue self-assembly method to create small diameter tissue-engineered blood vessels. First, a hanging drop method resulted in microtissues (spheroids) of human artery–derived fibroblasts (HAF), with 10,000 cells/drop seeded in an inverted 60-well plate. After two days, HUVEC were added to coat the spheroids and the final microtissues were harvested after five additional days of culture. The microtissues were collected in a bioreactor and cultured under either static or dynamic conditions (under pulsatile flow and circumferential mechanical stimulation). Accumulation of capillary-like structures occurred within 14 days. Under dynamic conditions, although a layered tissue structure was observed after 14 days of culture, it was not characteristic of native vasculature since EC were present inside the wall while SMA-positive fibroblasts lined the lumen.[73]

### 6.3.3 *In situ* Vascularization with Endogenous Cells

Instead of seeding scaffolds with EC prior to implantation, bio-functionalized materials have been used to mobilize endogenous cells such as EPC *in situ*. This approach differs from the use of GF-containing scaffolds in that it aims to stimulate de novo blood vessel formation through mobilization of vessel forming cells, instead of promoting GF-induced sprouting from existing blood vessels. This alternative *in situ* strategy exploits the patient's own cells, eliminating any immune response issues.

However, the definition of what constitutes an EPC cell is controversial, since there is no specific marker that is unique to this cell population, and different methods are currently used to isolate, identify, and culture these cells. As reviewed by Yoder and collaborators, there are currently several approaches to isolate EPC, which typically involve some form of *in vitro* cell culture of peripheral monocytes and the distinguishing of particular phenotypes as say early outgrowth or late outgrowth EPC.[74,75] Various cell marker combinations have been used ($CD34^+AC133^+KDR^+$ is a common one),[76] but none can specifically and uniquely identify EPC. In one study, two types of EPC were suspended in a mixture of collagen type I and fibronectin, and the gelled constructs were implanted in the flank of NOD-SCID mice. CD31 staining showed donor (human) derived blood vessels were only seen with EPC that had produced colonies of cobblestone shaped cells after *in vitro* culture. While both EPC expressed EC markers (such as CD31, VWF, UEA-1, and others), the EPC that did not form vessels also expressed monocyte/macrophage markers (CD14, CD115), as well as the hematopoietic marker CD45. The abundance of the two cell populations was also different, with only ~0.02 vessel forming EPC colonies/$10^6$ MNC (mononuclear cells), instead of ~4 colonies/$10^6$ cells of the non-functioning EPC.[77] In a different study, these functioning EPC cells were used to create a construct for bone tissue engineering applications using a porous biphasic calcium phosphate/bone morphogenetic protein-2 scaffold. Vasculogenesis by the EPC was shown four weeks after the implant, with donor (human) derived blood vessels containing host red blood cells.[27]

During ischemic injury, bone marrow derived ("hematopoietic") EPC have the ability to home to the site of injury and re-endothelialize the injured blood vessels. Designing a system that can mobilize and capture EPC is an attractive concept, but in addition to further clarifying what cells are actually being mobilized, several key issues must be addressed before its successful implementation. In order to increase the number of circulating EPC, a strategy will need to be devised to mobilize the EPC from the bone marrow, since only very few EPC are thought to be circulating in peripheral blood. Several factors are known to increase the number of circulating EPC, including SDF-1, IL-8, Ang-1, and granulocyte colony-stimulating factor (G-CSF).[78–82] The main challenge lies in creating a signal that is strong enough to attract EPC and that is above the "background noise" associated with inflammation and tissue injury associated with biomaterial implantation. Moreover, since the implanted biomaterial will not be lined with EC, strategies to inhibit thrombosis and prevent undesirable protein adsorption and cell adhesion will be required.

Alobaid and collaborators evaluated the potential of a nanocomposite biomaterial based on polyhedral oligomeric silsesquioxane nanocages with incorporated RGD sequences to promote capture of circulating EPC *in vitro*. When peripheral blood mononuclear cells containing 1–2% CD34+ cells were plated onto nanocomposite sheets incorporating the RGD sequence, or on control nanocomposite sheets without the RGD sequence, the RGD-nanocomposite biomaterial showed a significant increase in the number of attached EPC colonies, thus showing the potential of this approach for endothelialization with EPC.[83] However, in order for this method to be applicable *in vivo*, bioactive sequences other than RGD will need to be considered, since RGD is a universal cell adhesion peptide, and cells more abundant than EPC, such as monocytes, would most likely dominate the surface attachment sites.

SDF-1, a potent chemotactic cytokine, promotes angiogenesis (directly or by modulating the production of GF such as VEGF), and vasculogenesis by modulating cell homing to areas of neovascularization. In a study performed by Simock *et al.* using a nude rat animal model, a vascularized tissue was first created by placing an arteriovenous loop (AVL) inside a polycarbonate chamber in the groin area, leading to the formation of a highly vascular fibrous tissue (see Section 6.3.4).[84] SDF-1 was administered postoperatively into the chamber via a catheter connected to a pump, for a total of 1 mg of SDF-1 over one week. Human $CD34^+$ cells (collected by leukapheresis from patients after up to three days stimulation with G-CSF, sorted by magnetic affinity for CD34, and stored frozen until used) were injected intracardially six days after creating the AVL chamber and the tissue was explanted and studied two and seven days following cell injection. This approach was shown to be successful in delivering cells to the implant site, as the number of fluorescently pre-labeled $CD34^+$ cells was eight times higher for groups receiving SDF-1 treatment compared to the untreated group. However, these cells did not line the blood vessels at the AVL implant site, and there was no significant difference in the extent of vessel density or vascular maturity in the control versus the SDF-1 treated group. An increase in the number of leukocytes, including neutrophils and macrophages, was also observed when SDF-1 was administered. The concomitant influx of leukocytes and $CD34^+$ cells raised the concern over the non-specificity of EPC homing signals. Moreover, the increased number of leukocytes in the treated group also presents a challenge for further studies to decouple the effects due to the homing of $CD34^+$ cells from the effects potentially caused by the presence of other infiltrating cells.[84]

### 6.3.4 Scaffold Prevascularization

A relevant concern of scaffold vascularization schemes via GF and/or cell delivery is that the length of time it takes to form a functional vasculature (less than one week) is often too long maintain viability of cells co-implanted within the scaffold. Consequently, a vasculature may eventually form, but the therapeutic cells within the engineered construct will have already died from lack of oxygen and nutrients in the interim. To address this issue, scaffolds have been prevascularized prior to implantation at the target site *in vivo*, so the scaffold vessels need only anastomose with the host vasculature. The prevascularization can occur *in vitro*[55,85] or *in vivo*.[86,87] However, the co-culture of multiple cell types *in vitro* prior to implantation is not trivial, since different cell types typically have different culture requirements. Different cell culture media, as well as different cell ratios between the co-cultured cells can significantly impact the angiogenic activity and the maintenance of cell phenotypes.[47,55]

Hiscox and colleagues investigated the potential of a prevascularized collagen construct to enhance the survival of transplanted islets.[85,88] Islets are highly vascularized microtissues. However, the isolation process destroys much of the islet microvasculature, and prompt revascularization and perfusion of transplanted islets is essential to islet survival and function.[89] Freshly isolated microvessel fragments from rat adipose tissue were cultured within collagen gels and were shown to form a vascular network *in vitro* and rapidly integrated with the host vasculature *in vivo*.[90] Isolated islets were embedded within a thin collagen gel, which was then sandwiched between two layers of this prevascularized collagen gel. This construct was implanted subcutaneously into SCID mice, and was shown to improve islet survival and insulin production over the 28 days of the study, compared to islets implanted without the preformed vasculature, which had no viable islets or insulin production.[85] Furthermore, insulin-producing cells were detected and were co-localized with intra-islet EC illustrating an association between islet viability and the presence of intra-islet vasculature.[85] While further *in vivo* work is required to demonstrate the therapeutic potential of this approach, this prevascularization strategy significantly improved the viability of the implanted cells, and could be used to create prevascularized constructs for other target tissues.

*In vitro* prevascularization strategies were used for bone tissue engineering by long-term co-culture of EC and osteoblasts seeded on a variety of scaffolds.[91,92] Co-culture of primary human osteoblasts and human outgrowth EC (OEC) on silk fibroin scaffolds resulted in highly organized prevascular structures and mineralized matrix after four weeks in culture. Individual phenotype and functionality of both OEC and osteoblasts needs to be further assessed in this co-culture system since qPCR analysis indicated a decrease in some EC and osteogenic markers.[92] OEC, a subset of mononuclear cells of peripheral blood, can potentially serve as an easily accessible autologous EC source due to their minimally invasive isolation, high expansion potential, and differentiated EC phenotype similar to such cells as HUVEC.[93] When implanted subcutaneously, starch – poly(caprolactone) fiber meshes prevascularized for seven days with OEC-osteoblast cocultures had CD31-positive vessels in the scaffold within 48 hours of implant.[94] The presence of erythrocytes within the donor-cell lined vessels suggested rapid anastomoses of the prevascular structure with the host vasculature. The vessels persisted within the scaffold for the duration of the 14-day study.

*In vivo* prevascularization strategies use the host as a bioreactor to vascularize the scaffold, prior to implanting the construct at its target site. One strategy uses an AVL and vascular pedicles to generate a vascularized matrix within an enclosing chamber. In this approach, an AVL or a pedicle, which contains both an artery and a vein, is enclosed within a polymer chamber that contains a matrix (such as collagen or Matrigel) or can be left empty.[95] Within the chamber, a microvascular bed is formed within the enclosed matrix, or in the case of empty chambers, new vascularized granulation tissue is formed. Functional cells are generally incorporated once the prevascularized tissue is formed.[86,96] The newly vascularized tissue is then relocated to the desired target site or left in place.[97] This technique has been used to create a variety of vascularized tissues including adipose,[97,98] skeletal muscle,[99] pancreatic (islet),[86,96] and cardiac tissues.[100,101] An example of this strategy utilized prevascularized chambers to support the survival and function of transplanted islets.[86,96] A chamber containing GF-reduced Matrigel™ and supplemental bFGF was placed around the epigastric pedicle in the groin of a diabetic mouse and formed a highly vascularized adipose tissue.[86] After 21 days, isolated islets were transplanted into the prevascularized chamber and the implant was left in place. After another 3 weeks, there was a significant reduction in blood glucose levels and improved glycemic control as measured by a glucose tolerance test.[86] The improvement in blood glucose regulation was due to the transplanted islets, as the mice returned to the initial hyperglycemic state after the chambers were removed.[86,96]

The highly vascularized omentum is a natural "bioreactor" that has been used to vascularize tissue engineering scaffolds, such as prevascularized cardiac patches for the repair of the myocardial tissue following an infarct.[87] A mixture of neonatal cardiac cells with GF-reduced Matrigel™ and pro-survival and angiogenic factors (insulin-like growth factor 1, SDF-1, and VEGF) were seeded into an alginate scaffold containing alginate-sulfate groups to enhance binding and sustained release of the GF. After 48 hours *in vitro*, the patch was transplanted into the omentum and was allowed to mature and remodel. After seven days *in vivo*, the patch contained cardiac muscle and a network of perfused blood vessels with associated perivascular cells. The prevascularized cardiac patches were then transplanted onto infarcted rat hearts seven days after induction of myocardial infarction. The evaluation at 28 days post-implant showed that prevascularized cardiac patches were structurally and electrically integrated into the host myocardium. Moreover, the presence of the vascularized cardiac patch induced thicker tissue formation, prevented further dilation of the chamber and ventricular dysfunction. This study provided evidence that prevascularization of the cardiac patch resulted in better grafting of the patch and led to improved cardiac function after myocardial infarction.

### 6.3.5 Decellularized Scaffolds

While still having elements of cell transplantation (see Section 6.3.2), decellularized scaffolds take advantage of the pre-existing architecture of a tissue that has been rendered acellular. The scaffolds are produced by perfusing an organ with detergents and buffers that effectively remove all the immunogenic cellular components from the tissue while preserving the tissue architecture and matrix, including the vascular basement membrane. The advantage of using an organ is that there is a vascular inlet and outlet. This allows the organ to be connected to a perfusion system that can distribute the decellularizing agents throughout the entire organ. As the vascular architecture has been preserved, it is hypothesized that the entire vascular tree of the organ, including the capillaries, is re-endothelialized by perfusing the decellularized organ with EC. The scaffold is then seeded with functional cells, such as hepatocytes or cardiomyocytes, to create a vascularized tissue engineering construct.

In 2007, the Mertsching group engineered liver-like tissue using a decellularized porcine jejunal segment.[102] After removing the original cells, the preserved acellular vasculature was seeded with porcine microvascular EC, which were allowed to attach and grow for one week. The luminal surface of the decellularized intestine was then seeded with primary porcine hepatocytes suspended in a collagen gel, and the cells grew in multiple layers around the endothelialized capillaries. The 3D liver tissue construct was maintained for three weeks *in vitro*, and the hepatocytes retained the capacity to perform liver specific functions, as measured by urea synthesis.[102] In other studies, decellularized arterial[103] and ureter[104] scaffolds have been used to engineer vascular grafts.

Taylor and colleagues presented a seminal study in which they engineered a bioartificial heart using an intact, decellularized cadaveric heart.[105] The cadaveric heart was decellularized by perfusing a sodium dodecyl sulfate (SDS)-based detergent through the heart chambers for 12 hours. The optimized decellularization scheme fully removed all cellular components from the matrix, while maintaining the micro and macro structures within the heart, including, presumably, a perfusable vasculature. The heart tissue was repopulated with neonatal cardiac cells through intramural injections and was maintained in a specialized bioreactor that simulated the physiological perfusion of the heart and electrical stimulation. The construct was allowed to mature under simulated physiological conditions and after 8 days, developed into a contractile tissue capable of approximately 2% of the pump action of an adult heart.[105] The group also demonstrated that re-endothelialization of the decellularized whole-heart could be achieved by perfusion of EC into the vascular conduits. After 7 days, EC formed single layers in both larger and smaller vessels and within ventricular cavities. In a similar study, Petersen *et al.* decellularized adult rat lungs, then repopulated the airways with pulmonary epithelial cells and the vasculature with

EC.[106] The lungs were successfully transplanted, perfused, ventilated for two hours, and were shown to be functional and effective at gas exchange.[106]

### 6.3.6 Angiogenic Biomaterials

In some cases, biomaterials themselves may induce angiogenesis without the addition of bioactive components such as GF or cells. By comparison to other vascularization strategies, using biomaterials as agonists of angiogenesis may be associated with lower cost, ease of storage, and increased versatility (the same biomaterial can be used to induce vascularization in various circumstances). The Sefton group has demonstrated the angiogenic effect of two synthetic methacrylic acid-based copolymers: poly(methacrylic acid -co- methyl methacrylate) (MAA) beads and poly(butyl methacrylate -co- methacrylic acid) (BMA-MAA) scaffolds. The MAA beads (150–200 μm diameter; 45 mol% methacrylic acid, 64 mol% methyl methacrylate, and 1% ethylene glycol dimethacrylate (EGDMA)) are non-biodegradable, and have a negative surface charge and rough surface topography.[107] *In vivo*, MAA beads significantly improved the vascularization of skin grafts in rats,[107] and diabetic wounds in mice,[108] compared to control poly(methyl methacrylate) (PMMA, 100% methyl methacrylate, 150–200 μm diameter) beads or no biomaterial. For tissue engineering applications, a porous BMA-MAA scaffold (45 mol% methacrylic acid, 54 mol% n-butyl methacrylate, 1% EGDMA) was developed.[109] BMA-MAA scaffolds were implanted subcutaneously in mice and promoted aggressive tissue penetration. The tissue invading the BMA-MAA scaffolds had a higher microvessel density compared to control poly(butyl methacrylate) (BMA) scaffolds at 21 and 30 days.

The mechanisms of action of MAA beads and BMA-MAA scaffold are unclear. The pro-angiogenic effect of the MAA beads was attributed to the methacrylic acid content, as no increased angiogenesis was observed in response to PMMA beads (same diameter, no methacrylic acid), and in early studies MAA beads and gels with low methacrylic acid content had a reduced effect on angiogenesis. It had been presumed that the anionic charge of MAA beads and BMA-MAA scaffold might bind endogenous cationic GF then release them slowly, extending the duration of the pro-angiogenic signaling. Also, in the context of beads, the surface topography may also play a role; MAA beads are rough while PMMA beads are smooth.

To clarify the biological mechanism of MAA-induced angiogenesis, the group has used quantitative real-time PCR to identify the effect of MAA beads on HUVEC and macrophage-like cells (dTHP-1).[110] The MAA beads did not modify the gene expression of typical angiogenic genes (such as VEGF), but did modulate the expression of cytokines important in wound healing.[110] Gene expression analysis in MAA-treated cutaneous wounds in diabetic mice showed a significant increase in the amount of Shh (pro-angiogenic morphogen) mRNA at day 4 post-wounding, which was followed by increased microvessel density at day 7.[111] Despite these insights, the mechanism remains unclear.

Based on earlier work by Brauker and collaborators,[112] the Ratner group has used spherical, interconnected pores to enhance vascularization in poly(2-hydroxyethyl methacrylate) scaffolds.[113] They extended the use of this scaffold microtemplating strategy for cardiac tissue engineering by adding parallel channels to encourage the organization of aligned cardiomyocyte bundles in poly(2-hydroxyethyl methacrylate-co-methacrylic acid) scaffolds.[114] Histological examination demonstrated that the scaffold pores and the surrounding interface were filled with vascular granulation tissue, and that an optimal pore size of 30–40 μm maximized neovascularization while minimizing fibrous encapsulation. Other groups have developed self-assembling, peptide-based nanofiber scaffolds (reviewed in Ref. 115) that enhance EC survival and angiogenesis, despite using peptide sequences that are not naturally occurring and are presumed to be biological inactive.[115-119] Narmoneva and colleagues demonstrated regulation of EC activation and angiogenesis *in vivo* using injectable RAD16-II (AcN-RARADADARARADADA-CNH$^2$) peptide nanofibers.[116,120] In a diabetic wound healing mouse model, wound treatment with RAD16-II nanofibers significantly enhanced angiogenesis and improved healing at day 7 post-wounding.[116] The presence of nanofibers resulted in significant increases of VEGF protein levels in the wound tissue. Biotinylated lectin injections demonstrated that most of the vessels formed within the granulation tissue were anastomosed with the host vasculature by day 7. Addition of $\beta 1$ and $\beta 3$ integrin inhibitors completely abrogated both *in vitro* and *in vivo* effects caused by the nanofibers suggesting that nanofiber-induced angiogenesis was, at least in part, mediated by integrins despite the low-affinity binding kinetics of the integrin-nanofiber interactions.[116]

Nanofiber gels formed by self-assembled peptide amphiphiles (PA) have also demonstrated angiogenic activity *in vivo*.[117–119] PA consist of a hydrophobic alkyl tail and a hydrophilic peptide head, which contains a sequence to promote self-assembly and bioactive domains that can be customized for specific applications.[117-119] Heparin-binding PA (HBPA) was engineered to self-assemble in the presence of heparan sulfate or heparan sulfate-like glycosaminoglycans (HSGAG). These HBPA-HS nanostructures were originally designed to stimulate extensive vascularization by binding minute amounts of VEGF and bFGF.[118] In a serendipitous discovery during an *in vivo* biocompatibility study, the control nanofibers (without the addition of any GF) were found to promote angiogenesis and the formation of vascularized tissue in both murine subcutaneous implant and dorsal skinfold chamber models, in the absence of a fibrotic response.[117] However by day 60, macrophages had completely degraded the nanofiber gel and the neovasculature had regressed. The observed vascularization at earlier time points (day 10 and 30) was not noted in nanofiber gels prepared without heparan sulfate or for only heparan sulfate treatment, suggesting that the angiogenic effect was attributed specifically to the presence of heparan sulfate within the nanofiber gel. It is possible that the heparan sulfate presented on the surface of HBPA-HS binds, stabilizes and gradually releases GF secreted by infiltrating endogenous cells, thereby enhancing angiogenesis.[117]

Another example of an angiogenic biomaterial is a PA nanostructure designed to mimic the activity of VEGF.[119] The PA incorporated a synthetic oligopeptide that was designed to mimic the $\alpha$-helical receptor-binding domain of native VEGF, and was capable of binding and inducing the phosphorylation of the VEGF receptor.[119] During *in vitro* culture, the VEGF-mimetic PA (VEGF-PA) enhanced EC proliferation, survival, and migration compared to untreated control and to a higher degree than recombinant VEGF. In a chicken chorioallantoic membrane assay, VEGF-PA nanogels elicited a strong angiogenic response, with over two-fold increase in microvessel density at three days post-implantation, compared to the initial blood vessel density (day 0). To evaluate the therapeutic efficacy of the VEGF-mimetic nanofibers, they were injected into the ischemic muscle three days following the induction of critical hind-limb ischemia. The nanofibers decreased tissue necrosis and improved tissue perfusion, functional recovery, and limb salvage at days 21 and 28 compared to controls (VEGF peptide, mutant PA, and saline), and the nanofibers performed as well or better than the administration of recombinant VEGF protein. The most significantly affected measure was the increased CD31-positive capillary density that VEGF-PA induced in the ischemic muscle, compared to a 20 μg bolus injection of $VEGF_{165}$ and a saline control.[119]

### 6.3.7 Microfabrication Methods

Many of the strategies for creating vascularized scaffolds rely on manipulating the biology of vessel formation, using angiogenic factors, transplanted cells, or bioreactors to induce vessel formation within the scaffold. Another vascularization paradigm focuses on generating predefined, hierarchical pseudo-vasculatures *in vitro*, with a high degree of spatial control using technologies such as MicroElectroMechanical Systems and cell/protein printing. These approaches allow the vascular network to be designed, ensuring optimal blood flow and mass transport characteristics.[121,122] While this section will only highlight the application of micromachining methods to create vascular blueprints, an in-depth review of scaffold fabrication techniques by Neves *et al.* can be found in Chapter 11.

The Vacanti group pioneered the concept of engineering a vasculature *in vitro* using micromachining technologies to create a blueprint for a microvascular network within the scaffold.[123] While early experiments aimed at growing spatially patterned cell sheets that were layered to form a 3D construct,[123] the group eventually focused on forming polymer films from non-degradable poly(dimethyl siloxane) (PDMS)[124,125] and degradable poly(glycerol sebacate) (PGS).[121] These contained branching, vessel-like channels throughout the construct that were partially lined with EC to form a pseudo-vasculature *in vitro*. Zheng *et al.* used a microfabrication technique to create a scaffold with a well-defined microchannel network capable of modulating scaffold vascularization *in vivo*.[126] The micro-structured collagen templates were implanted subcutaneously and the void spaces within the collagen gels were rapidly infiltrated with endogenous cells (compared to the 2% collagen volume). After 14 days, vascularized host tissue had invaded the patterned pores, with vessels running along the length of the pores. The large vessels (20–

30 μm diameter) surrounded by SMA-positive cells and containing erythrocytes were found along the axis of the pores, while smaller capillaries were found penetrating the deepest regions of the pores and the lateral branches of invaded tissue.[126] While over time, the collagen scaffold would likely undergo remodeling and degradation, at early time points it demonstrated the potential to guide the initial cellular invasion and vascularization.

Micropatterning techniques can also be used to pattern surfaces with cell instructive ligands, to spatially direct a desired cell response. The West group used this approach to regulate the formation of capillary-like structures.[115] The surface of non-adhesive PEGDA hydrogels was patterned with a cell adhesive ligand, Arg-Gly-Asp-Ser (RGDS), in stripes of varying width and ligand concentration. The response was inhibited on wider stripes, and stripes containing a higher concentration of ligand, suggesting that both geometrical and biochemical cues were controlling EC morphogenesis.[115]

While the technology is still in its infancy, 3D bioprinting may offer an interesting means to creating thick tissue engineered constructs with pre-defined architecture. Nakamura and colleagues used an inkjet-based platform capable of printing cells and biomolecules, layer by layer to form 3D constructs (reviewed in Ref. 127). However, to date no efforts have been made to attempt to print the complex 3D architecture that would be required for a vascular network.

## 6.4 Limitations and the Road Ahead

Much effort has been devoted in the last decade to develop functional, vascularized tissue-engineered constructs. Despite significant progress, especially in terms of expanding our current understanding of the design requirements for creating a functional vasculature within the construct, there is still much more required. The vascularization process is typically too slow to maintain the viability of the therapeutic cells of interest, and only tissues with low vascularization requirements, such as thin skin, or cartilage, are currently available clinically.

Most of the current vascularization approaches have proven successful in inducing blood vessel formation *in vivo*, but many of these blood vessels are not fully functional (leaky or not connected to the host vasculature), and many regress over time. Where some of these approaches have resulted in a functional, persistent blood vessel network, the time required to achieve such functional vascularization is still too long to be clinically relevant (several weeks for a millimeter sized implant).[128] Without an immediate vascular supply, the majority of the therapeutic cells (ex. liver cells, islets) within the construct will not survive. Going forward, much effort will need to be directed toward accelerating the formation of a perfused (i.e., anastomosed) vascular network within the construct and ensuring the new vasculature matures and persists.

Another key lesson learned from previous attempts to generate vascularized tissues is that any vascular structures pre-formed *in vitro* prior to implantation will most likely be remodeled *in vivo*. For example, when EC-coated modular constructs were implanted in mouse or rat animal models, the EC migrated off the surface of the modules and formed blood vessels in the channels between the modules instead of staying on the modules and providing non-thrombogenic surface to the channels randomly formed between the modules, as was originally expected.[56] The implanted cells were presumed to be the drivers of the remodeling process, with endogenous cells invading and remodeling the construct.[59–61,129] Finding ways to predict and guide the remodeling process *in vivo* will likely emerge as a key feature for the next generation of vascularized scaffolds, and the focus of vascularization research will likely switch from building pre-vascularized structures to controlling how the vascular structures will be remodeled and integrated with the host vasculature *in vivo*.

## 6.5 Summary

The tissue-engineered constructs that are being developed currently have limited clinical usefulness due to the lack of inherent vasculature, which critically constrains the construct size. A vascular supply is needed to maintain the viability of a clinically relevant number of therapeutic cells (in clinically relevant tissue densities) in order to achieve a functional benefit upon transplantation. A broad range of vascularization approaches are being explored to promote blood vessel formation within the construct. Initial studies using single GF delivery from tissue engineering scaffolds were successful in inducing blood vessel formation at early time points, but failed to produce a lasting vascular network. Hence, it was reasoned that the use of multiple GF, with different functions in angiogenesis, would be more suitable to recapitulate all stages of vessel formation. Multiple GF delivery improved vessel maturation and functionality, although issues still remain in terms of delivering physiological doses of GF, in a spatio-temporal controlled manner. Scaffolds are also used to deliver (transplant) cells to a target site, or to mobilize a specific cell type *in situ*. Transplanted or mobilized vascular cells contribute to the formation of a vascular network both directly, by re-assembling into vessel-like structures, as well as indirectly, by secreting various GF and driving the construct remodeling process and host response. However, the time frame required to develop a perfusable, functional blood vessel network within the construct is on the order of weeks, which significantly hinders the clinical benefit of this approach since most functional cells will not be able to survive without a vascular supply for such an extended period of time.

To circumvent this last issue, scaffold prevascularization approaches have been developed. The scaffold can be seeded with vascular cells and cultured *in vitro* to create tube-like structures within the scaffold prior to implantation (*in vitro* prevascularization). Alternatively, avascular scaffolds can be implanted in a highly vascularized site *in vivo* (such as the omentum) to facilitate construct vascularization, followed by relocation and connection to the host vasculature at the desired target site once the construct has been fully vascularized. Decellularized scaffolds have been used to provide a natural template for recreating the vascular tree architecture. Tissues or whole organs are first decellularized (while preserving the tissue architecture), and then repopulated with new cells (EC to repopulate the vascular tree and the desired functional cell populations). Another approach to induce vascularization is the use of biomaterials that act as angiogenic agonists. The use of angiogenic biomaterials provides a simple, cost effective and versatile method of inducing angiogenesis, as the same biomaterial can potentially be used to induce vascularization of various tissues with minimal modification. Finally, microfabrication techniques are currently being explored to create predefined, hierarchically organized vascular structures *in vitro*. The remodeling that occurs upon implantation of these structures *in vivo* remains to be further clarified.

Overall, significant progress has been made in expanding our understanding of the design criteria for fabricating vascularized tissue-engineered constructs. However, there is still a limited number of clinically relevant tissue-engineered constructs available, and further work is required in resolving the issue of poor tissue construct vascularity. Moving forward, vascularization strategies need to focus on accelerating the vascularization process and on improving the functionality of the vascular network formed.

## References


1. Carmeliet, P. and Jain, R. K. (2000) Angiogenesis in cancer and other diseases. *Nature*, **407**(6801), 249–257.
2. Dejana, E. (2004) Endothelial cell-cell junctions: Happy together. *Nat. Rev. Mol. Cell. Biol.*, **5**(4), 261–270.
3. Ramcharan, K. S., Lip, G. Y. H., Stonelake, P. S. and Blann, A. D. (2011) The endotheliome: A new concept in vascular biology. *Thromb. Res.*, **128**(1), 1–7.
4. Jain, R. K. (2003) Molecular regulation of vessel maturation. *Nat. Med.*, **9**(6), 685–693.
5. Heinke, J., Patterson, C. and Moser, M. (2012) Life is a pattern: Vascular assembly within the embryo. *Front. Biosci. (Elite Ed.)*, **4**, 2269–2288.
6. Rossant, J. and Howard, L. (2002) Signaling pathways in vascular development. *Annu. Rev. Cell. Dev. Biol.*, **18**, 541–573.
7. Jin, S.W. and Patterson, C. (2009) The opening act: Vasculogenesis and the origins of circulation. *Arterioscler. Thromb. Vasc. Biol.*, **29**(5), 623–629.
8. Aitsebaomo, J., Portbury, A. L., Schisler, J. C. and Patterson, C. (2008) Brothers and sisters: Molecular insights into arterial-venous heterogeneity. *Circ. Res.*, **103**(9), 929–939.
9. Torres-Vazquez, J., Kamei, M. and Weinstein, B. M. (2003) Molecular distinction between arteries and veins. *Cell Tissue Res.*, **314**(1), 43–59.
10. Goel, S., Duda, D. G., Xu, L., Munn, L. L., Boucher, Y., Fukumura, D., *et al.* (2011) Normalization of the vasculature for treatment of cancer and other diseases. *Physiol. Rev.*, **91**(3), 1071–1121.
11. Aird, W. C. (2007) Phenotypic heterogeneity of the endothelium: I. Structure, function, and mechanisms. *Circ. Res.*, **100**(2), 158–173.

12. Aird, W. C. (2007) Phenotypic heterogeneity of the endothelium: II. Representative vascular beds. *Circ. Res.*, **100**(2), 174–190.

13. Käβmeyer, S., Plendl, J., Custodis, P. and Bahramsoltani, M. (2009) New insights in vascular development: Vasculogenesis and endothelial progenitor cells. *Anat. Histol. Embryol.*, **38**(1), 1–11.

14. Cooke, J. P. and Losordo, D. W. (2002) Nitric oxide and angiogenesis. *Circulation*, **105**(18), 2133–2135.

15. Maisonpierre, P. C., Suri, C., Jones, P. F., Bartunkova, S., Wiegand, S. J., Radziejewski, C., *et al.* (1997) Angiopoietin-2, a natural antagonist for Tie2 that disrupts *in vivo* angiogenesis. *Science*, **277**(5322), 55–60.

16. Holash, J., Maisonpierre, P. C., Compton, D., Boland, P., Alexander, C. R., Zagzag, D., *et al.* (1999) Vessel cooption, regression, and growth in tumors mediated by angiopoietins and VEGF. *Science*, **284**(5422), 1994–1998.

17. Reiss, Y., Droste, J., Heil, M., Tribulova, S., Schmidt, M. H. H., Schaper, W., *et al.* (2007) Angiopoietin-2 impairs revascularization after limb ischemia. *Circ. Res.*, **101**(1), 88–96.

18. Carmeliet, P. (2000) Mechanisms of angiogenesis and arteriogenesis. *Nat. Med.*, **6**(4), 389–395.

19. Gerhardt, H. (2003) VEGF guides angiogenic sprouting utilizing endothelial tip cell filopodia. *J. Cell. Biol.*, **161**(6), 1163–1177.

20. Geudens, I. and Gerhardt, H. (2011) Coordinating cell behaviour during blood vessel formation. *Development*, **138**(21), 4569–4583.

21. Carmeliet, P. and Jain, R. K. (2011) Molecular mechanisms and clinical applications of angiogenesis. *Nature*, **473**(7347), 298–307.

22. Adams, R. H. and Alitalo, K. (2007) Molecular regulation of angiogenesis and lymphangiogenesis. *Nat. Rev. Mol. Cell. Biol.*, **8**(6), 464–478.

23. Fantin, A., Vieira, J. M., Gestri, G., Denti, L., Schwarz, Q., Prykhozhij, S., *et al.* (2010) Tissue macrophages act as cellular chaperones for vascular anastomosis downstream of VEGF-mediated endothelial tip cell induction. *Blood*, **116**(5), 829–840.

24. Davis, G. E., Stratman, A. N., Sacharidou, A. and Koh, W. (2011) Molecular basis for endothelial lumen formation and tubulogenesis during vasculogenesis and angiogenic sprouting. *Int. Rev. Cell. Mol. Biol.*, **288**, 101–165.

25. Schuch, G., Kisker, O., Atala, A. and Soker, S. (2002) Pancreatic tumor growth is regulated by the balance between positive and negative modulators of angiogenesis. *Angiogenesis*, **5**(3), 181–190.

26. Tongers, J., Roncalli, J. G. and Losordo, D. W. (2010) Role of endothelial progenitor cells during ischemia-induced vasculogenesis and collateral formation. *Microvasc. Res.*, **79**(3), 200–206.

27. Levengood, S. K. L., Poellmann, M. J., Clark, S. G., Ingram, D. A., Yoder, M. C. and Johnson, A. J. W. (2011) Human endothelial colony forming cells undergo vasculogenesis within biphasic calcium phosphate bone tissue engineering constructs. *Acta Biomater.*, **7**(12), 4222–4228.

28. Velazquez, O. C. (2007) Angiogenesis and vasculogenesis: Inducing the growth of new blood vessels and wound healing by stimulation of bone marrow-derived progenitor cell mobilization and homing. *J. Vasc. Surg.*, **45**(Suppl A),A39–47.

29. Masuda, H. and Asahara, T. (2003) Post-natal endothelial progenitor cells for neovascularization in tissue regeneration. *Cardiovasc. Res.*, **58**(2), 390–398.

30. Carmeliet, P. and Jain, R. K. (2011) Principles and mechanisms of vessel normalization for cancer and other angiogenic diseases. *Nat. Rev. Drug. Discov.*, **10**(6), 417–427.

31. Ehrbar, M., Metters, A., Zammaretti, P., Hubbell, J. A. and Zisch, A. H. (2005) Endothelial cell proliferation and progenitor maturation by fibrin-bound VEGF variants with differential susceptibilities to local cellular activity. *J. Control Release*, **101**(1–3), 93–109.

32. Phelps, E. A., Landázuri, N., Thulé, P. M., Taylor, W. R. and García, A. J. (2010) Bioartificial matrices for therapeutic vascularization. *Proc. Natl. Acad. Sci. USA*, **107**(8), 3323–3328.

33. Simons, M., Bonow, R. O., Chronos, N. A., Cohen, D. J., Giordano, F. J., Hammond, H. K., *et al.* (2000) Clinical trials in coronary angiogenesis: Issues, problems, consensus: An expert panel summary. *Circulation*, **102**(11), E73–E86.

34. Zisch, A.H., Lutolf, M.P. and Hubbell, J.A. (2003) Biopolymeric delivery matrices for angiogenic growth factors. *Cardiovasc. Pathol.*, **12**(6), 295–310.

35. Doi, K., Ikeda, T., Marui, A., Kushibiki, T., Arai, Y., Hirose, K., *et al.* (2007) Enhanced angiogenesis by gelatin hydrogels incorporating basic fibroblast growth factor in rabbit model of hind limb ischemia. *Heart Vessels*, **22**(2), 104–108.

36. Ennett, A. B., Kaigler, D. and Mooney, D. J. (2006) Temporally regulated delivery of VEGF *in vitro* and *in vivo*. *J. Biomed. Mater. Res. A*, **79**(1), 176–184.

37. Perets, A., Baruch, Y., Weisbuch, F., Shoshany, G., Neufeld, G. and Cohen, S. (2003) Enhancing the vascularization of three-dimensional porous alginate scaffolds by incorporating controlled release basic fibroblast growth factor microspheres. *J. Biomed. Mater. Res. A*, **65**(4), 489–497.

38. Ehrbar, M., Zeisberger, S. M., Raeber, G. P., Hubbell, J. A., Schnell, C. and Zisch, A. H. (2008) The role of actively released fibrin-conjugated VEGF for VEGF receptor 2 gene activation and the enhancement of angiogenesis. *Biomaterials*, **29**(11), 1720–1729.

39. Zisch, A. H., Zeisberger, S. M., Ehrbar, M., Djonov, V., Weber, C. C., Ziemiecki, A., *et al.* (2004) Engineered fibrin matrices for functional display of cell membrane-bound growth factor-like activities: Study of angiogenic signaling by ephrin-B2. *Biomaterials*, **25**(16), 3245–3257.

40. Shen, Y. H., Shoichet, M. S. and Radisic, M. (2008) Vascular endothelial growth factor immobilized in collagen scaffold promotes penetration and proliferation of endothelial cells. *Acta Biomater.*, **4**(3), 477–489.

41. Moore, K., MacSween, M. and Shoichet, M. (2006) Immobilized concentration gradients of neurotrophic factors guide neurite outgrowth of primary neurons in macroporous scaffolds. *Tissue Eng.*, **12**(2), 267–278.

42. Ganeff, C., Remouchamps, C., Boutaffala, L., Benezech, C., Galopin, G., Vandepaer, S., *et al.* (2011) Induction of the alternative NF-κB pathway by lymphotoxin $\alpha\beta$ (LT$\alpha\beta$) relies on internalization of LT$\beta$ receptor. *Mol. Cell. Biol.*, **31**(21), 4319–4334.

43. Chen, R. R., Silva, E. A., Yuen, W. W. and Mooney, D. J. (2007) Spatio-temporal VEGF and PDGF delivery patterns blood vessel formation and maturation. *Pharm. Res.*, **24**(2), 258–264.

44. Hao, X., Silva, E. A., Månsson-Broberg, A., Grinnemo, K.-H., Siddiqui, A. J., Dellgren, G., *et al.* (2007) Angiogenic effects of sequential release of VEGF-A165 and PDGF-BB with alginate hydrogels after myocardial infarction. *Cardiovasc. Res.*, **75**(1), 178–185.

45. Richardson, T. P., Peters, M. C., Ennett, A. B. and Mooney, D. J. (2001) Polymeric system for dual growth factor delivery. *Nat. Biotechnol.*, **19**(11), 1029–1034.

46. Greenberg, J. I., Shields, D. J., Barillas, S. G., Acevedo, L. M., Murphy, E., Huang, J., *et al.* (2008) A role for VEGF as a negative regulator of pericyte function and vessel maturation. *Nature*, **456**(7223), 809–813.

47. Khan, O. F. and Sefton, M. V. (2011) Endothelialized biomaterials for tissue engineering applications *in vivo*. *Trends Biotechnol.*, **29**(8), 379–387.

48. de Mel, A., Jell, G., Stevens, M. M. and Seifalian, A. M. (2008) Biofunctionalization of biomaterials for accelerated *in situ* endothelialization: A review. *Biomacromolecules*, **9**(11), 2969–2979.

49. Kraehenbuehl, T. P., Ferreira, L. S., Zammaretti, P., Hubbell, J. A. and Langer, R (2009) Cell-responsive hydrogel for encapsulation of vascular cells. *Biomaterials*, **30**(26), 4318–4324.

50. Cooper, T. P. and Sefton, M. V. (2011) Fibronectin coating of collagen modules increases *in vivo* HUVEC survival and vessel formation in SCID mice. *Acta Biomater.*, **7**(3), 1072–1083.

51. Kaully, T., Kaufman-Francis, K., Lesman, A. and Levenberg, S. (2009) Vascularization - the conduit to viable engineered tissues. *Tissue Eng. Part B Rev.*, **15**(2), 159–169.

52. Lovett, M., Lee, K., Edwards, A. and Kaplan, D. L. (2009) Vascularization strategies for tissue engineering. *Tissue Eng. Part B Rev.*, **15**(3), 353–370.

53. Koike, N., Fukumura, D., Gralla, O., Au, P., Schechner, J. S. and Jain, R. K. (2004) Tissue engineering: Creation of long-lasting blood vessels. *Nature*, **428**(6979), 138–139.

54. Moon, J. J., Saik, J. E., Poché, R. A., Leslie-Barbick, J. E., Lee, S.-H., Smith, A. A., *et al.* (2010) Biomimetic hydrogels with pro-angiogenic properties. *Biomaterials*, **31**(14), 3840–3847.

55. Levenberg, S., Rouwkema, J., Macdonald, M., Garfein, E. S., Kohane, D. S., Darland, D. C., *et al.* (2005) Engineering vascularized skeletal muscle tissue. *Nat. Biotechnol.*, **23**(7), 879–884.

56. McGuigan, A. P. and Sefton, M. V. (2006) Vascularized organoid engineered by modular assembly enables blood perfusion. *Proc. Natl. Acad. Sci. USA*, **103**(31), 11461–11466.

57. Khan, O. F. and Sefton, M. V. (2010) Perfusion and characterization of an endothelial cell-seeded modular tissue engineered construct formed in a microfluidic remodeling chamber. *Biomaterials*, **31**(32), 8254–8261.

58. Corstorphine, L. and Sefton, M. V. (2011) Effectiveness factor and diffusion limitations in collagen gel modules containing HepG2 cells. *J. Tissue. Eng. Regen. Med.*, **5**(2), 119–129.

59. Chamberlain, M. D., Gupta, R. and Sefton, M. V. (2011) Chimeric vessel tissue engineering driven by endothelialized modules in immunosuppressed Sprague-Dawley rats. *Tissue Eng. Part A*, **17**(1–2), 151–160.

60. Chamberlain, M. D., Gupta, R. and Sefton, M. V. (2012) Bone marrow-derived mesenchymal stromal cells enhance chimeric vessel development driven by endothelial cell coated microtissues. *Tissue Eng. Part A*, **18**(3–4), 285–294.

61. Gupta, R. and Sefton, M. V. (2011) Application of an endothelialized modular construct for islet transplantation in syngeneic and allogeneic immunosuppressed rat models. *Tissue Eng. Part A*, **17**(15–16), 2005–2015.

62. Leung, B. M. and Sefton, M. V. (2010) A modular approach to cardiac tissue engineering. *Tissue Eng. Part A*, **16**(10), 3207–3218.

63. Ciucurel, E. C. and Sefton, M. V. (2010) A poloxamine-polylysine acrylate scaffold for modular tissue engineering. *J. Biomater. Sci. Polym. Ed.*, **22**(18), 2515–2528.

64. Bruzewicz, D. A., McGuigan, A. P. and Whitesides, G. M. (2008) Fabrication of a modular tissue construct in a microfluidic chip. *Lab. Chip.*, **8**(5), 663–671.

65. Du, Y., Ghodousi, M., Qi, H., Haas, N., Xiao, W. and Khademhosseini, A. (2011) Sequential assembly of cell-laden hydrogel constructs to engineer vascular-like microchannels. *Biotechnol. Bioeng.*, **108**(7), 1693–1703.

66. Cheng, D., Lo, C. and Sefton, M. V. (2008) Effect of mouse VEGF164 on the viability of hydroxyethyl methacrylate-methyl methacrylate-microencapsulated cells *in vivo*: Bioluminescence imaging. *J. Biomed. Mater. Res. A*, **87**(2), 321–331.

67. Cheng, D. and Sefton, M. V. (2009) Dual delivery of placental growth factor and vascular endothelial growth factor from poly(hydroxyethyl methacrylate-co-methyl methacrylate) microcapsules containing doubly transfected luciferase-expressing L929 cells. *Tissue Eng. Part A*, **15**(8), 1929–1939.

68. Jabbarzadeh, E., Starnes, T., Khan, Y. M., Jiang, T., Wirtel, A. J., Deng, M., *et al.* (2008) Induction of angiogenesis in tissue-engineered scaffolds designed for bone repair: A combined gene therapy-cell transplantation approach. *Proc. Natl. Acad. Sci. USA*, **105**(32), 11099–11104.

69. De Coppi, P., Delo, D., Farrugia, L., Udompanyanan, K., Yoo, J. J., Nomi, M., *et al.* (2005) Angiogenic gene-modified muscle cells for enhancement of tissue formation. *Tissue Eng.*, **11**(7–8), 1034–1044.

70. Yang, F., Cho, S.-W., Son, S. M., Bogatyrev, S. R., Singh, D., Green, J. J., *et al.* (2010) Genetic engineering of human stem cells for enhanced angiogenesis using biodegradable polymeric nanoparticles. *Proc. Natl. Acad. Sci. USA*, **107**(8), 3317–3322.

71. L'Heureux, N., Paquet, S., Labbe, R., Germain, L. and Auger, F. A. (1998) A completely biological tissue-engineered human blood vessel. *FASEB J.*, **12**(1), 47–56.

72. Sasagawa, T., Shimizu, T., Sekiya, S., Haraguchi, Y., Yamato, M., Sawa, Y., *et al.* (2010) Design of prevascularized three-dimensional cell-dense tissues using a cell sheet stacking manipulation technology. *Biomaterials*, **31**(7), 1646–1654.

73. Kelm, J. M., Lorber, V., Snedeker, J. G., Schmidt, D., Broggini-Tenzer, A., Weisstanner, M., *et al.* (2010) A novel concept for scaffold-free vessel tissue engineering: Self-assembly of microtissue building blocks. *J. Biotechnol.*, **148**(1), 46–55.

74. Richardson, M. R. and Yoder, M. C. (2011) Endothelial progenitor cells: Quo vadis? *J. Mol. Cell. Cardiol.*, **50**(2), 266–272.

75. Critser, P. J., Voytik-Harbin, S. L. and Yoder, M. C. (2011) Isolating and defining cells to engineer human blood vessels. *Cell. Prolif.*, **44**(Suppl 1),15–21.

76. Peichev, M., Naiyer, A. J., Pereira, D., Zhu, Z., Lane, W. J., Williams, M., *et al.* (2000) Expression of VEGFR-2 and AC133 by circulating human CD34(+) cells identifies a population of functional endothelial precursors. *Blood*, **95**(3), 952–958.

77. Yoder, M. C., Mead, L. E., Prater, D., Krier, T. R., Mroueh, K. N., Li, F., *et al.* (2007) Redefining endothelial progenitor cells via clonal analysis and hematopoietic stem/progenitor cell principals. *Blood*, **109**(5), 1801–1809.

78. Aoki, J., Serruys, P. W., van Beusekom, H., Ong, A. T., McFadden, E. P., Sianos, G., *et al.* (2005) Endothelial progenitor cell capture by stents coated with antibody against CD34: The HEALING-FIM (healthy endothelial accelerated lining inhibits neointimal growth-first in man) Registry. *J. Am. Coll. Cardiol.*, **45**(10), 1574–1579.

79. Blindt, R., Vogt, F., Astafieva, I., Fach, C., Hristov, M., Krott, N., *et al.* (2006) A novel drug-eluting stent coated with an integrin-binding cyclic Arg-Gly-Asp peptide inhibits neointimal hyperplasia by recruiting endothelial progenitor cells. *J. Am. Coll. Cardiol.*, **47**(9), 1786–1795.

80. Pislaru, S. V., Harbuzariu, A., Gulati, R., Witt, T., Sandhu, N. P., Simari, R. D., *et al.* (2006) Magnetically targeted endothelial cell localization in stented vessels. *J. Am. Coll. Cardiol.*, **48**(9), 1839–1845.

81. Suuronen, E. J., Zhang, P., Kuraitis, D., Cao, X., Melhuish, A., McKee, D., *et al.* (2009) An acellular matrix-bound ligand enhances the mobilization, recruitment and therapeutic effects of circulating progenitor cells in a hindlimb ischemia model. *FASEB J.*, **23**(5), 1447–1458.

82. Avci-Adali, M., Ziemer, G. and Wendel, H. P. (2010) Induction of EPC homing on biofunctionalized vascular grafts for rapid *in vivo* self-endothelialization--a review of current strategies. *Biotechnol. Adv.*, **28**(1), 119–129.

83. Alobaid, N., Salacinski, H. J., Sales, K. M., Ramesh, B., Kannan, R. Y., Hamilton, G., *et al.* (2006) Nanocomposite containing bioactive peptides promote endothelialisation by circulating progenitor cells: An *in vitro* evaluation. *Eur. J. Vasc. Endovasc. Surg.*, **32**(1), 76–83.

84. Simcock, J. W., Penington, A. J., Morrison, W. A., Thompson, E. W. and Mitchell, G. M. (2009) Endothelial precursor cells home to a vascularized tissue engineering chamber by application of the angiogenic chemokine CXCL12. *Tissue Eng. Part A*, **15**(3), 655–664.

85. Hiscox, A. M., Stone, A. L., Limesand, S., Hoying, J. B. and Williams, S. K. (2008) An islet-stabilizing implant constructed using a preformed vasculature. *Tissue Eng. Part A*, **14**(3), 433–440.

86. Hussey, A. J., Winardi, M., Han, X.-L., Thomas, G. P. L., Penington, A. J., Morrison, W. A., *et al.* (2009) Seeding of pancreatic islets into prevascularized tissue engineering chambers. *Tissue Eng. Part A*, **15**(12), 3823–3833.

87. Dvir, T., Kedem, A., Ruvinov, E., Levy, O., Freeman, I., Landa, N., *et al.* (2009) Prevascularization of cardiac patch on the omentum improves its therapeutic outcome. *Proc. Natl. Acad. Sci. USA*, **106**(35), 14990–14995.

88. Hoying, J. B., Boswell, C. A. and Williams, S. K. (1996) Angiogenic potential of microvessel fragments established in three-dimensional collagen gels. *In Vitro Cell Dev. Biol. Anim.*, **32**(7), 409–419.

89. Jansson, L. and Carlsson, P. O. (2002) Graft vascular function after transplantation of pancreatic islets. *Diabetologia*, **45**(6), 749–763.

90. Shepherd, B. R., Chen, H. Y., Smith, C. M., Gruionu, G., Williams, S. K. and Hoying, J. B. (2004) Rapid perfusion and network remodeling in a microvascular construct after implantation. *Arterioscler. Thromb. Vasc. Biol.*, **24**(5), 898–904.

91. Unger, R. E., Sartoris, A., Peters, K., Motta, A., Migliaresi, C., Kunkel, M., *et al.* (2007) Tissue-like self-assembly in cocultures of endothelial cells and osteoblasts and the formation of microcapillary-like structures on three-dimensional porous biomaterials. *Biomaterials*, **28**(27), 3965–3976.

92. Fuchs, S., Jiang, X., Schmidt, H., Dohle, E., Ghanaati, S., Orth, C., *et al.* (2009) Dynamic processes involved in the pre-vascularization of silk

fibroin constructs for bone regeneration using outgrowth endothelial cells. *Biomaterials*, **30**(7), 1329–1338.

93. Fuchs, S., Motta, A., Migliaresi, C. and Kirkpatrick, C. J. (2006) Outgrowth endothelial cells isolated and expanded from human peripheral blood progenitor cells as a potential source of autologous cells for endothelialization of silk fibroin biomaterials. *Biomaterials*, **27**(31), 5399–5408.

94. Ghanaati, S., Fuchs, S., Webber, M. J., Orth, C., Barbeck, M., Gomes, M. E., *et al.* (2011) Rapid vascularization of starch-poly(caprolactone) *in vivo* by outgrowth endothelial cells in co-culture with primary osteoblasts. *J. Tissue Eng. Regen. Med.*, **5**(6), e136–e143.

95. Cronin, K. J., Messina, A., Knight, K. R., Cooper-White, J. J., Stevens, G. W., Penington, A. J., *et al.* (2004) New murine model of spontaneous autologous tissue engineering, combining an arteriovenous pedicle with matrix materials. *Plast. Reconstr. Surg.*, **113**(1), 260–269.

96. Forster, N. A., Penington, A. J., Hardikar, A. A., Palmer, J. A., Hussey, A., Tai, J., *et al.* (2011) A prevascularized tissue engineering chamber supports growth and function of islets and progenitor cells in diabetic mice. *Islets*, **3**(5), 271–283.

97. Dolderer, J. H., Thompson, E. W., Slavin, J., Trost, N., Cooper-White, J. J., Cao, Y., *et al.* (2011) Long-term stability of adipose tissue generated from a vascularized pedicled fat flap inside a chamber. *Plast. Reconstr. Surg.*, **127**(6), 2283–2292.

98. Wiggenhauser, P. S., Muller, D. F., Melchels, F. P., Egana, J. T., Storck, K., Mayer, H., *et al.* (2012) Engineering of vascularized adipose constructs. *Cell Tissue Res.*, **347**(3), 747–757.

99. Borschel, G. H., Dow, D. E., Dennis, R. G. and Brown, D. L. (2006) Tissue-engineered axially vascularized contractile skeletal muscle. *Plast. Reconstr. Surg.*, **117**(7), 2235–2242.

100. Birla, R. K., Borschel, G. H., Dennis, R. G. and Brown, D. L. (2005) Myocardial engineering *in vivo*: Formation and characterization of contractile, vascularized three-dimensional cardiac tissue. *Tissue Eng.*, **11**(5–6), 803–813.

101. Morritt, A. N., Bortolotto, S. K., Dilley, R. J., Han, X., Kompa, A. R., McCombe, D., *et al.* (2007) Cardiac tissue engineering in an *in vivo* vascularized chamber. *Circulation*, **115**(3), 353–360.

102. Linke, K., Schanz, J., Hansmann, J., Walles, T., Brunner, H. and Mertsching, H. (2007) Engineered liver-like tissue on a capillarized matrix for applied research. *Tissue Eng.*, **13**(11), 2699–2707.

103. Zhao, Y., Zhang, S., Zhou, J., Wang, J., Zhen, M., Liu, Y., *et al.* (2010) The development of a tissue-engineered artery using decellularized scaffold and autologous ovine mesenchymal stem cells. *Biomaterials*, **31**(2), 296–307.

104. Derham, C., Yow, H., Ingram, J., Fisher, J., Ingham, E., Korrosis, S. A., *et al.* (2008) Tissue engineering small-diameter vascular grafts: Preparation of a biocompatible porcine ureteric scaffold. *Tissue Eng. Part A*, **14**(11), 1871–1882.

105. Ott, H. C., Matthiesen, T. S., Goh, S.-K., Black, L. D., Kren, S. M., Netoff, T. I., *et al.* (2008) Perfusion-decellularized matrix: Using nature's platform to engineer a bioartificial heart. *Nat. Med.*, **14**(2), 213–221.

106. Petersen, T. H., Calle, E. A., Zhao, L., Lee, E. J., Gui, L., Raredon, M. B., *et al.* (2010) Tissue-engineered lungs for *in vivo* implantation. *Science*, **329**(5991), 538–541.

107. Eckhaus, A. A., Fish, J. S., Skarja, G., Semple, J. L. and Sefton, M. V. (2008) A preliminary study of the effect of poly(methacrylic acid-co-methyl methacrylate) beads on angiogenesis in rodent skin grafts and the quality of the panniculus carnosus. *Plast. Reconstr. Surg.*, **122**(5), 1361–1370.

108. Martin, D. C., Semple, J. L. and Sefton, M. V. (2010) Poly(methacrylic acid-co-methyl methacrylate) beads promote vascularization and wound repair in diabetic mice. *J. Biomed. Mater. Res. A*, **93**(2), 484–492.

109. Butler, M. J. and Sefton, M. V. (2007) Poly(butyl methacrylate-co-methacrylic acid) tissue engineering scaffold with pro-angiogenic potential *in vivo*. *J. Biomed. Mater. Res. A*, **82**(2), 265–273.

110. Fitzpatrick, L. E., Chan, J. W. Y. and Sefton, M. V. (2011) On the mechanism of poly(methacrylic acid -co- methyl methacrylate)-induced angiogenesis: Gene expression analysis of dTHP-1 cells. *Biomaterials*, **32**(34), 8957–8967.

111. Fitzpatrick, L. E., Lisovsky, A. and Sefton, M. V. (2012) The expression of sonic hedgehog in diabetic wounds following treatment with poly(methacrylic acid-co-methyl methacrylate) beads. *Biomaterials*, **33**(21), 5297–5307.

112. Brauker, J. H., Carr-Brendel, V. E., Martinson, L. A., Crudele, J., Johnston, W. D. and Johnson, R. C. (1995) Neovascularization of synthetic membranes directed by membrane microarchitecture. *J. Biomed. Mater. Res.*, **29**(12), 1517–1524.

113. Fukano, Y., Usui, M. L., Underwood, R. A., Isenhath, S., Marshall, A. J., Hauch, K. D., *et al.* (2010) Epidermal and dermal integration into

sphere-templated porous poly(2-hydroxyethyl methacrylate) implants in mice. *J. Biomed. Mater. Res. A*, **94**(4), 1172–1186.

114. Madden, L. R., Mortisen, D. J., Sussman, E. M., Dupras, S. K., Fugate, J. A., Cuy, J. L., *et al.* (2010) Proangiogenic scaffolds as functional templates for cardiac tissue engineering. *Proc. Natl. Acad. Sci. USA*, **107**(34), 15211–15216.

115. Zhang, S. (2003) Fabrication of novel biomaterials through molecular self-assembly. *Nat. Biotechnol.*, **21**(10), 1171–1178.

116. Cho, H., Balaji, S., Sheikh, A. Q., Hurley, J. R., Tian, Y. F., Collier, J. H., *et al.* (2012) Regulation of endothelial cell activation and angiogenesis by injectable peptide nanofibers. *Acta. Biomater.*, **8**(1), 154–164.

117. Ghanaati, S., Webber, M. J., Unger, R. E., Orth, C., Hulvat, J. F., Kiehna, S. E., *et al.* (2009) Dynamic *in vivo* biocompatibility of angiogenic peptide amphiphile nanofibers. *Biomaterials*, **30**(31), 6202–6212.

118. Rajangam, K., Behanna, H. A., Hui, M. J., Han, X., Hulvat, J. F., Lomasney, J. W., *et al.* (2006) Heparin binding nanostructures to promote growth of blood vessels. *Nano. Lett.*, **6**(9), 2086–2090.

119. Webber, M. J., Tongers, J., Newcomb, C. J., Marquardt, K.-T., Bauersachs, J., Losordo, D. W., *et al.* (2011) Supramolecular nanostructures that mimic VEGF as a strategy for ischemic tissue repair. *Proc. Natl. Acad. Sci. USA*, **108**(33), 13438–13443.

120. Narmoneva, D. A., Oni, O., Sieminski, A. L., Zhang, S., Gertler, J. P., Kamm, R. D., *et al.* (2005) Self-assembling short oligopeptides and the promotion of angiogenesis. *Biomaterials*, **26**(23), 4837–4846.

121. Fidkowski, C., Kaazempur-Mofrad, M., Borenstein, J., Vacanti, J., Langer, R. and Wang, Y. (2005) Endothelialized microvasculature based on a biodegradable elastomer. *Tissue Eng.*, **11**(1/2), 302–309.

122. Kaazempur-Mofrad, M., Vacanti, J. and Kamm, R. D. (2001) in *Computational Fluid and Solid Mechanics*, Vol. 1 (K-J Bathe, ed.), Elsevier Science Ltd, pp. 864–867.

123. Kaihara, S., Borenstein, J., Koka, R., Lalan, S., Ochoa, E. R., Ravens, *et al.* (2000) Silicon micromachining to tissue engineer branched vascular channels for liver fabrication. *Tissue Eng.*, **6**(2), 105–117.

124. Borenstein, J., Terai, H., King, K., Weinberg, E., Kaazempur-Mofrad, M. and Vacanti, J. (2002) Microfabrication technology for vascularized tissue engineering. *Biomed. Microdevices*, **4**(3), 167–175.

125. Shin, M., Matsuda, K., Ishii, O., Terai, H., Kaazempur-Mofrad, M., Borenstein, J., *et al.* (2004) Endothelialized networks with a vascular

geometry in microfabricated poly(dimethyl siloxane). *Biomed. Microdevices*, **6**(4), 269–278.

126. Zheng, Y., Henderson, P. W., Choi, N. W., Bonassar, L. J., Spector, J. A. and Stroock, A. D. (2011) Microstructured templates for directed growth and vascularization of soft tissue *in vivo*. *Biomaterials*, **32**(23), 5391–5401.

127. Arai, K., Iwanaga, S., Toda, H., Genci, C., Nishiyama, Y. and Nakamura, M. (2011) Three-dimensional inkjet biofabrication based on designed images. *Biofabrication*, **3**(3), 034113.

128. Rouwkema, J., Rivron, N. C. and van Blitterswijk, C. A. (2008) Vascularization in tissue engineering. *Trends Biotechnol.*, **26**(8), 434–441.

129. Gupta, R., van Rooijen, N. and Sefton, M. V. (2009) Fate of endothelialized modular constructs implanted in an omental pouch in nude rats. *Tissue Eng. Part A*, **15**(10), 2875–2887.

# The Role of Osteoblasts in *in vitro* Vascularization

Fuchs *et al.*[1] have reported that human microvascular endothelial cells (HDMEC) co-cultured with primary human osteoblasts (pOB) on three-dimensional porous hydroxyapatite, nickel–titanium, calcium phosphate, and silk fibroin scaffolds, considered as bone biomaterials, rapidly form microcapillary-like structures containing a lumen. The formation of vascular structures was seen just in the case where pOB were cocultured with the endothelial populations, was not enhanced using endothelial cells precultured alone.[2] The obtained pre-vascularized fibroin scaffolds, seeded with HDMEC and pOB, were then implanted in nude mice, showing the rapid anastomosis between the host vasculature and the *in vitro* generated vessel network,[3] a phenomenon that was especially evident in the cocultured condition.[2] During the *in vitro* cocoltures, osteoblasts provided additional structural support for OEC-derived vessels, perhaps acting in a pericyte-like role, serving as stabilizing structural elements following implantation of pre-cultured vessels.[4] results demonstrated the potential of pOB as a potent source of growth factors and extracellular matrix (ECM) that provide co-cultured endothelial cells with the necessary signaling for successful vascularization of tissue constructs *in vivo*.[2] These cells secrete bone-specific ECM and are known to produce a variety of factors that act on endothelial cells and promote their recruitment, survival, proliferation, and capillary formation.[5] Those factors include basic fibroblast growth factors, osteopontin, VEGF, and collagen type-I. Another potential mechanism whereby osteoblasts could promote scaffold vascularization is through facilitating the permeation into the fibroin scaffold of host inflammatory cells,[6] especially macrophages, which also appear to play a role in the vascularization process by their production of pro-angiogenic growth factors. Macrophages and multinucleated giant cells, which are involved in the degradation of silk fibroin micronets,[4] produce angiogenic signaling molecules, such as VEGF, which could also contribute to scaffold vascularization.[7]

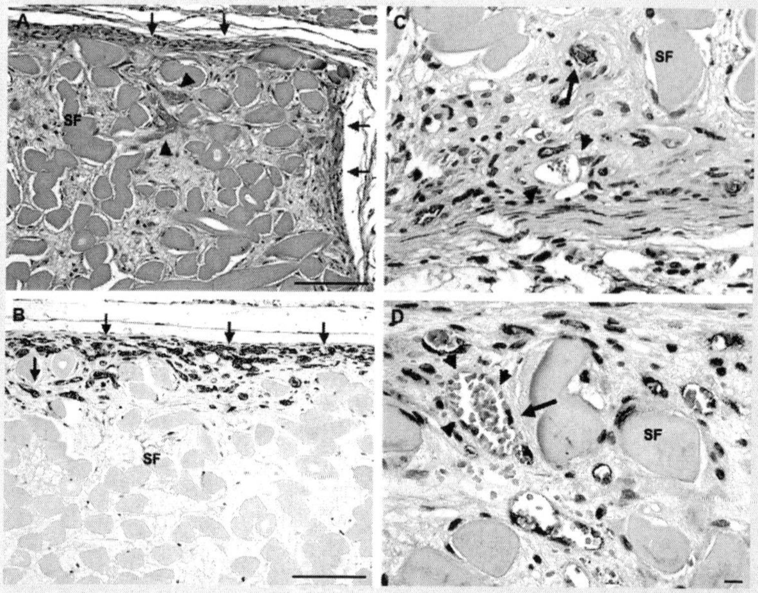


**Figure 1** Images of a representative HDMEC and HOS coculture on silk fibroin after 14 days *in vivo*. Image (A) is an HE-stained image of the implantation bed of the silk fibroin scaffold showing that the outer surface is surrounded by a large number of small lumen-containing vessels (arrows) and a few vessels with a larger lumen embedded within the fibers (arrowheads) both identifiable by their content, that is, red blood cells, indicating perfusion. Image (B), staining with human-specific vWF shows the contribution of the HDMEC to the vascularization. Images (C and D) show the presence of vWF-stained human-specific vessels (arrow) and unstained murine vessels (arrowheads). Image (D) shows chimeric vessels containing human endothelial cells stained with vWF (arrow) integrated into vessels made up of primarily murine endothelial cells (arrowhead). The vessels in (C) and (D) are filled with erythrocytes. (A and B: 200× magnification and C and D: 600× magnification; scale bar, 100 μm). Reprinted from Ref. 3, Copyright (2010), with permission from Elsevier.

The use of osteogenic cells could have implications in bone tissue engineering specifically, as the production of an osteogenic matrix would be concomitant with this induction of vascularization and could accelerate bone regeneration.

## References


1. Fuchs, S., Hofmann, A., and Kirkpatrick, C. J. (2007), Microvessel-like structures from outgrowth endothelial cells from human peripheral blood in 2-dimensional and 3-dimensional co-cultures with osteoblastic lineage cells, *Tissue Eng.*, **13**(10), 2577–2588.
2. Unger, R. E., Sartoris, A., Peters, K., Motta, A., Migliaresi, C., Kunkel, M., *et al.* (2007) Tissue-like self-assembly in cocultures of endothelial cells and osteoblasts and the formation of microcapillary-like structures on three-dimensional porous biomaterials, *Biomaterials*, **28**, 3965–3976.
3. Unger, R. E., Ghanaati, S., Orth, C., Sartoris, A., Barbeck, M., Halstenberg, S., Motta, A., Migliaresi, C., and Kirkpatrick, C. J. (2010), The rapid anastomosis between prevascularized networks on silk fibroin scaffolds generated *in vitro* with cocultures of human microvascular endothelial and osteoblast cells and the host vasculature, *Biomaterials*, **31**(27), 6959–6967.
4. Ghanaati, S., Fuchs, S., Webber, M. J., Orth, C., Barbeck, M., Gomes, M. E., Reis, R. L., and Kirkpatrick, J. C. (2011) Rapid vascularization of starch-poly(caprolactone) *in vivo* by outgrowth endothelial cells in co-culture with primary osteoblasts, *J. Tissue Eng. Regen. Med.*, **5**(6), 136–143.
5. Hirschi, K. K., Skalak, T. C., Peirce, S. M., and Little, C. D. (2002) Vascular assembly in natural and engineered tissues, *Ann. N. Y. Acad. Sci.*, **961**, 223–242.
6. Ghanaati, S., Unger, R. E., Webber, M. J., Barbeck, M., Orth, C., Kirkpatrick, J. A., Booms, P., Motta, A., Migliaresi, C., Sader, R. A., and Kirkpatrick, C. J. (2011) Scaffold vascularization *in vivo* driven by primary human osteoblasts with host inflammatory cells, *Biomaterials*, **32**, 8150–8160.
7. Ghanaati, S., Barbeck, M., Orth, C., Willershausen, I., Thimm, B. W., Hoffmann, C., Rasic, A., Sader, R. A., Unger, R. E., Peters, F., and Kirkpatrick, C. J. (2010) Influence of beta-tricalcium phosphate granule size and morphology on tissue reaction *in vivo*, *Acta Biomater.*, **6**, 4476–4487.

Chapter 7

# Engineering Biomimetic Scaffolds

**Nasim Annabi,[a,b,c] Nihal Engin Vrana,[a,b] Pinar Zorlutuna,[a,b] Fariba Dehghani,[d] and Ali Khademhosseini[a,b,c]**

[a]*Center for Biomedical Engineering, Department of Medicine, Brigham and Women's Hospital, Harvard Medical School, Cambridge, MA 02139, USA*

[b]*Harvard-MIT Division of Health Sciences and Technology, Massachusetts Institute of Technology, 77 Massachusetts Avenue, Cambridge, MA 02139, USA*

[c]*Wyss Institute for Biologically Inspired Engineering, Harvard University, Boston, MA 02115, USA*

[d]*School of Chemical and Biomolecular Engineering, University of Sydney, Sydney 2006, Australia*

alik@rics.bwh.harvard.edu

Tissue engineering has emerged as a growing field for repair and regeneration of damaged organs and tissues. The development of biomimetic scaffolds plays a crucial role in engineering tissue constructs by regulating cellular responses. Three dimensional (3D) scaffolds serve as a temporary extracellular matrix (ECM) that provide an environment with a desirable mechanical support for cell growth and tissue regeneration. Scaffolds that mimic the characteristics of natural ECM is ideal for tissue engineering as they direct different

*Scaffolds for Tissue Engineering: Biological Design, Materials, and Fabrication*
Edited by Claudio Migliaresi and Antonella Motta

ISBN 978-981-4463-20-1 (Hardcover), 978-981-4463-21-8 (eBook)
www.panstanford.com

cellular behaviors including cell migration, adhesion, proliferation, and differentiation. The architecture, physical characteristics, and biochemical properties of scaffolds play important roles in regulating cellular functions and subsequent tissue maturation. In this chapter, various aspects in the design of biomimetic scaffolds are discussed. The approaches developed for the modification of both surface and bulk properties of biomaterials are explained. These strategies include micro- or nanopatterning, 3D construction of scaffolds, and bulk modification such as rapid prototyping (RP), electrospinning, and phase separation. The effects of biophysical signals (e.g., mechanical stiffness and pore architecture) on cell behavior and differentiation are then elucidated. Finally, biochemical biomimicry in scaffolds based on incorporation of biomolecules such as proteins and growth factors are discussed.

## 7.1 Approaches to Engineer Biomimetic Constructs

Most native tissues have defined architectures, both in terms of their ECM structure and their cellular components, which usually provide certain functions or abilities. For example, in cornea, keratocytes are embedded within alternating layers of perpendicularly oriented collagen fibrils, which plays an important role in corneal transparency.[1] In blood vessels, smooth muscle cells form concentric layers perpendicular to pulsatile flow, which enables them to change the diameter of arteries with precision to control blood flow rate.[2] In bone, the organization of collagen fibrils and hydroxyapatite (HAP) crystals determines the anisotropic mechanical properties of the entire structure and defines in which axis it is the strongest.[3] Engineering constructs, which mimic the native tissue structures and microenvironment, is critical for creating functional tissues.[4,5] Various approaches have been applied to engineer biomimetic 3D scaffolds including RP, electrospinning, and phase separation. In addition, microscale technologies such as photolithography-based methods and nanoscale modification of surfaces (e.g., soft lithography, hot embossing lithography, and dip-pen nanolithography (DPN)) have been used as promising approaches to engineer micro- and nanopatterned substrates to guide cellular orientation via surface topography.

## 7.1.1 Fabrication of 3D Biomimetic Scaffolds

### 7.1.1.1 Rapid prototyping (RP)

Biomimetic structural features have been incorporated within 3D scaffolds by using various printing techniques, such as stereolithography and organ printing.[6,7] RP has been utilized to produce models for designing and manufacturing various prostheses including tailor-made prosthesis for individual patients.[8] This technique has also been used to fabricate scaffolds with desirable 3D shapes for any defect sites. The potential of this method has been recently identified for single cell and cell aggregate printing to create artificial organs.[9] RP has been primarily used for processing polymers that were stable at high temperatures (i.e., fuse deposition or particulate binding approaches).[10–12] Due to recent advances in photolabile polymer chemistry, RP method has been increasingly used to print hydrogels.[13] Stereolithography is one of the most effectively applied RP methods for hydrogels. In this technique, a layer-by-layer fabrication is achieved by using a photolabile polymer, a UV light source that can be highly focused, and a controllable stage. As the UV light (i.e., 325 nm wavelength) traces out each layer, it crosslinks the scaffold in the desired shape, and leaves excess areas in uncrosslinked form. The microcontrolled stage moves in a predetermined position upon fabrication of each layer and the deposition continues until all the layers are built. Stereolithography was previously used for the fabrication of micropatterned scaffolds from a blend of diethyl fumarate and poly(propylene fumarate) for bone tissue engineering applications.[14] Further, this technology was modified to fabricate cell-laden hydrogels by the introduction of a cytocompatible photoinitiator Irgacure 2959.[7,15–17]

### 7.1.1.2 Electrospinning

Micro- and nanosized fibers can be produced by electrospinning to fabricate biomimetic scaffolds. Since most of the ECM proteins such as collagen and elastin have a fibrous structure with fiber diameters ranging from nanometer to sub-micrometer scales, one of the widely employed methods for biomimetic scaffold fabrication is using approaches to create fibrillar scaffolds. Collagen is one of the most abundant ECM proteins in the body, which has fibrous structure with fiber diameters in the range of 50 nm to 500 nm.[18] Elastin is another example of fibrous ECM protein with fiber diameter of 10 nm to 12 nm.[19] Elastin fibers confer the properties of elasticity and resilience on elastic tissues such as arteries, lung, skin, vertebral ligamenta flava, vocal chords, and elastic cartilage.[20] Electrospinning has been extensively used for creating biomimetic scaffolds from a variety of materials, including synthetic and natural polymers.[21–25] Specifically, microfibers and nanofibers that are similar to native ECM proteins have been produced by this method for various tissue engineering applications.[21] Following cell seeding onto these biomimetic electrospun materials, the high surface area of scaffolds facilitates cell attachment and growth as well as regulates cell differentiation.[23]

In electrospinning, a high voltage is applied to a polymer solution, which overcomes surface tension and creates a fine charged jet. The charged polymer is then ejected, dried, and solidified onto an oppositely charged collector plate. The ejected polymer solutions repel each other during travel to the collector, which forms thin fibers after solvent evaporation.[26] The properties of the resultant materials such as fiber diameter, porosity, and morphology can be controlled by various parameters including applied voltage, viscosity, solution conductivity, and temperature.[27] This technique can be used to produce fibers ranging from 0.02 μm to 20 μm.[21] The resulting fibers are deposited randomly when using a stationary collector due to the random motion of the electrospinning polymer solution. However, the use of an electrically grounded rotating drum results in the formation of preferentially oriented fibers.[24,28]

Electrospinning has been applied for the fabrication of nanofibrous biomimetic scaffolds from natural polymers including chitosan,[29] collagen,[28] and elastin[30,31]; synthetic biodegradable polymers such as polyglycolic acid (PGA),[32] poly(lactic-co-glycolic acid) (PLGA),[33] poly(L-lactide) (PLLA), polycaprolactone (PCL),[34] and poly(glycerol sebacate) (PGS)[35]; and combinations of natural and synthetic polymers. These scaffolds were used for the formation of a variety of tissues such as bone and cardiac muscles. In one study, electrospinning technique was used to fabricate highly porous nano- and microfiber mats fabricated from poly(L-lactide-co-caprolactone) (PLCL) co-polymers with different compositions.[36] This polymer was used for the fabrication of fibers with diameters ranging from 0.3 μm to 7 μm and porosity between 56% and 63% (Fig. 7.1A–C). The constructs were then seeded with human umbilical vein endothelial cells (HUVECs) to investigate the effect of scaffold microstructures on cell behavior. The results of *in vitro* studies demonstrate that scaffolds comprised of small diameter fibers (0.3 μm to 1.2 μm) promoted HUVECs adhesion and proliferation, however, larger diameter fibers (e.g., 7 μm) impeded cell adhesion and spreading[36] (Fig. 7.1D–F). In another study, a combined electrospinning/salt leaching process was used to fabricate 3D porous hyaluronic acid/collagen scaffolds.[37] In this process, hyaluronic acid and collagen were first dissolved and mixed in NaOH/*N,N*-dimethyl formamide. The solution was then electrospun into a nanofiber structure. During electrospinning, salt particles were homogenously deposited into the fibers after which the fibers were stabilized through chemical crosslinking. Following salt leaching, a macroporous hyaluronic acid/collagen scaffold containing nanofibers was generated. The resultant composite was shown to support bovine chondrocyte adhesion and proliferation.[37]

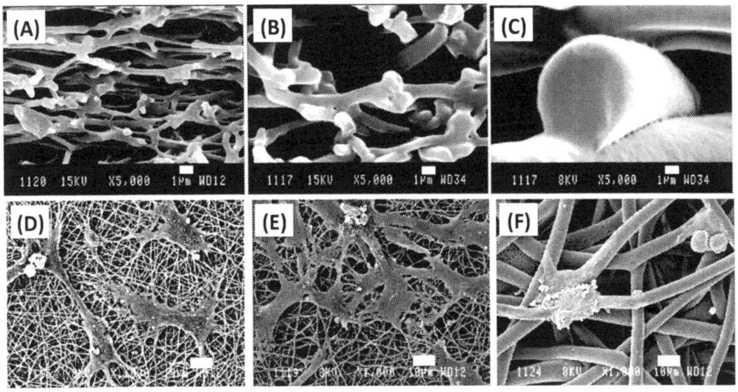


**Figure 7.1** (A–C) SEM images of unseeded and (D–F) HUVEC seeded; PLCL scaffolds with different diameter fibers: (A, D) 0.3 mm, (B, E) 1.2 μm, and (C, F) 7 μm after 7 days of culture. The scale bar for the top panel is 1 μm and for the bottom panel is 10 μm. Reprinted from Ref. 36, Copyright (2005), with permission from Elsevier.

#### 7.1.1.3 Phase separation

Phase separation process has been used as another powerful technique to produce biomimetic fibrous scaffolds for various tissue-engineering applications.[21,24] In this method, phase separation of a homogenous multicomponent system occurs to decrease the free energy of the system, when it becomes thermodynamically unstable. For instance, a polymer-lean phase with a low concentration of polymer, and a polymer-rich phase with a high polymer concentration are formed due to a temperature change or addition of a non-solvent.[24] After subsequent solvent removal and solidification of polymer-rich phase, a fibrous scaffold is formed. The morphologies and pore structure of the resultant scaffolds can be manipulated by changing the polymer system and phase separation conditions.[21] This technique has been used to produce porous biomimetic scaffolds using a variety of biodegradable polymers such as PLLA,[38] poly($\alpha$-hydroxyl acids)/HAP,[39] and nanohydroxyapatite (nHAP)/PLLA.[40,41] In one of the study, Ma *et al.* used a thermally induced phase separation process to fabricate PLLA scaffolds with nanofibrillar structure similar to that of natural collagen.[42] The nanofibrous scaffolds were prepared by thermally induced gelation of polymer solution followed by solvent exchange and freeze-drying. The fabricated fibrous scaffolds had 90% porosity and contained nanofibers in the range of 50 nm to 500 nm. It was also feasible to control pore morphology and average fiber diameters of the resultant scaffolds by changing the thermal annealing conditions, polymer concentration, solvent exchange, and freezing temperature.

Nanofibrous scaffolds with controlled fiber diameter and morphology can be produced by using phase separation technique. However, these scaffolds generally possess a low degree of pore interconnectivity. In a study, Wei *et al.* developed a method based on a combination of phase separation and porogen leaching to enhance pore interconnectivity in PLLA/nHAP.[41] In this method, a sugar template was first formed by binding water-soluble sugar spheres. The PPLA/nHAP solution in tetrahydrofuran was then cast onto the template and the temperature was decreased to induce phase separation.[41] The removal of sugar template by distilled water resulted in the formation of porous scaffolds with average pore sizes ranging from 180 μm to 250 μm and porosity between 97% and 98.2%.[41] This method enabled control of the resulting pore size and morphology by variation of sugar size and phase separation process parameters.[41] Liu *et al.* applied a similar process using paraffin spheres as template to produce nanofibrous gelatin and gelatin/apatite composites.[43,44] The fabricated scaffolds contained fibers with diameters in the range of 50 nm to 500 nm, pores in the range of 250 μm to 420 μm, and porosity of 97.5% (Fig. 7.2A,B).[43] In comparison to Gelfoam (a commercial gelatin foam) with similar pore size and porosity, the fabricated nanofibrous gelatin scaffolds exhibited much higher surface area and mechanical strength. *In vitro* studies demonstrated higher cell attachment and proliferation on nanofibrous gelatin scaffolds compared to Gelfoam during two weeks of culture. In addition, histology analysis exhibited uniform cell distribution and neo-tissue formation for cell seeded nanofibrous gelatin scaffolds after 14 days (Fig. 7.2C). However, cells formed a denser layer on the outer surface of Gelfoam compared to the centre of the scaffold, which could obstruct the nutrient diffusion within the scaffold.[43]

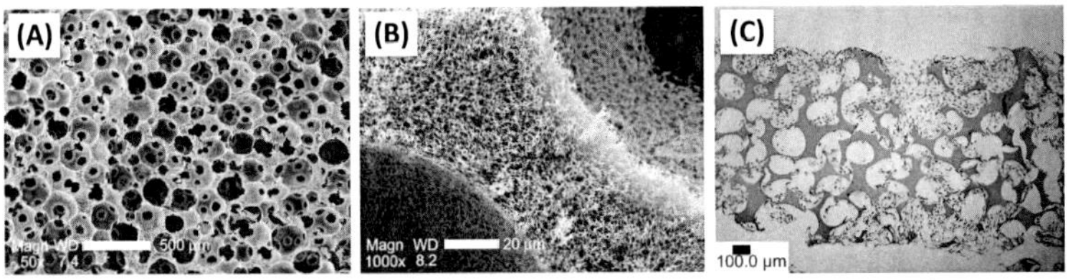


**Figure 7.2** SEM images of gelatin scaffold produced by using a combined phase separation/paraffin leaching process at (A) low (scale bar: 500 μm) and (B) high (scale bar: 20 μm) magnification, paraffin spheres ranging from 150 μm to 250 μm were used to prepare template; (C) H&E staining of MC3T3-E1 pre-osteoblast gelatin scaffold after 14 days of culture (scale bar: 100 μm). Reprinted from Ref. 43, Copyright (2009), with permission from Elsevier.

### 7.1.2 Effect of Surface Topography on Biomimetic Scaffolds

The surface architecture may have an impact on cellular attachment, phenotype, differentiation, and apoptosis. Biomimetic approaches have shown that patterned surfaces affect the gene and protein expression of cells.[45,46] For example, hybrid features of lines and toroids rather than line or toroid patterns were shown to regulate the degree of myoblasts differentiation[46] (Fig. 7.3). Nanoscale surface features have also been shown to enhance mechanical properties of vascular smooth muscle cell (VSMC) seeded collagen membranes[47] or retention of endothelial cells on surfaces upon the application of fluid shear stress.[17] Therefore, the exploitation of specific structural designs for controlling cellular response in tissue engineering is critical.[48] Various techniques such as photolithography approaches and nanoscale modification of surfaces (e.g., soft lithography, hot embossing lithography, and DPN) have been used as promising strategies to engineer micro- and nanopatterned substrates to guide cellular orientation via surface topography. Such substrates can be produced in free standing forms and be used for layer-by-layer production of 3D scaffolds.[49]

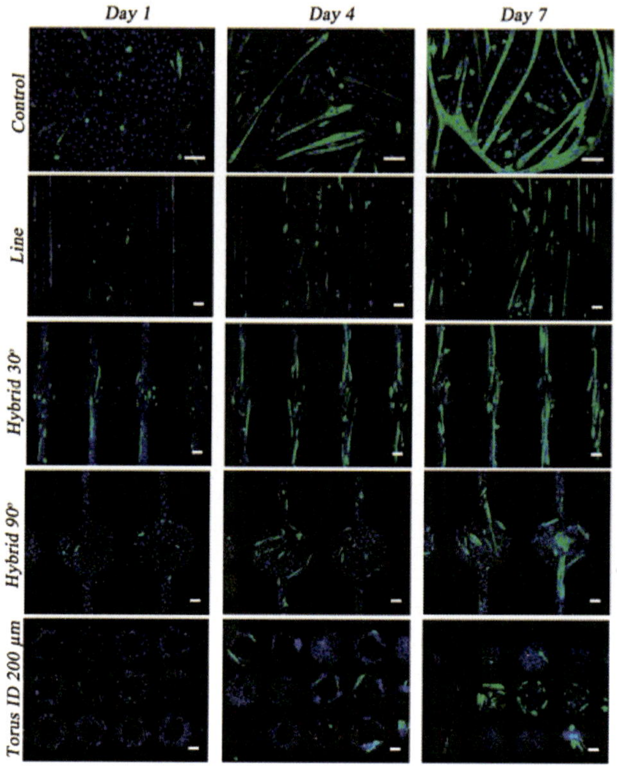


**Figure 7.3** Fluorescent images of the C2C12 cells on the different patterns. Column 1: C2C12 cells after one day; column 2: cells after four days; and column 3: cells after seven days, in the differentiation media. Row 1: C2C12 cells present on the unpatterned (control) surface; row 2: linear cell islands; rows 3 and 4: cells on hybrid patterns with different arc degrees (30° and 90°); and row 5: cells on toroid cell islands. Cells were stained for anti-MHC (green) and nucleus (blue) and these were used for the calculation of the fusion index and the maturation index (% myotubes ≥5 nuclei); Scale bar: 100 μm. Reprinted from Ref. 46, with permission from Royal Society of Chemistry, copyright (2011).

Cell contact guidance is usually achieved through microscale, and more recently, nanoscale patterned scaffolds where cellular orientation is achieved via surface topography.[50] The morphology of cells affects their functionality, therefore, it is critical to direct cells into a specific orientation that mimic the target tissue.[51–53] Photolithography is one of the most common methods for creating patterns on the surfaces of scaffolds. This method involves fabricating a template using a silicon wafer that has been produced in a clean room with a designed pattern. A solution of polymer is then added on the template that is generally located in a mould. The designed pattern can, therefore, be produced by methods such as solvent casting, physical, or chemical crosslinking.[54,55] Micropatterned scaffolds produced by this strategy have been used to study the effects of patterns on cell behavior, their attachment and orientation, and investigate how to improve cell–biomaterial function.[17,47,56,57]

The creation of nanopatterns on the surface of scaffold enables researchers to mimic the molecular level interactions of natural ECM components.[58,59] Therefore, nanopatterning promotes the biomimicry of scaffolds. Methods such as soft lithography, hot embossing, and DPN are used to create nanoscale features on polymer surfaces. Despite high resolution and pattern fidelity, each of these methods has certain limitations.[60] For example, nano-sized pattern as small as 10 nm can be produced using hot embossing lithography, however, heating and pressures used in this method is an obstacle to use this technique for heat labile biomaterials such as biopolymers. Soft lithography operates at moderate temperatures. In addition, sub-100 nm sized patterns can be created by using this technique.[61,62] DPN utilizes an atomic force microscope (AFM) in tapping mode to transfer the desired molecules from solution to a substrate surface according to a predetermined design. Features with sizes in the range of 30–50 nm were successfully created by this method.[63] AFM was also used to create 3D nanopatterns by scratching a surface in contact mode.[64,65] Alternatively, 3D nanopatterns were fabricated on silicone or acrylate-based polymers by using X-ray interference lithography, which provides periodic patterns within 30 nm range on large areas in short processing times.[66] Electron-beam lithography can be used to produce features with sizes less than 10 nm.[67]

Surface features by themselves are usually not sufficient to induce the desired changes in cellular behavior. Therefore, the introduction of other physical environment components (i.e., mechanical properties, architecture, etc.) and biochemical signals is crucial. Moreover, a major problem with the biomimicry of physical components is that the implementations of these properties are generally conducted in 2D.[68] Such 2D studies, although informative, are limited in their scope, as they do not mimic the 3D nature of the *in vivo* microenvironment. Therefore, studies using 3D microarchitectures and biomimetic fibrous or porous scaffolds are important for more in-depth studies and for engineering functional tissues. For example, in heart tissue the physical cues are an intricate match of electrical stimulation and the orientation of muscle layers, which is controlled by the ECM properties. Several studies have shown that a biomaterial with tunable mechanical properties is a key element in cardiac tissue, due to the variation of stiffness of heart muscle several folds between systole and diastole.[69] For heart tissue regeneration it is also critical to consider the excitation of cardiomyocytes, which depends on their alignment, and failure to mimic this function results in problems such as arrhythmia. To generate a more biomimetic cardiac scaffold, a 3D honeycomb patterned scaffold has been developed to withstand the continuous mechanical stress of heart tissue and to mimic its general anisotropy (Fig. 7.4).[70] However, architectural cues are often not enough to direct cell behavior and the other physical properties of the scaffold should mimic the target organ.

## 7.2 Controlling the Physical Properties of Biomimetic Scaffolds

The physical properties of tissue engineered scaffolds should ideally mimic the features of native ECM to regulate cell function. Thus, controlling the physical characteristics of scaffolds such as elasticity, mechanical stiffness, and porosity can be used to regulate cell survival, migration, and proliferation during the regeneration process.

### 7.2.1 Mechanical Properties

The biomechanical features of cellular matrix affect the cell growth and function by providing mechanical signals. These signals mainly come from the fibrous components of ECM such as elastin

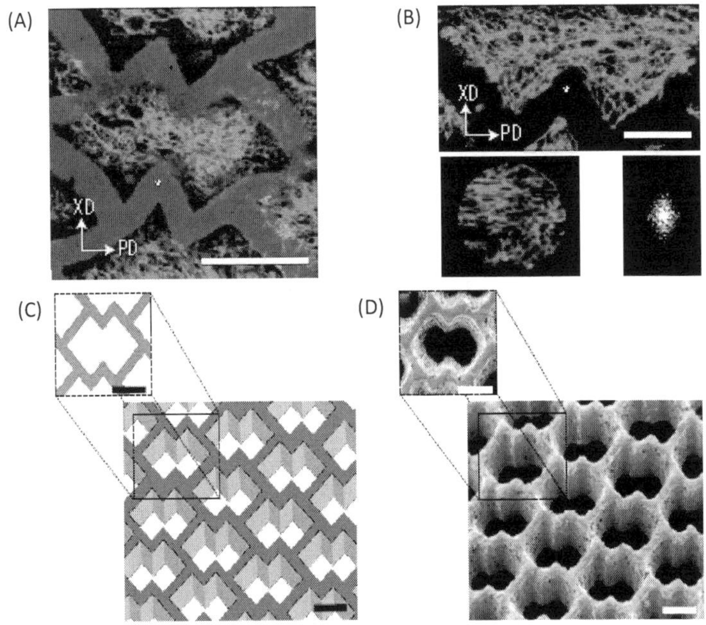


**Figure 7.4** Honeycomb shapes are able to imitate the anisotropic mechanical properties of the cardiac tissue solely by their architectural properties. They are also able to control the cell orientation in a way similar to the natural tissue. Confocal microscopy images from neonatal rat heart cells seeded honeycomb scaffolds; cells were labeled for (A, B) F-actin and stained for DNA; (C) architecture of scaffold which mimicked the directional properties of the heart; (D) SEM image showing high fidelity production of resultant accordion-like honeycomb PGS scaffold; Scale bars: (A) 200 μm; (B) 100 μm; (C) 200 μm; (D) 200 μm. Reprinted from Ref. 70, with permission of Nature Publishing Group, Copyright © 2008.

and collagen, which provide structural support and adequate stiffness.[71] Therefore, many researchers have attempted to mimic the mechanical characteristics of natural ECM to guide and regulate cellular function.[71] Tissues in the body have different mechanical properties with elastic modulus ranging from 10s of Pa to 100s of MPa. To engineer these tissues, it is required to design scaffolds with biomimetic mechanical properties, which possess similar elastic modulus and/or elasticity to that of native tissues. In a study, Engler *et al.* controlled the differentiation of mesenchymal stem cells (MSCs) into lineages corresponding to the stiffness of native environment by tuning the stiffness of the hydrogel surfaces.[72] MSCs were cultured on collagen-coated hydrogels with elastic modulus and elasticity similar to various tissues ranging from soft tissues (brain) to hard tissues (bone). It was demonstrated that MSCs differentiated to neurons when cultured on soft gels with modulus in the range of 0.1–1 kPa. The gels with stiffness and elasticity similar to muscle (i.e., 8–17 kPa) induced MSCs differentiation toward myogenic cells. Osteogenic differentiation was also observed when MSCs cultured on hydrogels, which had highest stiffness between 25–40 kPa and elasticity similar to bone.[72] Hu *et al.* studied the effect of material elasticity on osteogenic and myogenic differentiation of cells on silk/tropoelastin scaffolds.[73] Composite scaffolds with different ratios of tropoelastin and silk were fabricated by using a temperature-controlled water vapor annealing process and subsequently seeded by human bone marrow stem cells (hMSCs) and myoblasts. Cellular differentiation and growth were controlled by changing the polymer compositions. Increasing the concentration of tropoelastin enhanced the elasticity of scaffold, which resulted in proliferation and differentiation of hMSCs to osteoblast on the surfaces of composites with elastic modulus of 5 MPa. However, proliferation and myogenic differentiation of myoblasts were promoted by using stiffer silk–tropoelastin composites with the stiffness ranging from 20 MPa to 28 MPa.[73]

Biomaterial stiffness also affects cell proliferation, signaling, and differentiation. For some cells, stiff matrices improve cell attachment via changing focal adhesions and cytoskeleton organization, while soft substrates inhibit both cell proliferation and spreading. Ren *et al.* regulated the differentiation of skeletal muscle cells (C2C12) by changing the stiffness of poly (L-lysine)/hyaluronan (PLL/HA) films.[74] Composite films with varying stiffness were fabricated through chemical crosslinking using ethyl-3-(3-dimethylamino-propyl) carbodiimide (EDC) and *N*-hydrosulfosuccinimide (sulfo-NHS) as crosslinking agents. C2C12 myocytes were subsequently seeded on these films and the effect of material stiffness on cell differentiation was studied. It was shown that film stiffness affected the cell morphology, differentiation, and the morphology of myotubes (Fig. 7.5). Cellular differentiation was observed only for few days followed by cell aggregation and detachment after ~2 to 4 days when soft films were used. In addition, the myotubes had a short and thick morphology on the surfaces of soft substrates. However, cell elongation and formation of thin myotubes were observed by culturing the cells on stiff films (stiffness >320 kPa) for up to 2 weeks.[74]

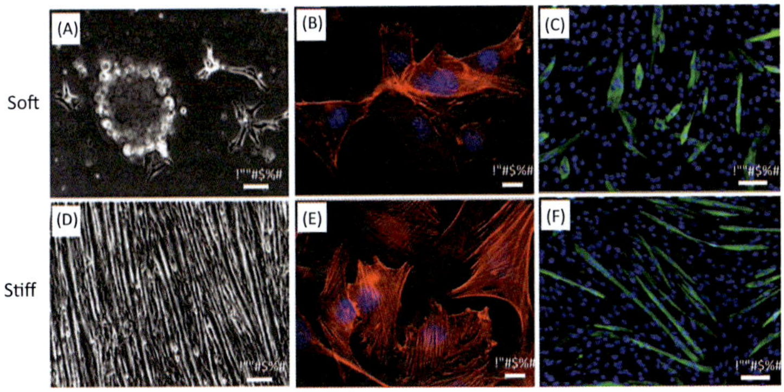


**Figure 7.5** Effect of material stiffness on C2C12 cell morphology, arrangement of cytoskeleton and focal adhesion proteins, and myotubes morphology on the surfaces of PLL/HA films. Phase contrast micrographs of C2C12 seeded on (A) soft and (D) stiff films on day 6 of culture (scale bar: 100 μm); Cytoskeletal actin staining of cells cultured on (B) soft and (E) stiff films on day 1 of culture (scale bar: 10 μm); Images of troponin (T–labeled in green and DNA-labeled in blue) of C2C12 cells cultured on (C) soft and (F) stiff films after 3 days of culture (scale bar: 100 μm). Reprinted with permission from Ref. 74, Copyright © 2008 Wiley-VCH Verlag GmbH & Co. KGaA, Weinheim.

In another study, the regeneration of cartilaginous components was directed by manipulating the mechanical properties of genipin-crosslinked polyethylene oxide (PEO)/chitin/chitosan composites.[75] The stiffness of composites was changed from 3 MPa to 12.5 MPa by tuning the polymer compositions. Increasing the weigh percentage of PEO reduced the elastic modulus and improved the extensibility or elasticity of the resultant composites. The optimal composition for chondrogenesis over 4 weeks of culture was reported to be 25% PEO, 25% chitin, and 50% chitosan.[75] In an attempt to regulate the spontaneous contraction of the construct, Shapira-Schweitzer *et al.* also modulated the stiffness of PEGylated fibrinogen hydrogels by using different ratios of fibrinogen and PEG.[76] To form the hydrogels, PEGylated fibrinogen precursor was mixed with neonatal rat cardiomyocytes and subsequently photo-crosslinked. The stiffness of hydrogels ranges from 8 kPa to 414 kPa, depending on the fibrinogen concentration and ratio of PEG to fibrinogen. The image analysis of the constructs showed an inverse correlation between hydrogels stiffness and amplitude of contraction.[76]

Biomaterials with gradients in mechanical properties have also been created to study the effect of substrate stiffness on cell behaviors including mobility, migration, signaling, and differentiation.[77–81] For example, Isenberg *et al.* investigated the effect of material stiffness on the migration of bovine aortic VSMCs on a hydrogel containing a gradient of mechanical properties with modulus ranging from 5 kPa to 80 kPa.[79] This gradient was created in collagen-incorporated polyacrylamide (PAAm) hydrogels by photopolymerizing the polymer precursor solution, containing a crosslinker and a photoinitiator, in a microfluidic device. The gradient of stiffness was generated in hydrogel by changing crosslinker concentration injected through one of the inlets. The VSMCs response to gradients in material stiffness (durotaxis) was characterized. Cell alignment was observed in the direction of the stiffness gradient and strengthened by increasing the magnitude of the gradient.[79] Similarly, Wong *et al.* showed that VSMCs migrated to stiffer regions of PAAm hydrogels when cultured on gradient hydrogels with modulus ranging from 5 kPa to 35 kPa.[78] In another experiment, the effect of PAAm stiffness on VSMCs spreading was studied by generating a gradient of elastic modulus in the range of 3 kPa to 40 kPa within the hydrogel.[77] It was found that cell spreading was significantly enhanced in the stiffer regions of hydrogel with modulus higher than 30 kPa after 18 h of culture.[77] Marklein *et al.* also fabricated a stiffness gradient in methacrylated hyaluronic acid gels to optimize the spreading and proliferation of hMSCs seeded on these gel.[80] Cells exhibited optimal spreading and proliferation in the regions of hydrogels with elastic modulus between 20 kPa and 90 kPa.[80] All these studies demonstrated that the material stiffness plays a crucial role in controlling various cell functions.

### 7.2.2 Pore Architectures

In addition to elasticity and stiffness, the microstructures of ECM affect the cell behavior and function. The pore architectures of ECM (e.g., porosity, pore interconnectivity, and pore size) significantly influence the extent of nutrient and oxygen diffusion for cell survival as well as removal of metabolic waste.[82] To function and grow properly, cells require highly porous and permeable environments, which provides adequate oxygen and nutrient delivery.[82] It is reported that cells can survive when they are located up to approximately 100 μm from a source of nutrient and oxygen, even in tissues containing highly vascularized networks.[83] There are many reports in literature on limited and peripheral tissue formation caused by inadequate nutrient and oxygen delivery into the 3D tissue constructs.[84,85] The formation of dense cell layers on the outer surfaces of these constructs reduces the supply of nutrients to the interior of tissues, which ultimately results in necrosis within the constructs.[86,87] Consequently, controlling the pore architectures of scaffolds can be utilized as a means of regulating cell survival, signaling, growth, and differentiation.

The pore size and porosity of biomaterials have been demonstrated to affect cell migration, proliferation, growth, and intercellular signaling. Generally, larger pores (>50 μm) allow for migration of cells into the scaffolds whereas small pores (<10 μm) improve cell adhesion properties.[88] Consequently, the presence of both microporosity and macroporosity within 3D scaffolds may facilitate cell attachment and growth, and ultimately tissue formation.[89] The optimal pore sizes for the proper function of various cells and tissues as suggested by various experiments are summarized in Table 7.1.

**Table 7.1** The effects of implant pore sizes on tissues regeneration

| Tissue formation/cell growth | Required pore size (μm) | References |
|---|---|---|
| Neovascularization | 5 | 90 |
| Fibroblast ingrowth | 5–15 | 91 |
| Regeneration of adult mammalian skin | 20–115 | 92, 93 |
| Regeneration of bone | 100–350 | 91, 94 |
| Osteoid ingrowth | 40–100 | 91 |
| Hepatocytes ingrowth | 20 | 95, 96 |
| Fibrovascular tissue | 500 | 97 |

The impact of scaffold pore architecture on cell growth and proliferation, and differentiation has been demonstrated by using various types of biomaterials. It has been reported that the porosity and average pore size of scaffold significantly affect the osteogenic cell proliferation as well as extent of osteogenesis.[98–101] In one study, non-woven polyethylene terephthalate (PET) scaffolds with varying porosities and fiber diameters were fabricated and used as substrates for rat MSCs growth and proliferation.[100] The rate of cell proliferation was enhanced when cells were cultured on scaffolds with higher porosity. On the other hand, lower porosity led to the higher osteogenic differentiation. Similarly, Mygind *et al.* reported increased proliferation rates of hMSCs cultured on HAP scaffolds with larger pores (500 μm) and faster rate of osteogenic differentiation on scaffold with smaller pore sizes (200 μm).[101] In another study, the effect of scaffold pore size on *in vivo* osteogenesis was investigated by implanting calcium aluminate cylindrical pellets with average pore sizes ranging from 10 μm to 200 μm and porosity of 46% in dog femorals for 12 weeks.[102] The results demonstrated that bone ingrowth and vascular formation promoted in implants with average pore size of 200 μm. Scaffolds with medium pore size (75–100 μm) supported the formation of unmineralized osteoid tissues and implants with small pores (<75 μm) facilitated the formation of fibrous connective tissues.[102]

Biomimetic scaffolds with gradients of pore size and porosity have also been created to study the effect of pores on cell behavior.[81,88,103,104] For example, Dubruel *et al.* produced two types of porous gelatin scaffolds with varying pore sizes and pore morphology, one of which contained conical pores and a gradient of pore sizes ranging from 20 μm to 330 μm and the other scaffold contained spherical pores with an average diameter of 135 μm.[103] Both scaffolds were cultured with various human cells such as epithelial, fibroblast, glial, and osteoblast to investigate the impact of pore characteristics on cell adhesion and spreading. The results of *in vitro* studies indicated that pore size and geometry had no significant impact on cell adhesion and proliferation and both scaffolds supported attachment and proliferation of cells.[103] Roy *et al.* created porosity gradient in a polylactic-co-glycolic acid/beta-tricalcium phosphate (PLGA/beta-TCP) scaffold to study the effect of porosity on *in vivo* bone formation.[105] The gradient scaffolds were fabricated by using a 3D printing technique and subsequently implanted in rabbit craniums. After 8 weeks of implantation, the regions of scaffold with higher porosity (up to 88%) exhibited increased bone tissue formation.[105] In another study, Oh *et al.* fabricated PCL scaffolds with gradients of porosity and pore sizes to investigate the *in vitro* cell interactions with the fabricated scaffold using different cells such as chondrocytes, osteoblasts, and fibroblasts.[88] The gradient scaffolds were produced by the centrifugation of a mold, which contained prefabricated fibril-like PCL, followed by heat-induced fibril bonding. The resultant scaffolds had gradients of pore size diameters ranging from 88 μm to 405 μm and porosity from 80% to 94%. *In vitro* studies demonstrated that different types of cells required different pore sizes for their optimal growth and proliferation; chondrocyte and osteoblast growth were promoted in the regions of scaffold with 380–405 μm pore size while fibroblasts grew faster in the scaffold region with pore size in the range of 186–200 μm. The *in vivo* tissue interactions with gradient material were also examined by using a rabbit model (skull bone defects). Following 4 weeks of implantation, faster bone formation was observed in scaffold region with pore sizes ranging from 290 μm to 310 μm.[88]

These studies together highlight the important role of physical properties of materials in dictating various cell behaviors such as migration, signaling, motility, and differentiation. Various fabrication techniques have been developed for controlling the physical properties of scaffolds to induce different cell behaviors. Some of these techniques were reviewed in sections 7.1 and 7.2. A more comprehensive review of these methods can be found in the literature.[81,106–108]

## 7.3 Biomimicry by Modification of Chemical Properties of Scaffolds

The biochemical aspect of biomimicry can be divided into three main areas: (1) mimicking the spatiotemporal distribution of biochemical factors; (2) mimicking the complexity of ECM structure to obtain cellular microenvironments; and (3) designing synthetic proteins or enzyme-sensitive synthetic polymers to control cells' remodeling activities.

### 7.3.1 Mimicking the Spatiotemporal Distribution of Biochemical Factors

Tissue engineering basically aims to bridge the gap between what the body can do during regeneration of tissues/organs and what it cannot in adulthood. Morphogenesis, the path of development in which a given organism assumes its shape, is strongly affected by the presence of physical and biochemical cues. Scaffolding can be viewed as a radical approach to morphogenesis, as it is used as a template for seeded cells to regenerate the final structure via its initial architecture and with the aid of bioactive agents. However, the distinct difference between the embryo where 3D cell aggregates turn into well-structured organs and the remodeling process of an artificial scaffold of cells to create a semblance of their *in vivo* environment should be taken into consideration.[109] Thus, it can be argued that instead of engineering the cellular compositions and the scaffold, it may be useful to engineer the actual process of the cellular remodeling of fabricated structure. Each step of scaffold remodeling then matches a developmental stage and biochemical signal plays a critical role in this process.[110] Growth factors both in liquid and solid phase have a direct control on the outcomes of biological events. Stem cell differentiation, phenotype changes, cell proliferation, angiogenesis, apoptosis, and many other cellular activities are directly controlled by biochemical signals. One approach to mimic this process is to use gradients of growth factors or other bioactive agents within tissue engineering scaffolds. Gradients of bioactive agents can be achieved by direct printing techniques such as inkjet printing,[111] inclusion in the certain parts of the scaffolds by micro- and nanopatterning techniques,[112] or by microfluidics methods.[113] A common method is the covalent conjugation of growth factors into the scaffold to mimic the actual structure that exists *in vivo* (i.e., bound to the ECM) (Fig. 7.6). This approach can be coupled to patterned surfaces to control the cellular differentiation by the gradients of different growth factors in microscale distances.[114] Compared to supplying growth factor exogenously, the approach of using covalent bonding and production of gradients of growth factors is beneficial as it prolongs their effects. This effect is mainly due to the fact that the growth factors in solution can easily diffuse away or can be degraded. The *in vivo* processing of growth factors is the topic of another line of study, which aims to exert control over temporal presence of bioactive signals. As growth factor stimulation also has a temporal aspect, methods that can mimic the removal of biochemical cues are in development. For example, recently a modified peptide sequence which contained both photo-degradable and photo-crosslinkable moieties, which can be incorporated into a hydrogel structure, has been used to control the distribution of biochemical signals.[115] With such a system, by using the light stimulation cell attachment and detachment can be controlled both in different areas of the scaffold and also at different time points.

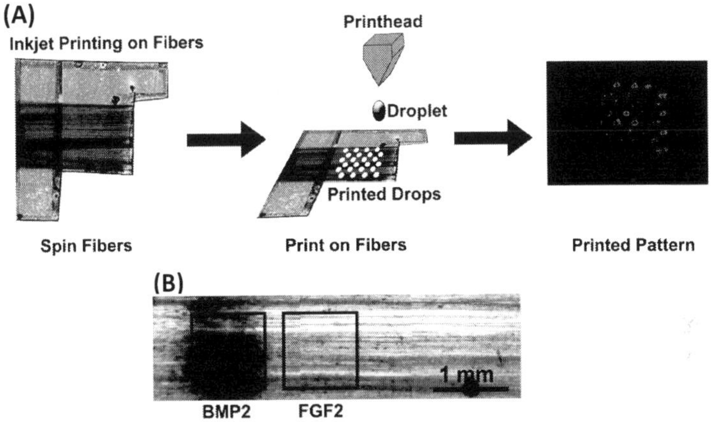


**Figure 7.6** Direct printing of bioactive molecules on aligned fibers to achieve biomimetic scaffolds that have spatial control over both on cell shape and differentiation. (A) The aligned fibers loaded with bone morphogenetic protein 2 (BMP2) or fibroblast growth factor 2 (FGF-2) can change the differentiation state of multipotent C3H10T1/2 cells to tenocytes or osteoblasts. (B) Different gradients can be used to locally differentiate cells to mimic the gradients in the musculoskeletal system. Local presence of BMP-2 increases ALP expression whereas printed FGF-2 or only fibers without growth factors have no effect, which demonstrates the importance of the local growth factor levels. Reprinted from Ref. 111, Copyright (2011), with permission from Elsevier.

Surface functionalization is the easiest and the most utilized way of adding bioactive factors to the scaffold, either by covalent bonding, deposition of polyelectrolyte multilayers, or absorption of nanoparticles.[116] Another possible method is the direct addition of growth factors into the prepolymer solution that is used for the fabrication of scaffold but this approach is viable for mild scaffold fabrication processes that perform at moderate temperatures and mild operating conditions. For example, for scaffolds fabricated using electrospinning direct addition of growth factors to prepolymer solution is generally not viable. Only more resistant bioactive agents such as dexamethasone (osteoblast differentiation) or heparin (anticoagulant surface coating for vascular tissue engineering) can be processed by eletrospinning.[117]

Generally, bioactive agents work in combinations or in a temporally sequential way. For example, a wound healing cascade is a synchronized chain of events where certain bioactive molecules are up or down regulated at different phases.[118] In the case of bone healing, bone morphogenic protein (BMP) family members appear at different stages of healing *in vivo*. One way of mimicking this property has been to integrate multiple growth factor delivery systems within biomaterials.[119] Different modes of delivery ensure the timely release of critical growth factors in the correct order. Delivery can be either at the cytokine level or at the gene level.[120] Gene activated matrices are scaffolds which have a plasmid DNA component that can be used to direct the cellular response via transfection of the incoming progenitor cells and their expression of the plasmid product in the scaffold *in vivo*. The transfection is achieved by adenovirus or adeno-associated virus.[121] This method can precisely mimic the sustained release profile of desired biochemical agents. However, as in all gene delivery systems, care must be taken to minimize side effects of the over-expression of specific genes. Gene delivery is a powerful method, and successful long-term expressions of specific genes have been demonstrated in bone defects.[122]

Finally, another widely employed method of mimicking the growth factor milieu of healing tissue is the utilization of platelet-enriched plasma. This system exploits the fact that platelets are small depots of several important cytokines such as vascular endothelial growth factor (VGF), platelet-derived growth factor (PDGF), transforming growth factor beta (TGF-β), and many others, which are relevant to healing process. Injection of platelet-enriched plasma or its utilization in combination with scaffolds such as fibrin gels have been shown to improve outcomes in animal models.[123] However, this approach is limited by the inherent problem of variability and lack of control over the final contents.

### 7.3.2 ECM Components as Biomimetic Scaffold Building Blocks

The *in vivo* physical surrounding of the cells is defined by their ECM, which consists of a complex mixture of different components such as proteins, glycosaminoglycans, and cytokines. ECM is the interface between the biophysical and biochemical effects as ECM molecules both have functional and structural importance in the cellular microenvironment.[124] This fact has become the basis of many efforts in biomimicking cellular environment, whether it is a stem cell niche or the ECM of a particular organ, to direct cell behavior. For proper function of stem cells, their microenvironment should be faithfully mimicked.[125] Failing to have control over stem cell behavior can result in loss of regenerative capacity or tumor formation *in vivo*. A possible biomimicry method for the optimal ECM microenvironment is the decellularization of organs or tissues such as small intestinal submucosa or decellularized bladder matrix. These structures, which are obtained by either mild detergent or enzymatic treatments, elicit fast cellular integration and they can be rendered anti-immunogenic. The most important advantage of these systems is retaining the inherent ECM content, such as the glycosaminoglycan content of the original tissue. This effect is especially beneficial if the target tissue is similar to the source tissue (i.e., using decellularized allogenic trachea for tracheal tissue engineering).[126] Moreover, they have the capacity to direct the differentiation of stem cells. For example, when MSCs are cultured on a decellularized microvascular endothelium matrix, they differentiated toward vascular lineage.[127] The limitations of this approach include the need for donors, donor variability and lack of precise control of these structures.

A more reliable and reproducible approach to harness the advantages of ECM signals is to use the native ECM components after isolating them from animal tissues or through producing them using recombinant DNA technology.[128] In conjunction with native ECM molecules, a commonly used method for biomimicry is to direct the assembly of structural ECM molecules such as collagen and glycoaminoglycans to achieve gel structures, which can exert control over biological events. One of the most versatile components for this purpose is hyaluronic acid, which has anti-inflammatory properties, induces elastin production[129] and possesses pluripotency supporting properties.[130] ECM molecules can be used alone or in the form of composites with other ECM molecules. For instance, combinatorial approaches showed that the best microenvironment for embryonic stem cells (ESC) renewal is composed of several ECM components.[131] Such findings had significant impact on the biomaterial-based stem cell microenvironment fabrication techniques, which are in essence biomimetic scaffolds for controlling cell behavior. In addition to micropatterning techniques that achieve physical guidance to cells, microscale patterning of ECM molecules is one of the major methods to mimic the structural guidance *in vivo*. For this purpose, microfluidics, soft lithography and stencils have been used. To achieve precision at nanoscale, direct printing methods based on AFM have been developed.[50] Another approach for creating microscale cellular environments is the development of cell aggregates in confined volumes. There is a substantial amount of research in the literature for developing collectible cell aggregates with controllable sizes.[132] To improve the rate of collection, thermo-responsive polymer-based platforms have been developed which facilitate the removal of the final products. The cell aggregate environment can be further controlled by addition of biomaterial particles which can also modulate stem cell differentiation.[133]

Mimicking the bulk properties of tissues are important, however, the effective mimicry of the properties commences when reproducing the properties of the interfaces between the cells and ECM. This can range from hydrophilicity to roughness to interfacial energy. For example, interface hydrophilicity has a cascade effect, where first it influences the adsorption of the proteins, which subsequently changes the cell behavior. One of the major problems with controlling biochemical and physical effects is that they are generally intertwined, that is, it is very difficult to convolute the contribution of a specific effect from the others. Some possible methods to resolve this issue is to find systems that change in one aspect without affecting other factors. Another route is confining biochemical signals in defined areas. For example utilization of fibronectin islands showed that the size of the island could be used to direct stem cells in different differentiation pathways. Small islands caused adipogenesis whereas bigger islands resulted in osteogenesis mainly through Rho/Rock pathway dependent manner.[134] For such systems, surfaces that can promote the absorption of proteins in a way that would preserve their functional conformation are also necessary. This is an example of the interplay between scaffold surface properties with the biochemical signaling. The production of artificial ECM by deconstruction of the components in the target tissue and their reconstruction with respect to the desired effects *in vitro* is an active area of research. One step further in this line of research is to mimic the ECM properties with synthetic structures, such as artificial proteins.

### 7.3.3 Engineering Biomolecules for Tissue Engineering

Even though, many tissue engineering products based on natural polymers, extracted from animals are currently in the market, their use has several disadvantages. Natural polymers generally possess good biochemical properties that ensure cell attachment and proliferation, but their evolutionary pathway limits their versatility. The range of mechanical properties that can be obtained or the production techniques are highly dependent on the structures of the natural polymers. For example, collagen is a suitable material for scaffold production, but its propensity to denature even under mild environmental changes hinders its more widespread use. These can be alleviated by the utilization of synthetic proteins or polypeptides, due to their versatility and ability to mimic natural conditions. Use of recombinant proteins instead of naturally occurring proteins is advantageous since it provides control over post-translational modifications in a way that ensures the desired functionality. This approach also overcomes long isolation steps, which does not always give a reproducible source of material both in composition and quantity. Moreover, it avoids disease transmission related to utilization of xenogenic materials thus improves the product safety. Some examples are human elastin-like polypeptide,[135] recombinant tropoelastin, ubiquitin fragments (antimicrobial peptides for protection of scaffolds against infection), and self-assembling polypeptide chains (3D hydrogel-based scaffolds).[136–138] The synthesis of elastin and tropoelastin by cell is generally difficult; therefore, they can be produced by recombinant technology. This product can be used for regeneration of organs with elastic properties such as skin, cardiac, heart valve, and vascular.

Self-assembly fibrillar synthetic proteins can be used to form gels in a versatile manner where several functionalities can be combined in a way that does not exist in nature. By this way bio-inspired multifunctional materials can be obtained as scaffold bases.[139] An example of this is biotinylation of synthetic peptides to achieve retention of growth factors to enable their controlled delivery similar to the native ECM controls the release of such agents *in vivo*. Such a system was used *in vivo* and was shown to improve cardiac function following injury.[140]

One of the methods to obtain multifunctionality is to design protein structures or short polypeptide chains from scratch that endows the necessary properties while at the same time making them more resilient to the physical conditions such as temperature or pH change. Moreover, a fully synthetic protein production system provides high purity, an ultimate control over all physical properties, and the ability to modify the material at a single amino acid level thus obtain subtle changes in 3D folding and assembly of the peptide chains.[141] One step further is the functionalization of synthetic or natural polymers with such designed molecules to produce hybrid materials with superior properties.

Engineered inclusion and exclusion of the signals of natural molecules makes it possible to study and exploit the effect of single signal on cells. Moreover, in this way the interaction of the material units can be controlled and self-assembly processes that result in complex structures can be predesigned. Another advantageous property of synthetic polypeptides is that they do not elicit strong immune responses, due to their natural structures. Moreover, their level of complexity can be brought to a level where they become an active part of signaling processes at different stages of their degradation. As *in vivo* degradation products of proteins are important messengers, the synthetic protein structures can be designed in a way that their degradation releases shorter polypeptides with other functionalities, hence enabling researchers to temporally control the remodeling process of the scaffolds.

Nanofibrous structure of ECM is one of the important properties; especially the nanofibrillar nature of basement membrane and how it affects the behavior of epithelial cells and endothelial cells are quite relevant for tissue engineering scaffolds.[142] Nanofibrous structures can be engineered by methods such as electrospinning, phase separation, nanopatterning as described in previous sections. Synthetic protein technology also provides systems that can form nanofibers of controllable size via self-assembly (Fig. 7.7).[143] This way not only the physical fiber structure are mimicked but also the presence of biochemical signal along the fibers can be faithfully reproduced.

Self-assembling polypeptides were first obtained from natural resources. For example, one of the first self-assembling peptide sequences, EAK16-II, was from zeotin, a yeast protein.[144] The repetitive structure, which contained hydrophobic, hydrophilic, and ionic amino acids, resulted in the spontaneous formation of beta-sheets. This led to more studies to create structures from repeating sequences, like block co-polymer systems used for synthetic polymers. This method proved to be highly versatile, as even simple changes like increasing the repeat chain numbers or changing the amino acids to control hydrophobicity had important effects on the final mechanical properties and the formation process of self-assembling scaffolds. These scaffolds have been used for a wide variety of applications such as nerve and cardiac tissue engineering and stem cell niche development or cell encapsulation.[143,145,146]

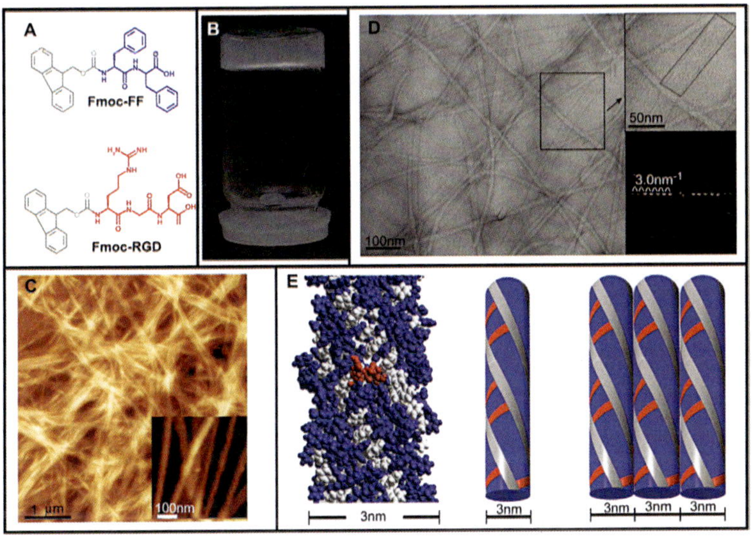


**Figure 7.7** Nanofibrillar scaffolds from short synthetic peptide derivatives (Fmoc-FF and Fmoc-RGD) that can self-assemble. Self-assembly process happens at 37°C and results in a hydrogel composed of nanofibers. Initially, 3 nm fibers are formed, later these fibers laterally self-assemble to form bigger fibers. The building blocks have RGD sequences (Fmoc-RGD) which is engineered in a way that upon self-assembly they are presented on the outer layers. With such a design, the interaction of cells with the scaffold can be finely controlled. Reprinted from Ref. 143, Copyright (2009), with permission from Elsevier.

*In vivo*, remodeling of tissues necessitates turnover of ECM components, which are cleaved by the enzymes secreted by the cells. One problem with the synthetic scaffolds, especially for the ones used for cell encapsulation is the lack of the necessary signals that can provide active degradation of the scaffold by the cells. This has significant effects on the ability of cells to proliferate and migrate after encapsulation. In order to mimic ECM degradation and cell response to ECM interaction by the release of matrix metalloproteinases (MMPs) and general proteases can be used. This is defined as dynamic reciprocity between cell and the matrix. To achieve this, synthetic polymers were modified with enzyme sensitive peptide sequences to produce hybrid, enzyme degradable scaffolds.[147] The sequences for MMP-mediated degradation or plasmin sensitive degradation are known and they have been successfully incorporated in either synthetic polypeptides or synthetic materials such as PEG to obtain more remodelable scaffolds. This way the remodeling process via degradation can be better mimicked and cell movement and spreading can be improved.[148] Moreover, the degradation process can be directly linked to the release of growth factors or dynamic control of cell adhesion, which provides an additional level of control and biomimicry. For example, when RGD functionalization of PEG gels linked with MMP-13 removal capability, the glycosaminoglycan production has been improved 10-fold.[149]

Most of the components in the body have multiple functions and their degradation products might have important functions. For example, ubiquitin which is a common protein, have proteolytic degradation products which can act as antifungal, antibacterial, or cell adhesion agents.[150] Next generation of biomaterials need to mimic this versatility in order to obtain long-term control over the fate of the scaffold not only *in vivo* initially but also long after its implantation. Together with the lessons learned from developmental biology, such as the ability to create the embryo's developmental environment for improved regeneration, in the future biomimicry of biochemical signals will become an even more important part of tissue engineering scaffold production efforts. In the long run, tissue engineering can be used for reverse engineering the biological processes to better understand developmental pathways without the ethical concerns related to the utilization of live embryos.

## 7.4 Summary and Future Directions

Significant efforts have been made in designing biomimetic scaffolds to regenerate clinically useful tissue replacements. The architecture, physical, and chemical characteristics of these scaffolds regulate cellular events including attachment, migration, proliferation, and differentiation. Various techniques have been employed including RP, electrospinning, and phase separation for creating 3D biomimetic scaffolds. These techniques provide control over the structural features of the scaffold to regulate cell functions. In addition to scaffold architecture, the physical properties of biomaterials such as mechanical stiffness and pore characteristics influence various cell behavior and controlling these properties may be beneficial in creating scaffolds with biomimetic properties. Another aspect of biomimetic scaffolds is biochemical properties, which can be controlled through the patterning of growth factors, fabrication of artificial ECM systems, and engineering synthetic proteins. Chemical stimuli by itself is usually not enough to induce the desired changes in cellular behavior, and the introduction of physical environment components (i.e., mechanical properties, architecture, etc.) is crucial. There is also limited capability of controlling cellular functions by only the scaffold structural features. Therefore, there is significant opportunity to advance the field further by controlling both physical architecture and chemical compositions of scaffolds in order to mimic certain characteristics of the natural ECM. The development of these biomimetic scaffolds may provide novel solutions in regenerative medicine with the ultimate goal of creating functional engineered tissues.

## References


1. West-Mays, J. A. and Dwivedi, D. J. (2006) The keratocyte: Corneal stromal cell with variable repair phenotypes, *Int. J. Biochem. Cell Biol.*, **38**, 1625–1631.
2. Wolinsky, H. and Glagov, S. (1964) Structural basis for the static mechanical properties of the aortic, *Circ. Res.*, **14**, 400–413.
3. Garnero, P., Borel, O., Gineyts, E., Duboeuf, F., Solberg, H., Bouxsein, M. L., Christiansen, C. and Delmas, P. D. (2006) Extracellular post-translational modifications of collagen are major determinants of biomechanical properties of fetal bovine cortical bone, *Bone*, **38**, 300–309.
4. Wheeldon, I., Farhadi, A., Bick, A. G., Jabbari, E. and Khademhosseini, A. (2011) Nanoscale tissue engineering: Spatial control over cell-materials interactions, *Nanotechnology*, **22**, 212001–212001.

5. Khademhosseini, A. and Langer, R. (2007) Microengineered hydrogels for tissue engineering, *Biomaterials*, **28**, 5087–5092.
6. Jeong, J. H., Chan, V., Cha, C., Zorlutuna, P., Dyck, C., Hsia, K. J., Bashir, R. and Kong, H. (2012) "Living" microvascular stamp for patterning of functional neovessels; orchestrated control of matrix property and geometry, *Adv. Mater.*, **24**, 58–63.
7. Chan, V., Zorlutuna, P., Jeong, J. H., Kong, H. and Bashir, R. (2010) Three-dimensional photopatterning of hydrogels using stereolithography for long-term cell encapsulation, *Lab Chip*, **10**, 2062–2070.
8. Bibb, R., Freeman, P., Brown, R., Sugar, A., Evans, P. and Bocca, A. (2000) An investigation of three-dimensional scanning of human body surfaces and its use in the design and manufacture of prostheses, *Proc. Inst. Mech. Eng. H*, **214**, 589–594.
9. Mironov, V., Boland, T., Trusk, T., Forgacs, G. and Markwald, R. R. (2003) Organ printing: Computer-aided jet-based 3D tissue engineering, *Trends Biotechnol.*, **21**, 157–161.
10. Zein, I., Hutmacher, D. W., Tan, K. C. and Teoh, S. H. (2002) Fused deposition modeling of novel scaffold architectures for tissue engineering applications, *Biomaterials*, **23**, 1169–1185.
11. Vozzi, G., Flaim, C., Ahluwalia, A. and Bhatia, S. (2003) Fabrication of PLGA scaffolds using soft lithography and microsyringe deposition, *Biomaterials*, **24**, 2533–2540.
12. Yilgor, P., Sousa, R. A., Reis, R. L., Hasirci, N. and Hasirci, V. (2008) 3D plotted PCL scaffolds for stem cell based bone tissue engineering, *Macromol. Symp.*, **269**, 92–99.
13. Zorlutuna, P., Jeong, J. H., Kong, H. and Bashir, R. (2011) Stereolithography-based hydrogel microenvironments to examine cellular interactions, *Adv. Funct. Mater.*, **21**, 3642–3651.
14. Cooke, M. N., Fisher, J. P., Dean, D., Rimnac, C. and Mikos, A.G. (2003) Use of stereolithography to manufacture critical-sized 3D biodegradable scaffolds for bone ingrowth, *J. Biomed. Mater. Res. B Appl. Biomater.*, **64B**, 65–69.
15. Arcaute, K., Mann, B. and Wicker, R. (2006) Stereolithography of three-dimensional bioactive poly(ethylene glycol) constructs with encapsulated cells, *Ann. Biomed. Eng.*, **34**, 1429–1441.
16. Arcaute, K., Mann, B. and Wicker, R. (2010) Stereolithography of spatially controlled multi-material bioactive poly(ethylene glycol) scaffolds, *Acta Biomater.*, **6**, 1047–1054.

17. Zorlutuna, P., Rong, Z., Vadgama, P. and Hasirci, V. (2009) Influence of nanopatterns on endothelial cell adhesion: Enhanced cell retention under shear stress, *Acta Biomater.*, **5**, 2451–2459.

18. Elsdale, T. and Bard, J. (1972) Collagen substrata for studies on cell behavior, *J. Cell Biol.*, **54**, 626–637.

19. Ramirez, F. (2000) Pathophysiology of the microfibril/elastic fiber system: Introduction, *Matrix Biol.*, **19**, 455–456.

20. Mithieux, S. M. and Weiss, A.S. (2005) Elastin, *Adv. Protein Chem.*, **70**, 437–461.

21. Wei, G. and Ma, P. X. (2008) Nanostructured biomaterials for regeneration, *Adv. Funct. Mater*, **18**, 3568–3582.

22. Zhang, Y., Venugopal, J. R., El-Turki, A., Ramakrishna, S., Su, B. and Lim, C. T. (2008) Electrospun biomimetic nanocomposite nanofibers of hydroxyapatite/chitosan for bone tissue engineering, *Biomaterials*, **29**, 4314–4322.

23. Li, M., Mondrinos, M. J., Gandhi, M. R., Ko, F. K., Weiss, A. S. and Lelkes, P. I. (2005) Electrospun protein fibers as matrices for tissue engineering, *Biomaterials*, **20**, 5999–6008.

24. Ma, P. X. (2008) Biomimetic materials for tissue engineering, *Adv. Drug Deliv. Rev.*, **60**, 184–198.

25. Chung, B. G., Kang, L. and Khademhosseini, A. (2007) Micro- and nanoscale technologies for tissue engineering and drug discovery applications, *Expert Opin. Drug Discov.*, **2**, 1653–1668.

26. Reneker, D. H. and Chun, I. (1996) Nanometer diameter fibers of polymer, produced by electrospinning, *Nanotechnology*, **7**, 216–223.

27. Pham, Q. P., Sharma, U. and Mikos, A. G. (2006) Electrospinning of polymeric nanofibers for tissue engineering applications: A review, *Tissue Eng.*, **12**, 1197–1211.

28. Matthews, J. A., Wnek, G. E., Simpson, D. G. and Bowlin, G. L. (2002) Electrospinning of Collagen Nanofibers, *Biomacromolecules*, **3**, 232–238.

29. Geng, X., Kwon, O.-H. and Jang, J. (2005) Electrospinning of chitosan dissolved in concentrated acetic acid solution, *Biomaterials*, **26**, 5427–5432.

30. Nivison-Smith, L., Rnjak, J. and Weiss, A. S. (2010) Synthetic human elastin microfibers: Stable cross-linked tropoelastin and cell interactive constructs for tissue engineering applications, *Acta Biomater.*, **6**, 354–359.

31. Rnjak, J., Li, Z., Maitz, P. K. M., Wise, S. G. and Weiss, A. S. (2009) Primary human dermal fibroblast interactions with open weave three-dimensional scaffolds prepared from synthetic human elastin, *Biomaterials*, **30**, 6469–6477.

32. Wulkersdorfer, B., Kao, K. K., Agopian, V. G., Ahn, A., Dunn, J. C., Wu, B. M. and Stelzner, M. (2010) Bimodal porous scaffolds by sequential electrospinning of poly(glycolic acid) with sucrose particles, *Int. J. Polym. Sci.*, doi:10.1155/2010/436178.

33. Li, W.-J., Laurencin, C. T., Caterson, E. J., Tuan, R. S. and Ko, F. K. (2002) Electrospun nanofibrous structure: A novel scaffold for tissue engineering, *J. Biomed. Mater. Res.*, **60**, 613–621.

34. Pham, Q. P., Sharma, U. and Mikos, A. G. (2006) Electrospun poly(e-caprolactone) microfiber and multilayer nanofiber/microfiber scaffolds: Characterization of scaffolds and measurement of cellular infiltration, *Biomacromolecules*, **7**, 2796–2805.

35. Sant, S., Hwang, C. M., Lee, S.-H. and Khademhosseini, A. (2011) Hybrid PGS–PCL microfibrous scaffolds with improved mechanical and biological properties, *J. Tissue Eng. Regen. Med.*, **5**, 283–291.

36. Keun Kwon, I., Kidoaki, S. and Matsuda, T. (2005) Electrospun nano- to microfiber fabrics made of biodegradable copolyesters: Structural characteristics, mechanical properties and cell adhesion potential, *Biomaterials*, **26**, 3929–3939.

37. Kim, T. G., Chung, H. J. and Park, T. G. (2008) Macroporous and nanofibrous hyaluronic acid/collagen hybrid scaffold fabricated by concurrent electrospinning and deposition/leaching of salt particles, *Acta Biomater.*, **4**, 1611–1619.

38. Chen, V. J. and Ma, P. X. (2006) The effect of surface area on the degradation rate of nano-fibrous poly(L-lactic acid) foams, *Biomaterials*, **27**, 3708–3715.

39. Zhang, R. and Ma, P. X. (1999) Poly(alpha-hydroxyl acids)/hydroxyapatite porous composites for bone-tissue engineering. I. Preparation and morphology, *J. Biomed. Mater. Res.*, **44**, 446–455.

40. Wei, G. and Ma, P. X. (2004) Structure and properties of nano-hydroxyapatite/polymer composite scaffolds for bone tissue engineering, *Biomaterials*, **25**, 4749–4757.

41. Wei, G. and Ma, P. X. (2006) Macroporous and nanofibrous polymer scaffolds and polymer/bone-like apatite composite scaffolds generated by sugar spheres, *J. Biomed. Mater. Res. A*, **78**, 306–315.

42. Ma, P. X. and Zhang, R. (1999) Synthetic nano-scale fibrous extracellular matrix. *J. Biomed. Mater. Res.*, **46**, 60–72.

43. Liu, X., Smith, L. A., Hu, J. and Ma, P. X. (2009) Biomimetic nanofibrous gelatin/apatite composite scaffolds for bone tissue engineering, *Biomaterials*, **30**, 2252–2258.

44. Liu, X. and Ma, P. X. (2009) Phase separation, pore structure, and properties of nanofibrous gelatin scaffolds, *Biomaterials*, **30**, 4094–4103.

45. Lee, M. H., Kang, J. H. and Lee, S. W. (2012) The significance of differential expression of genes and proteins in human primary cells caused by microgrooved biomaterial substrata, *Biomaterials*, **33**, 3216–3234.

46. Bajaj, P., Reddy, B., Millet, L., Wei, C. N., Zorlutuna, P., Bao, G. and Bashir, R. (2011) Patterning the differentiation of C2C12 skeletal myoblasts, *Integr. Biol.*, **3**, 897–909.

47. Zorlutuna, P., Elsheikh, A. and Hasirci, V. (2009) Nanopatterning of collagen scaffolds improve the mechanical properties of tissue engineered vascular grafts, *Biomacromolecules*, **10**, 814–821.

48. Grayson, W. L., Martens, T. P., Eng, G. M., Radisic, M. and Vunjak-Novakovic, G. (2009) Biomimetic approach to tissue engineering, *Semin. Cell Dev. Biol.*, **20**, 665–673.

49. Papenburg, B. J., Liu, J., Higuera, G. A., Barradas, A. M. C., de Boer, J., van Blitterswijk, C. A., Wessling, M. and Stamatialis, D. (2009) Development and analysis of multi-layer scaffolds for tissue engineering, *Biomaterials*, **30**, 6228–6239.

50. Curtis, A. S. G., Dalby, M. J. and Gadegaard, N. (2006) Nanoprinting onto cells, *J. R. Soc. Interface*, **3**, 393–398.

51. Jackman, R. J., Brittain, S. T., Adams, A., Prentiss, M. G. and Whitesides, G. M. (1998) Design and fabrication of topologically complex, three-dimensional microstructures, *Science*, **280**, 2089–2091.

52. Walboomers, X. F. and Jansen, J. A. (2001) Cell and tissue behavior on micro-grooved surfaces, *Odontology*, **89**, 0002–0011.

53. Shen, J. Y., Chan-Park, M. B., Zhu, A. P., Zhu, X., Beuerman, R. W., Yang, E. B., Chen, W. and Chan, V. (2006) Three-dimensional microchannels in biodegradable polymeric films for control orientation and phenotype of vascular smooth muscle cells, *Tissue Eng.*, **12**, 2229–2240.

54. Fukuda, J., Khademhosseini, A., Yeo, Y., Yang, X., Yeh, J., Eng, G., Blumling, J., Wang, C.-F., Kohane, D. S. and Langer, R. (2006) Micromolding of photocrosslinkable chitosan hydrogel for spheroid microarray and co-cultures, *Biomaterials*, **27**, 5259–5267.

55. Yeh, J., Ling, Y., Karp, J. M., Gantz, J., Chandawarkar, A., Eng, G., Blumling, J. I., Langer, R. and Khademhosseini, A. (2006) Micromolding of shape-

controlled, harvestable cell-laden hydrogels, *Biomaterials*, **27**, 5391–5398.

56. Vrana, E., Builles, N., Hindie, M., Damour, O., Aydinli, A. and Hasirci, V. (2008) Contact guidance enhances the quality of a tissue engineered corneal stroma, *J. Biomed. Mater. Res. A*, **84**, 454–463.
57. Aubin, H., Nichol, J. W., Hutson, C. B., Bae, H., Sieminski, A. L., Cropek, D. M., Akhyari, P. and Khademhosseini, A. (2010) Directed 3D cell alignment and elongation in microengineered hydrogels, *Biomaterials*, **31**, 6941–6951.
58. Craighead, H. G., James, C. D. and Turner, A. M. P. (2001) Chemical and topographical patterning for directed cell attachment, *Curr. Opin. Solid State Mater. Sci.*, **5**, 177–184.
59. Diehl, K. A., Foley, J. D., Nealey, P. F. and Murphy, C. J. (2005) Nanoscale topography modulates corneal epithelial cell migration, *J. Biomed. Mater. Res. A*, **75A**, 603–611.
60. Hasirci, V., Vrana, E., Zorlutuna, P., Ndreu, A., Yilgor, P., Basmanav, F. B. and Aydin, E. (2006) Nanobiomaterials: A review of the existing science and technology, and new approaches, *J. Biomater. Sci. Polym. Ed.*, **17**, 1241–1268.
61. Kane, R. S., Takayama, S., Ostuni, E., Ingber, D. E. and Whitesides, G. M. (1999) Patterning proteins and cells using soft lithography, *Biomaterials*, **20**, 2363–2376.
62. Choi, D. G., Jeong, J. H., Sim, Y. S., Lee, E. S., Kim, W. S. and Bae, B. S. (2005) Fluorinated organic-inorganic hybrid mold as a new stamp for nanoimprint and soft lithography, *Langmuir*, **21**, 9390–9392.
63. Wilson, D. L., Martin, R., Hong, S., Cronin-Golomb, M., Mirkin, C. A. and Kaplan, D. L. (2001) Surface organization and nanopatterning of collagen by dip-pen nanolithography, *Proc. Natl. Acad. Sci. USA*, **98**, 13660–13664.
64. Tang, Q., Shi, S. Q. and Zhou, L. M. (2004) Nanofabrication with atomic force microscopy, *J. Nanosci. Nanotechnol.*, **4**, 948–963.
65. Yang, M. L., Zheng, Z. K., Liu, Y. Q. and Zhang, B. L. (2006) Kinetics of atomic force microscope-based scanned probe oxidation on an octadecylated silicon(111) surface, *J. Phys. Chem. B*, **110**, 10365–10373.
66. Heyderman, L. J., Solak, H. H., David, C., Atkinson, D., Cowburn, R. P. and Nolting, F. (2004) Arrays of nanoscale magnetic dots: Fabrication by X-ray interference lithography and characterization, *Appl. Phys. Lett.*, **85**, 4989–4991.

67. Robinson, A. P. G., Palmer, R. E., Tada, T., Kanayama, T., Allen, M. T., Preece, J. A. and Harris, K. D. M. (1999) 10 nm scale electron beam lithography using a triphenylene derivative as a negative/positive tone resist, *J. Phys. D. Appl. Phys.*, **32**, L75–L78.

68. Tibbitt, M. W. and Anseth, K. S. (2009) Hydrogels as Extracellular Matrix Mimics for 3D Cell Culture, *Biotechnol. Bioeng.*, **103**, 655–663.

69. Prabhakaran, M. P., Venugopal, J., Kai, D. and Ramakrishna, S. (2011) Biomimetic material strategies for cardiac tissue engineering, *Mater. Sci. Eng. C Mater. Biol. Appl.*, **31**, 503–513.

70. Engelmayr, G. C., Cheng, M. Y., Bettinger, C. J., Borenstein, J. T., Langer, R. and Freed, L. E. (2008) Accordion-like honeycombs for tissue engineering of cardiac anisotropy, *Nat. Mater.*, **7**, 1003–1010.

71. Pizzo, A., Kokini, K., Vaughn, L., Waisner, B. and Voytik-Harbin, S. (2005) Extracellular matrix (ECM) microstructural composition regulates local cell-ECM biomechanics and fundamental fibroblast behavior: A multidimensional perspective, *J. Appl. Physiol.*, **5**, 1909–1921.

72. Engler, A. J., Sen, S., Sweeney, H. L. and Discher, D. E. (2006) Matrix elasticity directs stem cell lineage specification, *Cell*, **126**, 677–689.

73. Hu, X., Park, S.-H., Gil, E. S., Xia, X.-X., Weiss, A. S. and Kaplan, D. L. (2011) The influence of elasticity and surface roughness on myogenic and osteogenic-differentiation of cells on silk-elastin biomaterials, *Biomaterials*, **32**, 8979–8989.

74. Ren, K., Crouzier, T., Roy, C. and Picart, C. (2008) Polyelectrolyte multilayer films of controlled stiffness modulate myoblast cell differentiation, *Adv. Funct. Mater.*, **18**, 1378–1389.

75. Kuo, Y.-C. and Ku, I. N. (2008) Cartilage regeneration by novel polyethylene oxide/chitin/chitosan scaffolds, *Biomacromolecules*, **9**, 2662–2669.

76. Shapira-Schweitzer, K. and Seliktar, D. (2007) Matrix stiffness affects spontaneous contraction of cardiomyocytes cultured within a PEGylated fibrinogen biomaterial, *Acta Biomater.*, **3**, 33–41.

77. Zaari, N., Rajagopalan, P., Kim, S. K., Engler, A. J. and Wong, J. Y. (2004) Photopolymerization in microfluidic gradient generators: Microscale control of substrate compliance to manipulate cell response, *Adv. Mater.*, **16**, 2133–2137.

78. Wong, J. Y., Velasco, A., Rajagopalan, P. and Pham, Q. (2003) Directed movement of vascular smooth muscle cells on gradient-compliant hydrogels, *Langmuir*, **19**, 1908–1913.

79. Isenberg, B. C., DiMilla, P. A., Walker, M., Kim, S. and Wong, J. Y. (2009) Vascular smooth muscle cell durotaxis depends on substrate stiffness gradient strength, *Biophys. J.*, **97**, 1313–1322.

80. Marklein, R. A. and Burdick, J. A. (2010) Spatially controlled hydrogel mechanics to modulate stem cell interactions, *Soft Matter*, **6**, 136–143.

81. Sant, S., Hancock, M. J., Donnelly, J. P., Iyer, D. and Khademhosseini, A. (2010) Biomimetic gradient hydrogels for tissue engineering, *Can. J. Chem. Eng.*, **88**, 899–911.

82. Owen, S. C. and Shoichet, M. S. (2010) Design of three-dimensional biomimetic scaffolds, *J. Biomed. Mater. Res. A*, **94A**, 1321–1331.

83. McClelland, R. E. and Coger, R. N. (2000) Use of micropathways to improve oxygen transport in a hepatic system, *J. Biomech. Eng.*, **122**, 268–273.

84. Landman, K. A. and Cai, A. Q. (2007) Cell proliferation and oxygen diffusion in a vascularising scaffold, *Bull. Math. Biol.*, **69**, 2405–2428.

85. Sachlos, E. and Czernuszka, J. T. (2003) Making tissue engineering scaffolds work, *Eur. Cell Matter.*, **5**, 29–40.

86. Yu, H., VandeVord, P. J., Mao, L., Matthew, H. W., Wooley, P. H. and Yang, S. Y. (2009) Improved tissue-engineered bone regeneration by endothelial cell mediated vascularization, *Biomaterials*, **30**, 508–517.

87. Santos, M. I. and Reis, R. L. (2010) Vascularization in bone tissue engineering: Physiology, current strategies, major hurdles and future challenges, *Macromol. Biosci.*, **10**, 12–27.

88. Oh, S. H., Park, I. K., Kim, J. M. and Lee, J. H. (2007) *In vitro* and *in vivo* characteristics of PCL scaffolds with pore size gradient fabricated by a centrifugation method, *Biomaterials*, **28**, 1664–1671.

89. Karageorgiou, V. and Kaplan, D. (2005) Porosity of 3D biomaterial scaffolds and osteogenesis, *Biomaterials*, **26**, 5474–5491.

90. Brauker, J. H., Carr-Brendel, V. E., Martinson, L. A., Crudele, J., Johnston, W. D. and Johnson, R. C. (1995) Neovascularization of synthetic membranes directed by membrane microarchitecture, *J. Biomed. Mater. Res.*, **29**, 1517–1524.

91. Klawitter, J. J. and Hulbert, S. F. (1971) Application of porous ceramics for the attachment of load bearing internal orthopedic applications, *J. Biomed. Mater. Res.*, **5**, 161–229.

92. Gogolewski, S. and Pennings, A. J. (1983) An artificial skin based on biodegradable mixtures of polylactides and polyurethanes for full-

thickness skin wound covering. *Makromol. Chem. Rapid Commun.*, **4**, 675–680.

93. Yannas, I. V., Lee, E., Orgill, D. P., Skrabut, E. M. and Murphy, G. F. (1989) Synthesis and characterization of a model extracellular matrix that induces partial regeneration of adult mammalian skin, *Proc. Natl. Acad. Sci. USA*, **86**, 933–937.

94. Schwartz, I., Robinson, B. P., Hollinger, J. O., Szachowicz, E. H. and Brekke, J. (1995) Calvarial bone repair with porous D,L-polylactide, *J. Otolaryngol. Head Neck Surg.*, **112**, 707–713.

95. Zhu, X. H., Lee, L. Y., Jackson, J. S. H., Tong, Y. W. and Wang, C.-H. (2008) Characterization of porous poly(D,L-lactic-co-glycolic acid) sponges fabricated by supercritical $CO_2$ gas-foaming method as a scaffold for three-dimensional growth of Hep3B cells, *Biotechnol. Bioeng.*, **100**, 998–1009.

96. Yang, S., Leong, K. F., Du, Z. and Chua, C. K. (2001) The design of scaffolds for use in tissue engineering. Part I. Traditional factors, *Tissue Eng.*, **7**, 679–689.

97. Wake, M. C., Patrick, C. W., Jr. and Mikos, A. G. (1994) Pore morphology effects on the fibrovascular tissue growth in porous polymer substrates, *Cell Transplant*, **3**, 339–343.

98. Kuboki, Y., Takita, H., Kobayashi, D., Tsuruga, E., Inoue, M., Murata, M., Nagai, N., Dohi, Y. and Ohgushi, H. (1998) BMP-induced osteogenesis on the surface of hydroxyapatite with geometrically feasible and nonfeasible structures: Topology of osteogenesis, *J. Biomed. Mater. Res.*, **39**, 190–199.

99. Kasten, P., Beyen, I., Niemeyer, P., Luginbuhl, R., Bohner, M. and Richter, W. (2008) Porosity and pore size of beta-tricalcium phosphate scaffold can influence protein production and osteogenic differentiation of human mesenchymal stem cells: An *in vitro* and *in vivo* study, *Acta Biomater.*, **4**, 1904–1915.

100. Takahashi, Y. and Tabata, Y. (2004) Effect of the fiber diameter and porosity of non-woven PET fabrics on the osteogenic differentiation of mesenchymal stem cells, *J. Biomater. Sci., Polym. Ed.*, **15**, 41–57.

101. Mygind, T., Stiehler, M., Baatrup, A., Li, H., Zou, X., Flyvbjerg, A., Kassem, M. and Bunger, C. (2007) Mesenchymal stem cell ingrowth and differentiation on coralline hydroxyapatite scaffolds, *Biomaterials*, **28**, 1036–1047.

102. Hulbert, S. F., Young, F. A., Mathews, R. S., Klawitter, J. J., Talbert, C. D. and Stelling, F. H. (1970) Potential of ceramic materials as permanently implantable skeletal prostheses, *J. Biomed. Mater. Res.*, **4**, 433–456.

103. Dubruel, P., Unger, R., Van Vlierberghe, S., Cnudde, V., Jacobs, P. J. S., Schacht, E. and Kirkpatrick, C. J. (2007) Porous gelatin hydrogels: Porous gelatin hydrogels: 2. *In vitro* cell interaction study 2. *In vitro* cell interaction study, *Biomacromolecules*, **8**, 338–344.

104. Tripathi, A., Kathuria, N. and Kumar, A. (2009) Elastic and macroporous agarose–gelatin cryogels with isotropic and anisotropic porosity for tissue engineering, *J. Biomed. Mater. Res. A*, **90A**, 680–694.

105. Roy, T. D., Simon, J. L., Ricci, J. L., Rekow, E. D., Thompson, V. P. and Parsons, J. R. (2003) Performance of degradable composite bone repair products made via three-dimensional fabrication techniques, *J. Biomed. Mater. Res.*, **66**, 283–291.

106. Annabi N, Nichol, J. W., Zhong, X., Ji, C., Koshy, S., Khademhosseini, A. and Dehghani, F. (2010) Controlling the porosity and microarchitecture of hydrogels for tissue engineering, *Tissue Eng. Part B*, **16**, 371–380.

107. Barbetta, A., Rizzitelli, G., Bedini, R., Pecci, R. and Dentini, M. (2010) Porous gelatin hydrogels by gas-in-liquid foam templating, *Soft Matter*, **6**, 1785–1792.

108. Dehghani, F. and Annabi, N. (2011) Engineering porous scaffolds using gas-based techniques, *Curr. Opin. Biotechnol.*, **22**, 661–666.

109. Ingber, D. E., Mow, V. C., Butler, D., Niklason, L., Huard, J., Mao, J., Yannas, I., Kaplan, D. and Vunjak-Novakovic, G. (2006) Tissue engineering and developmental biology: Going biomimetic, *Tissue Eng.*, **12**, 3265–3283.

110. Lenas, P., Moos, M. and Luyten, F. P. (2009) Developmental engineering: A new paradigm for the design and manufacturing of cell-based products. Part II. From genes to networks: Tissue engineering from the viewpoint of systems biology and network science, *Tissue Eng. Part B Rev.*, **15**, 395–422.

111. Ker, E. D. P., Nain, A. S., Weiss, L. E., Wang, J., Suhan, J., Amon, C. H. and Campbell, P. G. (2011) Bioprinting of growth factors onto aligned sub-micron fibrous scaffolds for simultaneous control of cell differentiation and alignment, *Biomaterials*, **32**, 8097–8107.

112. Miller, E. D., Li, K., Kanade, T., Weiss, L. E., Walker, L. M. and Campbell, P. G. (2011) Spatially directed guidance of stem cell population migration by immobilized patterns of growth factors, *Biomaterials*, **32**, 2775–2785.

113. Chung, B. G. and Choo, J. (2010) Microfluidic gradient platforms for controlling cellular behavior, *Electrophoresis*, **31**, 3014–3027.

114. Ker, E. D. F., Chu, B., Phillippi, J. A., Gharaibeh, B., Huard, J., Weiss, L. E. and Campbell, P. G. (2011) Engineering spatial control of multiple

differentiation fates within a stem cell population, *Biomaterials*, **32**, 3413–3422.

115. DeForest, C. A. and Anseth, K. S. (2012) Photoreversible patterning of biomolecules within click-based hydrogels, *Angew. Chem. Int. Ed. Engl.*, **124**, 1852–1855. doi: 10.1002/anie.201106463.

116. Gabriel, D., Dvir, T. and Kohane, D. S. (2012) Delivering bioactive molecules as instructive cues to engineered tissues, *Expert Opin. Drug Deliv.*, **9**, 473–492.

117. Szentivanyi, A., Chakradeo, T., Zernetsch, H. and Glasmacher, B. (2011) Electrospun cellular microenvironments: Understanding controlled release and scaffold structure, *Adv. Drug Deliv. Rev.*, **63**, 209–220.

118. Chen, F. M., Zhang, M. and Wu, Z. F. (2010) Toward delivery of multiple growth factors in tissue engineering, *Biomaterials*, **31**, 6279–6308.

119. Yilgor, P., Tuzlakoglu, K., Reis, R. L., Hasirci, N. and Hasirci, V. (2009) Incorporation of a sequential BMP-2/BMP-7 delivery system into chitosan-based scaffolds for bone tissue engineering, *Biomaterials*, **30**, 3551–3559.

120. Dupont, K. M., Boerckel, J. D., Stevens, H. Y., Diab, T., Kolambkar, Y. M., Takahata, M., Schwarz, E. M. and Guldberg, R. E. (2012) Synthetic scaffold coating with adeno-associated virus encoding BMP2 to promote endogenous bone repair, *Cell Tissue Res.*, **347**, 575–588.

121. Evans, C. H., Palmer, G. D., Pascher, A., Porter, R., Kwong, F. N., Gouze, E., Gouze, J. N., Liu, F., Steinert, A., Betz, O., Betz, V., Vrahas, M. and Ghivizzani, S. C. (2007) Facilitated endogenous repair: Making tissue engineering simple, practical, and economical, *Tissue Eng.*, **13**, 1987–1993.

122. Bonadio, J., Smiley, E., Patil, P. and Goldstein, S. (1999) Localized, direct plasmid gene delivery *in vivo*: Prolonged therapy results in reproducible tissue regeneration, *Nat. Med.*, **5**, 753–759.

123. Anitua, E., Alkhraisat, M. H. and Orive, G. (2012) Perspectives and challenges in regenerative medicine using plasma rich in growth factors, *J. Control. Release*, **157**, 29–38.

124. Kim, B. S., Park, I. K., Hoshiba, T., Jiang, H. L., Choi, Y. J., Akaike, T. and Cho, C. S. (2011) Design of artificial extracellular matrices for tissue engineering, *Prog. Polym. Sci.*, **36**, 238–268.

125. Zhang, H., Dai, S., Bi, J. X. and Liu, K. K. (2011) Biomimetic three-dimensional microenvironment for controlling stem cell fate, *Interface Focus*, **1**, 792–803.

126. Haag, J., Baiguera, S., Jungebluth, P., Barale, D., Del Gaudio, C., Castiglione, F., Bianco, A., Comin, C. E., Ribatti, D. and Macchiarini, P. (2012)

Biomechanical and angiogenic properties of tissue-engineered rat trachea using genipin cross-linked decellularized tissue, *Biomaterials*, **33**, 780–789.

127. Lozito, T. P., Kuo, C. K., Taboas, J. M. and Tuan, R. S. (2009) Human mesenchymal stem cells express vascular cell phenotypes upon interaction with endothelial cell matrix, *J. Cell. Biochem.*, **107**, 714–722.

128. Werkmeister, J. A. and Ramshaw, J. A. M. (2012) Recombinant protein scaffolds for tissue engineering, *Biomed. Mater.*, **7**, 012002.

129. Huang, N. F. and Li, S. (2011) Regulation of the matrix microenvironment for stem cell engineering and regenerative medicine, *Ann. Biomed. Eng.*, **39**, 1201–1214.

130. Gerecht, S., Burdick, J. A., Ferreira, L. S., Townsend, S. A., Langer, R. and Vunjak-Novakovic, G. (2007) Hyaluronic acid hydrogen for controlled self-renewal and differentiation of human embryonic stem cells, *Proc. Natl. Acad. Sci. USA*, **104**, 11298–11303.

131. Yang, F., Cho, S.-W., Son, S. M., Hudson, S. P., Bogatyrev, S., Keung, L., Kohane, D. S., Langer, R. and Anderson, D. G. (2010) Combinatorial extracellular matrices for human embryonic stem cell differentiation in 3D, *Biomacromolecules*, **11**, 1909–1914.

132. Tekin, H., Anaya, M., Brigham, M. D., Nauman, C., Langer, R. and Khademhosseini, A. (2010) Stimuli-responsive microwells for formation and retrieval of cell aggregates, *Lab Chip*, **10**, 2411–2418.

133. Bratt-Leal, A. M., Carpenedo, R. L., Ungrin, M. D., Zandstra, P. W. and McDevitt, T. C. (2011) Incorporation of biomaterials in multicellular aggregates modulates pluripotent stem cell differentiation, *Biomaterials*, **32**, 48–56.

134. McBeath, R., Pirone, D. M., Nelson, C. M., Bhadriraju, K. and Chen, C. S. (2004) Cell shape, cytoskeletal tension, and RhoA regulate stem cell lineage commitment, *Dev. Cell*, **6**, 483–495.

135. Bozzini, S., Giuliano, L., Altomare, L., Petrini, P., Bandiera, A., Conconi, M. T., Fare, S. and Tanzi, M.C. (2011) Enzymatic cross-linking of human recombinant elastin (HELP) as biomimetic approach in vascular tissue engineering, *J. Mater. Sci. Mater. Med.*, **22**, 2641–2650.

136. Kozel, B. A., Ciliberto, C. H. and Mecham, R. P. (2004) Deposition of tropoelastin into the extracellular matrix requires a competent elastic fiber scaffold but not live cells, *Matrix Biol.*, **23**, 23–34.

137. Kieffer, A.-E., Goumon, Y., Ruh, O., Chasserot-golaz, S., Nullans, G., Gasnier, C., Aunis, D. and Metz-Boutigue, M.-H. (2003) The N- and

C-terminal fragments of ubiquitin are important for the antimicrobial activities, *FASEB J.*, **17**, 776–778.

138. Hartgerink, J. D., Beniash, E. and Stupp, S. I. (2002) Peptide-amphiphile nanofibers: A versatile scaffold for the preparation of self-assembling materials, *Proc. Natl. Acad. Sci. USA*, **99**, 5133–5138.

139. Gomes, S., Leonor, I. B., Mano, J. F., Reis, R. L. and Kaplan, D. L. (2012) Natural and genetically engineered proteins for tissue engineering, *Prog. Polym. Sci.*, **37**, 1–17.

140. Davis, M. E., Hsieh, P. C. H., Takahashi, T., Song, Q., Zhang, S. G., Kamm, R. D., Grodzinsky, A. J., Anversa, P. and Lee, R. T. (2006) Local myocardial insulin-like growth factor 1 (IGF-1) delivery with biotinylated peptide nanofibers improves cell therapy for myocardial infarction, *Proc. Natl. Acad. Sci. USA*, **103**, 8155–8160.

141. Romano, N. H., Sengupta, D., Chung, C. and Heilshorn, S. C. (2011) Protein-engineered biomaterials: Nanoscale mimics of the extracellular matrix, *Biochim. Biophys. Acta*, **1810**, 339–349.

142. Lutolf, M. P. and Hubbell, J. A. (2005) Synthetic biomaterials as instructive extracellular microenvironments for morphogenesis in tissue engineering, *Nat. Biotech.*, **23**, 47–55.

143. Zhou, M., Smith, A. M., Das, A. K., Hodson, N. W., Collins, R. F., Ulijn, R. V. and Gough, J. E. (2009) Self-assembled peptide-based hydrogels as scaffolds for anchorage-dependent cells, *Biomaterials*, **30**, 2523–2530.

144. Gelain, F., Horii, A. and Zhang, S. G. (2007) Designer self-assembling peptide scaffolds for 3-D tissue cell cultures and regenerative medicine, *Macromol. Biosci.*, **7**, 544–551.

145. Cigognini, D., Satta, A., Colleoni, B., Silva, D., Donega, M., Antonini, S. and Gelain, F. (2011) Evaluation of early and late effects into the acute spinal cord injury of an injectable functionalized self-assembling scaffold, *PLoS One*, **6**, E19782.

146. Kim, J. H., Jung, Y., Kim, S. H., Sun, K., Choi, J., Kim, H. C. and Park, Y. (2011) The enhancement of mature vessel formation and cardiac function in infarcted hearts using dual growth factor delivery with self-assembling peptides, *Biomaterials*, **32**, 6080–6088.

147. Ehrbar, M., Rizzi, S. C., Schoenmakers, R. G., San Miguel, B., Hubbell, J. A., Weber, F. E. and Lutolf, M. P. (2007) Biomolecular hydrogels formed and degraded via site-specific enzymatic reactions, *Biomacromolecules*, **8**, 3000–3007.

148. Sala, A., Hanseler, P., Ranga, A., Lutolf, M. P., Voros, J., Ehrbar, M. and Weber, F.E. (2011) Engineering 3D cell instructive microenvironments by rational assembly of artificial extracellular matrices and cell patterning, *Integr. Biol.*, **3**, 1102–1111.

149. Salinas, C. N. and Anseth, K. S. (2008) The enhancement of chondrogenic differentiation of human mesenchymal stem cells by enzymatically regulated RGD functionalities, *Biomaterials*, **29**, 2370–2377.

150. Metz-Boutigue, M. H., Kieffer, A. E., Goumon, Y. and Aunis, D. (2003) Innate immunity: Involvement of new neuropeptides, *Trends Microbiol.*, **11**, 585–592.

## Chapter 8

# Biomaterials for Scaffolds: Synthetic Polymers

**Luis Rojo,**[a,b] **Blanca Vázquez,**[a,b] **and Julio San Román**[a,b]
[a]*Institute of Polymer Science and Technology,*
*Consejo Superior de Investigaciones Científicas (CSIC),*
*Juan de la Cierva 3, 28006 Madrid, Spain*
[b]*Networking Biomedical Research Center in Bioengineering,*
*Biomaterials and Nanomedicine (CIBER-BBN), Spain*
jsroman@ictp.csic.es

Synthetic polymers offer enormous possibilities for the preparation of porous systems, controlled degradable matrices, and modulated scaffolds for tissue engineering and drug delivery. The election of monomeric species and the advanced possibilities of new synthetic routes have opened a wide panorama for the preparation of polymer systems and composites with specific properties, and the most important with the possibility of mimic the nature of the most representative macromolecular components of the human body. The macromolecular chemistry plays an important role in the design and preparation of systems with specific structures, morphologies, and properties, using molecules that give not only high molecular weight polymers, but also functionality, structural order, and orientation,

*Scaffolds for Tissue Engineering: Biological Design, Materials, and Fabrication*
Edited by Claudio Migliaresi and Antonella Motta

ISBN 978-981-4463-20-1 (Hardcover), 978-981-4463-21-8 (eBook)
www.panstanford.com

self-organization of microdomains, nano- and microcrystallinity, and modulated biodegradability.

## 8.1 Introduction

The increasing activity in regenerative medicine needs the contribution of specific and functional biomaterials to achieve the main objective of this new discipline. General methods in tissue engineering require a balanced combination of cell culture growth with biomaterials and the contribution of bioactive molecules to reach functional solutions with possibilities of application. A successful approach considers the replacement or repair of damaged tissues with viable ones by creation of an environment that promotes the capacity of cells to integrate, differentiate, and proliferate.[1–4] The creation of the adequate environment can be achieved by the implantation of a three-dimensional (3D) scaffold seeded with appropriate living cells secreting their own extracellular matrix (ECM).[1] The proliferation of cells is induced and activated by the addition of growth factors and other active components of the ECM. In a natural occurring scenario, the bioactive molecules should also promote the intercellular communication and attachment of cells to the scaffold. In addition, it is clear that vascularization is a fundamental requisite for the success of the regenerative process. In this context, the assistance of a porous and permeable polymeric scaffold is decisive to support and direct generation of a tissue by the seeded cells. The 3D polymeric scaffolds provide frameworks for cells to attach proliferate and form ECM. Additional bioactive molecules, such as adhesion receptors to cells, growth factors or their biomimetic functional activators, enzymes, and other regulators of cell attachment, proliferation, and differentiation, are very convenient and necessary to improve the regeneration of tissues in a biomimetic approach.[1–3] Natural and synthetic polymers with the specific functionality offer very interesting possibilities for design and development. The combination of microstructural design and preparation of molecular polymer architectures, synthetic pathways, and technological methodologies for the preparation of tissue engineering scaffolds and drug delivery devices, provides a powerful tool for regenerative medicine.

The application of biostable porous polymer systems is dependent on their impact to cellular responses. Two main features have to be considered: the 3D topography and the surface texture of the scaffold. The 3D topography can be conceptualized as the size, the shape, and the surface texture of the material; and it has been demonstrated clearly that bone cells are sensitive to the gross topography of the underlying material. Cells respond to the substrate topography by adapting their orientation, movement, and attachment kinetics.[5–10] New polymerization techniques and methodologies provide homo- and copolymer systems with very well-controlled molecular weight and size of the block sequences, and thus have made the preparation of polymeric systems with strict control of size and composition of the different blocks of the molecular architecture relatively easy. This provides an excellent control of the composition, microstructure, and topography of the polymer matrices designed for cell adhesion, proliferation as well as controlled delivery supports. Thanks to advanced polymerization methodologies such as atom transfer radical polymerization (ATRP) and the reversible addition fragmentation transfer (RAFT) that have been described by Matyjaszewski and Gregory *et al.* very well,[11,12] it is possible the preparation of well-defined low-polydispersity homopolymers and multiblock copolymers with specific composition and microstructure of hard and flexible sequences possible, which, in turn, has made the preparation of a wide range of scaffolds and supports for tissue engineering and controlled release of bioactive compounds possible. The preparation of multiblock copolymers with stiff blocks and soft sequences provides systems with modulated properties and self-organizing microdomains which can respond to specific stimulus in the point of application. This concept is being applied with a practical interest in the new strategies of regenerative medicine.

It is important to consider that cells can respond to the shape and topography of their environment.[13,14] *In vivo* cells interact directly with surface topography through protein folding and collagen banding. For instance, Dalby *et al.*[15] have demonstrated that the respond of mesenchymal stem cells (MSC) to micropatterned substrata, strongly depends on the distribution and size of the micropatterned spots generated by electron beam lithography. One of the initial events in substrate-induced stem cell differentiation is the ability of the cells to form mature fibrillar-like adhesions.[15] Chemical patterning and material hardness[16,17] will alter the adhesion formation in stem and other cell lines, and modulation of adhesion of fibroblasts in response to nanotopography has been shown.[18]

## 8.2 Polymers Hydrophilicity and Water Permeability

The application of synthetic biodegradable polymers has been a challenge for applications in tissue engineering, drug delivery, and for resorbable implants. Hydrophilic–hydrophobic balance plays an important role in biodegradable polymers whose degradation occurs through hydrolysis. This is the case of polyhydroxyalkanoates, and in particular poly(glycolic acid) (PGA), poly(lactic acid) (PLA) and their copolymers poly(lactic-co-glycolic acid) (PLGA). The higher hydrophilicity of PGA compared to the more hydrophobic PLA renders hydrolytic degradation of PGA faster than the PLA. For both, biodegradation is very sensitive to the diffusion of water that causes the hydrolysis of the carboxylic ester groups. PGA and PLA homopolymers are semicrystalline, this controlling the kinetics of the biodegradative process that depends on the permeation to water which causes the hydrolysis of the carboxylic ester groups. As it is shown in the scheme of Fig. 8.1, the amorphous domains of the systems are most sensitive to degradation than the crystalline domains, because of the higher water permeability of the disordered amorphous microstructures. Moreover, under the biodegradative process, the long polymer chains are broken in small fragments, that can crystallize with their higher mobility.

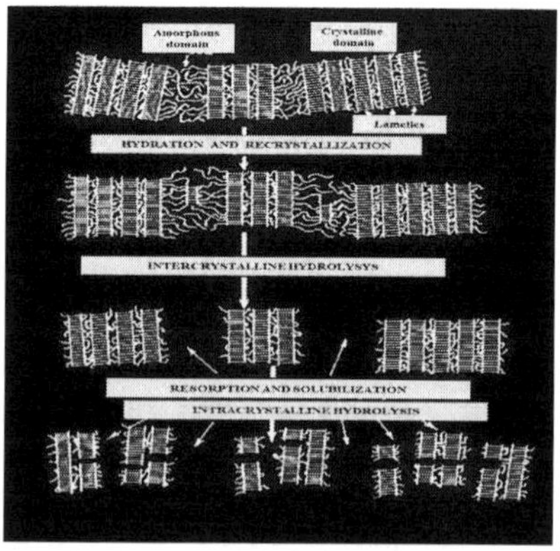


**Figure 8.1** Morphology and hydrolytic behavior of semicrystalline aliphatic polyesters (e.g., PGA, PLA, and PCL).

The biodegradative process depends on the molecular weight of the polymer, the molecular weight distribution and even the size and the shape of the polymer device. Micro- or nanoparticles degrade faster than plates, nails or screws, and porous systems are more sensitive to the degradative process than solid bulk implants and supports.

Random copolymers of glycolic and lactic acid are essentially amorphous, with very low degree of crystallinity and, therefore, are much more sensitive to the biodegradation than the corresponding homopolymers. Depending on the amount of water, processing temperatures can produce a noticeable thermal degradation of the original polymers, with the formation of lower molecular weight chains that can increase the polymer degradation kinetics and lower the device mechanical characteristics.[19] Although these polyesters could be prepared by the direct polycondensation of the corresponding hydroxy acid, this process provides relatively low molecular weight systems. The preparation of high molecular weight polymer systems occurs through the ring opening polymerization of the corresponding cyclic lactones, glycolide, or lactide. The ring opening polymerization is usually initiated by tin octoate in non-polar solvents. The FDA has accepted the use of this catalyst, provided that less than 0.1% of the catalyst is used in the synthesis process. The ring opening polymerization results in high molecular weight and low polydispersity polymer systems.

In this chapter we consider the description of synthetic polymer systems considering the characteristics and properties of them, better than the consideration of the chemical structure of the corresponding polymer components. Accordingly, we have grouped the systems in hydrophobic, hydrophilic, and amphiphilic polymers.

## 8.3 Hydrophobic Polymer Systems

Hydrophobic polymer systems are characterized by the presence of functional groups with low polarity and therefore, non-solubility in water and physiological fluids. The design of molecular structures is very wide and covers all kind of polymer systems from biostable amorphous and semicrystalline macromolecules up to biodegradable highly porous and stiff materials. Porous systems with interconnected and modulated porosity can be prepared using non-expensive, traditional methodologies and techniques such as salt leaching, precipitation, foaming with specific gases and supercritical fluids, self-foaming with selected reactions, dissolution of porogens such as sugar, glass beads, and gelatin, and many other procedures.

Considering biostable or non-biodegradable systems, there have been suggested excellent polyacrylic formulations with very good biomechanical stability and modulated stiffness.[20] They are based on different copolymers of methyl methacrylate (MMA), ethyl acrylate, alkyl acrylates, and methacrylates that have been shown to be highly biocompatible systems with good tissue integration and long-term stability. Their characteristics can be modulated by the molecular architecture of the polymers and copolymers. The insolubility in water makes these systems very attractive for tissue regeneration in bone repair applications, ocular implants, wound healing, and many other possibilities. Furthermore, they can be matrices for the controlled release system for bioactive molecules. After integration, the behavior of the systems is in general excellent and they are incorporated in the regenerated tissue without complications.

Thermoplastic elastomers and soft or stiff polymer-based scaffolds are being applied for replacement and reconstruction of soft and hard tissues. Silicones, polyesters like dacron (polyethylene terephthalate (PET)), polytetrafluorethylene (PTFE), polyethylene (PE), polypropylene (PP), and a variety of polyurethanes (PUs), which are relatively inert, biostable, and biocompatible have demonstrated a good behavior when implanted in the human body.[21] The formation of fibrous capsule around the implants, which require communication and interaction with the surrounding tissue or with the circulatory system, render it ineffective. It is essential for implants not only to be surrounded by a highly vascularized tissue, but also to be penetrated by it.[1]

Poly(ε-caprolactone) (PCL) is a highly hydrophobic and versatile aliphatic polyester with interesting applications in tissue engineering and drug delivery systems. It is characterized by low toxicity, excellent biocompatibility, and acceptable bioabsorbability *in vivo*. However, the low hydrophilicity and relative high crystallinity slow down its degradation rate thus, limiting its possible applications.[22–24] Other polyalkanoates obtained from bacteria, polyhydroxybutyrates (PHBV), have displayed excellent possibilities for processing very attractive designs and geometries, from nanoparticles to relatively big devices.[25,26]

The high hydrophobicity and solubility in organic solvents as well as the thermal conditions for processing, allow the preparation of a great variety of systems with several techniques such as porogen leaching, gas foaming, phase separation, fiber electrospinning, supercritical $CO_2$ technologies, microsphere sintering, and rapid prototyping. These techniques provide 3D scaffolds with different porosities and surface characteristics. The synergistic combination of macromolecular systems and cell therapy is becoming of great importance in regenerative medicine and tissue engineering. The application of MSC in combination with synthetic scaffolds in the regeneration of bone has been suggested since the beginning of the 1980s.[25] In these applications, the use of biodegradable synthetic polymers has been decisive. However, most of the *in vivo* models are based on the behavior of 3D-polymer scaffolds/cells applied in rodents, and the translation to humans cannot be predicted with accuracy. Therefore, new models and more precise methodologies are necessary to have a clear idea of the positive effects and limitations of the application of combined polymers/cell therapies for tissue regeneration. To reach an acceptable regenerative process the angiogenesis of the implanted scaffolds/tissue precursors is necessary. Since late vascularization of 3D-engineered constructs or scaffolds limits the transfer of tissue engineering from *in vitro* to *in vivo*.[25] Polymeric supports with well-interconnected porosity are necessary to have enough permeability and diffusion of oxygen, physiological fluids, and nutrients to cells. Scaffolds containing angiogenic factors such as vascular endothelial growth factor (VEGF) in combination with protein morphogenetic growth factor, BMP-2 and BMP-7, can be engineered *in vitro*, and then implanted *in vivo* with a prevascularized systems, which provides an active medium for tissue regeneration.[25]

There is increase in interest in the development of biomaterials that will allow the delivery of growth factors to enhance the integration in tissue engineering. Bone morphogenic proteins (BMPs) are able to induce bone formation and offer a potential means of enhancing bone generation in functionalized and seeded scaffolds. Saito *et al.*, have evaluated biodegradable PLA-based scaffolds and matrices as carriers of BMP and their results showed that the BMP-charged matrices induced ectopic bone formation on implantation into animal muscles.[27] The modification of the porosity and properties of the PLA-based scaffolds controlled the 3D growth process.[27]

Dendritic polymers including dendrimers and hyperbranched polymers, have a highly branched structure, intramolecular voids, small rheological volumes, make low viscosity solutions, and possess a high density of functional groups at the periphery.[22,28,29] Dendrimers synthesis requires step-by-step onerous purification procedures. However, hyperbranched polymers have a structure rather similar to dendritic systems, and can be synthesized with much easier procedures. These systems that can be based on biodegradable polyesters offer properties and functionality, which can be advantageously exploited for the preparation of scaffolds for tissue engineering and matrices for drug delivery. The functionality can be used in part for the incorporation of specific bioactive compounds such as growth factors, antioxidants, or anti-inflammatory agents, which can help the regenerative process.

Recently, biodegradable PUs incorporating amino acid–derived isocyanates have been proposed in tissue engineering for regenerating soft and hard tissues. Scaffolds have shown to possess excellent properties, such as good biodegradative control and modulated biomechanical and biochemical behavior. A good example has been published recently by Alperin and Woodhose with PU films seeded with stem cell derived cardiomyocytes for cardiac tissue regeneration[20,30,33,35–37] in the treatment of infarction.[35] The use of bioengineered cardiac grafts has been suggested to replace and reorganize infarcted cardiac tissue, and the application of seeded PU films coated with gelatine, laminin, or collagen IV has demonstrated a good cell adhesion and proliferative behavior, with the subsequent tissue regeneration.[30]

## 8.4 Hydrophilic Polymer Systems

Most common scaffolds used for soft tissue healing are based on hydrophilic polymers due to their capability to form highly defined and evolving 3D matrices that mimic the microstructure and physicochemical properties of ECM.[31,32] In these conditions, the cell population can experience appropriate biomechanical stimulation, thus enhancing their physical integration on highly hydrated body environments such as musculoskeletal,[33] myocardium,[34] cornea,[35] and other soft tissues.[36] Compared to natural polymers, hydrophilic synthetic polymers including esters, amides, urethanes, ethers, and phosphates can be mass produced with specific structures and molecular weights offering improved control of the chemical composition and matrix architecture.[37]

A wide range of hydrophilic polymer families can be synthesized to fit the desired biological, physical, chemical, and biomechanical properties of the tissue to be repaired. In general, these properties can be fine-tuned through variations in the chemical structure and functionality of the macromolecular chains.[32] The presence of carbonyl, carboxyl, ester, amino, amide, or phosphorous derived functional groups onto synthetic polymeric-based scaffolds increase the polar component of the free energy of the surface making these biomedical devices more wettable and therefore, more susceptible to adsorption of adhesion-mediated ECM biomolecules.[38] Highly hydrated engineered scaffolds are usually prepared from injectable formulations based on hydrophilic synthetic polymers which self-form hydrogels by physical or chemical crosslinking mechanisms.

Following are the most used synthetic polymers[39,40]:

- Poly(glutamic acid) or PGA
- poly(ethylene glycol) or PEG
- poly(propylene fumarate) or PPF
- poly(hydroxyethyl methacrylate) or PHEMA
- poly(vinyl alcohol) or PVA
- poly(*N*-isopropylacrylamide) or PNIPAM
- poly(aminoamides) or PAA
- poly(aminoacids)
- poly(phosphoesters) or PPE
- poly(phosphazenes) or PPZ

Table 8.1 shows the structure of the synthetic hydrophilic polymers and their targeted tissues.

Poly($\alpha$-hydroxy acids) constitute a family of polymeric systems widely spread for the preparation of tissue-engineered scaffolds.[41] PGA is one of the first bioabsorbable synthetic polymer that was approved for clinical applications and constitutes the simplest hydrophilic member of poly($\alpha$-hydroxy acids) family consisting of repeating glycolic acid ester linkages through the polymeric backbone.[19] PGA-based porous scaffolds have high initial mechanical strength and good biocompatibility. However, glycolic ester backbone suffers from fast hydrolytic degradation in aqueous media, releasing glycolic acid monomer or soluble oligomers that can be responsible of undesired inflammatory response.[42–44] This fact together with the significantly reduction on its mechanical properties over time have limited the applicability of PGA in biomedical applications. Nevertheless, due to the feasibility to obtain more hydrophobic, and thus slower hydrolyzable copolymers when synthetized with other poly($\alpha$-hydroxy acids) such as PLA, these PGA-based polymeric systems are still under investigation mainly for short-term applications in musculoskeletal, cardiovascular, vaginal, intestinal, lymphatic, and spinal soft tissues regeneration.[32]

**Table 8.1** Hydrophilic systems based on synthetic polymers

| System | Applications | Reference |
|---|---|---|
| PGA 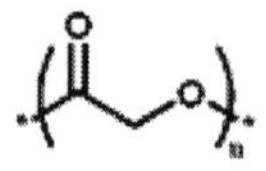 | Fabrication of mesh networks for musculoskeletal, cardiovascular, vaginal, intestinal, lymphatic, and spinal regeneration | 32 |
| PPF 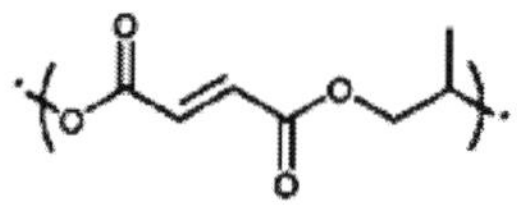 | Fabrication of peptide-modified PPF based 3D porous scaffolds applicable in bone tissue engineering | 41 |
| PEG 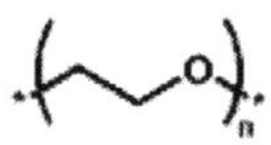 | Hydrogel-based scaffolds for musculoskeletal, vascular, dental pulp, and endothelial tissue regeneration | 34, 42–44 |
| Poly(*N*-isopropylacrylamide) (PEGDA) 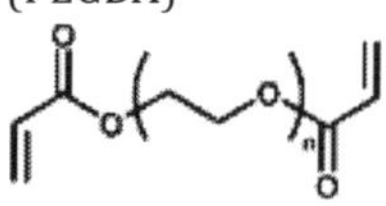 | Free radical self-crosslinkable hydrogel for musculoskeletal, vascular, dental pulp, and endothelial tissue regeneration | 34, 42–44 |
| Poly(*N*-isopropylacrylamide) (PNIPAM) 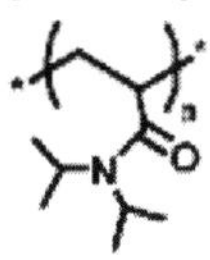 | Engineered methacrylate-based hydrogels for cartilage, myocardial, and cornea tissue healing | 45, 46 |
| PHEMA 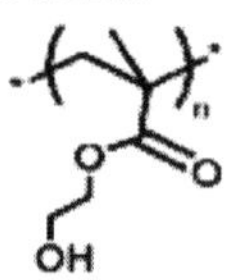 | Fabrication of keratoprothesis | 47 |

(*Continued*)

**Table 8.1** (*Continued*)

| System | Applications | Reference |
|---|---|---|
| PVA 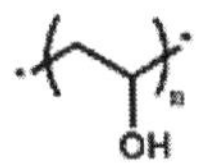 | Preparation of grafted peptide PVA porous scaffolds for *in vitro* regeneration of cartilage-like tissue | 48, 49 |
| Poly(L-glutamic acid) (PLG) 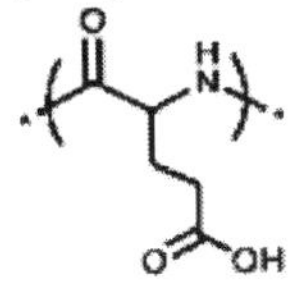 | Scaffolds for peripheral nerve and cartilage regeneration | 50 |
| PPE 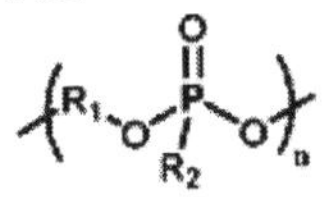 | Fabrication of biocompatible and biodegradable scaffolds for cartilage tissue engineering | 51, 52 |
| PPZ 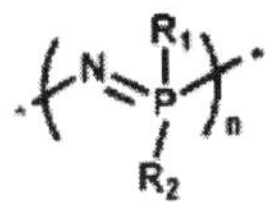 | Obtaining of biodegradable scaffolds for hard tissue engineering applications | 53 |

PPF is also a hydrophilic synthetic polyester widely explored for biomedical engineering. PPF is synthesized from diethyl fumarate and propylene glycol by a two-step transesterification reaction leading to an unsaturated backbone. Further, crosslinking reaction through the C=C double bound yields to covalent polymer networks that can be hydrolyzed in aqueous media leaching non-toxic degradation products, mainly fumaric acid and propylene glycol.[45] This polymer has been used in the preparation of *in situ* polymerizable biodegradable scaffolds for filling irregular shaped bone defects with minimal surgical intervention. For this application different vinylic and acrylic monomers have been used as crosslinking agents, vinyl pyrrolidone being the most representative.[46] Alternative and less toxic crosslinking agents based on macromonomers such as PPF-diacrylate[47] or hydrophilic methacrylate poly(ethers), that is, dimethacrylated PEG[48] have been used leading to more biocompatible injectable degradable networks. However, crosslinked PPF networks without the use of additional crosslinking agents was also achieved.[49] Additionally, the preparation of macroporous scaffolds was attained by incorporation of a porogen to the formulation.[50] In order to enhance osteoblasts adhesion and proliferation, the surface properties of a PPF/diethyl fumarate (DEF) scaffolds were altered with different peptide motif modifications, Arg-Gly-Asp (RGD), cyclo-RGD, and Arg-Gly-Asp Lys-Arg-Ser-Arg mixture, or RGD-KRSR mixture, the latter one giving the most promising results, suggesting that this macroporous scaffolds are optimal for musculoskeletal tissue engineering applications.[51]

Synthetically derived poly(ethers) and their copolymers are other highly biocompatible polymers that have been used for the preparation of hydrogel-based scaffolds in tissue engineering applications. The most successful results have been obtained with three types of poly(ethers), including poly(ethylene oxide) (PEO), poly(propylene oxide) (PPO), and poly(tetrahydrofuran) (PTF), however, a limited number of studies have been conducted using PTF.[52]

PEGs constitute a wide family of low-immunogenicity hydrophilic polymers. The oxyethylene backbone makes the water molecule soluble with a high hydration volume and molecular flexibility leading to a reduced cell and protein adsorption capacity.[53] Nevertheless, bioelimination mechanism of polyether chains is limited to the excretory system and thus, only low molecular weight PEGs are currently approved for medical applications.[32] In spite of this limitation, PEG remains as a widely spread system for the preparation of engineered scaffold due to its feasibility to be chemically modified by the substitution of terminal hydroxyl groups with either acrylates or methacrylates residues yielding to crosslinkable macromonomers.[54] Linear poly(ethylene glycol diacrylate) (PEGDA), dimethacrylate (PEGDMA), and multi-arm PEG acrylate can then form biocompatible oxyethylene-based hydrogels with tailored biomechanical properties depending on the concentration, molecular weight, and molecular architecture of the ethylene glycol precursor. In addition, hydrolytically labile PEG-based hydrogels can be obtained through copolymerization with other acrylated and methacrylated comonomers or with commonly degradable $\alpha$-hydroxy acids such as those derived from lactic acid, glutamic acid, or propylene fumarate, giving degradable composite systems for the preparation of bioactive scaffolds of musculoskeletal,[55,56] vascular,[34] dental pulp,[57] and endothelial tissues.[58]

The hydrophilic methacrylate-based polymers present the advantage of easy crosslinking with other comonomers through classic copolymerization methods obtaining tailored materials in large scale productions and with highly tunable properties that seem to be extremely suitable as 3D growth matrices. Additionally, polymethacrylate hydrogels containing surface reactive moieties such as carboxyl, hydroxyl, or amino functional groups might be chemically modified by grafting of ECM proteins or cell adhesion receptor peptide sequences in order to enhance the cell–matrix interactions and thus, improving the biofunctionality of the engineered scaffolds.[59] This is the case of Poly(*N*-isopropylacrylamide) and poly(acrylic acid) which have been successfully modified for the preparation of cartilage,[60] myocardial,[61,62] and ophthalmological-engineered scaffolds.[63] Other example of engineered methacrylate-based scaffold consists in the preparation of keratoprothesis (artificial cornea) from sponged PHEMA composites. Earlier studies on synthetic PHEMA showed encouraging results in human clinical trials and later keratoprothesis prototypes based on PEG/PHEMA interpenetrated networks proved their optimal features in terms of wettability, water uptake capacity (>80% w/w), and mechanical properties for ophthalmological tissue engineering applications.[63–65]

Vinyl polymers constitute other class of synthetic hydrophilic materials that show very important advantages in this field. PVA formed by ethylene backbone with pendant hydroxyl groups is one of the simplest members of the group. However, PVA-based hydrogels are distinguished by their good biomechanical features and the ability to retain large amount of water (>70%) within the structure, ensuring a prolonged highly hydrated environment and a correct diffusivity of ions and biomolecules through the material.[66] Crosslinking mechanism of hydrogels forming networks can take place physically through freeze–thaw cycling and phase separation from theta solutions or chemically with crosslinking agents such as glutharaldehyde, formaldehyde, or epichlorohydrin.[67] PVA hydroxy functional groups can be also chemically modified with a variety of substituents. In this sense, methacrylated PVA chains also conjugated with cell adhesion peptide sequences (e.g., integrins) have been evaluated for the fabrication of porous scaffolds that provide the appropriate stimuli and biomechanical environment for *in vitro* regeneration of cartilage-like soft tissues.[68,69]

As mentioned above, soft tissue regeneration implicates not only high cell matrix interactions but also tunable elasticity and nutrient permeability within a highly moistly tridimensional supporting structural scaffold. However, it is rare to find a material that possesses all the desired characteristics. In this sense, synthetic PAAs have been gaining an increased attention in the last few years. They constitute a family of biocompatible and biodegradable polymers containing carboxyl groups along with amino groups in their repeating units that possess low interfacial tension and can be prepared as soft hydrogels with tunable mechanical properties to be used as implantable cell supporting structural devices.[70,71] These PAAs are synthetized by polyaddition of amines to bisacrylamides through Michael-type reactions leading to regularly arranged *ter*-amino and amide group macromolecular chains, in which chemical properties can be tuned by using functionalized monomers containing additional side *ter*-amino, carboxyl, hydroxyl, and allyl groups.[72] Recent studies on novel guanidine containing amphoteric PAA-based hydrogels have shown very promising *in vivo* results when used as implantable scaffold for peripheral nerve regeneration.[73]

Very close macromolecular architecture to PAA polymers can be obtained from direct polycondensation of $\alpha$-amino acids, yielding to homo or copoly(amino acids) linked together through amide bonds with active functional groups on the side.[40] Due to their feasible side functionalization together with the high similarity with naturally occurring proteins, synthetic poly(amino acids) have been extensively evaluated for the preparation of biomedical materials.[40] However, only poly(L-glutamic acid) (PLG) has been found promising for the preparation of soft tissue engineered scaffolds.[74] As an alternative approach, peptide bonds within synthetic poly(amino acids) backbones have been replaced by non-amide linkages ester, leading to pseudo-poly(amino acids) that are also been applied as scaffolds in the field of tissue engineering.[74,75] In fact, the search for structural architectures similar to the natural occurring biomacromolecules has constituted a permanent challenge for polymer science specialists. In this sense, synthetic PPE include an attractive family of biocompatible and biodegradable nucleic acid mimetic polymers consisting of repeating phosphoester monomeric units in the backbone[76] such as those of poly(2-aminoethyl propylene phosphate) (PPE-EA),[77] poly(6-aminohexyl propylene phosphate) (PPE-HA),[78] and poly(2-(2-oxo-1,3,2-dioxaphospholoyloxy) ethyl methacrylate) (OPEMA)[79] that have been used for the preparation of bone tissue engineering scaffolds.

There are also other phosphorous-based polymers that have also been investigated for biomedical applications.[80] Poly(phosphazenes), also called poly(phosphonitrile), are hybrid organic–inorganic high molecular weight polymers consisting of phosphorus–nitrogen backbone bonded through alternating double and single bonds that offer good versatility of being chemically functionalized by esterification, etherification, or amidification of the two side phosphorus units with a variety of biomolecules appropriate to be used for the design and preparation of hard tissue engineered scaffold.[81]

As it has been reviewed during this section, many hydrophilic synthetic polymers can be used for the preparation of engineered scaffold promoting and enhancing the replacement and/or regeneration of soft tissues and their biological functions by mimicking not only the natural occurring tissue supporting matrices but also the mechanical, structural requirements, and molecular interactions with cells that promote tissue healing. However, despite all the advances achieved during the last decades in polymer science, ideal hydrophilic matrices which combine all desired features in one engineered system do not exist and perhaps, they might not ever exist because different tissues require different biofunctional niches to be developed. New approaches based on the combination of polymers hold promise for the successful design of polymer-assisted soft tissue healing therapies.

## 8.5 Amphiphilic Systems

Amphiphilic systems are heterogeneous by nature and consist of both hydrophilic and hydrophobic parts. Their main feature is the ability to form intermolecular and/or intramolecular self-assembly that leads to different nanostructures and architectures. For example, intermolecular self-assembly of amphiphilic molecules leads to the formation of micelles in aqueous milieu and intramolecular assembly occurs in protein folding. Furthermore, molecules can have the ability to form supramolecular assembles. As self-assembly phenomenon occurs in nature spontaneously in many biological systems, in latest years, huge investigation has been directed to produce scaffolds for tissue regeneration based on this phenomenon. At present, this field is growing up especially due to the positive results of investigated systems and to their formidable potential for cell proliferation and differentiation.

### 8.5.1 Block Copolymer Scaffolds

Block copolymerization of monomers that have very different hydrophilicity is the most usual method to obtain synthetic polymeric amphiphilic systems. Block copolymers constitute a versatile source to design nanoscale scaffolds. Thus, the morphology of block copolymers can be tuned with composition, molecular weight of the individual blocks, and the spatial relationship among them. By changing the molecular weight and spatial arrangement of the blocks, different morphologies such as spheres, cylinders, lamellae, and bicontinuous conformations, can be obtained.[82] Another advantage of using block copolymers is their potential to form nanostructures on non-flat surfaces, such as spheres and irregular 3D structures.

Triblock copolymers of PEO and PPO is a family of commercially available amphiphilic systems (Pluronic® or Poloxamer®) that have gelation properties associated with their micelle-forming capability in water.[83] Aqueous solutions of these copolymers exhibit phase transition from sol to gel and gel to sol with increase in temperature when the polymer concentration is upon a critical value.[84] The mechanism of gelation is still under debate but is commonly attributed to the association of micelles and formation of various lyotropic liquid crystalline phases.[85] It has been established that the stability of gels is related with the shape and packaging symmetry of the ordered structures and depends on the ratio of the lengths of the hydrophilic and hydrophobic segments. The X-shaped copolymers formed by four PPO and PEO block chains bonded to an ethylene diamine central group (Tetronic®) are commercial copolymers known as poloxamines and their ability to form micelles in aqueous solutions and gelation has more recently demonstrated.[86] The properties of the core of these micelles can strongly differ from those of the Pluronic micelles since the hydrophilic diamine group connects the hydrophobic PPO segments. In addition, the pH sensitiveness of Tetronic micelles offer interesting features compared to the micelles of non-ionic block copolymers. Cytotoxicity screening of various poloxamine varieties have been reported revealing Tetronic 908, 1107, 1301, and 1307 solutions as the most cytocompatible[87] for tissue engineering scaffolds.

Pluronic® and Tetronic® are mainly used as injectable gel scaffolds. Gelation temperature depends on polymer composition and solution concentration and the copolymer Pluronic F-127 that forms gels at

37°C has been the most employed so far as injectable scaffold, for example, in burn treatment and in the formation of a cartilage layer on the host bone.[88] This poloxamer has been also used to produce tissue-engineered cartilage in conjunction with distraction osteogenesis in the reconstruction of a condylar osteochondral defect.[89] The scaffold consisted of a mixture of chondrocytes and Pluronic F-127 injected on the notched surface of a transport disc which was implanted *in vivo*, producing the simultaneous rehabilitation and regeneration of condylar bone and cartilage. Pluronic F-127 has been employed to produce a nanostructuring hydrogel scaffold made from crosslinked PEG-fibrinogen conjugates suitable to control cellular morphogenesis through defined cell–matrix interactions. It was observed that structural modifications induced by the Pluronic F-127 micelles without changing stiffness were the dominant factor that influenced the cell morphology response of human foreskin fibroblasts cells.[90]

A new class of injectable and erodible hydrogels exhibiting highly robust gel strength at body temperature was fabricated by enzyme-mediated crosslinking between Pluronic copolymer micelles. Micelles in aqueous solution consist of tyramine-conjugated Pluronic F-127 at two terminal ends of PEO. Tyrosinase was employed to convert these micelles to highly reactive catechol-conjugated micelles that could further crosslink individual Pluronic copolymer micelles to form a highly stable gel structure. The resulting hydrogel presented a lower critical gelation concentration and enhanced strength at the gel state along with tissue adhesive properties to porcine stomach mucin due to the presence of unreacted catechol groups.[91]

Engineering scaffolds based on Pluronic in combination with other polymers were reported. Thus, a scaffold made from mixtures of polycaprolactone (PCL) and Pluronic F-127 has been proposed as a new type of cylindrical scaffold with gradually increasing growth factor concentrations. This scaffold has the required characteristics for *in vitro* investigations of 3D chemotaxis/haptotaxis for the control of specific biological processes.[92] Other researchers have prepared a polydioxanone/Pluronic F-127 (PDO/Pluronic F127) scaffold for bone formation using periosteal-derived cells in presence of adipose tissue-derived endothelial-like cells with positive results.[93]

Poloxamines were later used in tissue engineering. The copolymers Tetronic 1107 and 1307 gave viscoelastic gels at 37°C and they were evaluated both as an active osteogenic component and as a vehicle for rhBMP-2 injectable implants. Both poloxamines presented intrinsic osteogenic activity (not observed for Pluronic F127) and Tetronic 1307, solely or loaded with rhBMP-2, caused differentiation of MSC to both osteoblasts and adipocytes,[87] offering interesting perspectives for bone regeneration using minimally invasive procedures. An injectable hydrogel based on a modified 4-arm PPO–PEO block copolymer that was conjugated to RGD and crosslinked *in situ*, was a novel approach for tissue engineering medicine.[94] Hydrogels were successfully obtained by simple mixing of RGD-Y in a tyramine-conjugated Tetronic polymer solution in presence of horseradish peroxidase (HRO) and hydrogen peroxide ($H_2O_2$). *In vitro* results of osteoblasts (MC3T3-E1) cultured on the hydrogel surfaces for 24 h, revealed good viability and adequate morphology of the adhered cells, demonstrating that the *in situ* conjugation of RGD peptides to tyrosine residues enhanced the attachment and spreading of osteoblasts in a surface density-dependent manner.

In other approaches, Tetronic 1107 was utilized for the preparation of semi-synthetic matrices in combination with collagen.[95] The poloxamine was first functionalized with methacrylate groups and subsequently photopolymerized in the presence of collagen. The resulting gel had a sharp increase in stiffness and HepG2 and bovine aortic endothelial cells adhered to and spread on the surface of these matrices suggesting good potential in tissue regeneration. A summary table (Table 8.2) displayed the described systems based on commercial poloxamers and their applications in tissue regeneration.

Injectable gels scaffolds with improved biodegradability and permanence of the gel in the body are based on diblock and triblock copolymers of PEO and poly(L-lactic acid) (PLLA), PEO-PLLA, and PEO-PLLA-PEO, respectively. These systems form micelles at low concentrations which are driven by hydrophobic interactions of the PLLA blocks. They exhibit a sol–gel transition with increasing temperature which is determined by the length of PLLA block when PEO block is constant as well as by concentration. Thus, specific compositions of the triblock copolymer form a gel upon subcutaneous injection in rat.[96] Accordingly, aqueous solutions of triblock copolymers of PEO and PLGA, PEO-PLGA-PEO, become a gel at the body temperature showing good mechanical strength and integrity of the gel persisted longer than one month.[97] Injectable gels of PLA-PEO-PLA triblock copolymers have been proposed for tissue

**Table 8.2** Amphiphilic systems based on poloxamers

| System | Application | Reference |
|---|---|---|
| Poloxamer 407 mixed with chondrocytes | Formation of cartilage on the host bone | 88 |
| Mixture of Pluronic F-127 and chondrocytes injected on the notched surface of a disk | Scaffold for reconstruction of condylar osteochondral defect | 89 |
| Crosslinked PEG-fibrinogen modified with Pluronic F-127 | Scaffold for control cellular morphogenesis of human foreskin fibroblasts cells | 90 |
| Enzyme-mediated crosslinking of tyramine-conjugated Pluronic F-127 | Erodible hydrogels with muco-adhesive properties to porcine stomach mucin | 91 |
| Mixtures of Pluronic F-127–PCL | Surface area gradient scaffold for growth factors, i.e., heparin, VEGF; model to study chemotaxis/ haptotaxis in biological processes | 92 |
| Mixtures of PDO–Pluronic F-127 | Scaffold for bone formation using periosteal-derived cells and endothelial-like cells | 93 |
| Tetronic 1107 and 1307 loaded with rhBMP-2 | Injectable implants with osteogenic activity; differentiation of MSC; bone regeneration | 87 |
| RGD-conjugated Tetronic-tyramine | Hydrogel crosslinked *in situ* for osteoblasts cultures; bone regeneration | 94 |
| Semi-synthetic collagen-Tetronic 1107 matrices | Scaffolds for adhesion and spread of HepG2 bovine aortic endothelial cells | 95 |

engineering applications.[98] In order to enhance their mechanical properties photo-crosslinked PLA-PEO-PLA hydrogels were synthesized by the use of a novel two-step approach. In the first step, the end-functionalized PLA-PEO-PLA triblock was self-assembled into a physical hydrogel through hydrophobic micelle network junctions, and in the second step, the network was stabilized by photo-crosslinking the terminal acrylate groups. The resulting gels demonstrated an extended degradation time and elastic modulus that can be tuned from 5 to 400 kPa by controlling the polymer concentration. This tunability will allow for better matching between the hydrogel scaffold material and soft target tissues.[99]

Self-crosslinkable and biodegradable amphiphilic copolymers of poly(caprolactone fumarate) (PCLF) and poly(ethylene glycol fumarate) (PEGF), PEGF-co-PCLF, have been developed for tissue engineering applications using a novel synthesis method. They are essentially copolymers of fumaryl chloride, which contain double bonds for *in situ* crosslinking with PCL or/and PEG that has a flexible chain to facilitate self-crosslinking via photo-initiation. These copolymers were proposed to prepare polymer networks and scaffolds with controlled hydrophilicity, swelling, and mechanical properties for applications in drug release and micro-encapsulation of cells.[100]

Supramolecular hydrogels self-assembled by $\alpha$-cyclodextrin and methoxypolyethylene glycol-poly(caprolactone)-(dodecanedioic acid)-poly(caprolactone)-methoxypolyethylene glycol (MPEG-PCL-MPEG) triblock polymers were prepared and characterized *in vitro* and *in vivo*. ECV304 cells and marrow MSC were encapsulated and cultured in the hydrogels. The *in vitro* cell viability studies and the *in vivo* histological studies demonstrated that MSC maintained the spread morphologies in the hydrogel although the hydrogel network degraded at some degree. The rapid gelation kinetics is attractive because of its entrapment of biologically active additives and cells contribute to the tissue regeneration. The *in vivo* histological studies demonstrated that degradation products of the hydrogels did not produce signs of cytotoxicity.[101]

Amphiphilic triblock copolymers of methoxy-PEO (MPEO), PLLA, and poly(L-lysine) (PLL), MPEO-PLLA-PLL, were obtained to create scaffolds with surface properties suitable for cellular adhesion. Thin films of the triblock copolymer promoted human osteoblast adhesion and proliferation compared with pure PLLA and MPEG-PLLA diblock copolymer attributed to the presence of free amino groups, which self-segregate on the surface of the polyester. This triblock copolymer can potentially be used to immobilize cell recognition ligands or growth factors to induce a specific and desired cellular interactions what can be of relevance in the preparation of functional scaffolds for tissue engineering.[102] In other works, the effects of immobilizing peptides in amphiphilic triblock copolymers of poly(ε-caprolactone-co-lactide) (PCLA) and PEO, PCLA-PEO-PCLA, on the cell-binding efficacy have been studied. These block copolymers self-assembled into hydrogels forming a percolated micelle network, showing a sol–gel transition temperature between room temperature and body temperature. Immobilization of RGD in hydrophilic blocks effectively enhanced cell adhesion of chondrocytes on the corresponding hydrogel surface much more significantly than that in hydrophobic blocks because of the RGD ligands in hydrophilic blocks exposed on the micellar surface.[103]

The morphology and protein adsorption behavior of diblock and triblock copolymers of MMA and acrylic acid (AA), MMA-AA and AA-MMA-AA respectively, were investigated using atomic force microscopy. They present distinct nanomorphologies at the same chemical composition of the blocks. In the diblock copolymers PMMA and PAA cylinders were observed whereas in the triblock copolymers ring-like features were observed. Differences in surface properties affected bovine serum albumin adsorption, critical for understanding biocompatibility, in the sense that triblock copolymer exhibited higher adhesion force, and should be considered for studies on cellular adhesion. A further conclusion of this study suggests that it is possible to tune protein adsorption by changing the arrangement of the corresponding block components through chemical synthetic techniques, which subsequently leads to different surface nanomorphologies.[104] Table 8.3 shows the systems based on the synthetic copolymers with amphiphilic properties described above.

### 8.5.2 Peptide-Based Scaffolds

In the latest years great interest has emerged on the use of totally degradable small molecules capable of self-assembly giving supramolecular nanostructures as scaffolds for tissue regeneration. Small peptides have attracted much attention due to their functionality and cell signaling capacity. The versatility of the chemical composition based on amino acid sequences leads to a variety of possible secondary, tertiary, and quaternary structures through folding and hydrogen bond formation.[105] Entanglement of peptide-based nanostructures can lead to the formation of 3D networks in nanofibers that mimic the native ECM that could function as highly biomimetic artificial matrices.

**Table 8.3** Amphiphilic systems based on synthetic copolymers

| System | Application | Reference |
|---|---|---|
| Photo-crosslinked PLA-PEO-PLA triblock copolymers | Hydrogel scaffolds with properties for soft tissues | 99 |
| Self-crosslinkable PEGF-co-PCLF copolymers | Scaffolds with controlled properties for drug release and micro-encapsulation of cells | 100 |
| Self-assembled systems of $\alpha$-cyclodextrins and MPEG-PCL-MPEG triblock copolymers | Supramolecular hydrogels for drug delivery and stem cells encapsulation | 101 |
| MPEO-PLLA-PLL triblock copolymers | Scaffolds adequate for osteoblasts cultures. Bone formation | 102 |
| PCLA-PEO-PCLA triblock copolymers with RGD immobilized | Scaffolds for encapsulation of chondrocytes | 103 |
| MMA-AA and AA-MMA-AA diblock and triblock copolymers | Scaffolds with different nanomorphology for cell–polymer interactions | 104 |

Systems based on peptide amphiphiles (PAs) have been extensively studied by Stupp *et al.*[106] Basically, it consists of short peptide sequences attached to a hydrophobic tail through an amide bond. The PA includes one or more charged residues to aid in solubility and gelation. Self-assembly into cylindrical nanofibers takes place through hydrophobic collapse of the aliphatic tails along with intermolecular hydrogen bonding formation, usually $\beta$-sheet formation, so that the bioactive residue of the PA is displayed on the surface of the nanofiber whereas, the hydrophobic tails pack in the core. Depending on the sequence composition the assembly can be controlled by pH or addition of multivalent cations. The effect of hydrogen bonding on the self-assembly nanostructures and the mechanical properties of the assembled gel[107] was investigated using a series of 26 PA derivatives, including 19 *N*-methylated variants and 7 alanine mutants; results revealed the relevant role of key hydrogen bonds.

For regenerative medicine, usually the PA molecules are combined with bioactive principles such as growth factors, glycosaminoglycans among others, or even cells, forming a liquid with low viscosity which can be injected into the tissue where the PAs self-assemble due to the presence of the electrolytes of the physiological medium giving nanofibers and subsequent formation of a gel network. The control of the kinetics of gelation could be critical to tailor these systems for specific clinical applications. The gelation kinetics of a series of PAs presenting different peptide sequences without changing the bioactive segment was studied.[108] The pre-existence of hydrogen-bonded aggregates of more hydrophobic PAs in the solution appeared to accelerate gelation kinetics. Mutation of the peptide sequence to include more hydrophilic and bulky amino acids suppressed formation of these nuclei and effectively slowed down gelation through self-assembly of the nanofiber network.

PAs can be modified by addition of a bioactive sequence to signal encapsulated cells, that is, the Arg-Gly-Asp-Ser (RGDS) peptide epitope.[109] The Ile-Lys-Val-Ala-Val (IKVAV) peptide sequence derived from laminin has been incorporated to PAs for neural regeneration. Neural progenitor cells encapsulated *in vitro* within the 3D network differentiated rapidly into neurons, while discouraging the development of astrocytes.[110] These results suggested the potential of this system in the treatment of spinal cord injury (SCI) where formation of a glial scar prevents axonal regeneration after injury.[111] In fact, when the IKVAV-modified system was applied to a mouse SCI model, it was demonstrated that the PA-reduced astrogliosis and increased the number of oligodendroglia at the injury site. Furthermore, the nanofibers promoted regeneration of both descending motor fibers and ascending sensory fibers through the lesion site.[112]

To generate systems with proangiogenic activity, heparin was used to nucleate the self-assembly of nanostructures from a heparin-binding PA (HBPA). The rigid nanofibers obtained displayed heparin chains capable of capturing many potent signaling proteins, such as fibroblast growth factor 2 (FGF-2) and VEGF, both very important in angiogenesis processes. *In vivo* studies demonstrated that the nanostructures stimulated extensive new blood vessel formation using nanogram amounts of growth factor proteins that by themselves did not induce any detectable angiogenesis.[113] Biocompatibility studies revealed that the material was biodegraded and quickly remodeled into a well-vascularized connective tissue without the addition of any exogenous growth factors.[114] Supramolecular nanofibers formed by self-assembly of a heparin-binding PA and heparan sulfate-like glycosaminoglycans were evaluated using a dorsal skinfold chamber model to dynamically monitor the interaction between the nanofiber gel and the microcirculation.[114] Results obtained advocate the use of this system for the healing of chronic wounds or to enhance efficacy of skin grafts.

Hartgerink *et al.*,[106] designed a PA which contained four consecutive cysteine amino acids, a phosphoserine residue, and the bioactive sequence RGD in order to prepare composites for bone regeneration. The self-assembled of the PA molecules was induced by pH giving nanofibers reversibly crosslinked. After crosslinking, the fibers were able to direct mineralization of hydroxyapatite in such a way that the crystallographic c axes of hydroxyapatite were aligned with the long axes of the fibers, as that observed between collagen fibrils and hydroxyapatite crystals in bone. PAs-based scaffolds have been investigated for enamel regeneration which is still a major challenge in regenerative medicine. For this application, branched PA molecules containing the peptide motif RGD (abbreviated BRGD-PA) hydrogels were cultured with ameloblast-like cells and primary enamel organ epithelial cells.[115] Results showed an enhancement in proliferation with expression of amelogenin and ameloblastin, two proteins secreted by ameloblasts during enamel formation. In other works, PAs-based scaffolds have been employed for dental stem cells differentiation. PA hydrogel scaffolds that incorporated the cell adhesion sequence RGD and an enzyme-cleavable site were used to encapsulate living cells. Stem cells from human exfoliated deciduous teeth (SHED) secrete a soft collagen matrix and dental pulp stem cells (DPSC) differentiate into an osteoblast phenotype and deposit mineral.[116] Other studies were devoted to develop a new biomaterial in which synthetic peptides self-assembled to form a hydrogel. Results showed feasibility of this material to serve as a matrix for osteogenic differentiation of co-cultured bone marrow cells *in vitro* and *in vivo*.[117]

Following a mimimalist approach small amphiphilic blocks of peptides were selected for the preparation of nanofiber scaffolds that form macroscopic hydrogels at neutral pH. Thus, amphiphilic blocks of mixtures of dipeptides linked to fluorenylmethoxycarbonyl (Fmoc) have been prepared by Jayawarna *et al.* (group of Ulijn).[118] Seven Fmoc-dipeptides made up of combinations of the amino acids; glycine, alanine, leucine, and phenylalanine were used. The self-assembly of the Fmoc-dipeptides is driven by hydrogen bonding and $\pi$–$\pi$ interactions between $\pi$ electrons in the aromatic fluorenyl rings. It was found that variations in the molecular architecture of blocks provoked drastic changes in the self-assembly behavior. Specific compositions gave gels stable at physiological pH and resulted suitable to support the growth of bovine chondrocytes. Based on these systems a biomimetic hydrogel was obtained using a mixture of the two aromatic short peptides Fmoc-FF (Fmoc-diphenylalanine) and Fmoc-RGD as the simplest self-assembling moieties reported up till then for the construction of small-molecule-based bioactive hydrogels.[119] The self-assembly of FF and RGD peptide sequences into $\beta$-sheets generated cylindrical nanofibers. This rapid gelling material promoted adhesion, spreading, and proliferation of encapsulated dermal fibroblasts through specific RGD-integrin binding. A recent review on the progress of minimal self-assembling peptides and functionalized amino acids to create hydrogels has been reported.[120] The article deals with important topics such as the mechanism of self-assembly and the relationship between the molecular structure with their viscoelastic and biochemical properties, focusing on the outlook for the use of these modular and dynamic materials as robust networks for tissue engineering.

A new class of ionic self-complementary oligopeptides was devised by Zhang *et al.*[121] They were formed by primary sequences of alternating hydrophilic and hydrophobic amino acids where sequences of charged amino acid residues include alternating positive and negative charges. This group described the use of hexadecapeptides to generate highly hydrated gels (over 99% water content) and proposed their application as scaffolds that mimic ECM suitable for mammalian cell attachment in tissue repair. The scaffolds could support neuronal cell attachment and differentiation as well as extensive neurite outgrowth[122] and they result successful for cartilage repair in which encapsulated chondrocytes retained their morphology and expressed a stable chondrocyte phenotype.[123] The functionalization of one of the peptide AcN-RADARADARADARADA-$CONH_2$, by direct solid phase synthesis extension at the amino terminal with three short-sequence motifs that are present in proteins of the basement membrane laminin 1 (YIGSR, RYVVLPR) and collagen IV (TAGSCLRKFSTM), was attempted.[124] Human aortic endothelial cells cultured in this hydrogel expressed endothelial phenotype and promoted basement membrane deposition which is a prerequisite for successful expansion of this type of cells.

Peptide-based hydrogels consisting of oligopeptide sequences that fold intramolecularly into amphiphilic, asymmetric $\beta$-hairpins and intermolecularly into fibrillar nanostructures are another type of systems with multiple possibilities for tissue replacement. These systems are known as $\beta$-hairpin peptides. They can be prepared from monomers of alternating hydrophilic and hydrophobic character, namely lysine and valine, respectively. These peptides adopt a random conformation in water, however, they fold into its $\beta$-hairpin conformation in the presence of ions of the physiological fluid or changes in pH.[125] Subsequent self-assemble of the individual hairpins due to hydrophobic interactions and intramolecular hydrogen bonds produces a hydrogel structure. The lysine residue can alternatively be replaced by a negatively charged glutamic acid.[126] A variety of this type of diblock copolypeptide amphiphiles containing charged and hydrophobic segments have been studied. Dilute solutions of these copolypeptides formed hydrogels that retain their mechanical strength up to temperatures of about 90°C and recovered rapidly after stress. Gelation depends not only on the amphiphilic nature of the polypeptides, but also on chain conformations-$\alpha$-helix, $\beta$-strand, or random coil, indicating that shape-specific supramolecular assembly is essential for the gelation process.[127]

Physically crosslinked and fibrillar hydrogel networks were developed by the self-assembly of $\beta$-hairpin peptide molecules with varying degrees of strand asymmetry. The system was based on the MAX1 [$(VK)_4$-VDPPT-$(KV)_4$-$NH_2$] consisting of two strands of alternating valine (V) and lysine (K) residues.[128] The peptide chemistry in the self-assembled state was used to generate fibrils with twisting, non-twisting, or laminated morphology. The mass density of the networks varied significantly, and it could be directly related to the local fibrillar morphology. The density of the network was dependent of changes in the peptide concentration. Bulk rheological properties of the hydrogels correlated with the fibrillar nanostructure. Stiffer and laminated fibrils formed networks with higher storage modulus as compared to the flexible and singular fibrillar networks.

Kumar *et al.*[129] have reported a review on the recent patents concerning the molecular self-assembly of peptides, PAs, and peptidomimetics through nanoarchitectures to give hydrogels. The design rules of this rapidly growing field described in the review are centered mainly on the construction of different types of peptides, peptide-based block copolymers as well as polymer peptide conjugates. The applicability of these systems in cell culture scaffolds for tissue engineering, drug, and gene delivery and as templates for nanofabrication and biomineralization is also discussed. As in previous section, Table 8.4 displays the different peptide-based systems proposed for tissue engineering purposes.

**Table 8.4** Peptide-based amphiphilic systems

| System | Application | Reference |
|---|---|---|
| PA with IKVAV peptide sequence motif | 3D network of nanofibers for neural progenitor cells encapsulation; treatment of SCI | 110–112 |
| Heparin-binding PA (HBPA) nucleated by heparin | Nanostructures with proangiogenic activity | 113 |
| Systems of HBPA and heparin sulfate-like glycosaminoglycans | Supramolecular nanostructures for angiogenesis; treatment of chronic wounds or as skin grafts | 114 |
| PA containing cysteine and phosphoserine residues and RGD sequence | Preparation of composites for bone regeneration | 106 |
| Branched PA molecules containing the peptide motif RGD (BRGD-PA) | Hydrogel scaffolds for ameloblast-like cells and primary enamel organ epithelial cells cultures; enamel regeneration | 115 |

(*Continued*)

**Table 8.4** (*Continued*)

| System | Application | Reference |
|---|---|---|
| PA incorporating the sequence RGD and an enzyme-cleavable site | Hydrogel scaffolds for dental stem cells differentiation | 116 |
| Systems based on mixtures of Fmoc-FF and Fmoc-RGD | A biomimetic hydrogel for encapsulation of dermal fibroblasts | 119 |
| Hexadecapeptides with alternating charged and hydrophobic amino acids | Highly hydrated gels as scaffolds for neuronal regeneration and cartilage formation | 122, 123 |
| Peptide AcN-RADARADARADARADA-$CONH_2$ incorporating the three sequence motif YIGSR, RYVVLPR, and TAGSCLRKFSTM | Scaffold for endothelial cells encapsulation | 124 |
| $\beta$-Hairpin peptides | Hydrogel scaffolds which retain and recover their mechanical properties | 127 |
| $\beta$-Hairpin peptide molecules with varying degrees of strand asymmetry based on the MAX1 [$(VK)_4$-VDPPT-$(KV)_4$-$NH_2$] system | Hydrogel network structure with different local fibril nanostructure | 128 |

## References


1. Jagur-Grodzinski J (2006) Polymers for tissue engineering, medical devices, and regenerative medicine. Concise general review of recent studies. *Polymer for Advanced Technologies* **17**, 395–418.
2. Lavik E, Langer R (2004) Tissue engineering: Current state and perspectives. *Applied Microbiology and Biotechnology* **65**, 1–8.
3. Stupp SI (2005) Biomaterials for regenerative medicine. *Materials Research Society Bulletin* **30**, 546–553.
4. Wang YK, Yong T, Ramakrishna S (2005) Nanofibres and their influence on cells for tissue regeneration. *Australian Journal of Chemistry* **58**, 704–712.

5. Meyer U, Loos U, Joos U *et al.* (2004) Biological and biophysical principles in extracorporal bone tissue engineering. Part I. *International Journal of Oral & Maxillofacial Surgery* **33**, 325–332.

6. Dalby MJ, Di Silvio L, Davies GW *et al.* (2000) Surface topography and HA filler volume effect on primary human osteoblasts *in vitro*. *Journal of Materials Science-Materials in Medicine* **11**, 805–810.

7. Hubbell JA (1999) Bioactive biomaterials. *Current Opinion in Biotechnology* **10**, 123–129.

8. Lysaght MJ, Reyes J (2001) The growth of tissue engineering. *Tissue Engineering* **7**, 485–493.

9. Meyer U, Szulczewski DH, Moller K *et al.* (1993) Attachment kinetics and differentiation of osteoblasts on different biomaterials. *Cells and Materials* **3**, 129–140.

10. Palecek SP, Loftus JC, Ginsberg MH *et al.* (1997) Integrin-ligand binding properties govern cell migration speed through cell-substratum adhesiveness. *Nature* **385**, 537–540.

11. Matyjaszewski K, Tsarevsky NV (2009) Nanostructured functional materials prepared by atom transfer radical polymerization. *Nature Chemistry* **1**, 276–288.

12. Gregory A, Stenzel MH (2012) Complex polymer architectures via RAFT polymerization: From fundamental process to extending the scope using click chemistry and nature's building blocks. *Progress in Polymer Science* **37**, 38–105.

13. Curtis A, Wilkinson C (2001) Nantotechniques and approaches in biotechnology. *Trends in Biotechnology* **19**, 97–101.

14. Stevens MM, George JH (2005) Exploring and engineering the cell surface interface. *Science* **310**, 1135–1138.

15. Dalby MJ, Gadegaard N, Tare R *et al.* (2007) The control of human mesenchymal cell differentiation using nanoscale symmetry and disorder. *Nature Materials* **6**, 997–1003.

16. McBeath R, Pirone DM, Nelson CM *et al.* (2004) Cell shape, cytoskeletal tension, and RhoA regulate stem cell lineage commitment. *Developmental Cell* **6**, 483–495.

17. Curtis A, Sokolikova-Csaderova L, Aitchison G (2007) Measuring cell forces by a photoelastic method. *Biophysical Journal* **92**, 2255–2261.

18. Dalby MJ, Biggs MJP, Gadegaard N *et al.* (2007) Nanotopographical stimulation of mechanotransduction and changes in interphase centromere positioning. *Journal of Cellular Biochemistry* **100**, 326–338.

19. Albertsson (2003) Recent developments in ring opening polymerization of lactones for biomedical applications. *Biomacromolecules* **4**, 1466–1486.

20. Pérez-Merino P, Parra F, Ibares-Frías L *et al.* (2010) Clinical and pathological effects of different acrylic intracorneal ring segments in corneal additive surgery. *Acta Biomaterialia* **6**, 2572–2579.

21. Puskas JE, Chen YH (2004) Biomedical application of commercial polymers and novel polyisobutylene-based thermoplastic elastomers for soft tissue replacement. *Biomacromolecules* **5**, 1141–1154.

22. Wang T, Li M, Gao H *et al.* (2011) Nanoparticle carriers based on copolymers of poly(epsilon-caprolactone) and hyperbranched polymers for drug delivery. *Journal of Colloid and Interface Science* **353**, 107–115.

23. Zhu Z, Li Y, Li X *et al.* (2010) Paclitaxel-loaded poly(N-vinylpyrrolidone)-b-poly(epsilon-caprolactone) nanoparticles: Preparation and antitumor activity *in vivo*. *Journal of Controlled Release* **142**, 438–446.

24. Zhang W, Li Y, Liu L *et al.* (2010) Amphiphilic toothbrushlike copolymers based on poly(ethylene glycol) and poly(epsilon-caprolactone) as drug carriers with enhanced properties. *Biomacromolecules* **11**, 1331–1338.

25. Stevens MM (2008) Biomaterials for bone tissue engineering. *Materials Today* **11**, 18–25.

26. Xing Z-C, Chae W-P, Baek J-Y *et al.* (2010) *In vitro* assessment of antibacterial activity and cytocompatibility of silver-containing PHBV nanofibrous scaffolds for tissue engineering. *Biomacromolecules* **11**, 1248–1253.

27. Saito N, Takaoka K (2003) New synthetic biodegradable polymers as BMP carriers for bone tissue engineering. *Biomaterials* **24**, 2287–2293.

28. Gao C, Yan D (2004) Hyperbranched polymers: From synthesis to applications. *Progress in Polymer Science* **29**, 183–275.

29. McKee MG, Unal S, Wilkes GL *et al.* (2005) Branched polyesters: Recent advances in synthesis and performance. *Progress in Polymer Science* **30**, 507–539.

30. Alperin C, Zandstra PW, Woodhouse KA (2005) Polyurethane films seeded with embryonic stem cell-derived cardiomyocytes for use in cardiac tissue engineering applications. *Biomaterials* **26**, 7377–7386.

31. Kim BS, Park IK, Hoshiba T *et al.* (2011) Design of artificial extracellular matrices for tissue engineering. *Progress in Polymer Science* **36**, 238–268.

32. Ulery BD, Nair LS, Laurencin CT (2011) Biomedical applications of biodegradable polymers. *Journal of Polymer Science, Part B: Polymer Physics* **49**, 832–864.

33. Varghese S, Elisseeff JH (2006) Hydrogels for musculoskeletal tissue engineering. In: C Werner, JH Eisseeff, C Fischbach, T Freier, AJ Garcia, DJ Mooney, T Pompe, K Salchert, S Varghese, S Zhang and X Zhao, editors. *Polymers for Regenerative Medicine* (pp. 95–144). Berlin: Springer-Verlag Berlin.

34. Prabhakaran MP, Venugopal J, Kai D *et al.* (2011) Biomimetic material strategies for cardiac tissue engineering. *Materials Science & Engineering C-Materials for Biological Applications* **31**, 503–513.

35. Grinstaff M (2007) Dendritic macromers for hydrogel formation: Tailored materials for ophthalmic, orthopedic, and biotech applications. *Journal of Polymer Science Part A, Polymer Chemistry* **46**, 383–400.

36. Gentsch R, Borner HG (2011) Designing three-dimensional materials at the interface to biology. In: HG Borner and JF Lutz, editors. *Bioactive Surfaces* (vol. 240, pp. 163–192). Berlin: Springer-Verlag Berlin.

37. Place ES, George JH, Williams CK *et al.* (2009) Synthetic polymer scaffolds for tissue engineering. *Chemical Society Reviews* **38**, 1139–1151.

38. Bacakova L, Filova E, Parizek M *et al.* (2011) Modulation of cell adhesion, proliferation and differentiation on materials designed for body implants. *Biotechnology Advances* **29**, 739–767.

39. Tan H, Marra KG (2010) Injectable, biodegradable hydrogels for tissue engineering applications. *Materials* **3**, 1746–1767.

40. Tian H, Tang Z, Zhuang X *et al.* (2012) Biodegradable synthetic polymers: Preparation, functionalization and biomedical application. *Progress in Polymer Science* **37**, 237–280.

41. Pok S, Jacot JG (2011) Biomaterials advances in patches for congenital heart defect repair. *Journal of Cardiovascular Translational Research* **4**, 646–654.

42. Benoit DSW, Durney A, Anseth K (2006) Manipulations in hydrogel degradation behavior enhance osteoblast function and mineralized tissue formation. *Tissue Engineering* **12**, 1663–1673.

43. Berkland C, Pollauf E, Varde N *et al.* (2007) Monodisperse liquid-filled biodegradable microcapsules. *Pharmaceutical Research* **24**, 1007–1013.

44. Bernd HE, Kunze C, Freier T *et al.* (2009) Poly(3-hydroxybutyrate) (PHB) patches for covering anterior skull base defects - an animal study with minipigs. *Acta Oto-Laryngol* **129**, 1010–1017.

45. Kasper FK, Tanahashi K, Fisher J *et al.* (2009) Synthesis of poly(propylene fumarate). *Nature Protocols* **4**, 518–525.

46. Peter SJ, Kim P, Yasko AW *et al.* (1999) Crosslinking characteristics of an injectable poly(propylene fumarate)/β-tricalcium phosphate paste and mechanical properties of the crosslinked composite for use as a biodegradable bone cement. *Journal of Biomedical Materials Research* **44**, 314–321.

47. Timmer MD, Ambrose CG, Mikos AG (2003) *In vitro* degradation of polymeric networks of poly(propylene fumarate) and the crosslinking macromer poly(propylene fumarate)-diacrylate. *Biomaterials* **24**, 571–577.

48. He S., Yaszemski MJ, Yasko AW *et al.* (2000) Injectable biodegradable polymer composites based on poly(propylene fumarate) crosslinked with poly(ethylene glycol)-dimethacrylate. *Biomaterials* **21**, 2389–2394.

49. Fisher J, Tirnmer MD, Timmer M *et al.* (2003) Photoinitiated cross-linking of the biodegradable polyester poly(propylene fumarate). Part I. Determination of network structure. *Biomacromolecules* **4**, 1327–1334.

50. Fisher J, Holland T, Dean D *et al.* (2003) Photoinitiated cross-linking of the biodegradable polyester poly(propylene fumarate). Part II. *In vitro* degradation. *Biomacromolecules* **4**, 1335–1342.

51. Shin JH, Lee JW, Jung JH *et al.* (2011) Evaluation of cell proliferation and differentiation on a poly(propylene fumarate) 3D scaffold treated with functional peptides. *Journal of Materials Science* **46**, 5282–5287.

52. Pomel C, Leborgne C, Cheradame H *et al.* (2008) Synthesis and evaluation of amphiphilic poly(tetrahydrofuran-b-ethylene oxide) copolymers for DNA delivery into skeletal muscle. *Pharmaceutical Research* **25**, 2963–2971.

53. Spiller KL, Maher SA, Lowman AM (2011) Hydrogels for the repair of articular cartilage defects. *Tissue Engineering Part B: Reviews* **17**, 281–299.

54. Hubbell JA (1998) Synthetic biodegradable polymers for tissue engineering and drug delivery. *Current Opinion in Solid State & Materials Science* **3**, 246–251.

55. Elisseeff J, Anseth K, Sims D *et al.* (1999) Transdermal photopolymerization of poly(ethylene oxide)-based injectable hydrogels for tissue-engineered cartilage. *Plastic and Reconstructive Surgery* **104**, 1014–1022.

56. Fitzgerald J, Jin M, Dean D *et al.* (2004) Mechanical compression of cartilage explants induces multiple time-dependent gene expression patterns and involves intracellular calcium and cyclic AMP. *Journal of Biological Chemistry* **279**, 19502–19511.

57. Yuan Z, Nie H, Wang S *et al.* (2011) Biomaterial selection for tooth regeneration. *Tissue Engineering Part B: Reviews* **17**, 373–388.

58. Leslie-Barbick JE, Moon JJ, West JL (2009) Covalently-immobilized vascular endothelial growth factor promotes endothelial cell tubulogenesis in poly(ethylene glycol) diacrylate hydrogels. *Journal of Biomaterials Science: Polymer Edition* **20**, 1763–1779.

59. Vasita R (2008) Improved biomaterials for tissue engineering applications: Surface modification of polymers. *Current Topics in Medicinal Chemistry* **8**, 341–353.

60. Ibusuki S, Iwamoto Y, Matsuda T (2003) System-engineered cartilage using poly(N-isopropylacrylamide)-grafted gelatin as *in situ*-formable scaffold: *In vivo* performance. *Tissue Engineering* **9**, 1133–1142.

61. Ozturk N, Girotti A, Kose GT *et al.* (2009) Dynamic cell culturing and its application to micropatterned, elastin-like protein-modified poly(N-isopropylacrylamide) scaffolds. *Biomaterials* **30**, 5417–5426.

62. Singelyn JM, Christman KL (2011) Injectable materials for myocardial tissue engineering myocardial tissue engineering. In: AR Boccaccini and SE Harding, editors (vol. 6, pp. 133–163). Berlin, Heidelberg: Springer.

63. Shimmura S, Doillon CJ, Griffith M *et al.* (2003) Collagen-poly(N-isopropylacrylamide)-based membranes for corneal stroma scaffolds. *Cornea* **22**, S81–S88.

64. Park S, Nam SH, Koh WG (2012) Preparation of collagen-immobilized poly(ethylene glycol)/poly(2-hydroxyethyl methacrylate) interpenetrating network hydrogels for potential application of artificial cornea. *Journal of Applied Polymer Science* **123**, 637–645.

65. Chirila TV (2001) An overview of the development of artificial corneas with porous skirts and the use of PHEMA for such an application. *Biomaterials* **22**, 3311–3317.

66. Lanza R, Langer R, Vacanti J, editors (2007) Principles of tissue engineering, 3th ed, *Tissue Engineering Intelligence Unit.* Hardbound: Elsevier Academic Press.

67. Peppas NA, Hilt JZ, Khademhosseini A *et al.* (2006) Hydrogels in biology and medicine: From molecular principles to bionanotechnology. *Advanced Materials* **18**, 1345–1360.

68. Bryant SJ, Davis Arehart KA, Luo N *et al.* (2004) Synthesis and characterization of photopolymerized multifunctional hydrogels: Water-soluble poly(vinyl alcohol) and chondroitin sulfate macromers for chondrocyte encapsulation. *Macromolecules* **37**, 6726–6733.

69. Schmedlen KH, Masters KS, West JL (2002) Photocrosslinkable polyvinyl alcohol hydrogels that can be modified with cell adhesion peptides for use in tissue engineering. *Biomaterials* **23**, 4325–4332.

70. Ferruti P, Bianchi S, Ranucci E *et al.* (2005) Novel agmatine-containing poly(amidoamine) hydrogels as scaffolds for tissue engineering. *Biomacromolecules* **6**, 2229–2235.

71. Ferruti P, Bianchi S, Ranucci E *et al.* (2005) Novel poly(amido-amine)-based hydrogels as scaffolds for tissue engineering. *Macromolecular Bioscience* **5**, 613–622.

72. Ferruti P, Marchisio MA, Duncan R (2002) Poly(amido-amine)s: Biomedical applications. *Macromolecular Rapid Communications* **23**, 332–355.

73. Magnaghi V, Conte V, Procacci P *et al.* (2011) Biological performance of a novel biodegradable polyamidoamine hydrogel as guide for peripheral nerve regeneration. *Journal of Biomedical Materials Research - Part A* **98A**, 19–30.

74. Nair LS, Laurencin CT (2007) Biodegradable polymers as biomaterials. *Progress in Polymer Science* **32**, 762–798.

75. Khan W, Muthupandian S, Farah S *et al.* (2011) Biodegradable polymers derived from amino acids. *Macromolecular Bioscience* **11**, 1625–1636.

76. Zhao Z, Wang J, Mao HQ *et al.* (2003) Polyphosphoesters in drug and gene delivery. *Advanced Drug Delivery Reviews* **55**, 483–499.

77. Lu Z-Z, Wu J, Sun T-M *et al.* (2008) Biodegradable polycation and plasmid DNA multilayer film for prolonged gene delivery to mouse osteoblasts. *Biomaterials* **29**, 733–741.

78. Li Q, Wang J, Shahani S *et al.* (2006) Biodegradable and photocrosslinkable polyphosphoester hydrogel. *Biomaterials* **27**, 1027–1034.

79. Du JZ, Sun TM, Weng SQ *et al.* (2007) Synthesis and characterization of photo-cross-linked hydrogels based on biodegradable polyphosphoesters and poly(ethylene glycol) copolymers. *Biomacromolecules* **8**, 3375–3381.

80. Ilia G (2009) Phosphorus containing hydrogels. *Polymers for Advanced Technologies* **20**, 707–722.

81. Morozowich N, Weikel A, Nichol J *et al.* (2011) Polyphosphazenes containing vitamin substituents: Synthesis, characterization, and hydrolytic sensitivity. *Macromolecules* **44**, 1355–1364.

82. Sperling LH (2011) *Introduction to physical polymer science.* 3rd ed. New York, NY: Wiley.

83. Alexandridis P, Alan Hatton T (1995) Poly(ethylene oxide) poly(propyleneoxide)poly(ethyleneoxide)blockcopolymersurfactants in aqueous solutions and at interfaces: Thermodynamics, structure, dynamics, and modeling. *Colloids and Surfaces A: Physicochemical and Engineering Aspects* **96**, 1–46.

84. Malmsten M, Lindman B (1992) Self-assembly in aqueous block copolymer solutions. *Macromolecules* **25**, 5440–5445.

85. Zhang K, Khan A (1995) Phase behavior of poly(ethylene oxide)-poly(propylene oxide)-poly(ethylene oxide) triblock copolymers in water. *Macromolecules* **28**, 3807–3812.

86. Alvarez-Lorenzo C, Gonzalez-Lopez J, Fernandez-Tarrio M *et al.* (2007) Tetronic micellization, gelation and drug solubilization: Influence of pH and ionic strength. *European Journal of Pharmaceutics and Biopharmaceutics* **66**, 244–252.

87. Rey-Rico A, Silva M, Couceiro J *et al.* (2011) Osteogenic efficiency of *in situ* gelling poloxamine systems with and without bone morphogenetic protein-2. *European Cells and Materials* **21**, 317–340.

88. Gutowska A, Jeong B, Jasionowski M (2001) Injectable gels for tissue engineering. *Anatomical Record* **263**, 342–349.

89. Yu H, Yang X, Cheng J *et al.* (2011) Distraction osteogenesis combined with tissue-engineered cartilage in the reconstruction of condylar osteochondral defect. *Journal of Oral and Maxillofacial Surgery* **69**, e558–e564.

90. Frisman I, Seliktar D, Bianco-Peled H (2011) Nanostructuring PEG-fibrinogen hydrogels to control cellular morphogenesis. *Biomaterials* **32**, 7839–7846.

91. Lee SH, Lee Y, Lee SW *et al.* (2011) Enzyme-mediated cross-linking of Pluronic copolymer micelles for injectable and *in situ* forming hydrogels. *Acta Biomaterialia* **7**, 1468–1476.

92. Oh SH, Kim TH, Lee JH (2011) Creating growth factor gradients in three dimensional porous matrix by centrifugation and surface immobilization. *Biomaterials* **32**, 8254–8260.

93. Lee JH, Kim JH, Oh SH *et al.* (2011) Tissue-engineered bone formation using periosteal-derived cells and polydioxanone/pluronic F127

scaffold with pre-seeded adipose tissue-derived CD146 positive endothelial-like cells. *Biomaterials* **32**, 5033–5045.

94. Park KM, Jun I, Joung YK *et al.* (2011) *In situ* hydrogelation and RGD conjugation of tyramine-conjugated 4-arm PPO-PEO block copolymer for injectable bio-mimetic scaffolds. *Soft Matter* **7**, 986–992.
95. Sosnik A, Sefton MV (2005) Semi-synthetic collagen/poloxamine matrices for tissue engineering. *Biomaterials* **26**, 7425–7435.
96. Jeong B, Bae YH, Lee DS *et al.* (1997) Biodegradable block copolymers as injectable drug-delivery systems. *Nature* **388**, 860–862.
97. Jeong B, Bae YH, Kim SW (2000) *In situ* gelation of PEG-PLGA-PEG triblock copolymer aqueous solutions and degradation thereof. *Journal of Biomedical Materials Research* **50**, 171–177.
98. Saffer EM, Tew GN, Bhatia SR (2011) Poly(lactic acid)-poly(ethylene oxide) block copolymers: New directions in self-assembly and biomedical applications. *Current Medicinal Chemistry* **18**, 5676–5686.
99. Sanabria-DeLong N, Crosby AJ, Tew GN (2008) Photo-cross-linked PLA-PEO-PLA hydrogels from self-assembled physical networks: Mechanical properties and influence of assumed constitutive relationships. *Biomacromolecules* **9**, 2784–2791.
100. Wang S, Lu L, Gruetzmacher JA *et al.* (2006) Synthesis and characterizations of biodegradable and crosslinkable poly(ε-caprolactone fumarate), poly(ethylene glycol fumarate), and their amphiphilic copolymer. *Biomaterials* **27**, 832–841.
101. Wu DQ, Wang T, Lu B *et al.* (2008) Fabrication of supramolecular hydrogels for drug delivery and stem cell encapsulation. *Langmuir* **24**, 10306–10312.
102. Peng H, Xiao Y, Mao X *et al.* (2009) Amphiphilic triblock copolymers of methoxy-poly(ethylene glycol)-b-poly(L-lactide)-b-poly(L-lysine) for enhancement of osteoblast attachment and growth. *Biomacromolecules* **10**, 95–104.
103. Zhang Z, Lai Y, Yu L *et al.* (2010) Effects of immobilizing sites of RGD peptides in amphiphilic block copolymers on efficacy of cell adhesion. *Biomaterials* **31**, 7873–7882.
104. Palacio M, Schricker S, Bhushan B (2010) Morphology and protein adsorption characteristics of block copolymer surfaces. *Journal of Microscopy* **240**, 239–248.
105. Hartgerink JD, Zubarev ER, Stupp SI (2001) Supramolecular one-dimensional objects. *Current Opinion in Solid State and Materials Science* **5**, 355–361.

106. Hartgerink JD, Beniash E, Stupp SI (2001) Self-assembly and mineralization of peptide-amphiphile nanofibers. *Science* **294**, 1684–1688.

107. Paramonov SE, Jun HW, Hartgerink JD (2006) Self-assembly of peptide-amphiphile nanofibers: The roles of hydrogen bonding and amphiphilic packing. *Journal of the American Chemical Society* **128**, 7291–7298.

108. Niece KL, Czeisler C, Sahni V *et al.* (2008) Modification of gelation kinetics in bioactive peptide amphiphiles. *Biomaterials* **29**, 4501–4509.

109. Guler MO, Hsu L, Soukasene S *et al.* (2006) Presentation of RGDS epitopes on self-assembled nanofibers of branched peptide amphiphiles. *Biomacromolecules* **7**, 1855–1863.

110. Silva GA, Czeisler C, Niece KL *et al.* (2004) Selective differentiation of neural progenitor cells by high-epitope density nanofibers. *Science* **303**, 1352–1355.

111. Silver J, Miller JH (2004) Regeneration beyond the glial scar. *Nature Reviews Neuroscience* **5**, 146–156.

112. Tysseling-Mattiace VM, Sahni V, Niece KL *et al.* (2008) Self-assembling nanofibers inhibit glial scar formation and promote axon elongation after spinal cord injury. *Journal of Neuroscience* **28**, 3814–3823.

113. Rajangam K, Behanna HA, Hui MJ *et al.* (2006) Heparin binding nanostructures to promote growth of blood vessels. *Nano Letters* **6**, 2086–2090.

114. Ghanaati S, Webber MJ, Unger RE *et al.* (2009) Dynamic *in vivo* biocompatibility of angiogenic peptide amphiphile nanofibers. *Biomaterials* **30**, 6202–6212.

115. Huang Z, Sargeant TD, Hulvat JF *et al.* (2008) Bioactive nanofibers instruct cells to proliferate and differentiate during enamel regeneration. *Journal of Bone and Mineral Research* **23**, 1995–2006.

116. Galler KM, Cavender A, Yuwono V *et al.* (2008) Self-assembling peptide amphiphile nanofibers as a scaffold for dental stem cells. *Tissue Engineering - Part A* **14**, 2051–2058.

117. Ozeki M, Kuroda S, Kon K *et al.* (2011) Differentiation of bone marrow stromal cells into osteoblasts in a self-assembling peptide hydrogel: *In vitro* and *in vivo* studies. *Journal of Biomaterials Applications* **25**, 663–684.

118. Jayawarna V, Ali M, Jowitt TA *et al.* (2006) Nanostructured hydrogels for three-dimensional cell culture through self-assembly of fluorenylmethoxycarbonyl-dipeptides. *Advanced Materials* **18**, 611–614.

119. Zhou M, Smith AM, Das AK *et al.* (2009) Self-assembled peptide-based hydrogels as scaffolds for anchorage-dependent cells. *Biomaterials* **30**, 2523–2530.

120. Ryan DM, Nilsson BL (2012) Self-assembled amino acids and dipeptides as noncovalent hydrogels for tissue engineering. *Polymer Chemistry* **3**, 18–33.

121. Zhang S, Holmes TC, DiPersio CM *et al.* (1995) Self-complementary oligopeptide matrices support mammalian cell attachment. *Biomaterials* **16**, 1385–1393.

122. Holmes TC, De Lacalle S, Su X *et al.* (2000) Extensive neurite outgrowth and active synapse formation on self-assembling peptide scaffolds. *Proceedings of the National Academy of Sciences of the United States of America* **97**, 6728–6733.

123. Kisiday J, Jin M, Kurz B *et al.* (2002) Self-assembling peptide hydrogel fosters chondrocyte extracellular matrix production and cell division: Implications for cartilage tissue repair. *Proceedings of the National Academy of Sciences of the United States of America* **99**, 9996–10001.

124. Genové E, Shen C, Zhang S *et al.* (2005) The effect of functionalized self-assembling peptide scaffolds on human aortic endothelial cell function. *Biomaterials* **26**, 3341–3351.

125. Schneider JP, Pochan DJ, Ozbas B *et al.* (2002) Responsive hydrogels from the intramolecular folding and self-assembly of a designed peptide. *Journal of the American Chemical Society* **124**, 15030–15037.

126. Haines-Butterick L, Rajagopal K, Branco M *et al.* (2007) Controlling hydrogelation kinetics by peptide design for three-dimensional encapsulation and injectable delivery of cells. *Proceedings of the National Academy of Sciences of the United States of America* **104**, 7791–7796.

127. Nowak AP, Breedveld V, Pakstis L *et al.* (2002) Rapidly recovering hydrogel scaffolds from self-assembling diblock copolypeptide amphiphiles. *Nature* **417**, 424–428.

128. Hule RA, Nagarkar RP, Hammouda B *et al.* (2009) Dependence of self-assembled peptide hydrogel network structure on local fibril nanostructure. *Macromolecules* **42**, 7137–7145.

129. Kumar P, Pillay V, Modi G *et al.* (2011) Self-assembling peptides: Implications for patenting in drug delivery and tissue engineering. *Recent Patents on Drug Delivery and Formulation* **5**, 24–51.

Chapter 9

# Biomaterials for Scaffolds: Natural Polymers

**Lindsay S. Wray and David L. Kaplan**
*Science and Technology Center, Tufts University, Medford, MA 02155, USA*
david.kaplan@tufts.edu

Scaffold materials are critical components of tissue engineering. There are two classes of polymer materials for biodegradable scaffold fabrication: synthetic and natural. Synthetic materials are derived from man-made synthesis and have been extensively investigated for the myriad of tissue engineering applications. Natural polymers are derived from natural processes of growth and metabolism and offer many advantages over synthetic polymers, in particular, their biocompatibility and similarity to the natural extracellular matrix (ECM). Overall, natural polymers offer a wide spectrum of material properties accounting to tunable scaffold assembly and chemical manipulation methods. More work into coordinated blending of natural polymers and translation of natural polymer synthesis to microbial fermentation production has the potential to advance the use of natural polymers in tissue engineering.

---

*Scaffolds for Tissue Engineering: Biological Design, Materials, and Fabrication*
Edited by Claudio Migliaresi and Antonella Motta

ISBN 978-981-4463-20-1 (Hardcover), 978-981-4463-21-8 (eBook)
www.panstanford.com

## 9.1 Introduction

In tissue engineering, scaffold materials play a significant role in directing functional tissue outcomes. The scaffold material should be selected based on the requirements of the tissue engineering study or application. Depending on which tissue is being engineered, some materials are more appropriate than others, but in general, all scaffolding material must satisfy the following criteria:

- Purity (i.e., free of microorganisms and toxic chemicals)
- Minimal adverse immunogenic response
- Biocompatibility
- Biodegradation and/or bioerosion
- Structural stability (i.e., tensile, compressive, shear strength)

Additionally, these materials must allow for porous scaffold assembly for cellular migration and proliferation, surface properties and chemical functionalization that promote desired cell functions, growth factor storage and release for controlling cell behavior, and mechanical properties that mimic the structural characteristics of human tissue.

There are a variety of synthetic and naturally derived (ND) polymeric materials that satisfy these criteria. Synthetic polymers are synthesized by human-directed polymerization processes. In contrast, ND polymer materials are synthesized by animal, plant, and microorganism metabolism and growth (Fig. 9.1). The various synthetic polymer materials available for tissue engineering applications have been reviewed elsewhere in this book. This chapter will present ND polymeric materials that are commonly implemented in scaffolds for tissue engineering.

The most commonly used synthetically derived polymers and the ND polymers presented in this chapter are polyglycolic acid (PGA), polycaprolactone (PCL), polyethylene glycol (PEG), polylactic acid (PLA), glycosaminoglycans (GAGs), and polyhydroxyalkanoates (PHA).

### 9.1.1 Benefits and Limitations

Natural polymers are well suited for tissue engineering applications because they are biocompatible and can be resorbed by the host tissue upon implantation. Compared to synthetic polymers, natural polymers are advantageous for tissue engineering because they resemble the protein and polysaccharide extracellular matrix (ECM) components. Cells recognize and possess the necessary binding domains for the amino acid and saccharide sequences. In contrast, the elastomeric backbone of synthetic polymers typically requires surface modifications or chemical functionalization in order to promote selective cell binding.

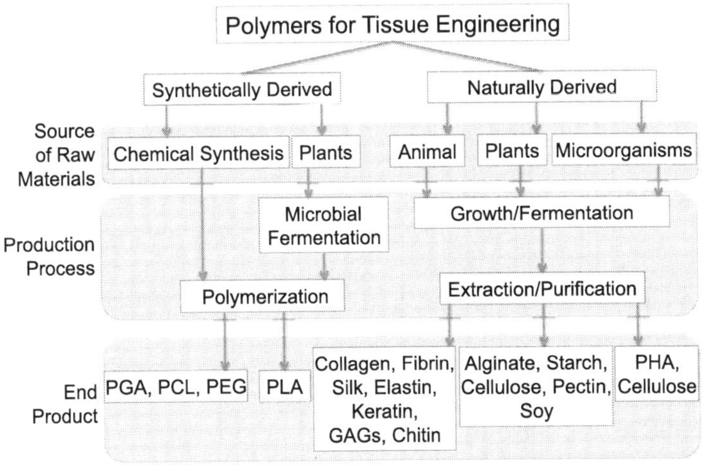


**Figure 9.1** Polymers commonly used in tissue engineering scaffolds can be derived by man-made chemical synthesis or natural pathways of metabolism and tissue growth. (*Note:* Cellulose is shown twice because it is commonly produced by both animal and microorganism sources.)

Given the strict medical regulatory standards for material purity and consistent quality, scaffold materials should have controllable and reproducible properties. ND polymer structure and chemical components can vary widely between natural sources.[1–3] For example alginate, which is derived from marine algae, can have variable material properties depending on when it was harvested during the year due to changes in temperature and sunlight exposure, as well as dependent on different species of algae.[2] Consequentially, material properties from batch-to-batch may not be reproducible and quality assurance/quality control (QA/QC) manufacturing practices are currently challenging to implement. In contrast, synthetic polymer material properties are reproducible from batch-to-batch because they are determined by tunable processing parameters and the QA/

QC practices are well established. In this regard, ND polymers are less desirable compared to synthetic polymers. However, with high demand from the consumer and food industries for biodegradable polymer material that is derived from renewable sources,[4] processes for reproducible ND polymer production are being investigated.[5] Furthermore, the establishment of tissue-engineered medical product standards will serve to guide ND polymer production processes on meeting quality benchmarks for clinical applications.[6]

### 9.1.2 Naturally Derived Polymer Diversity

As they exist in nature, ND polymers play important roles in tissue structural support, cell adhesion and migration, and chemokine storage and release. This makes ND polymers exemplary materials for scaffolds, as they should provide similar functions for engineered tissues. ND polymers predominantly used in tissue engineering are proteins and polysaccharides (Table 9.1). Polyesters produced by bacterial fermentation are also being investigated for engineering tissue equivalents. Sources for ND polymers include animals, plants, and microorganisms. Animal sources provide limited materials in some cases, such as elastin derived from bovine aorta, while in other cases larger amounts are available as with chitin derived from crustacean exoskeletons. Animal sources for ND polymer-based scaffolds also include animal or human cadaver tissue decellularization. Plant-derived polymers are abundant and the infrastructure for isolating these materials has been established by agriculture and food industries. Microorganisms produce natural polymers, which can be extracted from the native organisms or produced via fermentation. Altering fermentation parameters allows a degree of control over the features of the resulting material. Material abundance, ease of extracting, and purification of the material determines the extent to which the ND polymer can be used in clinical applications. Given the limited availability of some ND polymers, efforts have been made to generate these materials with recombinant microorganisms.

As seen in Table 9.1, ND polymers have been implemented in engineering many different human tissues, ranging from soft tissues such as neural and adipose to mechanically tough tissues such as cartilage and bone. Collectively, ND polymers have a wide range of material properties due to the diversity of proteins and polysaccharides found throughout nature, allowing for their use in

**Table 9.1** Summary of ND polymers, sources, and tissue engineering applications

| Polymer | Primary structural components | Predominant sources used in engineering tissues | Primary tissue engineering uses/ applications |
|---|---|---|---|
| Proteins | | | |
| Collagen/Gelatin | Monomer: three polypeptide $\alpha$ chains assembled into a triple helix.<br>Repeating unit: tripeptide of G-X-Y in which G is glycine and X is often proline and Y is often hydroxyproline or hydroxylysine. | Collagen—bovine tendon, rat tail<br>Gelatin—bovine (bones and skin), porcine (skin), fish, and recombinant bacteria | skin, wound dressing, cartilage, cardiovascular |
| Fibrin | Monomer: fibrinogen which has been enzymatically cleaved by thrombin. | Bovine plasma<br>Human plasma | tissue adhesive, cardiovascular, microvascular |
| Silk | Monomer: Heavy chain which contains alternating hydrophobic and hydrophilic blocks.<br>Repeating unit: G-X where G is glycine X is often alanine or serine. | Silk worm cocoons<br>Recombinant bacteria | bone, cartilage, ligament, vascular graft |
| Elastin/Tropoelastin | Monomer: tropoelastin composed of hydrophobic and hydrophilic domains. The hydrophobic domains contain predominantly G, V, A, and P amino acids and the hydrophilic domains contain primarily K and A amino acids. | Bovine aorta and ligament<br>Recombinant bacteria | skin, cardiovascular, wound dressing |

(Continued)

**Table 9.1** (*Continued*)

| Polymer | Primary structural components | Predominant sources used in engineering tissues | Primary tissue engineering uses/ applications |
|---|---|---|---|
| Keratin | Exists as three families of proteins: $\alpha$, $\beta$, and $\gamma$. | Human hair<br>Wool | nerve regeneration, wound dressing |
| Soy protein | Components: conglycinin and glycinin. Glycinin is predominantly made up of the amino acids N and E. | Soybean | bone filler, wound dressing |
| **Polysaccharides** | | | |
| Glycosaminoglycans | Linear chains of repeating disaccharide monomers.<br>Hyaluronic acid: D-glucuronic acid and *N*-acetyl-D-glucosamine with alternating $\beta$-(1,3) and $\beta$-(1,4) glycosidic linkages.<br>Chondroitin sulfate: D-glucuronic acid and *N*-acetyl-D-galactosamine-sulfate with $\beta$-(1,3) glycosidic linkages. | Hyaluronic acid: rooster comb, bovine vitreous humor<br>Chondroitin sulfate: bovine and poultry cartilage | cartilage, skin, neural, adipose |
| Chitin/Chitosan | Linear polymer of *N*-acetyl-D-glucosamine linked by $\beta$-(1,4) glycosidic bonds. | Crustacean shells | bone, cartilage, skin, neural |
| Alginate | Linear block copolymer of $\beta$-D-mannuronic acid and $\alpha$-L-guluronic acid linked at the 1,4 position. | Brown seaweed | cartilage |

| Polymer | Primary structural components | Predominant sources used in engineering tissues | Primary tissue engineering uses/ applications |
|---|---|---|---|
| Starch | D-glucose linked with $\alpha$-(1,4) glycosidic bonds. May contain side chains of D-glucose oligosaccharides via an $\alpha$-(1,6) glycosidic bond. | Potato<br>Corn<br>Sweet potato | bone |
| Cellulose | Linear polysaccharide of $\beta$-(1,4) linked D-glucose. | Wood pulp<br>Bacteria | cartilage, wound dressing |
| Pectin | Primarily composed of (1,4)-linked $\alpha$-D-galacturonic acid interrupted with regions of rhamnogalacturonan I and II. | Fruit pomace | material coatings and microspheres for cell stabilization |
| **Polyesters** | | | |
| Polyhydroxyalkanoates | Linear polymer of ester containing monomers with varying numbers of backbone carbon atoms.<br>Polymer backbone may contain methyl, ethyl, propyl, or nonyl side chains. | Bacteria | heart valve, bone, retinal pigment epithelium, vascular graft, neural conduit |

diverse tissue engineering applications. The pool of available ND polymers offer a wide spectrum of biodegradation rates, mechanical characteristics, and potential for bioactive modification that are needed to mimic the tissues found throughout the human body. For example, elastin is often used for soft tissue engineering because of its elastic quality, while silk is appropriate for hard tissue engineering because of its unparalleled toughness.[7] Moreover, ND polymers blended with other ND polymers have been extensively exploited to achieve an even wider range and complexity of scaffold properties.[8]

ND polymers can also be assembled into many different material architectures and formats. These include porous scaffolds, hydrogels, fibrous scaffolds, and microspheres.[9] ND polymers have also been adapted to microfabrication[10] and solid free form printing techniques.[11] These assembly methods are often aqueous-based, which is an advantage over synthetic assembly methods that often require toxic solvents and extreme processing parameters (i.e., temperature, pressure).

This chapter focuses on topics relevant to ND polymer selection and scaffold fabrication and reviews the sources, structure, tissue engineering applications, material formats, chemical modifications, mechanical properties, biodegradation dynamics, *in vivo* immune responses, and clinical implementation. We focus on ND polymers that have been significantly studied for their use in tissue engineering. It should be noted that there are many ND polymers that have yet to be extensively investigated and that the information found here should not be considered exhaustive.

## 9.2 Proteins

### 9.2.1 Collagen and Gelatin

#### 9.2.1.1 Collagen

Collagen is the most abundant protein in the human body and is the dominant ECM component of most tissues. Collagen is not only primarily involved in providing cells and tissues with structural support but also plays a significant role in directing cellular behavior and chemokine storage and release.[12] A collagen building block is a right-handed triple helix of three polypeptide chains, termed $\alpha$

chains, held together in the core by hydrogen bonding. The $\alpha$ chains that make up the triple helix generally follow a tripeptide repeating unit of G-X-Y where, G is glycine, X is often proline, and Y is often hydroxyproline or hydroxylysine (Fig. 9.2A). When assembled into the triple helix building block, the glycine residues face in toward the helix core and the proline in the second position allows the polypeptide chain to twist.[13,14] It is estimated that at least 25 different types of collagen exist.[13] The amino acid compositions of the $\alpha$ chains determine the type of collagen. For example, collagen type I is composed of one $\alpha 2$ chain and two $\alpha 1$ chains, while collagen type II is composed of three $\alpha 1$ chains. Of all the collagen types, type I and II are most abundant in the human body.[15]

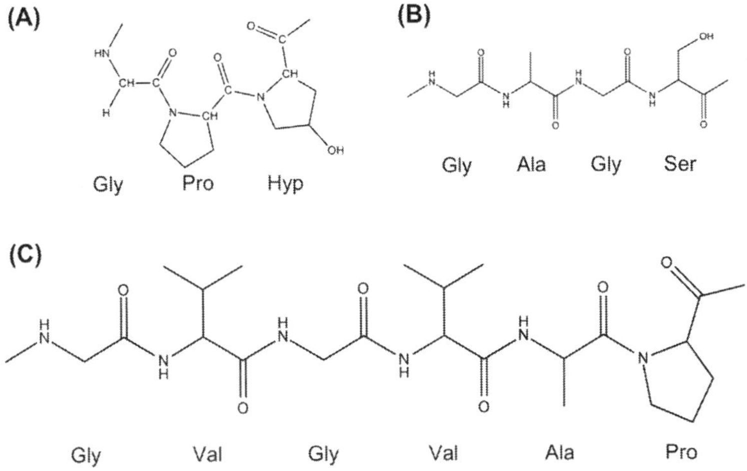


**Figure 9.2** Chemical structures of repeating peptide units for (A) collagen, (B) silk, and (C) elastin.

Collagen-based scaffolds have been implemented in many tissue engineering applications. Collagen can be assembled into all forms of scaffold architectures and is amenable to being blended with other ND polymers.[16–18] Collagen is primarily degraded by matrix metalloproteinases but is also degraded by cysteine and serine proteinases.[19] Upon implantation, collagen scaffolds undergo rapid degradation on the order of days to weeks depending on the type of scaffold and implant location.[20,21] Crosslinking of collagen is commonly performed with glutaraldehyde because it significantly stabilizes scaffolds against degradation.[22] However, glutaraldehyde crosslinking has been shown to compromise collagen scaffold biocompatibility.[23,24] Alternative crosslinking options that are more biocompatible than glutaraldehyde include chemical treatments (i.e., cyanamide, 1-ethyl-3-(3-dimethyl aminopropyl) carbodiimide (EDC), and hexamethylene diisocyanate), physical treatments (e.g. ultraviolet, heat irradiation), and enzymatic treatments (e.g. transglutamase, lysil oxidase).[22,25,26] Collagen crosslinking is also used to improve scaffold mechanical properties.[25] Collagen is inherently bioactive so chemical functionalization is not necessary, but collagen can be chemically modified to change its surface charge. This has been implemented to alter the solubility of the collagen, improve crosslinking, and repel bacterial cell attachment.[27,28]

Collagen scaffolds have been widely used clinically for regenerating ophthalmic, dermal, vascular, and orthopedic tissues. Collagen-based scaffolds are clinically available as corneal shields for enhanced drug delivery and wound healing.[20] The collagen is obtained from porcine dermis and shaped to fit the curvature of the cornea. The shields are crosslinked to varying extents to obtain different dissolution times, typically between 12 and 72 hours. Collagen hydrogels have been used clinically for injectable dermal fillers[29,30] and tissue adhesives.[31,32] For dermal fillers, collagen is obtained from bovine and porcine tissue or from human fibroblast cultures. Non-crosslinked products last between 2 and 4 months while crosslinked products can last longer, in some cases up to 12 months.[33] For tissue adhesives, bovine collagen is combined with bovine thrombin and in some products also with autologous plasma.[31] These products provided hemostasis by establishing a clot and attracting fibrinogen to the wound site.[32] Collagen scaffolds are used for regenerating skin to heal wounds such as burns and skin ulcers.[34] These scaffolds are made from reconstituted collagen or from decellularized skin. Collagen-based living skin equivalents are composed of reconstituted bovine skin and allogenic cells that are cultured together *ex vivo* to recreate the stratified layers of normal skin. This technology is used to treat partial and full thickness wounds and ulcers that would otherwise be difficult to heal. Porous collagen scaffolds functionalized with recombinant human bone morphogenic protein-2 are also used to treat orthopedic injuries and defects.[35]

#### 9.2.1.2 Gelatin

Gelatin is a derivative of collagen, where the hydrogen bonds between the triple helix $\alpha$ chains have been broken. Compared to collagen, gelatin has similar biocompatibility but reduced mechanical properties and faster degradation.[36] Gelatin is highly soluble in aqueous solutions and forms a thermally reversible hydrogel. Gelatin can also be assembled into three-dimensional porous scaffolds.[37] Gelatin is often blended with other ND polymer materials as a way to improve scaffold biological activity.[38,39] Like collagen, gelatin is crosslinked with glutraldehyde and EDC to increase mechanical stiffness and prolong degradation time.[40] Regardless of crosslinking, gelatin has very low mechanical properties and fast degradation time. Gelatin sponges, powders, and pastes are clinically used as hemostatic agents. For gelatin-based hemostatic gels, gelatin granules are combined with thrombin granules that expand when applied to the bleeding surface. The expansion of the hemostatic agent helps to close the wound and stop bleeding.[41] Depending on where the gelatin product is applied, complete resorption of material occurs between two days and six weeks.[42]

### 9.2.2 Fibrin

Fibrin is a glycoprotein that facilitates clotting of blood and wound healing.[43] Upon tissue trauma and breaching of the vascular wall, signaling cascades activate the enzyme thrombin, which cleaves the plasma protein, fibrinogen. Fibrinogen is composed of two sets of A$\alpha$, B$\beta$, and $\gamma$-polypeptide chains, which are held together by disulfide bridges. Activated thrombin cleaves the *N*-terminus of the A$\alpha$ and B$\beta$ chains, which exposes a region of the fibrinogen molecule that enables spontaneous self-assembly into a branching fibril network.[44] This network is stabilized by non-covalent and electrostatic forces and by covalent crosslinking, which is initiated by the factor XIII enzyme. Fibrin clots provide a fibrous network over which new tissue can form and after the tissue has been regenerated, the fibrin clot is quickly degraded by plasmin.

Predominantly, fibrin has been used to build hydrogel scaffolds for engineering cardiovascular tissue,[45] neural tissue engineering,[46] and microvascular network formation.[47] Fibrin can also be formed into porous scaffolds by mixing with calcium phosphate granules.[48]

When forming fibrin hydrogels, the pH, ionic strength, and temperature affect the mechanical strength and degradation profile of the end product. Due to the rapid degradation rate of fibrin *in vivo*, methods for extending the degradation rate via chemical crosslinking have been investigated using genipin[49] and glutaraldehyde.[50] Fibrin is clinically available as a hemostatic agent and tissue adhesive and significantly improves hemostasis when used to secure skin grafts, as well as improves graft integration with host tissue.[51]

### 9.2.3 Silk

Silks are generally defined as extruded protein fibers in which the process of spinning silk involves conversion of an aqueous protein solution into an insoluble filament.[52] In nature, silk is spun by a variety of arthropods for a variety of different purposes such as in the construction of webs, cocoons, and nests. In *Bombyx mori* cocoon silk, the peptide chain has alternating hydrophobic and hydrophilic blocks in which the hydrophobic blocks typically contain a repeating sequence of G-X where, X is A 65% and S 23%[53] (Fig. 9.2B). This polypeptide structure leads to *β*-sheet formation and extensive hydrogen bonding between the hydrophobic domains. This extensive hydrogen bonds accounts for high strength and toughness of the silk. Silks obtained from silk worm cocoons are predominantly used to assemble scaffolds for tissue engineering applications.

Silk scaffolds can take on many material formats such as porous membranes,[54] hydrogels,[55] and porous sponges.[56] Silk scaffolds have been implemented for engineering anterior cruciate ligament,[57] bone,[58] cartilage,[59] adipose,[21] and vascular grafts.[60] Silk has also been investigated as a tissue adhesive.[61] The biological properties of silk scaffolds can be augmented through bulk loading, surface decoration, or formation of composite materials.[62] Silk scaffolds can be chemically functionalized with bioactive agents such as the arginine–glycine–aspartic acid (RGD) integrin binding domain and surface hydrophilicity can be altered with diazonium coupling chemistry.[63] Silk has robust mechanical properties and tunable degradation properties, which can be matched to the tissue engineering application. Silk scaffolds can be assembled without the use of chemical crosslinking by inducing *β*-sheet crystallinity via temperature-controlled water annealing.[64] Silk induces minimal immune response and exhibits slow degradation *in vivo* ranging from weeks to years to resorb completely.[65]

### 9.2.4 Elastin

Elastin is a mammalian ECM protein that plays a role in controlling vascular smooth muscle proliferation and supporting endothelial cell growth.[66] Elastin also provides tissues with elasticity and is found in ligaments, large arteries, lung, and skin. Elastogenesis occurs during the late fetal and early neonatal times and is generally not regenerated thereafter or the process is exceedingly slow. The elastin building block, termed tropoelastin, is composed of hydrophobic and hydrophilic domains. The hydrophilic domains contain primarily lysine and alanine amino acids and the hydrophobic domains contain predominantly glycine, valine, alanine, and proline amino acids (Fig. 9.2C). Types of elastin for scaffold fabrication include solubilized elastin from animal sources and recombinant tropoelastin and elastin-like sequences.[67–69]

Elastin has been assembled into electrospun fibrous mats,[70] hydrogels,[71] and porous scaffolds[72] and is primarily used to engineer dermal tissue, cardiovascular tissue, and vascular conduits.[72] Elastin is commonly incorporated into scaffolds as a blend with other ND biomaterials, most notably collagen[70] and silk.[73] Elastin scaffolds are capable of being assembled without a chemical crosslinker[74] but to further stabilize elastin constructs, crosslinking agents have been employed.[72] Implanted elastin hydrogels elicit a moderate foreign body reaction and support tissue regeneration.[74] Elastin products are available for clinical use as a component of decellularized tissue matrix products.

### 9.2.5 Keratin

Keratin is a protein found in skin, hair, and nails. It exists as three families of proteins: $\alpha$, $\beta$, and $\gamma$. The $\alpha$-keratin is the dominant keratin component and provides structural support to the hair shaft, while $\beta$-keratins play a protective role and form the hair cuticle. The $\gamma$-keratins function as disulfide crosslinkers and hold the hair shaft structure together. Keratin has been fabricated into hydrogels[75] and porous scaffolds.[76] Porous scaffolds can be formed using freeze-drying methods or keratin powder can be mixed with NaCl porogens and compression-molded.[77] These keratin biomaterials possess cell binding sites such as RGD and leucine-aspartic acid-valine (LDV)[78] and demonstrate good *in vitro* cytocompatibility.[79] To improve keratin mechanical properties photopolymerizing crosslinking agents have been investigated.[80] Implanted keratin porous scaffolds display minimal inflammation, support host tissue vascularization, and degrade slowly.[81] Keratin hydrogels have also been used as hemostatic agents and *in vivo* performed as well as commercially available hemostatic bandages.[82]

### 9.2.6 Soy Protein

Soy protein comes from the soybean and functions as energy supply during seedling germination. The protein has two components, conglycinin (20 kDa), which makes up 35% and glycinin (~350 kDa), which makes up 52% of the total protein found in the soybean.[83] Soy protein has been assembled into hydrogels,[84] fibers,[85] electrospun mats,[86] and freeze-dried porous scaffolds.[87] Common tissue applications of soy protein have been primarily bone fillers[88] and wound dressings.[89] In order to control scaffold mechanical properties and degradation rates, crosslinking agents have been implemented that include adipic/acetic anhydride,[83] glyceraldehydes and glutaraldehyde,[90] and transglutaminase.[87] Soy protein incorporated into freeze-dried cellulose sponges exhibited better biocompatibility and increased degradation rate compared to cellulose scaffolds.[91] Soy protein granules were tested *in vitro* and supported osteoblast differentiation and suppressed proinflammatory cytokines released by mononuclear cells.[92] These soy protein granules have also been implanted into bone defects and supported mature trabeculae after eight weeks post-operation compared to sham operation controls.[88]

## 9.3 Polysaccharides

### 9.3.1 Glycosaminoglycans: Hyaluronate and Chondroitin Sulfate

Glycosaminoglycans (GAGs) are major components of tissue ECM and perform a myriad of roles in maintaining tissue structure and directing cell function. GAGs are found in cartilage tissue where they play a role in resisting compressive forces and lubricating articular joints and in basement membranes where they are involved in cell binding. GAGs are linear chains of repeating disaccharide monomers.

The predominant types of GAG chains found in the body are hyaluronan (HA), chondroitin sulfate (CS), dermatan sulfate, heparan sulfate, and keratan sulfate. In tissue engineering, the primary GAGs used for scaffold construction are HA and CS (Fig. 9.3A). HA consists of a repeating disaccharide of D-glucuronic acid and *N*-acetyl-D-glucosamine with alternating *β*-(1,3) and *β*-(1,4) glycosidic linkages. CS consists of a repeating disaccharide of D-glucuronic acid and *N*-acetyl-D-galactosamine sulfate with *β*-(1,3) glycosidic linkages. The *N*-acetyl-D-galactosamine sulfate can be sulfated on either or both of the 4 and 6 position hydroxyl groups.

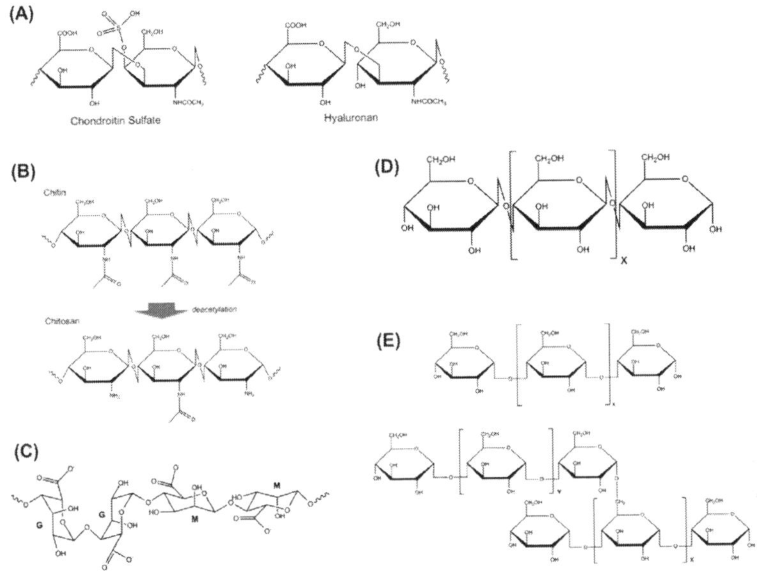


**Figure 9.3** Chemical structure of polysaccharide materials: (A) glycosaminoglycans, (B) chitin and chitosan, (C) alginate, (D) cellulose, and (E) amylose and amylopectin.

GAGs have been used in a variety of material formats for various tissue engineering applications due to their ability to direct cell behavior toward the desired tissue function. The use of GAGs in engineering dermal,[93] brain,[94] and adipose tissues[95] has been reported, but they are predominantly fabricated into hydrogels for cartilage engineering. GAGs can be functionalized through the carboxyl and hydroxyl groups on the saccharide residues and on the *N*-acetyl group following deamidation.[96] Using both aqueous- and organic solvent-based chemistry GAGs have been modified with amine and carboxyl groups, which affects biological and degradation properties.[97] GAGs have been crosslinked with agents such as dialdehydes using carbodiimide chemistry[98] and by photopolymerization.[99] When implanted, uncrosslinked GAGs degrade rapidly via hyaluronidases and produce oligosaccharide fragments.[100] Tuning GAG degradation properties using HA chemically modified with methacrylate groups has been successful.[101] GAGs are available clinically as components of decellularized tissue matrices and as dermal fillers for correcting deep lines and wrinkles.[102]

### 9.3.2 Chitin and Chitosan

Chitin is found in crustacean exoskeletons and in fungal mycelia. Chitin is a linear polysaccharide of *N*-acetyl-D-glucosamine linked by $\beta$-(1,4) glycosidic bonds. Chitosan is derived from chitin by a deacetylation process thereby resulting in a heteropolymer of D-glucosamine and *N*-acetyl-D-glucosamine monomers (Fig. 9.3B). Chitin and chitosan are commonly characterized by their degree of deacetylation (DD), which is the ratio of 2-acetamido-2-deoxy-D-glucopyranose to 2-amino-2-deoxy-D-glucopyranose structural units. For chitin the DD is greater that 90–95% and for chitosan the DD is less than 35%.[103]

Chitosan has been implemented as a scaffold material for a variety of tissue engineering applications including bone,[104] cartilage,[105] skin,[106] and neural tissue.[107] Chitosan can generate different material formats including hydrogels,[105] electrospun mats,[104] and microspheres,[108] and porous scaffolds.[109] To improve cell binding, chitosan can be functionalized at the C2 amino group, the C6 primary hydroxyl group, and the C3 secondary hydroxyl group.[110] Chitin and chitosan have been functionalized with sulfation to improve coagulant properties, antimicrobial strength, and solubility in water.[111] To bolster mechanical properties and decrease *in vivo* degradation rate, a variety of crosslinking mechanisms have been investigated including the use of glutaraldehyde[112] or genipin.[113] The degradation process is primarily mediated by lysozyme hydrolysis and the products are non-toxic saccharide derivatives.[114] Implanted chitosan elicits a minimal foreign body response and has inherent antimicrobial and thrombogenic properties.[115] Chitosan bandages have been clinically implemented as wound dressings.[116]

### 9.3.3 Alginate

Alginate is found in the cell walls of brown algae such as *Laminaria hyperborea*, *Macrocystis pyrifera*, and *Laminaria digitata* where it functions as flexible structural support.[2] Alginate is also produced by bacteria such as *Pseudomonas aeruginosa* as an exopolysaccharaide.[117] Alginate is a linear, unbranched block copolymer of $\beta$-D-mannuronic acid and $\alpha$-L-guluronic acid linked at the 1,4 position (Fig. 9.3C). The blocks can be pure mannuronic acid (M block), purely guluronic acid (G block), or a mixture of both mannuronic and guluronic acid (MG block). The ratio of mannuronic acid to guluronic acid depends on the source and time that the alginate is harvested.[2]

Alginate has been primarily investigated in tissue engineering as a hydrogel scaffold for cartilage tissue engineering.[118] In addition to hydrogels, alginate can also be electrospun[119] and fabricated into porous sponges[120] by adding divalent cations ($Ca^{2+}$, $Mg^{2+}$, $Ba^{2+}$, $Sr^{2+}$) to alginate solutions. Divalent cations can be added directly or incorporated as a buffered divalent cation so that it is released into the alginate solution via a pH trigger.[121] Alginate binding with divalent ions takes place in the G blocks of the polymer, so the higher the G block content, the stronger the gel. Hydrogels can also be formed by lowering the pH below the pKa of the alginate monomers so that intermolecular hydrogen bonding occurs.[121] Because of the cytocompatible gelling regime, alginate has been used for the delivery of encapsulated cells.[122,123] Divalent cation crosslinking quickly loses integrity due to leaching of the cation,[124] so additional crosslinking methods are utilized. These include covalent crosslinking agents such as dialdehydes[125] and photopolymerizable functional groups.[126] Alginate lacks cell binding domains and due to its hydrophilic nature resists protein adsorption, so it is commonly functionalized on the carboxyl groups with moieties such as RGD peptides[127] and galactose.[128] Alginate has also been sulfated for improvement of blood compatibility.[129]

### 9.3.4 Cellulose

Cellulose is the most abundant organic matter found on the planet and exists as a structural component of plant cell walls. Cellulose is also produced by bacteria as an extracellular protective envelope. It is composed of a linear polysaccharide of $\beta$-(1,4)-linked D-glucose that assembles in to $\beta$-sheets with extensive hydrogen bonding[130] (Fig. 9.3D). Within the plant cell walls, cellulose forms nanofibrils that bundle together to form larger microfibers.[131] Bacterial cellulose (BC) exists as a nanofibrous mesh.[132]

Cellulose and has been formed into a variety of material formats such as porous membranes,[132] hydrogels,[133] and porous scaffolds[134] and can be patterned with submicron features.[135] BC scaffolds have been produced in pellicle form and cultured with chondrocytes. These porous, fibrous scaffolds promoted cellular proliferation compared to tissue culture plastic without activating proinflammatory cytokines.[136] Non-woven cellulose fabrics activated with $Ca(OH)_2$ solution and coated with calcium phosphate supported chondrocyte proliferation and collagen type I and type II matrix production.[137] Cellulose degradation *in vivo* is very slow because mammals lack enzymes that cleave the $\beta$-(1,4) linkage. Subcutaneous implantation of cellulose sponges exhibited decreased size and increased scaffold fragments after 16 weeks but complete dissolution of the sponge was not observed during the 60-week study.[138] Cellulose *in vivo* degradation rate can be increased by pretreating the scaffold with cellulase[135] and by lowering the material crystallinity.[139] Upon implantation cellulose has demonstrated mild foreign body reactions and cellulose dermal wound healing dressings are clinically available to treat lesions, burns, and abrasions.[140]

### 9.3.5 Starch: Amylose and Amylopectin

Starch is a polysaccharide produced by plants for energy storage and is localized in plant cell organelles called amyloplasts. Starch exists as either a linear polysaccharide called amylose, or as a highly branched polysaccharide called amylopectin. Amylose is a polysaccharide of D-glucose linked with $\alpha$-(1,4) glycosidic bonds. Amylopectin is also composed of $\alpha$-(1,4)-linked D-glucose but is also linked to side chains of D-glucose oligosaccharides via an $\alpha$-(1,6) glycosidic bond (Fig. 9.3E).

Starches can be rendered into porous, three-dimensional scaffolds using a variety of methods including freeze-drying,[141] compression molding with salt particles,[142] and extrusion with a citric acid–based blowing agent.[143] Rapid prototyped starch-based scaffolds have been produced with microporous morphology and highly controllable macroscopic design features.[11] Starch hydrogels have been fabricated by functionalizing amylopectin with phenolic hydroxyl groups that gel upon exposure to horseradish peroxidase.[144] This gel material is cytocompatible and the degradation rate can be controlled by altering horseradish peroxidase and phenolic hydroxyl concentration. *In vitro* exposure of starch scaffolds to $\alpha$-amylase, which is found in human saliva and blood, generated glucose degradation products.[145] Amylose chains are more tightly packed than amylopectin due to the lack of side chains, and is therefore, more resistant to degradation via hydrolysis compared to amylopectin. Otherwise, starch is highly resistant to enzymatic and acid hydrolysis.[146] Starch-based scaffolds implanted in the subcutaneous and intramuscular space did not elicit adverse immune reactions[147] and supported host tissue integration.[148]

### 9.3.6 Pectin

Pectin is found in terrestrial plants and plays a structural role in plant cell walls and in bonding plant cells together. Pectin can be found in high abundance in citrus fruit peels, apple pomace, and beet pulp, however, it can also be found to a lesser degree in a variety of other fruit pomaces.[149] Pectins are a class of heteropolysaccharides with a range of molecular weights and side chain content. Pectin primarily exists as a polymer of (1,4)-linked $\alpha$-D-galacturonic acid, in which some of the carboxyl groups are methyl esterified.[150] The galacturonate polysaccharide backbone may be interrupted with regions referred to as rhamnogalacturonan I and II (RG-I and RG-II). RG-I regions consist of alternating galacturonate and rhamnose residues that support branching side chains of oligosaccharides rich in arabinose and galactose.[151]

Pectins have been investigated for biomedical applications primarily in the form of material coatings[152] and microspheres.[153] Pectin blended with chitosan is amenable to three-dimensional scaffold assembly by freeze-drying. Pectins rich with the RG-I moiety derived from apples have been used to coat surfaces on which fibroblasts, macrophages, and osteoblasts were cultured.[152] Pectins were also fabricated into microspheres using an electrostatic bead generation system and $CaCl_2$ as a gelling agent. The microspheres were successfully coupled with RGD using carbodiimide chemistry and cultured with preosteoblasts. The preosteoblasts differentiated to osteoblasts and produced mineralized matrix within the microspheres. Without RGD functionalization the preosteoblasts did not adhere and had a rounded morphology.[153] Macrophages cultured *in vitro* on pectin coatings and *in vivo* implantation of pectin microspheres suggests that pectin is biocompatible and holds potential for a variety of tissue engineering applications.[152,154]

## 9.4 Polyhydroxyalkanoates

PHAs are linear polyesters naturally synthesized by a variety of bacteria (both gram-positive and gram-negative) under limited nitrogen, phosphorous, or oxygen conditions.[155] There are many different types of PHAs but the types most widely studied for tissue engineering include poly 3-hydroxybutyrate (PHB) and 3-hydroxyvalerate (PHV) (Fig. 9.4). PHAs accumulate in granules within the cell cytoplasm and function as a source of energy. PHAs are classified by the number of carbons in the monomeric units and chemical composition. Monomers with 3–5 carbons are classified as short chain length (scl-PHAs), monomers with 6–14 carbons are classified as medium chain length (mcl-PHAs), and monomers with more than 14 carbons are classified as long chain length (lcl-PHAs). PHAs with short chain lengths tend to be more crystalline while longer chain length PHAs tend to be more elastomeric. The monomeric unit can also have different functional side chains such as methyl, ethyl, propyl, or nonyl groups.

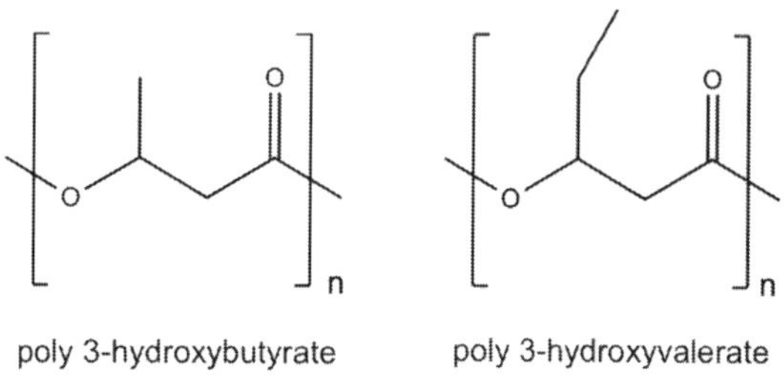


**Figure 9.4** Polyhydroxyalkanoates.

PHAs have been assembled into films, scaffolds, and hydrogels for bone, retinal pigment epithelium, blood vessel substitutes, neural conduits, and most notably heart valve tissue engineering.[156–158] PHAs exhibit a wide range of mechanical properties and biodegradation rates, which are determined by factors such as monomer length, scaffold configuration, PHA blend components and ratios, and chemical functionalization. The degradation product, D-3-hydroxybutyric acid, is a non-toxic and naturally occurring metabolite of higher living organisms.[159] PHAs have been surface modified with other ND polymers such as collagen and gelatin.[160,161] By including functionalized starting material in the fermentation step, PHAs can be produced with a diversity of chemical functionalization such as aryl, olefin, halogen, cyano, and epoxide groups.[158,159] PHAs are naturally very hydrophobic so to improve their hydrophilicity, they can be plasma treated.[162] PHAs exhibit resistance to enzymatic degradation with minimal loss of tensile strength and are degraded primarily by surface erosion.[163] PHAs have been investigated *in vivo* in the form of heart valves, which resist degradation up to 24 weeks and experience cellular infiltration by cells, no thrombus formation, and mild valvular regurgitation.[164]

## 9.5 Conclusions

For the last two decades, there has been growing interest in engineering tissues using ND polymers most notably for their similarity to the extracellular components of the human body. The ability to tune degradation rates and mechanical properties via controlled processing, offers additional utility to these polymers. The alternatives to ND polymers are synthetically derived degradable polymers such as polyglycolic acid, polylactic acid, and polycaprolactone, which have been widely investigated for their ability to regenerate human tissues and organs. ND polymers can be harvested from sustainable industries such as agriculture, sericulture, livestock, and fish farming. Additionally, with the advent of recombinant DNA technology, many ND polymers can be produced via microorganisms with tailorable features. There is no single ideal ND polymer that can satisfy every desired scaffold characteristic. Fortunately, these materials can be blended, which has been extensively exploited to obtain scaffolds with desired composition and characteristics. Given the diversity of ND polymers, future research efforts could focus on augmenting the advantages of ND polymers with other ND polymers instead of incorporating a synthetically derived biopolymer. Further, progress in the use of genetic engineering should also propel these polymers into the next generation of highly tailored biomaterials, due to the ability to control chirality, block chemistries, and molecular weight with precision; providing a path toward designer polymers for specific applications.

Future technology for the production of tissue engineered products will need to consider the sustainability of the raw materials required to support this advancing field. ND polymers derived from renewable and abundant resources offer a desirable alternative to synthetically derived materials. Polymerization of ND materials is conducted by living organisms as a part of their growth. In contrast, polymerization of synthetically derived materials requires high-energy chemical processing. While synthetically derived materials have been successfully implemented as tissue engineering scaffolds, ND polymers are anticipated to play an increasingly important role in the field. Improved QC and understanding of biological interactions will be needed to match advances in the uses of these polymers in biomaterials, tissue engineering, and regenerative medicine.

## References

1. Angele, P., Abke, J., Kujat, R., Faltermeier, H., Schumann, D., Nerlich, M., Kinner, B., Englert, C., Ruszczak, Z. and Mehrl, R. (2004) Influence of different collagen species on physico-chemical properties of crosslinked collagen matrices, *Biomaterials*, **25**(14), 2831–2841.
2. Draget, K. I., Smidsrød, O. and Skjåk-Bræk, G. (2005) Alginates from algae. In: Steinbüchel A. (Ed.). *Biopolymers Online* (pp. 1–30). New York: Wiley.
3. Zeugolis, D., Paul, R. and Attenburrow, G. (2008) Factors influencing the properties of reconstituted collagen fibers prior to self-assembly: Animal species and collagen extraction method, *J. Biomed. Mater. Res.*, **86**(4), 892–904.
4. Gross, R. A. and Kalra, B. (2002) Biodegradable polymers for the environment, *Science*, **297**(5582), 803–807.
5. U.S. Congress, Office of Technology Assessment. (1993) *Biopolymers: Making Materials Nature's Way-Background Paper, OTA-BP-E-102*, Washington, DC: U.S. Government Printing Office.
6. Dornish, M., Kaplan, D. and Skaugrud, O. (2001) Standards and guidelines for biopolymers in tissue-engineered medical products, *Ann. N. Y. Acad. Sci.*, **944**(1), 388–397.

7. Gosline, J., Lillie, M., Carrington, E., Guerette, P., Ortlepp, C. and Savage, K. (2002) Elastic proteins: Biological roles and mechanical properties, *Phil. Trans. R. Soc. Lond. B*, **357**(1418), 121–132.

8. Silva, S. S., Mano, J. F. and Reis, R. L. (2010) Potential applications of natural origin polymer-based systems in soft tissue regeneration, *Crit. Rev. Biotechnol.*, **30**(3), 200–221.

9. Brahatheeswaran, D., Yasuhiko, Y. and Toru, M. (2011) Polymeric scaffolds in tissue engineering application: A review, *Int. J. Polym. Sci.*, **2011**, 1–19.

10. Golden, A. P. and Tien, J. (2007) Fabrication of microfluidic hydrogels using molded gelatin as a sacrificial element, *Lab Chip*, **7**(6), 720–725.

11. Lam, C. X. F., Mo, X., Teoh, S. H. and Hutmacher, D. (2002) Scaffold development using 3D printing with a starch-based polymer, *Mater. Sci. Eng. C*, **20**(1), 49–56.

12. Kanematsu, A., Marui, A., Yamamoto, S., Ozeki, M., Hirano, Y., Yamamoto, M., Ogawa, O., Komeda, M. and Tabata, Y. (2004) Type I collagen can function as a reservoir of basic fibroblast growth factor, *J. Control. Release*, **99**(2), 281–292.

13. Gelse, K., Pöschl, E. and Aigner, T. (2003) Collagens: Structure, function, and biosynthesis, *Adv. Drug Deliv. Rev.*, **55**(12), 1531–1546.

14. Kadler, K. E., Baldock, C., Bella, J. and Boot-Handford, R. P. (2007) Collagens at a glance, *J. Cell Sci.*, **120**(12), 1955–1958.

15. Kadler, K. E., Hill, A. and Canty-Laird, E. G. (2008) Collagen fibrillogenesis: Fibronectin, integrins, and minor collagens as organizers and nucleators, *Curr. Opin. Cell Biol.*, **20**(5), 495–501.

16. Lee, C. H., Singla, A. and Lee, Y. (2001) Biomedical applications of collagen, *Int. J. Pharm.*, **221**(1), 1–22.

17. Zhong, S., Teo, W. E., Zhu, X., Beuerman, R., Ramakrishna, S. and Yung, L. Y. (2005) Formation of collagen-glycosaminoglycan blended nanofibrous scaffolds and their biological properties, *Biomacromolecules*, **6**(6), 2998–3004.

18. Wang, X., Sang, L., Luo, D. and Li, X. (2011) From collagen-chitosan blends to three-dimensional scaffolds: The influences of chitosan on collagen nanofibrillar structure and mechanical property, *Colloids Surf. B Biointerfaces*, **82**(1), 233–240.

19. Everts, V., van der Zee, E., Creemers, L. and Beertsen, W. (1996) Phagocytosis and intracellular digestion of collagen, its role in turnover and remodelling, *Histochem. J.*, **28**(4), 229–245.

20. Willoughby, C. E., Batterbury, M. and Kaye, S. B. (2002) Collagen corneal shields, *Surv. Ophthalmol.*, **47**(2), 174–182.

21. Mauney, J. R., Nguyen, T., Gillen, K., Kirker-Head, C., Gimble, J. M. and Kaplan, D. L. (2007) Engineering adipose-like tissue *in vitro* and *in vivo* utilizing human bone marrow and adipose-derived mesenchymal stem cells with silk fibroin 3D scaffolds, *Biomaterials*, **28**(35), 5280–5290.

22. Rault, I., Frei, V., Herbage, D., Abdul-Malak, N. and Huc, A. (1996) Evaluation of different chemical methods for cross-linking collagen gel, films and sponges, *J. Mater. Sci. Mater. Med.*, **7**(4), 215–221.

23. Van Wachem, P., Van Luyn, M., Damink, L., Dijkstra, P., Feijen, J. and Nieuwenhuis, P. (1994) Biocompatibility and tissue regenerating capacity of crosslinked dermal sheep collagen, *J. Biomed. Mater. Res.*, **28**(3), 353–363.

24. Petite, H., Duval, J. L., Frei, V., Abdul-Malak, N., Sigot-Luizard, M. F. and Herbage, D. (1995) Cytocompatibility of calf pericardium treated by glutaraldehyde and by the acyl azide methods in an organotypic culture model, *Biomaterials*, **16**(13), 1003–1008.

25. Elbjeirami, W. M., Yonter, E. O., Starcher, B. C. and West, J. L. (2003) Enhancing mechanical properties of tissue-engineered constructs via lysyl oxidase crosslinking activity, *J. Biomed. Mater. Res.*, **66**(3), 513–521.

26. Parenteau-Bareil, R., Gauvin, R. and Berthod, F. (2010) Collagen-based biomaterials for tissue engineering applications, *Materials*, **3**(3), 1863–1887.

27. Tiller, J. C., Bonner, G., Pan, L. C. and Klibanov, A. M. (2001) Improving biomaterial properties of collagen films by chemical modification, *Biotechnol. Bioeng.*, **73**(3), 246–252.

28. Brinkman, W. T., Nagapudi, K., Thomas, B. S. and Chaikof, E. L. (2003) Photo-cross-linking of type I collagen gels in the presence of smooth muscle cells: Mechanical properties, cell viability, and function, *Biomacromolecules*, **4**(4), 890–895.

29. Goldberg, D. J. (2009) Breakthroughs in US dermal fillers for facial soft-tissue augmentation, *J. Cosmet. Laser Ther.*, **11**(4), 240–247.

30. Lorenc, Z. P., Nir, E. and Azachi, M. (2010) Characterization of physical properties and histologic evaluation of injectable Dermicol-p35 porcine-collagen dermal filler, *Plast. Reconstr. Surg.*, **125**(6), 1805–1813.

31. CoStasis Multi-center Collaborative Writing Committee. (2001) A novel collagen-based composite offers effective hemostasis for multiple

surgical indications: Results of a randomized controlled trial, *Surgery*, **129**(4), 445–450.

32. Reece, T. B., Maxey, T. S. and Kron, I. L. (2001) A prospectus on tissue adhesives, *Am. J. Surg.*, **182**(2), S40–S44.
33. Narins, R. S., Brandt, F. S., Lorenc, Z. P., Maas, C. S., Monheit, G. D. and Smith, S. R. (2008) Twelve-month persistency of a novel ribose-cross-linked collagen dermal filler, *Dermatol. Surg.*, **34** (Suppl. 1), S31–S39.
34. Ruszczak, Z. (2003) Effect of collagen matrices on dermal wound healing, *Adv. Drug Deliv. Rev.*, **55**(12), 1595–1611.
35. McKay, W. F., Peckham, S. M. and Badura, J. M. (2007) A comprehensive clinical review of recombinant human bone morphogenetic protein-2 (INFUSE Bone Graft), *Int. Orthop.*, **31**(6), 729–734.
36. Ratanavaraporn, J., Damrongsakkul, S., Sanchavanakit, N., Banaprasert, T. and Kanokpanotn, S. (2006) Comparison of gelatin and collagen scaffolds for fibroblast cell culture, *JOM*, **16**, 31–36.
37. Kang, H. W., Tabata, Y. and Ikada, Y. (1999) Fabrication of porous gelatin scaffolds for tissue engineering, *Biomaterials*, **20**(14), 1339–1344.
38. Chang, C. H., Liu, H. C., Lin, C. C., Chou, C. H. and Lin, F. H. (2003) Gelatin-chondroitin-hyaluronan tri-copolymer scaffold for cartilage tissue engineering, *Biomaterials*, **24**(26), 4853–4858.
39. Huang, Y., Onyeri, S., Siewe, M., Moshfeghian, A. and Madihally, S. V. (2005) *In vitro* characterization of chitosan-gelatin scaffolds for tissue engineering, *Biomaterials*, **26**(36), 7616–7627.
40. Kuijpers, A., Engbers, G., Feijen, J., De Smedt, S., Meyvis, T., Demeester, J., Krijgsveld, J., Zaat, S. and Dankert, J. (1999) Characterization of the network structure of carbodiimide cross-linked gelatin gels, *Macromolecules*, **32**(10), 3325–3333.
41. Cappabianca, P., Esposito, F., Esposito, I., Cavallo, L. M. and Leone, C. A. (2009) Use of a thrombin-gelatin haemostatic matrix in endoscopic endonasal extended approaches: Technical note, *Acta Neurochir. (Wien)*, **151**(1), 69–77.
42. Weaver, F. A., Hood, D. B., Zatina, M., Messina, L. and Badduke, B. (2002) Gelatin-thrombin-based hemostatic sealant for intraoperative bleeding in vascular surgery, *Ann. Vasc. Surg.*, **16**(3), 286–293.
43. Ahmed, T. A., Dare, E. V. and Hincke, M. (2008) Fibrin: A versatile scaffold for tissue engineering applications, *Tissue Eng. Part B. Rev.*, **14**(2), 199–215.
44. Doolittle, R. F. (1984) Fibrinogen and fibrin, *Annu. Rev. Biochem.*, **53**, 195–229.

45. Ye, K. Y., Sullivan, K. E. and Black, L. D. (2011) Encapsulation of cardiomyocytes in a fibrin hydrogel for cardiac tissue engineering, *J. Vis. Exp.*, (55), e3251, doi:10.3791/3251.

46. Johnson, P. J., Parker, S. R. and Sakiyama-Elbert, S. E. (2010) Fibrin-based tissue engineering scaffolds enhance neural fiber sprouting and delay the accumulation of reactive astrocytes at the lesion in a subacute model of spinal cord injury, *J. Biomed. Mater. Res.*, **92**(1), 152–163.

47. Chen, X., Aledia, A. S., Ghajar, C. M., Griffith, C. K., Putnam, A. J., Hughes, C. C. W. and George, S. C. (2008) Prevascularization of a fibrin-based tissue construct accelerates the formation of functional anastomosis with host vasculature, *Tissue Eng. Part A*, **15**(6), 1363–1371.

48. Le Nihouannen, D., Guehennec, L. L., Rouillon, T., Pilet, P., Bilban, M., Layrolle, P. and Daculsi, G. (2006) Micro-architecture of calcium phosphate granules and fibrin glue composites for bone tissue engineering, *Biomaterials*, **27**(13), 2716–2722.

49. Schek, R., Michalek, A. and Latridis, J. (2011) Genipin-crosslinked fibrin hydrogels as a potential adhesive to augment intervertebral disc annulus repair, *Eur. Cell Mater.*, **21**, 373–383.

50. Brown, E. E., Laborie, M. P. G. and Zhang, J. (2012) Glutaraldehyde treatment of bacterial cellulose/fibrin composites: Impact on morphology, tensile and viscoelastic properties, *Cellulose*, **19**(1), 127–137.

51. Currie, L. J., Sharpe, J. R. and Martin, R. (2001) The use of fibrin glue in skin grafts and tissue-engineered skin replacements: A review, *Plast. Reconstr. Surg.*, **108**(6), 1713–1726.

52. Sutherland, T. D., Young, J. H., Weisman, S., Hayashi, C. Y. and Merritt, D. J. (2010) Insect silk: One name, many materials, *Annu. Rev. Entomol.*`**55**, 171–188.

53. Zhou, C. Z., Confalonieri, F., Jacquet, M., Perasso, R., Li, Z. G. and Janin, J. (2001) Silk fibroin: Structural implications of a remarkable amino acid sequence, *Proteins*, **44**(2), 119–122.

54. Gil, E. S., Park, S. H., Marchant, J., Omenetto, F. and Kaplan, D. L. (2010) Response of human corneal fibroblasts on silk film surface patterns, *Macromol. Biosci.*, **10**(6), 664–673.

55. Kim, U. J., Park, J., Li, C., Jin, H. J., Valluzzi, R. and Kaplan, D. L. (2004) Structure and properties of silk hydrogels, *Biomacromolecules*, **5**(3), 786–792.

56. Kim, U. J., Park, J., Joo Kim, H., Wada, M. and Kaplan, D. L. (2005) Three-dimensional aqueous-derived biomaterial scaffolds from silk fibroin, *Biomaterials*, **26**(15), 2775–2785.

57. Altman, G. H., Diaz, F., Jakuba, C., Calabro, T., Horan, R. L., Chen, J., Lu, H., Richmond, J. and Kaplan, D. L. (2003) Silk-based biomaterials, *Biomaterials*, **24**(3), 401–416.

58. Hofmann, S., Hagenmüller, H., Koch, A. M., Müller, R., Vunjak-Novakovic, G., Kaplan, D. L., Merkle, H. P. and Meinel, L. (2007) Control of *in vitro* tissue-engineered bone-like structures using human mesenchymal stem cells and porous silk scaffolds, *Biomaterials*, **28**(6), 1152-1162.

59. Wang, Y., Blasioli, D. J., Kim, H. J., Kim, H. S. and Kaplan, D. L. (2006) Cartilage tissue engineering with silk scaffolds and human articular chondrocytes, *Biomaterials*, **27**(25), 4434–4442.

60. Lovett, M., Cannizzaro, C., Daheron, L., Messmer, B., Vunjak-Novakovic, G. and Kaplan, D. L. (2007) Silk fibroin microtubes for blood vessel engineering, *Biomaterials*, **28**(35), 5271–5279.

61. Leisk, G. G., Lo, T. J., Yucel, T., Lu, Q. and Kaplan, D. L. (2010) Electrogelation for protein adhesives, *Adv. Mater*, **22**(6), 711–715.

62. Pritchard, E. M. and Kaplan, D. L. (2011) Silk fibroin biomaterials for controlled release drug delivery, *Expert Opin. Drug Deliv.*, **8**(6), 797–811.

63. Murphy, A. R., John, P. S. and Kaplan, D. L. (2008) Modification of silk fibroin using diazonium coupling chemistry and the effects on hMSC proliferation and differentiation, *Biomaterials*, **29**(19), 2829–2838.

64. Hu, X., Shmelev, K., Sun, L., Gil, E. S., Park, S. H., Cebe, P. and Kaplan, D. L. (2011) Regulation of silk material structure by temperature-controlled water vapor annealing, *Biomacromolecules*, **12**(5), 1686–1696.

65. Wang, Y., Rudym, D. D., Walsh, A., Abrahamsen, L., Kim, H. J., Kim, H. S., Kirker-Head, C. and Kaplan, D. L. (2008) *In vivo* degradation of three-dimensional silk fibroin scaffolds, *Biomaterials*, **29**(24–25), 3415–3428.

66. Waterhouse, A., Wise, S. G., Ng, M. K. and Weiss, A. S. (2011) Elastin as a nonthrombogenic biomaterial, *Tissue Eng. Part B. Rev.*, **17**(2), 93–99.

67. Soskel, N. T., Wolt, T. B. and Sandberg, L. B. (1987) Isolation and characterization of insoluble and soluble elastins, *Methods Enzymol.*, **144**, 196–214.

68. Bellingham, C., Woodhouse, K., Robson, P., Rothstein, S. and Keeley, F. (2001) Self-aggregation characteristics of recombinantly expressed human elastin polypeptides, *Biochim. Biophys. Acta*, **1550**(1), 6–19.

69. Martin, S. L., Vrhovski, B. and Weiss, A. S. (1995) Total synthesis and expression in Escherichia coli of a gene encoding human tropoelastin, *Gene*, **154**(2), 159–166.

70. Buttafoco, L., Kolkman, N., Engbers-Buijtenhuijs, P., Poot, A., Dijkstra, P., Vermes, I. and Feijen, J. (2006) Electrospinning of collagen and elastin for tissue engineering applications, *Biomaterials*, **27**(5), 724–734.

71. Annabi, N., Mithieux, S. M., Weiss, A. S. and Dehghani, F. (2009) The fabrication of elastin-based hydrogels using high pressure CO2, *Biomaterials*, **30**(1), 1–7.

72. Daamen, W. F., Nillesen, S. T. M., Wismans, R. G., Reinhardt, D. P., Hafmans, T., Veerkamp, J. H. and van Kuppevelt, T. H. (2008) A biomaterial composed of collagen and solubilized elastin enhances angiogenesis and elastic fiber formation without calcification, *Tissue Eng. Part A*, **14**(3), 349–360.

73. Hu, X., Park, S. H., Gil, E. S., Xia, X. X., Weiss, A. S. and Kaplan, D. L. (2011) The influence of elasticity and surface roughness on myogenic and osteogenic-differentiation of cells on silk-elastin biomaterials, *Biomaterials*, **32**(34), 8979–8989.

74. Mithieux, S. M., Tu, Y., Korkmaz, E., Braet, F. and Weiss, A. S. (2009) *In situ* polymerization of tropoelastin in the absence of chemical cross-linking, *Biomaterials*, **30**(4), 431–435.

75. Sierpinski, P., Garrett, J., Ma, J., Apel, P., Klorig, D., Smith, T., Koman, L. A., Atala, A. and Van Dyke, M. (2008) The use of keratin biomaterials derived from human hair for the promotion of rapid regeneration of peripheral nerves, *Biomaterials*, **29**(1), 118–128.

76. Srinivasan, B., Kumar, R., Shanmugam, K., Sivagnam, U. T., Reddy, N. P. and Sehgal, P. K. (2010) Porous keratin scaffold–promising biomaterial for tissue engineering and drug delivery, *J. Biomed. Mater. Res. B Appl. Biomater.*, **92**(1), 5–12.

77. Katoh, K., Tanabe, T. and Yamauchi, K. (2004) Novel approach to fabricate keratin sponge scaffolds with controlled pore size and porosity, *Biomaterials*, **25**(18), 4255–4262.

78. Tachibana, A., Furuta, Y., Takeshima, H., Tanabe, T. and Yamauchi, K. (2002) Fabrication of wool keratin sponge scaffolds for long-term cell cultivation, *J. Biotechnol.*, **93**(2), 165–170.

79. Reichl, S. (2009) Films based on human hair keratin as substrates for cell culture and tissue engineering, *Biomaterials*, **30**(36), 6854–6866.

80. Sando, L., Kim, M., Colgrave, M. L., Ramshaw, J. A., Werkmeister, J. A. and Elvin, C. M. (2010) Photochemical crosslinking of soluble wool keratins produces a mechanically stable biomaterial that supports cell adhesion and proliferation, *J. Biomed. Mater. Res. A.*, **95**(3), 901–911.

81. Hill, P., Brantley, H. and Van Dyke, M. (2010) Some properties of keratin biomaterials: Kerateines, *Biomaterials*, **31**(4), 585–593.

82. Aboushwareb, T., Eberli, D., Ward, C., Broda, C., Holcomb, J., Atala, A. and Van Dyke, M. (2009) A keratin biomaterial gel hemostat derived from human hair: Evaluation in a rabbit model of lethal liver injury, *J. Biomed. Mater. Res. B Appl. Biomater.*, **90**(1), 45–54.

83. Paetau, I., Chen, C. Z. and Jane, J. (1994) Biodegradable plastic made from soybean products. 1. Effect of preparation and processing on mechanical properties and water absorption, *Ind. Eng. Chem. Res.*, **33**(7), 1821–1827.

84. Maltais, A., Remondetto, G. E. and Subirade, M. (2010) Tabletted soy protein cold-set hydrogels as carriers of nutraceutical substances, *Food Hydrocoll.*, **24**(5), 518–524.

85. Reddy, N. and Yang, Y. (2009) Soyprotein fibers with high strength and water stability for potential medical applications, *Biotechnol. Prog.*, **25**(6), 1796–1802.

86. Vega-Lugo, A. C. and Lim, L. T. (2009) Controlled release of allyl isothiocyanate using soy protein and poly (lactic acid) electrospun fibers, *Food Res. Int.*, **42**(8), 933–940.

87. Chien, K. B. and Shah, R. N. (2011) Novel porous soy protein scaffolds: Materials characterization and effects on human mesenchymal stem cell growth and proliferation, *Acta Biomater.*, **8**(2), 694–703.

88. Merolli, A., Nicolais, L., Ambrosio, L. and Santin, M. (2010) A degradable soybean-based biomaterial used effectively as a bone filler *in vivo* in a rabbit, *Biomed. Mater.*, **5**(1), 15008.

89. Peles, Z. and Zilberman, M. (2012) Novel soy protein wound dressings with controlled antibiotic release: Mechanical and physical properties, *Acta Biomater.*, **8**(1), 209–217.

90. Caillard, R., Remondetto, G., Mateescu, M. and Subirade, M. (2008) Characterization of amino cross-linked soy protein hydrogels, *J. Food Sci.*, **73**(5), C283–C291.

91. Luo, L. H., Zhang, Y. F., Wang, X. M., Wan, Y., Chang, P. R., Anderson, D. P. and Chen, Y. (2010) Preparation, characterization, and *in vitro* and *in vivo* evaluation of cellulose/soy protein isolate composite sponges, *J. Biomater. Appl.*, **24**(6), 503–526.

92. Santin, M., Morris, C., Standen, G., Nicolais, L. and Ambrosio, L. (2007) A new class of bioactive and biodegradable soybean-based bone fillers, *Biomacromolecules*, **8**(9), 2706–2711.

93. Wang, T. W., Wu, H. C., Huang, Y. C., Sun, J. S. and Lin, F. H. (2006) Biomimetic bilayered gelatin-chondroitin 6 sulfate-hyaluronic acid biopolymer as a scaffold for skin equivalent tissue engineering, *Artif. Organs*, **30**(3), 141–149.

94. Wang, T. W. and Spector, M. (2009) Development of hyaluronic acid-based scaffolds for brain tissue engineering, *Acta Biomater.*, **5**(7), 2371–2384.

95. Tan, H., Ramirez, C. M., Miljkovic, N., Li, H., Rubin, J. P. and Marra, K. G. (2009) Thermosensitive injectable hyaluronic acid hydrogel for adipose tissue engineering, *Biomaterials*, **30**(36), 6844–6853.

96. Burdick, J. A. and Prestwich, G. D. (2011) Hyaluronic acid hydrogels for biomedical applications, *Adv. Mater.*, **23**(12), H41–H56.

97. Oh, E. J., Park, K., Kim, K. S., Kim, J., Yang, J. A., Kong, J. H., Lee, M. Y., Hoffman, A. S. and Hahn, S. K. (2010) Target specific and long-acting delivery of protein, peptide, and nucleotide therapeutics using hyaluronic acid derivatives, *J. Control. Release*, **141**(1), 2–12.

98. Wang, W., Zhang, M., Lu, W., Zhang, X., Ma, D., Rong, X., Yu, C. and Jin, Y. (2009) Cross-linked collagen–chondroitin sulfate–hyaluronic acid imitating extracellular matrix as scaffold for dermal tissue engineering, *Tissue Eng. Part C Methods*, **16**(2), 269–279.

99. Baier Leach, J., Bivens, K. A., Patrick, C. W., Jr. and Schmidt, C. E. (2003) Photocrosslinked hyaluronic acid hydrogels: Natural, biodegradable tissue engineering scaffolds, *Biotechnol. Bioeng.*, **82**(5), 578–589.

100. Stern, R., Kogan, G., Jedrzejas, M. J. and Soltés, L. (2007) The many ways to cleave hyaluronan, *Biotechnol. Adv.*, **25**(6), 537–557.

101. Chung, C., Beecham, M., Mauck, R. L. and Burdick, J. A. (2009) The influence of degradation characteristics of hyaluronic acid hydrogels on *in vitro* neocartilage formation by mesenchymal stem cells, *Biomaterials*, **30**(26), 4287–4296.

102. Yeom, J., Bhang, S. H., Kim, B. S., Seo, M. S., Hwang, E. J., Cho, I. H., Park, J. K. and Hahn, S. K. (2010) Effect of cross-linking reagents for hyaluronic acid hydrogel dermal fillers on tissue augmentation and regeneration, *Bioconjug. Chem.*, **21**(2), 240–247.

103. Ravi Kumar, M. N. V. (2000) A review of chitin and chitosan applications, *React. Funct. Polym.*, **46**(1), 1–27.

104. Yilgor, P., Tuzlakoglu, K., Reis, R. L., Hasirci, N. and Hasirci, V. (2009) Incorporation of a sequential BMP-2/BMP-7 delivery system into chitosan-based scaffolds for bone tissue engineering, *Biomaterials*, **30**(21), 3551–3559.

105. Jin, R., Moreira Teixeira, L., Dijkstra, P., Karperien, M., Van Blitterswijk, C., Zhong, Z. and Feijen, J. (2009) Injectable chitosan-based hydrogels for cartilage tissue engineering, *Biomaterials*, **30**(13), 2544–2551.

106. Adekogbe, I. and Ghanem, A. (2005) Fabrication and characterization of DTBP-crosslinked chitosan scaffolds for skin tissue engineering, *Biomaterials*, **26**(35), 7241–7250.

107. Scanga, V. I., Goraltchouk, A., Nussaiba, N., Shoichet, M. S. and Morshead, C. (2010) Biomaterials for neural-tissue engineering–Chitosan supports the survival, migration, and differentiation of adult-derived neural stem and progenitor cells, *Can. J. Chem.*, **88**(3), 277–287.

108. Kucharska, M., Walenko, K., Butruk, B., Brynk, T., Heljak, M. and Ciach, T. (2010) Fabrication and characterization of chitosan microspheres agglomerated scaffolds for bone tissue engineering, *Mater. Lett.*, **64**(9), 1059–1062.

109. Di Martino, A., Sittinger, M. and Risbud, M. V. (2005) Chitosan: A versatile biopolymer for orthopaedic tissue-engineering, *Biomaterials*, **26**(30), 5983–5990.

110. Subramanian, A., Vasanthan, K. S., Krishnan, U. M. and Sethuraman, S. (2011) Chitosan and its derivatives in clinical use and applications. In: Domb, A. J., Kumar, N., and Ezra, A. (Eds.). *Biodegradable Polymers in Clinical Use and Clinical Development* (pp.111–135). New York: Wiley.

111. Jayakumar, R., Nwe, N., Tokura, S. and Tamura, H. (2007) Sulfated chitin and chitosan as novel biomaterials, *Int. J. Biol. Macromol.*, **40**(3), 175–181.

112. Ji, C., Khademhosseini, A. and Dehghani, F. (2011) Enhancing cell penetration and proliferation in chitosan hydrogels for tissue engineering applications, *Biomaterials*, **32**(36), 9719–9729.

113. Wang, G., Zheng, L., Zhao, H., Miao, J., Sun, C., Ren, N., Wang, J., Liu, H. and Tao, X. (2011) *In vitro* assessment of the differentiation potential of bone marrow-derived mesenchymal stem cells on genipin-chitosan conjugation scaffold with surface hydroxyapatite nanostructure for bone tissue engineering, *Tissue Eng. Part A*, **17**(9-10), 1341–1349.

114. Hirano, S., Tsuchida, H. and Nagao, N. (1989) N-acetylation in chitosan and the rate of its enzymic hydrolysis, *Biomaterials*, **10**(8), 574–576.

115. Ishihara, M., Fujita, M., Kishimoto, S., Hattori, H. and Kanatani, Y. (2012) Biological, chemical, and physical compatibility of chitosan and biopharmaceuticals. In: Sarmento, B., and das Neves, J. (Eds.). *Chitosan-Based Systems for Biopharmaceuticals: Delivery, Targeting and Polymer Therapeutics* (pp. 93–106) New York: Wiley.

116. Mi, F. L., Shyu, S. S., Wu, Y. B., Lee, S. T., Shyong, J. Y. and Huang, R. N. (2001) Fabrication and characterization of a sponge-like asymmetric

chitosan membrane as a wound dressing, *Biomaterials*, **22**(2), 165–173.

117. Leid, J. G., Willson, C. J., Shirtliff, M. E., Hassett, D. J., Parsek, M. R. and Jeffers, A. K. (2005) The exopolysaccharide alginate protects Pseudomonas aeruginosa biofilm bacteria from IFN-$\gamma$-mediated macrophage killing, *J. Immunol.*, **175**(11), 7512–7518.

118. Stojkovska, J., Bugarski, B. and Obradovic, B. (2010) Evaluation of alginate hydrogels under *in vivo*–like bioreactor conditions for cartilage tissue engineering, *J. Mater. Sci. Mater. Med.*, **21**(10), 2869–2879.

119. Park, S. A., Park, K. E. and Kim, W. D. (2010) Preparation of sodium alginate/poly (ethylene oxide) blend nanofibers with lecithin, *Macromol. Res.*, **18**(9), 891–896.

120. Yen, C. N., Lin, Y. R., Chang, M. D., Tien, C. W., Wu, Y. C., Liao, C. J. and Hu, Y. C. (2008) Use of porous alginate sponges for substantial chondrocyte expansion and matrix production: Effects of seeding density, *Biotechnol. Prog.*, **24**(2), 452–457.

121. Draget, K., Skjåk Bræk, G. and Smidsrød, O. (1994) Alginic acid gels: The effect of alginate chemical composition and molecular weight, *Carbohydr. Polym.*, **25**(1), 31–38.

122. Tan, W. H. and Takeuchi, S. (2007) Monodisperse alginate hydrogel microbeads for cell encapsulation, *Adv. Mater.*, **19**(18), 2696–2701.

123. Dhollander, A. A. M., Verdonk, P. C. M., Lambrecht, S., Verdonk, R., Elewaut, D., Verbruggen, G. and Almqvist, K. F. (2012) Midterm results of the treatment of cartilage defects in the knee using alginate beads containing human mature allogenic chondrocytes, *Am. J. Sports Med.*, **40**(1), 75–82.

124. Lansdown, A. B. and Payne, M. J. (1994) An evaluation of the local reaction and biodegradation of calcium sodium alginate (Kaltostat) following subcutaneous implantation in the rat, *J. R. Coll. Surg. Edinb.*, **39**(5), 284–288.

125. Jejurikar, A., Seow, X. T., Lawrie, G., Martin, D., Jayakrishnan, A. and Grondahl, L. (2012) degradable alginate hydrogels crosslinked by the macromolecular crosslinker alginate dialdehyde, *J. Mater. Chem.*, **22**, 9751–9758.

126. Chou, A. I. and Nicoll, S. B. (2009) Characterization of photocrosslinked alginate hydrogels for nucleus pulposus cell encapsulation, *J. Biomed. Mater. Res. A*, **91**(1), 187–194.

127. Chang, J. C., Hsu, S. and Chen, D. C. (2009) The promotion of chondrogenesis in adipose-derived adult stem cells by an RGD-chimeric protein in 3D alginate culture, *Biomaterials*, **30**(31), 6265–6275.

128. Yang, J., Goto, M., Ise, H., Cho, C. S. and Akaike, T. (2002) Galactosylated alginate as a scaffold for hepatocytes entrapment, *Biomaterials*, **23**(2), 471–479.

129. Ronghua, H., Yumin, D. and Jianhong, Y. (2003) Preparation and *in vitro* anticoagulant activities of alginate sulfate and its quaterized derivatives, *Carbohydr. Polym.*, **52**(1), 19–24.

130. Nishiyama, Y., Langan, P. and Chanzy, H. (2002) Crystal structure and hydrogen-bonding system in cellulose I$\beta$ from synchrotron X-ray and neutron fiber diffraction, *J. Am. Chem. Soc.*, **124**(31), 9074–9082.

131. Eichhorn, S., Dufresne, A., Aranguren, M., Marcovich, N., Capadona, J., Rowan, S., Weder, C., Thielemans, W., Roman, M. and Renneckar, S. (2010) Review: Current international research into cellulose nanofibres and nanocomposites, *J. Mater. Sci.*, **45**(1), 1–33.

132. Czaja, W. K., Young, D. J., Kawecki, M. and Brown, R. M., Jr. (2007) The future prospects of microbial cellulose in biomedical applications, *Biomacromolecules*, **8**(1), 1–12.

133. Sannino, A., Demitri, C. and Madaghiele, M. (2009) Biodegradable cellulose-based hydrogels: Design and applications, *Materials*, **2**(2), 353–373.

134. Pajulo, O., Viljanto, J., Hurme, T., Saukko, P., Lönnberg, B. and Lönnqvist, K. (1996) Viscose cellulose sponge as an implantable matrix: Changes in the structure increase the production of granulation tissue, *J. Biomed. Mater. Res.*, **32**(3), 439–446.

135. Entcheva, E., Bien, H., Yin, L., Chung, C. Y., Farrell, M. and Kostov, Y. (2004) Functional cardiac cell constructs on cellulose-based scaffolding, *Biomaterials*, **25**(26), 5753–5762.

136. Svensson, A., Nicklasson, E., Harrah, T., Panilaitis, B., Kaplan, D., Brittberg, M. and Gatenholm, P. (2005) Bacterial cellulose as a potential scaffold for tissue engineering of cartilage, *Biomaterials*, **26**(4), 419–431.

137. Muller, F. A., Muller, L., Hofmann, I., Greil, P., Wenzel, M. M. and Staudenmaier, R. (2006) Cellulose-based scaffold materials for cartilage tissue engineering, *Biomaterials*, **27**(21), 3955–3963.

138. Martson, M., Viljanto, J., Hurme, T., Laippala, P. and Saukko, P. (1999) Is cellulose sponge degradable or stable as implantation material? An *in vivo* subcutaneous study in the rat, *Biomaterials*, **20**(21), 1989–1995.

139. Miyamoto, T., Takahashi, S., Ito, H., Inagaki, H. and Noishiki, Y. (1989) Tissue biocompatibility of cellulose and its derivatives, *J. Biomed. Mater. Res.*, **23**(1), 125–133.

140. Helenius, G., Backdahl, H., Bodin, A., Nannmark, U., Gatenholm, P. and Risberg, B. (2006) *In vivo* biocompatibility of bacterial cellulose, *J. Biomed. Mater. Res. A.*, **76**(2), 431–438.

141. Nakamatsu, J., Torres, F. G., Troncoso, O. P., Min-Lin, Y. and Boccaccini, A. R. (2006) Processing and characterization of porous structures from chitosan and starch for tissue engineering scaffolds, *Biomacromolecules*, **7**(12), 3345–3355.

142. Gomes, M., Godinho, J., Tchalamov, D., Cunha, A. and Reis, R. (2002) Alternative tissue engineering scaffolds based on starch: Processing methodologies, morphology, degradation and mechanical properties, *Mater. Sci. Eng. C*, **20**(1), 19–26.

143. Salgado, A., Coutinho, O. and Reis, R. (2004) Novel starch-based scaffolds for bone tissue engineering: Cytotoxicity, cell culture, and protein expression, *Tissue Eng.*, **10**(3–4), 465–474.

144. Sakai, S., Liu, Y., Matsuyama, T., Kawakami, K. and Taya, M. (2012) On-demand serum-degradable amylopectin-based *in situ* gellable hydrogel, *J. Mater. Chem.*, **22**(5), 1944–1949.

145. Azevedo, H. S., Gama, F. M. and Reis, R. L. (2003) *In vitro* assessment of the enzymatic degradation of several starch based biomaterials, **4**(6), 1703–1712.

146. Jane, J., Wong, K. and McPherson, A. E. (1997) Branch-structure difference in starches of A-and B-type X-ray patterns revealed by their Naegeli dextrins, *Carbohydr. Res.*, **300**(3), 219–227.

147. Santos, T. C., Marques, A. P., Silva, S. S., Oliveira, J. M., Mano, J. F., Castro, A. G., van Griensven, M. and Reis, R. L. (2010) Chitosan improves the biological performance of soy-based biomaterials, *Tissue Eng. Part A*, **16**(9), 2883–2890.

148. Mendes, S. C., Reis, R., Bovell, Y. P., Cunha, A., van Blitterswijk, C. A. and de Bruijn, J. D. (2001) Biocompatibility testing of novel starch-based materials with potential application in orthopaedic surgery: A preliminary study, *Biomaterials*, **22**(14), 2057–2064.

149. BeMiller, J. N. (1986) An introduction to pectins: Structure and properties. In: Fishman, M. L., and Jen, J.J. (Eds.). *Chemistry and Function of Pectins* (pp. 1–12). Washington, USA: ACS Symposium Series, American Chemical Society.

150. Coimbra, P., Ferreira, P., de Sousa, H. C., Batista, P., Rodrigues, M. A., Correia, I. J. and Gil, M. H. (2011) Preparation and chemical and biological characterization of a pectin/chitosan polyelectrolyte

complex scaffold for possible bone tissue engineering applications, *Int. J. Biol. Macromol.*, **48**(1), 112–118.

151. Perez, S., Rodriguez-Carvajal, M. A. and Doco, T. (2003) A complex plant cell wall polysaccharide: Rhamnogalacturonan II. A structure in quest of a function, *Biochimie*, **85**(1–2), 109–121.

152. Bussy, C., Verhoef, R., Haeger, A., *et al.* (2008) Modulating *in vitro* bone cell and macrophage behavior by immobilized enzymatically tailored pectins, *J. Biomed. Mater. Res. A.*, **86**(3), 597–606.

153. Munarin, F., Guerreiro, S. G., Grellier, M. A., Tanzi, M. C., Barbosa, M. A., Petrini, P. and Granja, P. L. (2011) Pectin-based injectable biomaterials for bone tissue engineering, *Biomacromolecules*, **12**(3), 568–577.

154. Giunchedi, P., Conte, U., Chetoni, P. and Saettone, M. F. (1999) Pectin microspheres as ophthalmic carriers for piroxicam: Evaluation *in vitro* and *in vivo* in albino rabbits, *Eur. J. Pharm. Sci.*, **9**(1), 1–7.

155. Keshavarz, T. and Roy, I. (2010) Polyhydroxyalkanoates: Bioplastics with a green agenda, *Curr. Opin. Microbiol.*, **13**(3), 321–326.

156. Chen, G. Q. and Wu, Q. (2005) The application of polyhydroxyalkanoates as tissue engineering materials, *Biomaterials*, **26**(33), 6565–6578.

157. Sabir, M. I., Xu, X. and Li, L. (2009) A review on biodegradable polymeric materials for bone tissue engineering applications, *J. Mater. Sci.*, **44**(21), 5713–5724.

158. Hazer, D. B., Kılıçay, E. and Hazer, B. (2012) Poly (3-hydroxyalkanoate) s: Diversification and biomedical applications: A state of the art review, *Mater. Sci. Eng. C*, **4**, 637–647

159. Freier, T., Kunze, C., Nischan, C., Kramer, S., Sternberg, K., Sass, M., Hopt, U. T. and Schmitz, K. P. (2002) *In vitro* and *in vivo* degradation studies for development of a biodegradable patch based on poly(3-hydroxybutyrate), *Biomaterials*, **23**(13), 2649–2657.

160. Guzmán, D., Kirsebom, H., Solano, C., Quillaguamán, J. and Hatti-Kaul, R. (2011) Preparation of hydrophilic poly (3-hydroxybutyrate) macroporous scaffolds through enzyme-mediated modifications, *J. Bioact. Compat. Polym.*, **26**(5), 452–463.

161. Baek, J. Y., Xing, Z. C., Kwak, G., Yoon, K. B., Park, S. Y., Park, L. S. and Kang, I. K. (2012) Fabrication and characterization of collagen-immobilized porous PHBV/HA nanocomposite scaffolds for bone tissue engineering, *J. Nanomater.*, **2012**, 1–11.

162. Wang, Y., Lu, L., Zheng, Y. and Chen, X. (2006) Improvement in hydrophilicity of PHBV films by plasma treatment, *J. Biomed. Mater. Res. A.*, **76**(3), 589–595.

163. Shishatskaya, E. I., Volova, T. G., Gordeev, S. A. and Puzyr, A. P. (2005) Degradation of P(3HB) and P(3HB-co-3HV) in biological media, *J. Biomater. Sci. Polym. Ed.*, **16**(5), 643–657.

164. Sodian, R., Sperling, J. S., Martin, D. P., Egozy, A., Stock, U., Mayer, J. E., Jr. and Vacanti, J. P. (2000) Fabrication of a trileaflet heart valve scaffold from a polyhydroxyalkanoate biopolyester for use in tissue engineering, *Tissue Eng.*, **6**(2), 183–188.

# Chapter 10

# Hydrogels: Characteristics and Properties

**Rolando Barbucci[a,b] and Daniela Pasqui[a,b]**
[a]*C.R.I.S.M.A. University of Siena, Via G. Matteotti 15/16, 53034 Colle di Val d'Elsa (Si), Italy*
[b]*Department of Biotechnologies, Chemistry and Pharmacy, University of Siena, Via A. Moro 2, 53100 Siena, Italy.*
barbucci@unisi.it

Hydrogels are a particular class of materials, consisting of hydrophilic polymer chains and water. Water is the major constituent of the hydrogels and could reach more than 90% of the weight. However, the presence and the chemical nature of the polymer chain affect largely the arrangement and the interaction of the water molecules which can form several layers on the macromolecule. The nature of hydrogels makes them particularly suitable to be used as biomaterials, especially in soft tissue engineering. In this chapter, mainly hydrogels consisting of polysaccharides are described. The strategy of synthesis is discussed as well as the solvent used in the synthesis which can determine "different" properties of the "same" hydrogel. The main methods used to characterize the hydrogels are also discussed. On the basis of the physicochemical and mechanical

*Scaffolds for Tissue Engineering: Biological Design, Materials, and Fabrication*
Edited by Claudio Migliaresi and Antonella Motta

ISBN 978-981-4463-20-1 (Hardcover), 978-981-4463-21-8 (eBook)
www.panstanford.com

properties of hydrogels, applications ranging from osteoarthritic therapy to the controlled release of drugs are shown.

## 10.1 Generality

A hydrogel is a material generally constructed by a hydrophilic polymer mixture and water that exhibits both solid-like and liquid-like properties. The ability to hold water depends on the hydrophilicity of the polymer, and can range from 10% up to several thousand times the value of the dry weight.[1,2] Water occupies the interstitial spaces that are formed in the three-dimensional structural framework constituted by the crosslinked polymeric chains (Fig. 10.1). The three-dimensional structure is usually described as a mesh, which holds the fluid in place providing the solidity of the hydrogel. The crosslinked polymer network may be formed by varying interactions among the polymeric chains, for instance, by hydrogen bonding, van der Waals interaction, covalent bonds, electrostatic interactions, or physical entanglements (Fig. 10.1b).

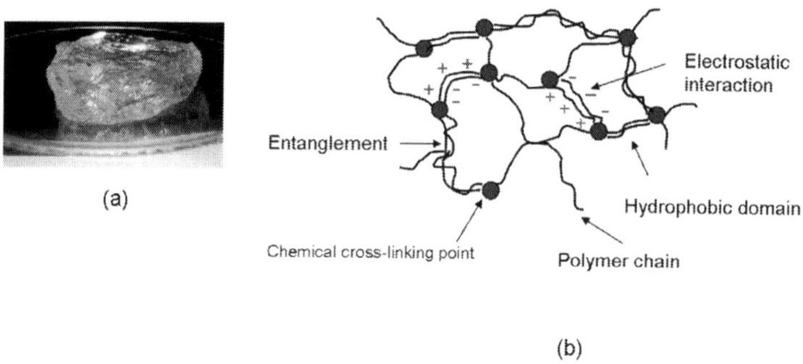


**Figure 10.1** (a) Image of a polysaccharide-based hydrogel at its maximum swelling degree. (b) Schematic drawing of a chemically crosslinked hydrogel showing the main interactions among the polymer chains.

The basic categories of structural hydrogels can be schematically represented by (i) thermoset hydrogels in which the polymeric chains are covalently crosslinked (also called chemical hydrogels), and which are stable in the presence of water but show weak and brittle properties at high water content; and (ii) thermoplastic hydrogels, which are physically crosslinked showing limited shape stability (physical hydrogel).[3,4] The chemical hydrogels always present physical interaction as well, thus both the characteristics are present in the chemical hydrogels (chemical bonds and physical interactions). The contrary, of course, does not occur.

In the physical hydrogels, the network is held together by molecular entanglement (temporary constraints) and/or secondary forces such as ionic, hydrophobic, and hydrogen bonds (Fig. 10.2). The physical hydrogels are not homogeneous, since groups of entanglements, or hydrophobically, or ionically associated domains, create dishomogeneity and do not present the characteristics of structural repeatability.[5–7] The chemical hydrogels are mainly characterized by a covalent crosslinked network and can be obtained by the chemical crosslinking of polymers or with the addition of metal ions which act as crosslinking agents among the polymeric chains, binding to coordinating groups present on the polymer[8] (Fig. 10.3).

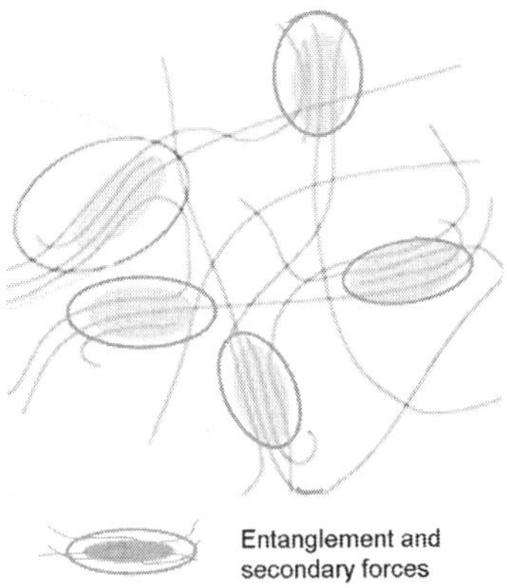


**Figure 10.2** Schematic drawing of a physically crosslinked hydrogel.

Also the chemical hydrogels are not homogeneous (although sometimes the crosslinker can be evenly distributed along the chain, so that we can speak about stoichiometric crosslinking) because the presence of physical interactions affect the homogeneity of the network. All this occurs due to hydrophobic aggregation of crosslinking agents and/or hydrophobic residues of the polymers that leads to a high local density of crosslinks, or to a possible phase separation, or formation of voids or macropores. As proof of this, high molecular weights usually promote more numerous intermolecular interactions among the polymer chains.

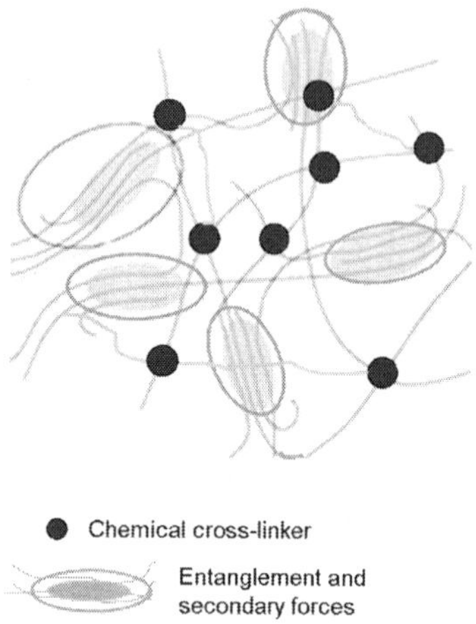


**Figure 10.3** Schematic drawing of a chemically crosslinked hydrogel.

In the fluid phase of the hydrogel, water generally fills up the interstitial pores of the network and makes the hydrogel wet and soft rendering it similar to biological tissues in some respects. Thus, there is space inside the hydrogel network that is available to accommodate water, determining one of the main characteristics of hydrogel, the swelling capacity. Also, the presence of ionizable groups onto the polymer chains affects the water uptake of the hydrogel.[2,9] The ionizable group may dissociate in solution completely or partially, depending on the dissociation constant, and the charged groups remain fixed to the polymer chains of the hydrogel network. The ionic charges in the polymer backbone repel each other in an aqueous solution and generate a significant expansion in space. A number of mobile ions that include counterions and co-ions, due to the presence of electrolytes, surround the hydrogel. Two examples are shown in Fig. 10.4.[10,11]

The difference in the ion concentration within the gels and the outside solution generates osmosis, and the greater the difference in the ion concentration is, the stronger the osmotic pressure becomes. Some solutes may also penetrate into the hydrogel network and somehow interact with network segments. This will probably

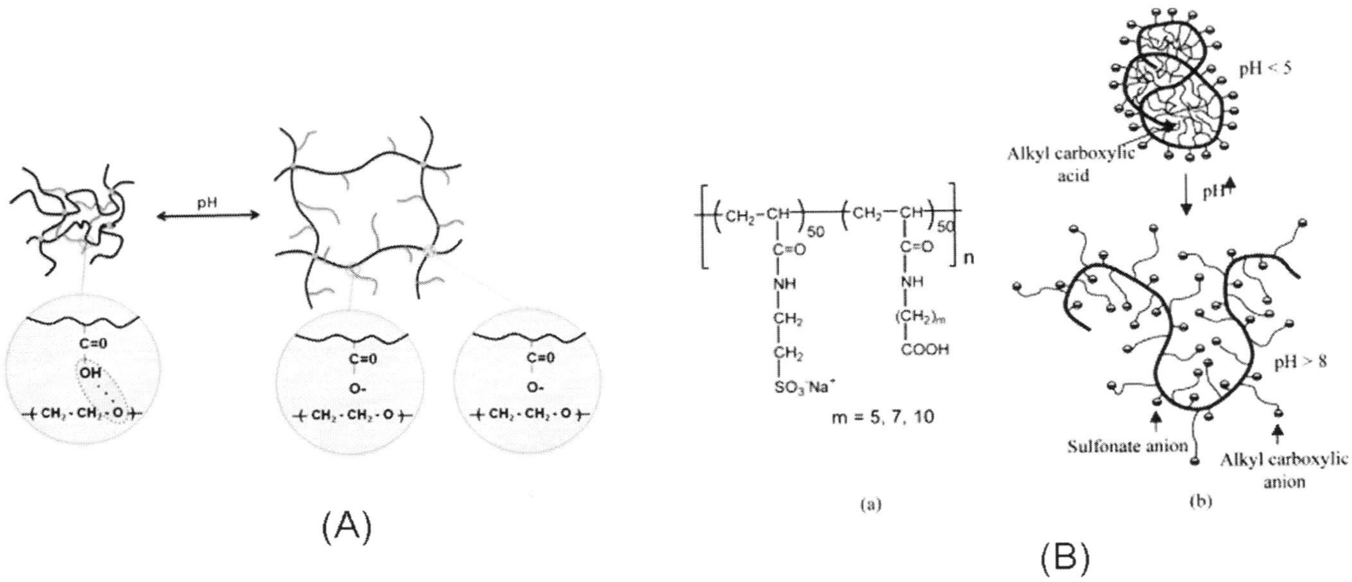


**Figure 10.4** Example of the influence of ionizable groups along the polymer chains in the water uptake of the hydrogel. Fig. A reprinted from Ref. 10, Copyright 2011 and Fig. B reprinted from Ref. 11, Copyright 2004, with permission from Elsevier.

affect the contractibility of the network. Other solutes are too bulky to penetrate into the network. Generally, inorganic salts or large or small organic molecules decrease the degree of equilibrium swelling compared with that achieved in pure distilled water. A typical example is represented by some neutral hydrogels for which equilibrium swelling had been reached in distilled water. They are dehydrated in a dextran solution even if no penetration of dextran molecules into the hydrogel network occurs; hence, the dehydration is a result of the balancing of the water chemical potential throughout the system.[12]

### 10.1.1 Water in the Hydrogel

Water is the main component of the hydrogel and can be classified into three types as shown in Fig. 10.5. Water in the outermost layer is called free water and can be easily removed from the hydrogel under mild conditions, such as by an absorbing paper. The interstitial water is the type of water, which is not attached to the hydrogel network, but physically trapped in between the hydrated polymer chains. Thus, free and interstitial water are the waters, which do not take part in bonds with the polymeric molecules. They show similar transition temperature, enthalpy, and DSC curves as the pure water. The bound water is directly attached to the polymer chain through the hydration of functional groups or ions. The interaction of this type of water is different according to the different functional groups present in the macromolecule skeleton. Generally, this bound water remains as an integral part of the hydrogel structure and can hardly be separated, even at high temperature. Although other layers of water can be accommodated into the hydrogel structure, this schematic view can be considered as a simple picture of water in the hydrogel[13] (Fig. 10.5). The type of water in a hydrogel determines the permeation of chemical and biological compounds in the gel.[14] When a hydrogel in the dry state absorbs water, the first molecules, which enter the matrix, will hydrate the polar groups (bound water). As soon as the polar groups are hydrated, the network swells and the hydrophobic groups will begin to aggregate amongst themselves, finding a hostile environment, which is the aqueous one. After this first process, the network will absorb a further amount of water due to osmotic pressure that pushes the polymer chains to an infinite dilution (interstitial water). This water goes to fill the space between the polymeric chains of the network, already hydrated with bound water. This swelling driving force is balanced by the covalent and physical crosslinkings, which give rise to the elastic force of the network, so that the swelling of the hydrogel will reach an equilibrium state. The interstitial water puts both the water and the hydrogel in contact with the external environment because it links with the free water, which keeps the characteristics of the water in the bulk.

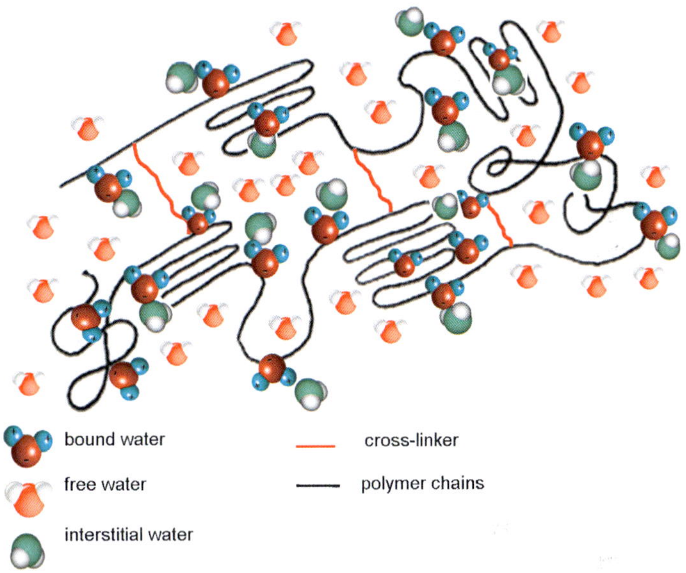


**Figure 10.5** Schematic drawing of the different types of water inside a chemically crosslinked hydrogel.

The dehydration of the hydrogel determines the loss of the free and interstitial water with a shrinking of the hydrogel, which drastically reduces its volume. By heating or lyophilizing the hydrogel, a further reduction of volume occurs because of the partial loss of the bound water. A complete removal of water molecules never occurs.

When the dry hydrogel is re-swollen in water, schematically the opposite process occurs, starting from the hydration of the polymeric chains and subsequently, the establishment of interstitial and free water layers. However, this process does not determine the same situation and disposition of waters inside the polymeric network. In fact, the volume of the hydrogel results different from earlier result.

The conformation of the polymeric chains, their entanglements, and physical interactions are different and the re-swelling process occurs differently whenever a dried gel is rehydrated (Fig. 10.6).

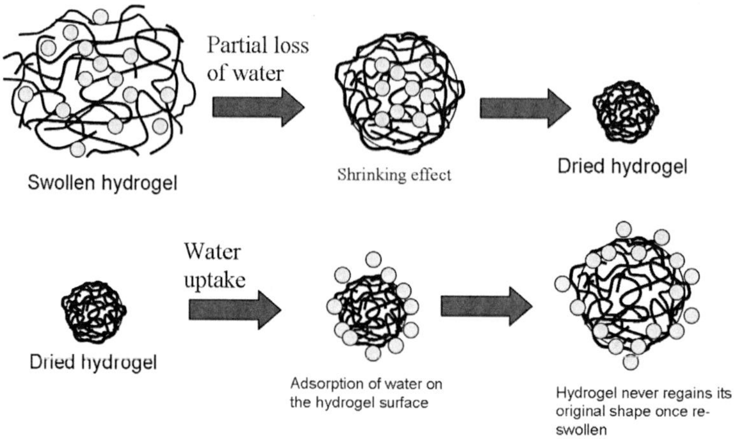


**Figure 10.6** Mechanism of water loss and subsequent re-swelling process.

### 10.1.2 Swelling Behavior

The rate of the hydrogel swelling is mainly determined by how fast the polymer chains can relax: schematically, the penetration of water into the polymer network can be described by three steps: (1) diffusion of water molecules; (2) relaxation of polymeric chains due to the hydration; and (3) expansion of the polymer network. When the diffusion rate is much faster than the relaxation rate, the relaxation process becomes the main determining step. This results in a continuous renewable interface between the swollen hydrogel region and the non-hydrated internal polymer network (glassy core). This interface moves continuously from the external region to the internal core as the solvent uptake proceeds until the formation of a fully swollen hydrogel[13] (Fig. 10.7). As shown in Fig. 10.8, the absorption process in highly crosslinked hydrogels resembles that occurring for metal mesh. It works like a single diffusion process as polymer chain movement is limited by the high-crosslinked density, as occurs for metal mesh. Generally, the more a hydrogel is crosslinked, the lower is its water uptake[15,16] (Fig. 10.9).

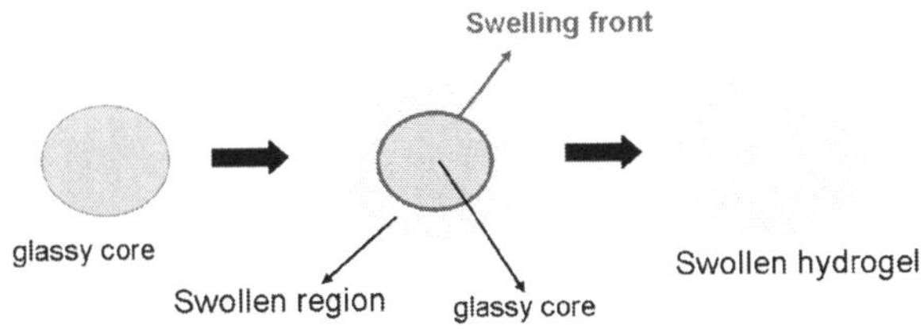


**Figure 10.7** Transport for swelling solvent from front moving into a glassy polymer interior.

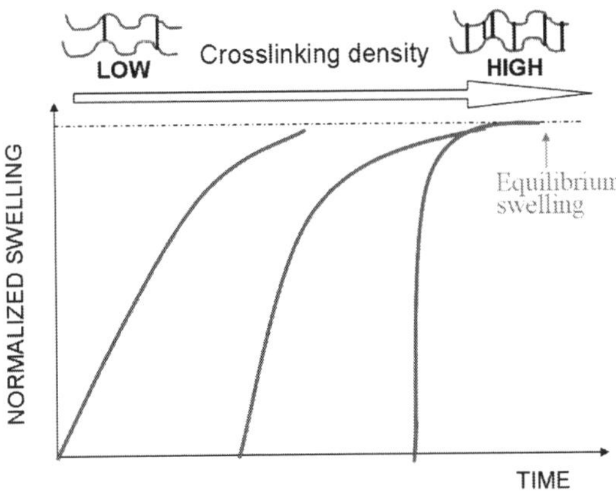


**Figure 10.8** Swelling kinetics.

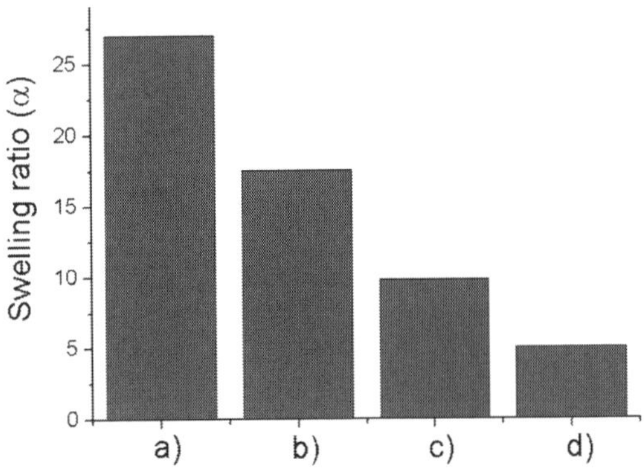


**Figure 10.9** Swelling ratio ($\alpha$) for a hydrogel of poly(*N*-isopropylacrylamide-co-2-hydroxypropyl-methacrylate) containing different increasing amounts of the crosslinker ethylenglycol dimethacrylate (EGDM): (a) 1 mol%, (b) 1.5 mol%, (c) 2 mol%, and (d) 5 mol%.

In order to measure the uptake of water molecules and their absorbing rate, we can measure the swelling degree of the hydrogel:

$$\%Sw = (m_t - m_d)/m_d \times 100$$

where Sw is the percentage of swelling at time $t$, $m_t$, weight of the swollen hydrogel at time $t$, and $m_d$, weight of the dry hydrogel. To measure the swelling kinetics, several values of Sw are measured at their corresponding swelling time. The swelling degree for hydrogel can also be expressed in other ways, for instance by estimating the variation in the height of each hydrogel from the dry to the swollen state.[17] The swelling degree (Sh) of each hydrogel[18] can be calculated according to the equation:

$$\%Sh = (h_s - h_0)/h_0 \times 100$$

where, $h_s$ is the height of the swollen hydrogel and $h_0$, initial height of the hydrogel.

Hydrogels can be compared through their swelling rate factor defined as the time to reach a certain percentage of the swelling capacity (Fig. 10.10).[18] For strongly hydrophilic polymers, such as polysaccharides that contain many hydrophilic groups in the backbone, the achievement of the maximum value of the swelling may occur in a very short time. In that case, in order to differentiate from one another, a second parameter can be taken into consideration: the de-swelling factor.[19] Deswelling in hydrogels is always a diffusion process and is related to the desorption of water molecules. Depending on the environment and how wet the surrounding environment is, the desorption process takes place at a slow or fast rate, ranging from a linear to an exponential trend (Fig. 10.11). From the slope of the curve we can envisage different water molecules present in the hydrogel. The first water molecules to be removed are those belonging to free water and the last ones are the bound water. When the bound water is removed the network collapses and the macromolecular chains may organize to form amorphous and/or crystalline domains.

Also the mechanical strength of a hydrogel is often derived almost entirely from the crosslinks in the system. The use of rheological analysis to investigate the mechanical properties of hydrogels is well known and has continued to be one of the most commonly used techniques in hydrogel characterization.[20] An example of

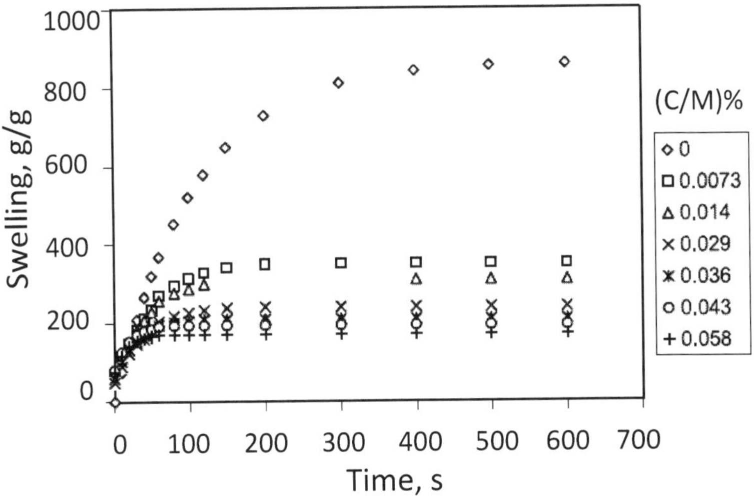


**Figure 10.10** Water swelling kinetics of acrylic hydrogels at different amounts of MBA (*N,N′*-methylenebisacrylamide) crosslinker. C/M is the molar ratio between crosslinker and monomer. Reprinted from Ref. 18, Copyright 2003, with permission from Elsevier.

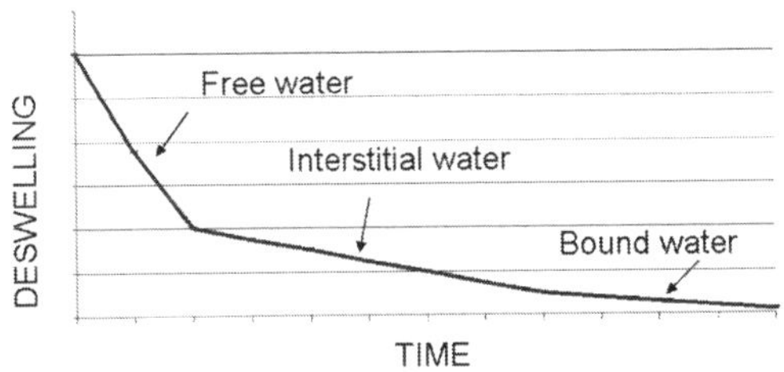


**Figure 10.11** General trend of de-swelling for a hydrogel.

the effects of swelling on the elastic modulus is presented in Fig. 10.12 for polyacrylamide gels crosslinked with *N,N*-methylene bisacrylamide.[21] These results clearly show that as the volume degree of swelling doubles, the elastic modulus decreases by nearly two orders of magnitude. The water swelling the copolymer is dramatically plasticizes the polymer and reduces the mechanical strength. For given water content, as crosslink density is increased, the ratio of bound to unbound water increases. Increase in the crosslink density provides more rigidity to the structure and thus, the moduli values increase through the relationship:

$$G'_R = r_0RT/M_c$$

where, $G'_R$ is the relaxed rubbery modulus; $r_o$, density, R, universal gas constant; $T$, temperature; and $M_c$, molecular weight between crosslinks.

An alternative form is

$$G'_R = n_iRT$$

where, $n_i$ is the number of crosslink sites per unit volume.

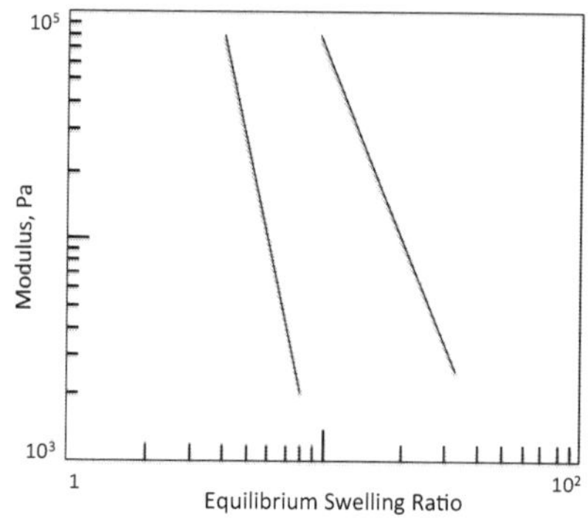


**Figure 10.12** The elastic modulus as a function of the equilibrium degree of swelling for polyacrylamide gels swollen in solutions of ionic strengths of 1.0 and $10^{-4}$ mol/l. Reprinted from Ref. 21, Copyright 1996, with permission from Elsevier.

By increasing the crosslinking density of the hydrogel, the network and thus its mechanical stiffness increase. The increase in the crosslinker ratio leads to an increase in the storage modulus of gel. These simple relationships do not account for contributions from trapped entanglements, which may be introduced during the crosslinking reaction.

### 10.1.3 Thixotropy

Thixotropy is a characteristic of some pseudo-plastic materials that leads to a progressive decrease in their viscosity when these materials are subjected to a specific mechanical stress, followed by the recovery of its rheological properties after a more or less prolonged period of rest.[22] Under stress conditions, an infinitesimal layer of fluid slides onto the adjacent layer, leading to the settling of a laminar flux. Hence, the fluid transforms from the gel state to the liquid state, representing a reversible process of gel–sol isothermal transformation. The thixotropic systems usually contain asymmetric particles, which, by means of numerous points of contact, constitute a certain crosslinked structure. The state of rest gives the system rigidity similar to that of a gel, and then, when applying a shear force, the structure begins to break as it reduces the contact points and the particles align in the flow direction, and the system passes from gel to sol with a reduction in viscosity. Once stopped the application of stress, the crosslinked structure slowly begins to recover as a consequence of the Brownian motion of particles[23] (Fig. 10.13). The most common substance that has this unique property is ketchup. If the container is left standing the sauce has a semi-solid appearance. But when the container is shaken, the sauce becomes smooth to the point of easy escape. The thixotropy also characterizes margarine, which is easily spreadable, and OME paints. In fact the thixotropy properties are exploited in the preparation of special "non-drip" paints: in fact, until the paint is brushed on, it is fluid, when the painting is over it thereby solidifies, avoiding the formation of drips.

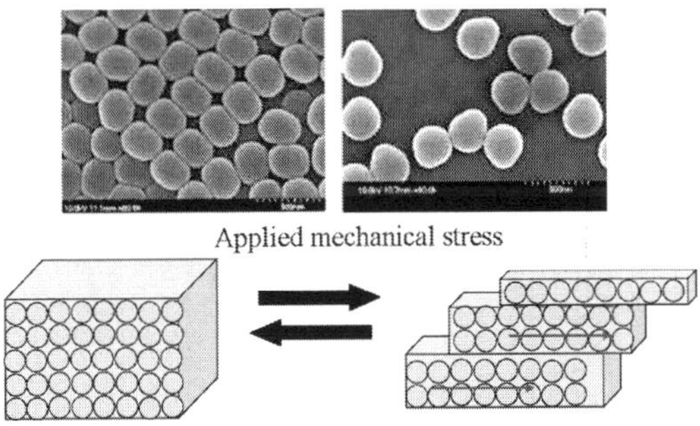


**Figure 10.13** Schematic drawing showing the behavior of a thixotropic material under a stress application.

The capability of the polysaccharide hydrogel to absorb a very large amount of water renders the hydrogel very similar to a highly viscous solution allowing for the thixotropic behavior of the polysaccharide also in the hydrogel form. Some examples are represented by carboxymethylcellulose (CMC), hyaluronane (Hyal), alginic acid amidated (AAA), and CMC amidated (CMCA).[24–26] This points out the unique property of these hydrogels to be injectable and is very useful for biomedical applications.[17] In fact, the hydrogel can be considered not as a whole structure but formed by several macro- or microparticles connected to each other by water molecules. Water clings to the hydrogel particles binding them together with strength. However, when an appropriate stress is applied, the particles slide on to the water layer, which connects them, and so under stress the system behaves like a fluid. Once the stress is removed, the hydrogel network forms again and the structure resumes its three-dimensional form. This characteristic of the hydrogel is time dependent and appears evident by a double rheological curve, in rheometric analysis, in which a rising and a descending curve can be distinguished (Fig. 10.14). The figure shows the viscosity trend versus shear rate of the guar gum (GG) hydrogel in a steady state flow test.[27] The two curves (forward and backward) are not overlapped, meaning that the material morphology has changed after the rheometer stress. The area within the two curves is called the hysteresis loop and it represents the energy loss required to obtain the sol–gel transition. The hysteresis loop also represents the time necessary for the rearrangement of the material. Since the energy loss, which is measured by the area of the hysteresis loop, changes linearly with the logarithm of the recovery time, the greater the area is, the longer it takes for the rearrangement of the material.[28] The presence of the hysteresis loop demonstrates the thixotropic nature of the GG hydrogel.[27]

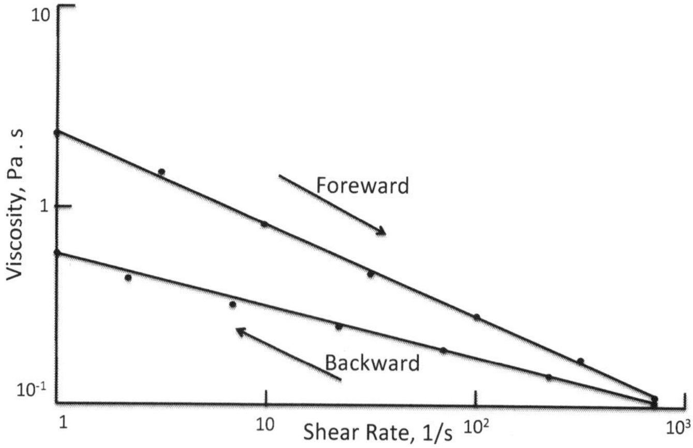


**Figure 10.14** Viscosity versus shear rate for the GG hydrogel in the steady state flow test. Reprinted from Ref. 27, Copyright 2008, with permission from Elsevier.

Furthermore, Fig. 10.15 shows the trend of strain percentage versus the shear stress of the GG hydrogel immediately after the rheometer test and of the same gel after 24 h of settling. This test represents another method to verify the thixotropic nature of a hydrogel. When the two loops are overlapped, it means that after the stress the hydrogel has regained its consistency and structure, at least partially, and that the material possesses a thixotropic nature.

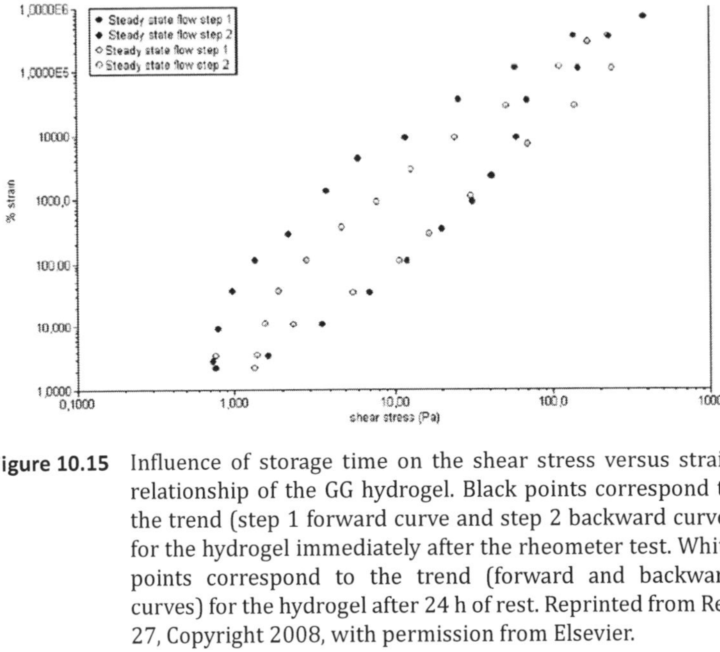


**Figure 10.15** Influence of storage time on the shear stress versus strain relationship of the GG hydrogel. Black points correspond to the trend (step 1 forward curve and step 2 backward curve) for the hydrogel immediately after the rheometer test. White points correspond to the trend (forward and backward curves) for the hydrogel after 24 h of rest. Reprinted from Ref. 27, Copyright 2008, with permission from Elsevier.

The elastic G′ and viscous G″ moduli give us further information on the characteristics of a hydrogel. The G′ and G″ moduli trend region versus oscillation stress is reported in Fig. 10.16. Within the linear viscoelastic region (LVR) the G′ values are always greater than the G″ values, indicating that the GG hydrogel shows more of an elastic character than a viscous one. G′ and G″ do not change within the range of the LVR suggesting that the structure of the material is not broken. As a further test, FTIR and NMR analysis of the same sample before and after the mechanical stress was performed. The analysis demonstrated that the two spectra are practically superimposed. Outside the LVR, for stress values between 30 and 801 Pa, the trend of G′ began to decrease, while G″ values remained more or less constant, meaning that the viscosity of the material increased with respect to its elasticity. At the crossover point (also called critical stress), at 801 Pa, the point at which the G′ and G″ curves intersected, the viscosity dropped drastically because the material collapsed and the transition from gel to sol occurred. After four days of settling the G′ and G″ regained the same values of the native hydrogel demonstrating the thixotropic nature of the GG hydrogel and the recovery of its original morphology. The value of the crosspoint represents the strength that permits the sol–gel transformation[17,27] (See Table 10.1).

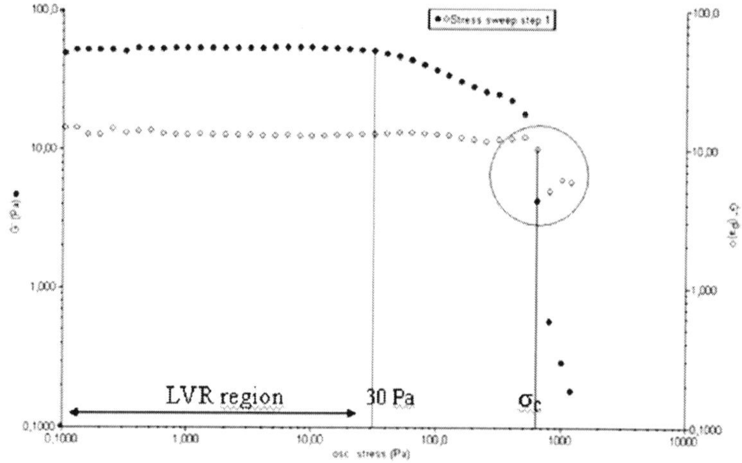


**Figure 10.16** Elastic and viscous moduli versus oscillation stress for the GG hydrogel in the stress sweep test. The LVR is between 0 and 30 Pa and the crossover point ($\sigma$) (the point at which the sol–gel transition occurs) is 801 Pa. Reprinted from Ref. 27, Copyright 2008, with permission from Elsevier.

**Table 10.1** Elastic (G′) and viscous (G″) moduli for different types of hydrogel (CMC, carboxymethylcellulose; CHT, chitosan; Hyal, Hyaluronan; and GG, guar gum. GG hydrogel in the stress sweep test. In the last column, the crosspoint value is reported)

| Hydrogel type | G′ (Pa) | G″ (Pa) | C.P. (Pa) |
|---|---|---|---|
| CMC | 550 ± 100 | 60 ± 30 | 200 ± 20 |
| CHT | 600 ± 120 | 44 ± 30 | 300 ± 20 |
| Hyal | 500 ± 50 | 60 ± 10 | 200 ± 30 |
| GG | 167 ± 40 | 20 ± 12 | 160 ± 37 |

## 10.2 Synthesis of Polysaccharide-Based Hydrogels

The long-standing use of polysaccharides as starting polymers for the preparation of hydrogels is due to the critical roles played by saccharide containing moieties in cell signaling pathways and to their versatile chemical structure, which allows several chemical modifications in the polymer backbone. The polysaccharides are based on the carbohydrate structures with the principal common feature, from the chemical reactivity point of view, of the presence of large numbers of functional groups. Thus, the reactions of the hydroxyl-group containing polymer are legion. While in the majority of synthetic polymers, the crosslinking reaction to obtain the hydrogels takes place starting from a monomer possibly using bulk, solution and reverse dispersion techniques, regarding polysaccharides, the crosslinking reaction occurs on the already formed polymer, and exploits precisely the functional groups that are located on the polymer chain. Thus, it becomes important to choose the crosslinking reaction so as not to damage the polysaccharide chain. Many reviews are present in literature on the subject and we limit ourselves only to the case of hydrogels of the polysaccharides that we use for biomedical applications. Hereafter, the structure of the repeating unit of CMC, hyaluronic acid (Hyal), CHT, alginic acid (AA), and GG are reported (Fig. 10.17).

The synthesis route used to obtain these hydrogels is shown in Figs. 10.18 and 10.19 and it can be traced to two different methods, one in an organic solvent, the other in water.[27,29–31]

The chemical route in organic solvent requires the polymer to be in the form of tetrabutylammonium salt for the solubilization in dymethylformammide[29] or in the acid form to be dissolved in dimethyl sulfoxide (DMSO). This implies a very long procedure including time-consuming steps.

Moreover, the hydrogel synthesized in organic solvent needs to be thoroughly washed in water for a long time to eliminate all traces of the organic solvent. However, the complete removal of the organic solvent is not easy as demonstrated by NMR spectra where some traces of organic solvent are revealed in the hydrogel at the swollen state even after an extensive washing. Therefore, even if the two types of hydrogels (synthesized in water or in organic solvent) present the

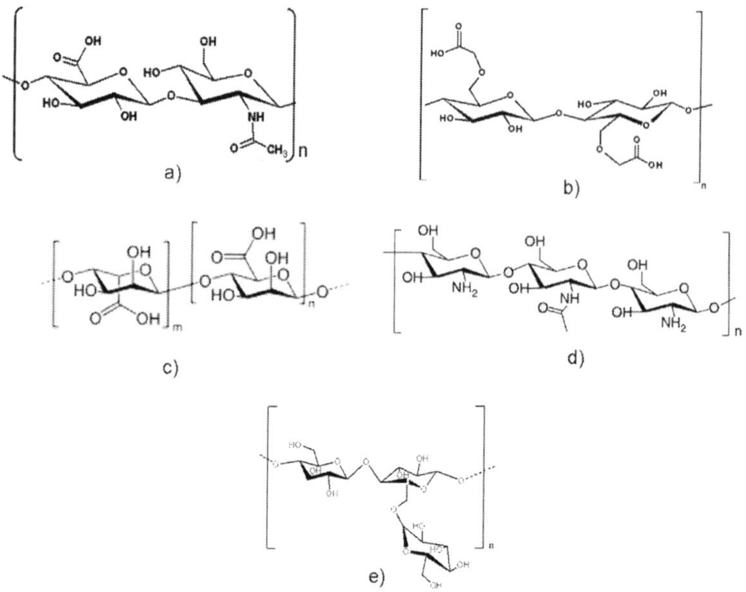


**Figure 10.17** Chemical structure of the repeating unit of (a) Hyal, (b) CMC, (c) AA, (d) CHT, and (e) GG.[31]

same chemical structure, the existence of traces of organic solvent may determine some different properties such as swelling degree or mechanical properties. The presence of the organic solvent in the water affects the conformation of the hydrogels and the hydrogen bonding between the hydrogels and water.[31] Sometimes, this may also induce a substitution of the interaction of the polymer with water with that of the organic solvent. This produces a change of polymer-polymer, water-solvent, and solvent-solvent interaction that determines a change in the swelling degree or in the mechanical properties of the hydrogel. It is known that the mixing ratio of $H_2O$–DMSO has a significant effect on the tensile strength and elongation at the break of the PVA hydrogel.[32]

### 10.2.1 Chemical Modifications on the Backbone of Polysaccharide Chains

Different chemical groups have been inserted on polysaccharide polymers with pretty simple chemical routes before being crosslinked

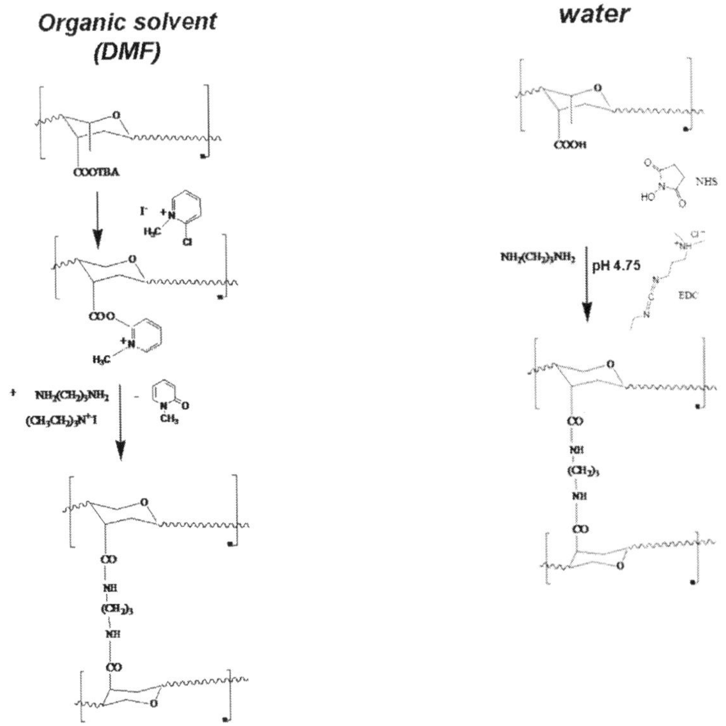


**Figure 10.18** Chemical route of the synthesis of polysaccharide-based hydrogels (A) in organic solvent, dimethylformammide (DMF) and (B) water.[31]

to form the hydrogel with the aim of tuning the properties of the hydrogel.

The introduction of sulfate groups on the Hyal polymer or other polysaccharides like CMC or AA makes them heparin-like molecules with anti-coagulant properties. Commercial heparin does not present a regular sequence of repeating units and the anticoagulant activity is given only by a peculiar sequence able to bind to the site of anti-thrombin III. This implies that only about one-third of commercial heparin preparation is really effective as an anti-coagulant. Sulfate groups have been inserted along Hyal polymer chains at different ratios ranging from one to four sulfate groups for disaccharide units. The procedure involves the introduction of sulfate groups on

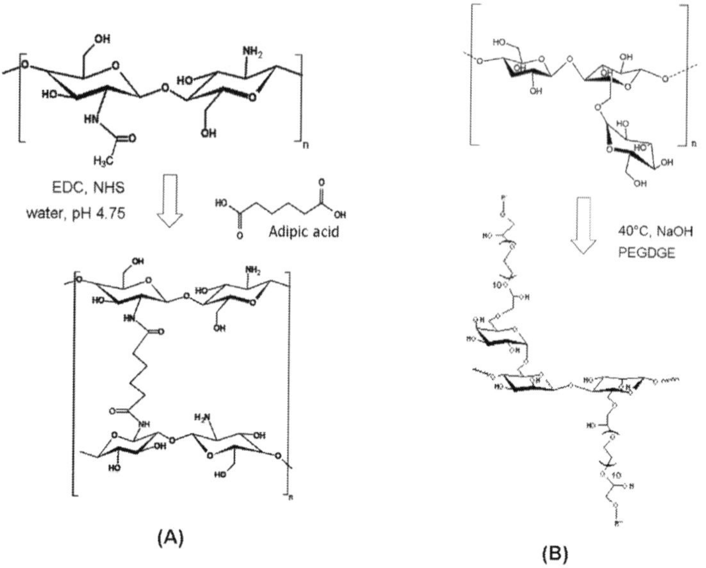


**Figure 10.19** Chemical route of the syntheses of (A) CHT by the addition of adipic acid as crosslinker and (B) GG by using polyethylene glycol diglycidyl ether (PEGDGE) under basic conditions. Both syntheses are carried out in water. Adapted from Refs. 27 and 30.

the alcoholic (where present) and oxidrile groups of the glycosidic rings of the polysaccharide. The reaction is carried out in organic solvent DMF or DMSO starting from the polymer in the form of tetrabutylammonium salt with the addition of a sulfation agent that is, $SO_3$-pyridine[33,34] (Fig. 10.20) (patent WO 95/25751). The sulfation degree was tuned by varying the molar ratio of the –OH groups of the polymer and sulfur trioxide pyridine (Table 10.2).

**Table 10.2** Number of $-SO_3^-$ groups contained in each disaccharide group of Hyal in function of the number of –OH groups available for sulfation

| Molar ratio –OH/$SO_3$ pyridine | Number of sulfate groups per repeating unit of Hyal |
|---|---|
| 1:1 | 1 |
| 1:2 | 2 |
| 1:5 | 2.5–3.0 |
| 1:8 | 3.5–4.0 |

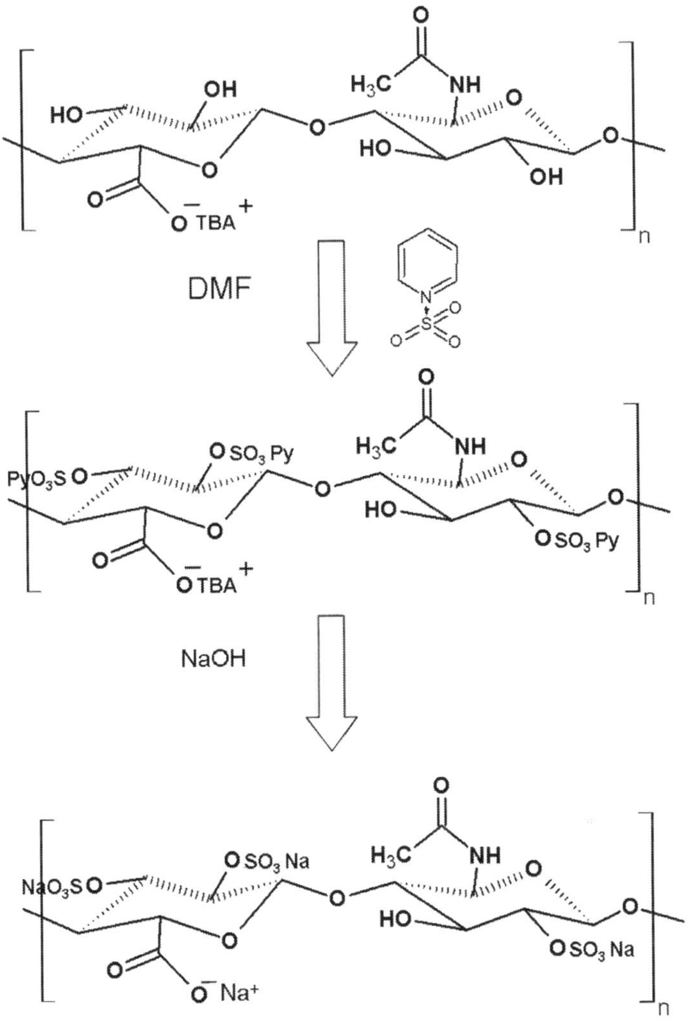


**Figure 10.20** Reaction scheme of the synthesis of sulfated Hyal. The same chemical route is employed for CMC and AA polymers.

The modified polymers (Hyals, CMCs, and AAs) especially those containing three and four sulfate groups for disaccharide units, demonstrated that they have satisfying anti-coagulant properties comparable to heparin and prevent platelet aggregation[34,35] (Table 10.3).

**Table 10.3** Thrombin time of the HyalS (with 3 $SO_3$ groups) solutions at different concentrations compared with the positive control that is, human plasma without the addition of the HyalS solution

| Samples | Thrombin time (TT sec ± S.E.) |
|---|---|
| Positive control | 12 ± 2 |
| HyalS solution (1 mg/ml) | >120 |
| HyalS solution ($2 \times 10^{-2}$ mg/ml) | 30 ± 2 |

These polymers were used for the preparation of new hydrogels, too. First attempts were made by inserting sulfate groups on a pre-synthetized hydrogel, but, since the reaction was carried out in organic solvent the hydrogel was shrunk and sulfate groups were found only on the hydrogel surface. On the contrary, starting from a sulfated polysaccharide, the synthesis of a new class of hydrogels bearing sulfate groups both inside and on the surface was successful.[36,37] The presence of additional negative charges, due to the sulfate groups, increases the swelling capability of all the hydrogels because of the increased hydrophilicity of the system after sulfation as well as the insertion of further negative charges which increase the repulsion among the polymer chains. Sulfated hydrogels showed anti-coagulant properties just as the linear starting polymers. No platelet adhesion was observed on sulfated Hyal hydrogel even under stressed conditions by scanning electron microscopy (SEM)[38] Sulfated hydrogels can also be effectively bound to the surface of catheters by a photoimmobilization reaction involving the free carboxylic acid groups once derivatized with a photosensitive molecule.

The total esterification of carboxylic acid groups of Hyal with benzyl alcohol allows for the obtainment of a non-crosslinked polymer, which is insoluble in aqueous solution (US Patents 4 851 521, 4 965 353). This modified Hyal, called Hyaff-11, spontaneously hydrolyses with time, releasing Hyal and benzylic alcohol. The appearance is a pad made of polymer fibers with an average diameter of 40 μm on which it is possible to culture chondrocytes or other types of cells. Moreover, a sulfated derivative of Hyaff with a degree of esterification equal to 50% has been modified by the insertion of sulfate groups on the remaining free carboxylic acid (HyaffS) (patent WO 95/25751). Hyaff-11 and also its sulfated derivative demonstrated to be excellent support for *in vitro* chondrocyte cultures generating a functional cartilage able to express specific genes for collagen type II and aggrecan.[39]

The possibility of introducing amide groups along the polymer chains of CMC or AA demonstrated to be an effective strategy to render these polysaccharides chemically similar to Hyal, but with a longer life in the implant site. An amidated derivative of CMC and AA can be used instead of a Hyal polymer to treat some pathological degenerative processes like osteoarthritis. Today, the injection of a Hyal solution still represents a possible therapy for the treatment of osteoarthritis, but several intra-articular injections of Hyal are usually required in patients because of the fast degradation of the polymer, which is caused by several enzymes.[40] An ideal substitute is a polymer with the same physicochemical characteristics of Hyal but with a slower degradability. The amide group can be easily introduced on CMC or AA polymers by the formation of an amide bond between the free carboxylic acid groups and the primary amine of methylamine with the help of a carbodiimide (CMCA, AAA)[41–43] (Fig. 10.21).

The reaction can be carried out in an organic solvent or in water. The degree of amidation defined as the number of carboxylic groups bound to the methylamine can be tuned and determined by NMR analysis and potentiometric titration. The remaining free carboxylic groups can be crosslinked to form an amidated hydrogel starting from CMCA and the AAA polymer. The presence of amide groups affects the mechanical properties and the water uptake of the hydrogel compared to the same hydrogel without the amide group. Basically, the storage modulus of CMCA and AAA hydrogels are lower than those of CMC and AA, respectively; and consequently, the water uptake at their maximum swelling degree is higher.[42,44]

Among the different strategies adopted to turn polymeric biomaterials with low or no tendency to induce osteintegration into mineralizing ones, the introduction of phosphate groups has been reported as a successful approach because they act as nucleating centers since anions attract and sequester $Ca^{2+}$ ions. The presence of phosphate groups along the polymer backbone improves the polysaccharide performance in terms of osteintegration. For this reason, phosphate groups have been bound to the carboxylate groups of the polymers by the formation of an amide bond with 2-aminoethyl-phosphonic acid by exploiting the well-known EDC chemistry[45] (Fig. 10.22) (patent EP07425286).

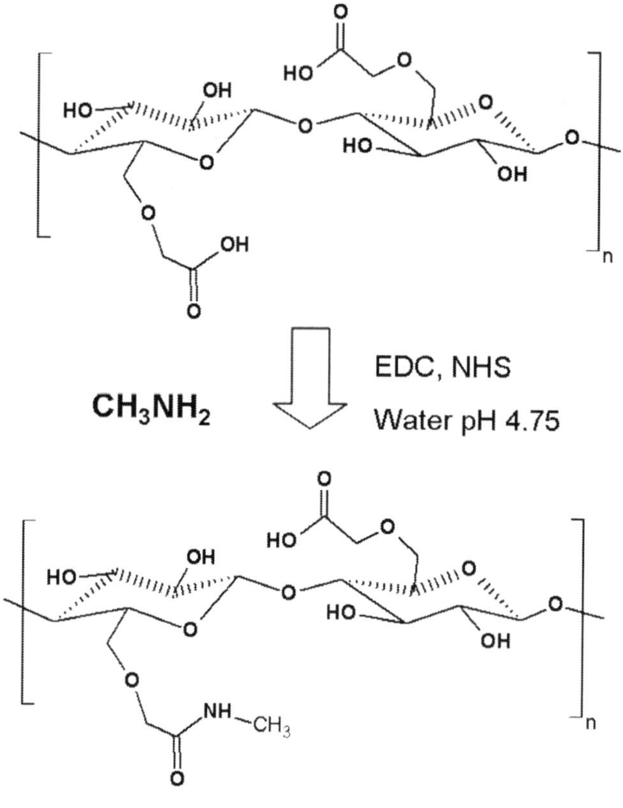


**Figure 10.21** Reaction scheme for the insertion of amide groups on the CMC polysaccharide.

The phosphonated polysaccharides were used both for the functionalization surfaces and the preparation of hydrogels. Biological tests (i.e., production of collagen type I, osteocalcin, and alkaline phosphatase) performed using osteoblast-like cells cultured on such materials emphasized the good performance of phosphonated CMC (both as a coating and as a 3D scaffold) toward bone-like cells (Fig. 10.23).[46,47]

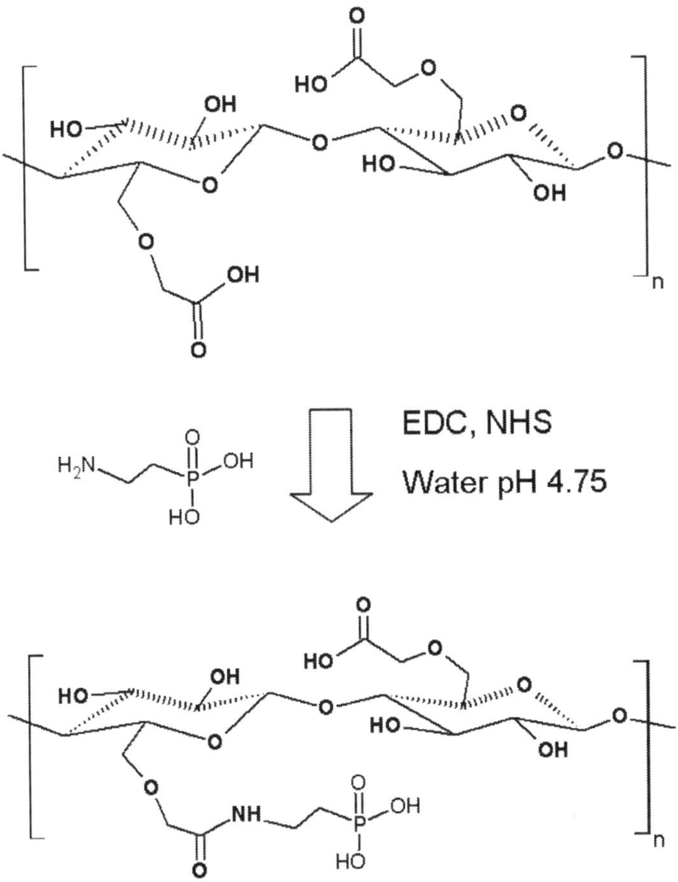


**Figure 10.22** Reaction scheme for the insertion of phosphate groups on the CMC polymer.

It is widely known that polysaccharides are capable of coordinating several metal ions assuming new biological properties according to the ions used for the complex formation. Among them, the complex with Cu(II) and Ag(I) are those showing the most promising application with their angiogenetic and antibacterial properties, respectively. Hyal and HyalS can easily form a stable complex with $Cu^{2+}$ thanks to the presence of free (not crosslinked) carboxylate groups.[48] Hyal and HyalS-Cu complexes improve endothelial cell functions relevant for vascularization such as adhesion, migration, and spreading. Moreover, in the case of HyalS-Cu complex, the

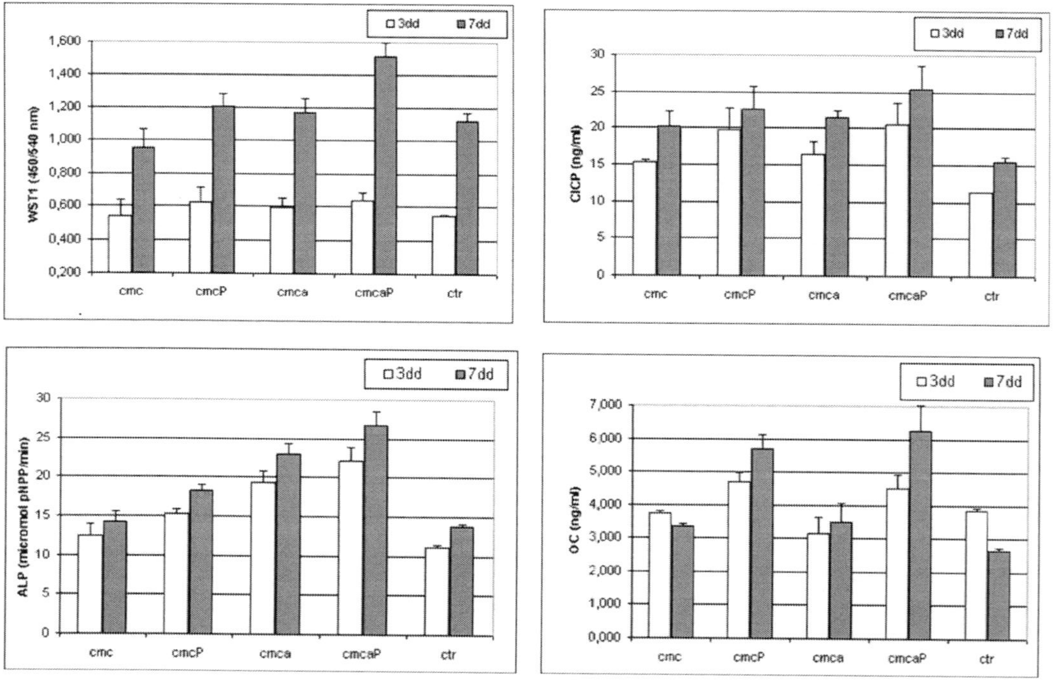


**Figure 10.23** Osteoblast proliferation, differentiation, and synthetic activity after 3 days (white column), 7 days (gray column) of culture with the different hydrogels, and the tissue culture polystyrene (ctr). WST-1: colorimetric assay for the quantification of cellular proliferation; CICP (C-propeptide of type I procollagen): biochemical indicator of collagen production and growth; ALP (alkaline phosphatase): assay for osteogenic differentiation; OC: osteocalcin.

presence of sulfate groups prevents the enzymatic degradation of the polymer and extends the angiogenetic effect of the complex on the surrounding tissue.[49,50] The treatment with silver as coating or added to different kinds of formulations is an efficient strategy to reduce bacteria adhesion. Hyal-based hydrogels combined with Cu(II) and Ag(I) ions were developed in order to design hydrogels able to satisfy peculiar requirements. Hyal hydrogels with a crosslinking degree of 50% are able to coordinate Cu (II) ions through the amidic and carboxylate groups and Ag(I) only through the carboxylate ones which remain free after the crosslinking reaction.[51]

## 10.3 Applications in Medicine

Polysaccharide-based hydrogels show peculiar properties among which the two most significant are

- capability to adsorb a huge amount of water once dipped in an aqueous solution or biological fluids, assuming a soft consistence similar to that of an extracellular matrix or soft tissue
- thixotropic behavior

These fundamental properties make them suitable materials for several biomedical applications such as scaffolds for tissue engineering, fillers and controlled drug release systems, etc.[42,52]

### 10.3.1 Scaffolds for Tissue Engineering: The Role of Thixotropy and Microporosity

Polysaccharide-based hydrogels have been tested as scaffolds for the growth of different kinds of cells: endothelial cells, fibroblasts, chondrocytes, pancreatic cells, etc. Generally, Hyal hydrogel demonstrates to be the most effective scaffold to promote the adhesion and proliferation of cells whereas, the AA hydrogel induces the lowest cell adhesion and the CMC one shows an intermediate behavior. One of the main critical points for the use of hydrogels as scaffolds for cell cultures deals with the seeding of cells on the scaffold. Given that the hydrogels are made up of water for more than 90% of their weight, cells hardly adhere on their surface and often slide along the wall of the gel down to the bottom of the Petri dish, reducing the number of cells inside the material. In order to overcome this limitation, a new strategy called the Amber effect was developed. It consisted of entrapping cells inside the hydrogel, exploiting its thixotropy, with the help of a syringe with a 19 gauge needle. This method demonstrated to be effective in keeping the cells inside the hydrogel matrix, reducing at the same time the number of cells that otherwise fall down and adhere to the Petri dish. Cells are well-distributed inside the hydrogel matrix as shown by cell counts made on different slide sections of a single piece of gel[53] (Fig. 10.24).

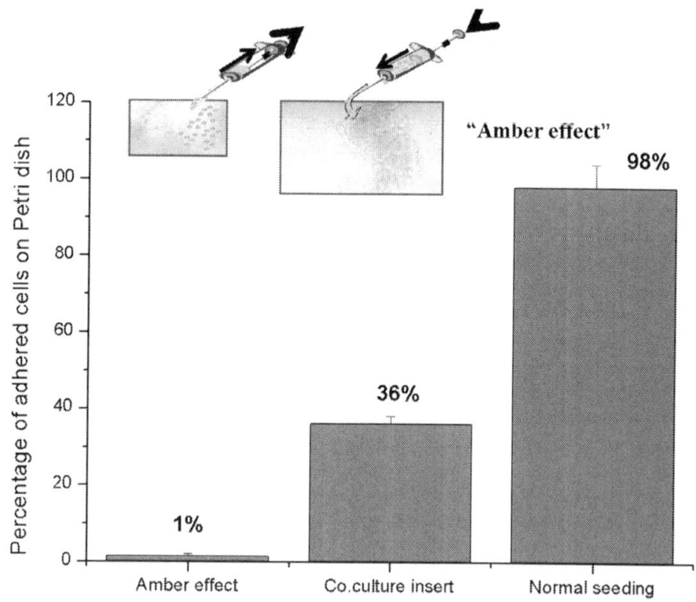


**Figure 10.24** The plot reports the number of adhered cells on the Petri dish seeded in three different ways: (1) via the "amber effect" which consists in squeezing the hydrogel together with cells inside the Petri well; (2) by using a co-culture insert on which cells are seeded; and (3) normal seeding of cells on the hydrogel surface. Adapted from Ref. 52.

Due to the thixotropic behavior of polysaccharide-based hydrogels, it is possible to obtain mixed hydrogels, namely interpenetrating hydrogels (IPHs) that can be easily injected *in situ*. In other words, presynthetized hydrogels can be squeezed through a syringe needle, mixed together and injected as a new single biomaterial containing the native hydrogels components. The physicochemical and mechanical properties of the IPHs are different from those of each native component. According to the type and the relative percentage of each individual component of the mix, it is possible to predict and tune the final properties of the IPH by taking into consideration the application of the final product. For example, by mixing hydrogels carrying opposite charges, that is, CMC (negatively charged) and CHT (positively charged), the electrostatic interaction between the opposite charges, evidenced by FTIR, induces an increase in the mechanical properties of the mixed compound with respect to those of the single components[31] (Fig. 10.25). On the contrary, by mixing hydrogels carrying the same charge that is, Hyaluronan (Hyal) and CMC, both negatively charged, repulsion occurs and the mechanical properties of the mix are drastically lower than those of each single component of the mixture. All IPHs systems demonstrated to be good scaffolds to promote and support cell survival, but the CMC-CHT IPH appeared to be the best one for cell growth. The advantage of these IPHs with respect to similar systems is related to the fact that no chemical reaction is necessary to obtain the mixture. Polysaccharide-based IPHs show promising applications for mini-invasive surgery for the fact that they can be *in situ* injected in relative percentages according to the desired properties.

An ideal cell scaffold ought to have a sufficient porosity to guarantee the exchange of nutrients with the surrounding medium and an adequate clearance of waste without losing the macroscopic mechanical properties of the material. The morphology of polysaccharide-based hydrogels itself consists of lamellae of different thicknesses and dimensions interconnected to each other. The method to obtain a microporous structure is based on stratifying the hydrogel on a filter with a defined porosity of 40, 70, or 100 μm. Once a layer of swollen hydrogel is spread on the filter, some drops of chloroform are added to prevent an excessive swelling of the hydrogel. The filter is place in a beaker containing a bicarbonate salt. On adding HCl solution, $CO_2$ bubbles form and pass through the filter and the gel allowing the formation of pores.[54] The presence of regular pores with dimensions matching the diameter of the filter mesh was observed under SEM (Fig. 10.26).

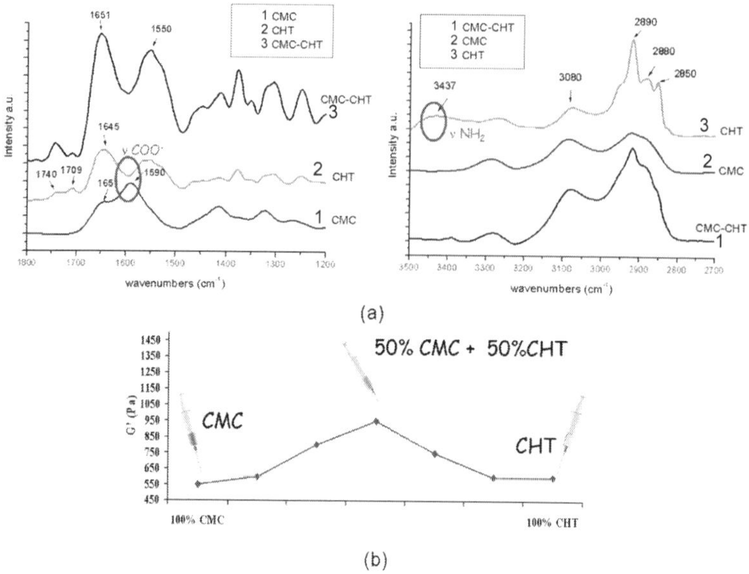


**Figure 10.25** (a) Infrared spectra of CMC-CHT IPH compared with the single CMC and CHT hydrogel. The disappearance of the peak ascribed to carboxylate groups of CMC and that of the peak relative to $NH_2$ of CHT evidenced the interaction between the two polymers. (b) Storage modulus G′ of the CMC-CHT mix prepared at different relative ratios of the two components. The maximum increase of G′ is observed when each component is present at 50%.[31]

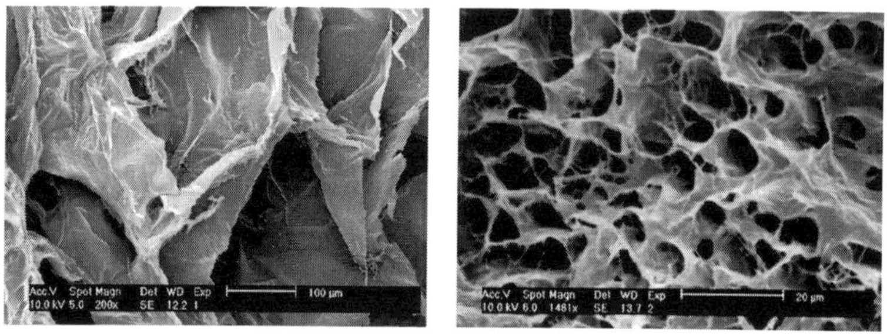

**Figure 10.26** SEM images showing the AA hydrogel before and after pore formation.

The presence of pores affects the swelling degree and the rheological properties of the hydrogels. Microporous hydrogels show a lower water uptake and increased mechanical properties compared to the same hydrogels without pores (Table 10.4).[55] This effect can be explained by considering that the presence of $CO_2$ induces the protonation of free carboxylate groups and as a consequence the reduction of the charge density and repulsion among the polymer chains. Moreover, infrared (IR) analysis highlighted the formation of hydrogen bonds responsible for a more compact structure of the hydrogel.[55] The presence of pores enhances the diffusion coefficient as ascertained by fluorescence recovery after a photobleaching (FRAP) test using labeled serum albumin.[54]

### 10.3.2 Drug Delivery

Polysaccharide-based hydrogels offer promising applications as delivery systems of drugs. Thanks to their capability to respond to pH, ionic strength, temperature, and in general to all variations of the surrounding environment by modifying their state that is, swollen or shrunken. As a consequence, the amount of the released drug can be controlled according to the type of hydrogel in relation to the characteristics of the medium in which it is dipped. The drug can be loaded inside the hydrogel matrix simply by dipping the dried hydrogel in a solution containing a defined amount of the drug. By using this method, Hyal and CMC hydrogels both in the native state and as microporous matrices have been loaded with ibuprofen-lysine, an anti-inflammatory drug usually used for osteoarthritis therapy.[55] Generally, as reported in literature, the presence of pores inside a hydrogel speeds up the release of the drug compared to the same hydrogels without pores. On the contrary, by increasing the pore dimensions of a CMC-based hydrogel, the release of ibuprofen-lysine was slower and reached a plateau level after five days (Fig. 10.27).[54] The presence of pores inside a polysaccharide-based hydrogel slows down the drug release. The motivation of this behavior can be due to the fact that the presence of pores decreases the swelling degree of the hydrogels, working against the delivery of the drug.

A peculiar use of polysaccharide-based hydrogels as delivery systems relies on fixing the hydrogel to an electrode bar of an electroporator.[54] Electrochemotherapy is a technique used for tumor ablation, which provides the delivery of a drug that is not able to permeate the cell membrane. It is based on the local application of short and intense electric pulses, which open the membrane pore and let the drug to enter and kill the tumor cell (Fig. 10.28). In our

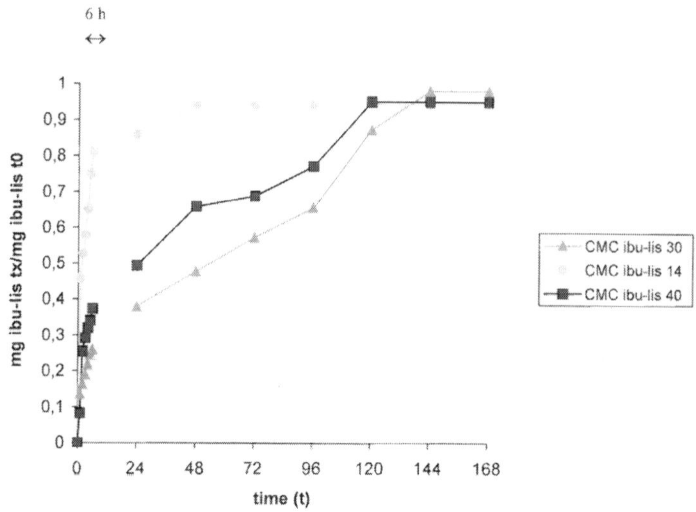


**Figure 10.27** Release kinetics of ibuprofen-lysine from the CMC hydrogel with a crosslinking degree of 50% having different pore sizes 30, 40, and 14 μm.

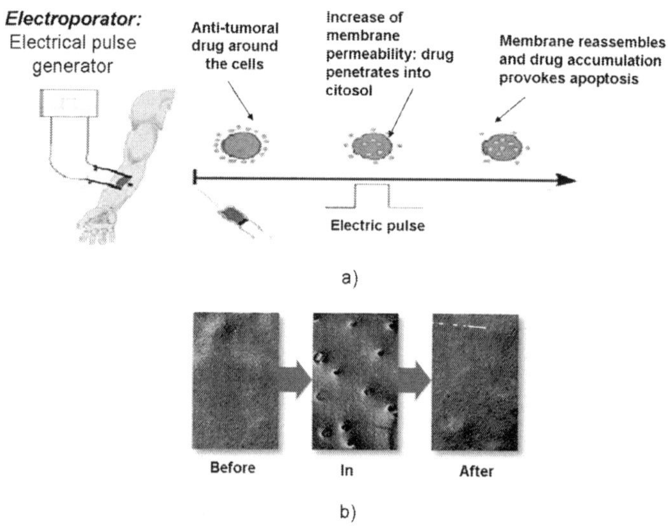


**Figure 10.28** (a) Scheme of local application of electrochemotherapy. (b) Transmission electron microscope (TEM) images showing cell membrane before and after the application of an electric pulse.

case, a neutral GG hydrogel loaded with a positively charged drug, bleomicin, an antibiotic inducing damages to the DNA of tumor cells, was attached to the positive electrode of the electroporator and the release of the drug was analyzed following the application of an electric field. The presence of the hydrogel allows for an effective release of bleomicin, which is able to selectively destroy cancer cells[55] (Fig. 10.29).

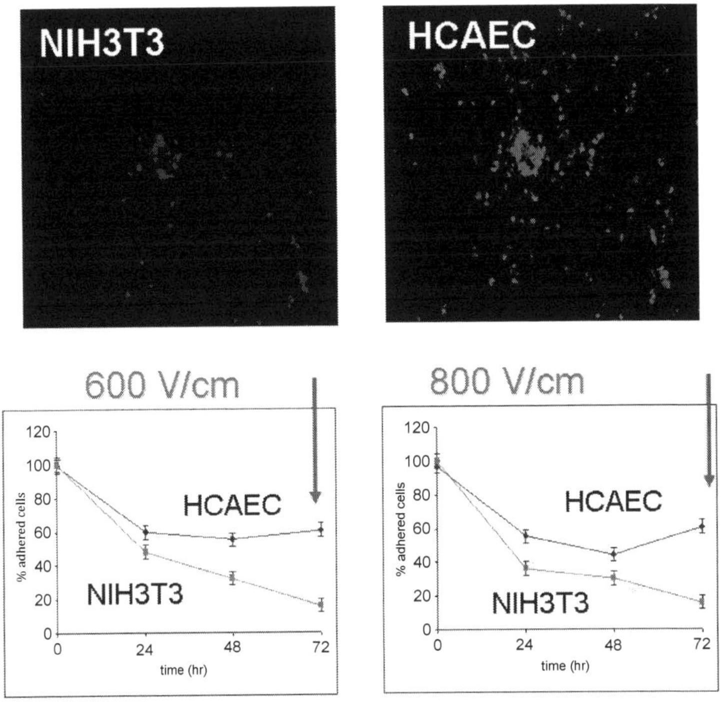


**Figure 10.29** Co-culture of a cancer fibroblast (line NIH-3T3) and endothelial cells (HCAEC line) treated with 30 μg of bleomicin. The applied electrical field at 600 V/cm and 800 V/cm was efficient enough to allow a sufficient permeation of bleomicin able to kill cancer cells without causing a large decrease in the endothelial cells.

## 10.3.3 Tissue Engineering Applications

Osteoarthritis is a degenerative joint disease clinically characterized by joint pain and instability. Current treatment is aimed at maintaining the mobility of the joint by physiotherapy and drugs to relieve the pain. A Hyal polymer is often used as local treatment by the injection of solutions at different concentrations or as systemic therapy. It provides relief to the patient with its lubricating and cushioning actions, showing some anti-inflammatory effects. However, the Hyal disappearance is very rapid, becomes quickly degraded by enzymatic or hydrolytic reactions. The use of a hydrogel matrix insoluble in water instead of using a polymer solution offers big advantages for the local treatment. The fact that such polysaccharide-based hydrogels and their thixotropic properties can be easily injected *in situ* makes the matter easier. Moreover, the Hyal hydrogel can be sterilized with ethylene oxide without being chemically or physically altered. Animals treated with the injection of the Hyal hydrogel revealed a repairing mechanism that lead to a partial healing of the damaged cartilage. At the end of the experimental time (50 days), the growth of a hyaline-like cartilage was observed in the knee joint of rabbits[52] (Fig. 10.30).

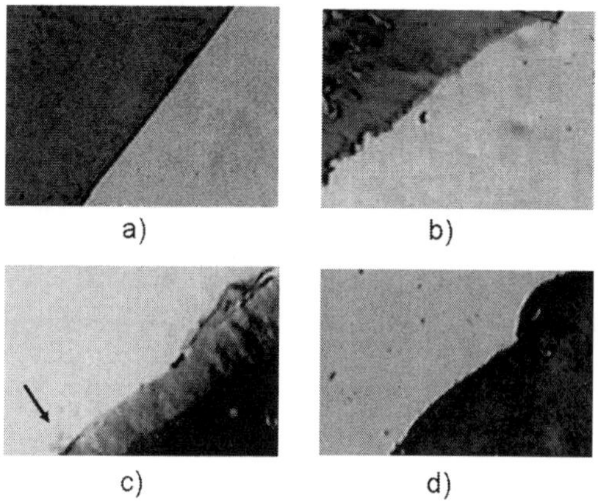


**Figure 10.30** Effect of the hyal hydrogel on damaged cartilage of rabbit: (a) normal cartilage tissue, (b) damaged cartilage tissue, (c, d) damaged cartilage filled with Hyal hydrogel with a crosslinking degree of 50% after 50 days. The arrow marked chondrocytes piling up on each other. The regenerated tissue is very similar to the healthy one.

This new tissue filled the bone defect presenting a smooth regular surface. Generally, osteoarthritis promotes a reduction of bone density inducing pain for the patient. Instead, the presence of a Hyal hydrogel inside the bone defect has proved to show only a minor reduction of bone density relieving the pain for the patient (Table 10.4).[56]

**Table 10.4** Mean percentage decrease of bone mineral density (BMD) of treated versus not treated side at each examined areas (femurs and tibiae) of rabbits

| | 30 days | | 50 days | |
|---|---|---|---|---|
| | Hyal | Hyal + Ibu | Hyal | Hyal + Ibu |
| F1 (whole femur) | −12.8% | −10.2% | −24.8% | −20.9% |
| T1 (whole tibia) | −13.6% | −14.8% | −29.2% | −20.7% |
| F2 (distal femoral epiphysis) | −17.8% | −17.4% | −34.5% | −27.7% |
| T2 (proximal femoral epiphysis) | −17.9% | −12.8% | −31.9% | −26.0% |

*Note:* The treatment consists of injections of Hyal hydrogel (Hyal) and Hyal hydrogel loaded with Ibuprofen-lysisne (Hyal+Ibu) inside induced bone defects at femoral medial condyle site of rabbits. After 30 days, no significant differences were found between the Hyal-Ibu treated group and the Hyal treated group in terms of reduction of the treated site, even if a clear trend in the reduction of the percentage of the defect was observed. At 50 days, a significant difference was found between the Hyal-Ibu group and the Hyal group in terms of reduction of the treated tibial site.

A critical point for tissue repair is the formation of a vessel network that helps to supply nutrients to the new tissue. Angiogenesis is a morphogenetic process, which leads to the development of new blood vessels, which requires the migration of endothelial cells, as well as their proliferation and differentiation into tube-like structures. Copper is known as an element, able to induce angiogenesisis *in vivo*. In order to test the real angiogenetic performance of hydrogel-copper(II) complexes, *in vivo* tests were performed by implanting sterilized Hyal-Cu(II) hydrogels in the subscapular panniculus adiposus of male rats and left for 21 days. The explanted hydrogel samples appeared to be highly hydrated with a remarkable neovascularization in the surrounding tissue. This effect was not found in the rats treated with Hyal hydrogel without the Cu(II) complex. Moreover, Hyal-Cu(II) complexes were mixed with freeze-dried non-demineralized bone and inserted in subcutaneous pockets of male rats in order to test their capability to promote the growth of new vessels.[58] The hydrogel-Cu(II) complex demonstrated to be effective in promoting the formation of new vessels. This method suggests the advantage offered by the combined use of a well-known biocompatible material (Hyal hydrogel) and Cu(II) ions, whose role in stimulating tissue vascularization is well known (Fig. 10.31).

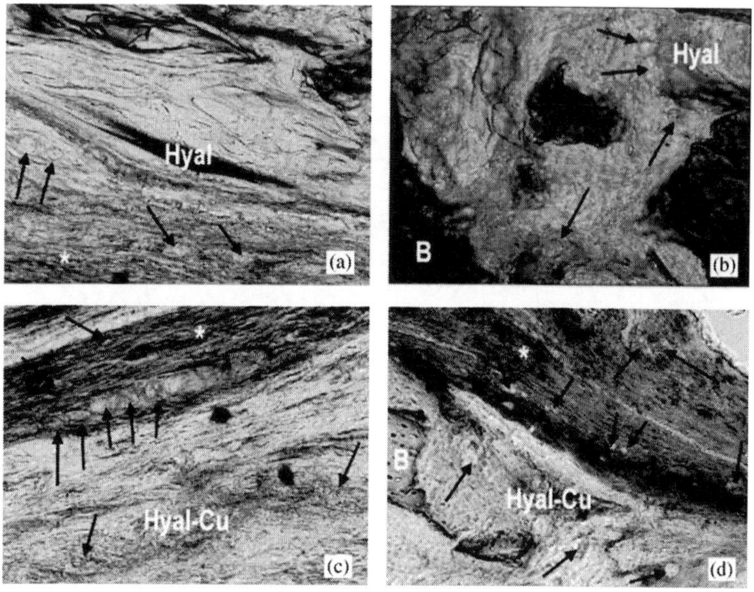


**Figure 10.31** Light microscopy of subcutaneous material implants at 4 weeks (magnification 20×): (a) Hyal-50%; (b) Hyal-50% plus freeze-dried bone; (c) Hyal-50%-$Cu^{2+}$; (d) Hyal-50%-$Cu^{2+}$ plus freeze-dried bone. Network of newly formed capillaries (arrows) in the connective matrix (*); and (B) bone. Reprinted from Ref. 58, Copyright 2005, with permission from Elsevier.

Generally, bacteria adhere to biomedical device surfaces through several different interactions. The bacteria adhesion is followed by a biofilm formation often resistant to antibiotic treatments and host defense mechanisms. This biofilm acts as a continuous focus of infections, which may cause implant rejection. Silver is a well-known antimicrobial agent added to many different commercial formulations. The Hyal-Ag(I) hydrogel complex shows antibacterial properties by preventing the adhesion and proliferation of bacteria (Staphylococcus epidermis strain) even at low concentrations of Ag(I) ($4.2 \times 10^{-3}$ mg Ag/mg dry Hyal) which is, at higher concentrations, toxic for cells.[51] Such hydrogel complexes can be stuck to different surfaces like polyurethane or other plastics, reducing the adhesion of bacteria.

## References


1. Hoffman, A. S. (2001) Hydrogels for biomedical applications, *Ann. N.Y. Acad. Sci.*, **944**, 62–73.
2. Peppas, N. A., Slaughter, B. V., Kanzelberger, M. A. (2012) Hydrogels. In: Matyjaszewski, K. and Möller, M. (Eds.), *Polymer Science: A Comprehensive Reference* (pp. 385–395). Amsterdam: Elsevier.
3. Peppas, N. A., Barr-Howell, B. D (1986) Characterization of the crosslinked structure of hydrogels. In: Peppas, N. A. (Ed.) *Hydrogels in Medicine and Pharmacy* (pp. 27–56). Boca Raton, FL: CRC Press.
4. Schacht, E. H. (2004) Polymer chemistry and hydrogel systems, *J. Phys. Conf. Ser.*, **3**, 22–28.
5. Guenet, J. M. (1992) *Thermoreversible gelation of polymers and biopolymers*. New York, NY: Academic Press.
6. Percec, V., Bera, T. K., Butera, R. J. (2002) A new strategy for the preparation of supramolecular neutral hydrogels, *Biomacromol.*, **3**, 272–279.
7. Kuo, C. K. and Ma, P. X. (2001) Ionically crosslinked alginate hydrogels as scaffolds for tissue engineering: Part I. Structure, gelation rate and mechanical properties, *Biomaterials*, **22**, 511–521.
8. Trabbic-Carlson, K., Setton, L. A., Chilkoti, A. (2003) Swelling and mechanical behaviors of chemically cross-linked hydrogels of elastin-like polypeptides, *Biomacromol.*, **4**, 572–580.
9. Philippova, O. E., Khokhlov, A. R. (2012) Polymers gels. In: Philippova, O. E., Khokhlov, A. R. (Eds.). *Polymer Science: A Comprehensive Reference* (Vol. 1, pp. 339–366). Amsterdam: Elsevier.
10. Liechty, W. B., Caldorera-Moore, M., Phillips, M. A., Schoener, C., Peppas, N. A. (2011) Advanced molecular design of biopolymers for transmucosal and intracellular delivery of chemotherapeutic agents and biological therapeutics, *J. Contr. Rel.*, **155**, 119–127.
11. Gil, E. S., Hudson, S. M. (2004) Stimuli-responsive polymers and their bioconjugates, *Prog. Polym. Sci.*, **29**, 1173–1222.
12. Ratner, B. D., Hoffman, A. S. (1976) Synthetic hydrogels for medical and related applications. In: Andrade, J. D. (Ed.). *Hydrogels for Medical and Related Applications, ACS Symposium Series, No. 31* (pp. 1–36). Washington, DC: American Chemical Society.
13. Omidian, H., Park, K. (2010) Introduction to hydrogels. In: Ottenbrite, R. M. (Ed.). *Biomedical Applications of Hydrogels Handbook* (pp. 1–16). London, UK: Springer.

14. Pasqui, D., De Cagna, M., Barbucci, R. (2012) Polysaccharide-based hydrogels: The key role of water in affecting mechanical properties, *Polymers,* **4**, 1517–1534.

15. Ilić-Stojanović Snežana, S., Nikoli, L., Nikolić, V., Stanković, M., Stamenković, J., Mladenović-Ranisavljević, I., Petrović, S. (2012) Influence of monomer and crosslinker molar ratio on the swelling behaviour of thermosensitive hydrogels, *Chem. Ind. Chem. Eng. Q.*, **18**, 1–9.

16. Li, H., (2009), *Smart Hydrogel Modelling*, Springer-Verlag, Berlin.

17. Barbucci, R., Giardino, R., De Cagna, M., Golini, L.; Pasqui, D. (2010) Inter-penetrating hydrogels (IPHs) as a new class of injectable polysaccharide hydrogels with thixotropic nature and interesting mechanical and biological properties, *Soft Matter,* **6**, 3524–3532.

18. Kabiri, B., Omidian, H., Hashemi, S.A., Zohuriaan-Mehr, M.J. (2003) Synthesis of fast-swelling superabsorbent hydrogels: effect of crosslinker type and concentration on porosity and absorption rate, *Eur. Pol. J.,* **39**, 1341–1348.

19. Okano, T. (ed.) (1998) *Biorelated polymers and gels: Controlled release and applications in biomedical engineering (polymers, interfaces and biomaterials)*, San Diego, CA: Academic Press.

20. Jones, D. S. (1999) Dynamic mechanical analysis of polymeric systems of pharmaceutical and biomedical significance, *Int. J. Pharm.,* **179**, 167–178.

21. Anseth, K. S., Bowman, C. N., Brannon-Peppas, L. (1996) Mechanical properties of hydrogels and their experimental determination, *Biomaterials*, **17**, 1647–1657.

22. Popa, L., Chica, M. V., Moisescu, S. (2003) Thixotropy of Na CMC hydrogels with indomethacin, *Ovidius Univ. Ann. Med. Sci. Pharm.*, 1, 222–226.

23. Frendlich, H., Juliiusberg, F. (1935) Thixotropy, influenced by the orientation of anisometric particles in sols and suspensions, *Trans. Faraday Soc.,* **31**, 920–921.

24. Barbucci, R., Leone, G., Lamponi, S. (2006) Thixotrophy property of hydrogels to evaluate the cell growing on the inside of the material bulk (Amber effect), *J. Biomed. Mater. Res. B Appl. Biomater.*, **76**, 33–40.

25. Mercuri, D., Leone, G., Barbucci, R., Favaloro, R., Signori, F., Bronco, S., Ciardelli, F. (2008) An artificial disc: Chemical and biomechanical analysis, *Macromol. Symp.*, **266**, 74–80.

26. Leone, G., Fini, M., Torricelli, P., Giardino, R., Barbucci, R. (2008) An amidated carboxymethylcellulose hydrogel for cartilage regeneration, *J. Mater. Sci. Mater. Med.*, **19**, 2873–2880.

27. Barbucci, R., Pasqui, D., Favaloro, R., Panariello, G. (2008) A thixotropic hydrogel from chemically cross-linked guar gum: Synthesis, characterization and rheological behavior, *Carbohydr. Res.*, **343**, 3058–3065.

28. Ker, R. F. (1996) The time-dependent mechanical properties of the human heel pad in the context of locomotion, *J. Exp. Biol.*, **199**, 1501–1508.

29. Barbucci, R., Magnani, A., Consumi, M. (2000) Swelling behavior of carboxymethylcellulose hydrogels in relation to cross-linking, pH, and charge density, *Macromolecules*, **33**, 7475–7480.

30. Pasqui, D., De Cagna, M., Barbucci, R. (2012) Polysaccharide-based hydrogels: The key role of water in affecting mechanical properties, *Polymers*, **4**, 1517–1534.

31. Barbucci, R., Giardino, R., De Cagna, M., Golini, L., Pasqui, D. (2010) Inter-penetrating hydrogels (IPHs) as a new class of injectable polysaccharide hydrogels with thixotropic nature and interesting mechanical and biological properties, *Soft Matter*, **6**, 3524–3532.

32. Li, H., Zhang, W., Xu, W., Zhang, X. (2000) Hydrogen bonding governs the elastic properties of poly(vinyl alcohol) in water: Single-molecular force spectroscopic studies on PVA, *Macromolecules*, **33**, 465–469.

33. Barbucci, R., Magnani, A., Casolaro, M., Marchettini, N, Rossi, C., Bosco, C. (1995) Modification of Hyaluronic acid by sulphate groups insertion to obtain a heparin-like molecule. Part I. Characterisation and behaviour in aqueous solution towards H+ and Cu2+ ions, *Gazz. Chim Ital.*, **125**, 169–180.

34. Magnani, A., Albanese, A, Lamponi, S., Barbucci, R. (1996) Blood-interaction performance of differently sulphated hyaluronic acids, *Thromb. Res.*, **81**, 383–385.

35. Barbucci, R., Lamponi, S., Magnani, A., Renier, D. (1998) The influence of molecular weight on the biological activity of heparin like sulphated hyaluronic acids, *Biomaterials*, **19**, 801–806.

36. Barbucci, R., Leone, G., Chiumiento, A., Di Cocco, M. E., D'Orazio, G., Gianferri, R., Delfini, M. (2006) Low- and high-resolution nuclear magnetic resonance (NMR) characterisation of hyaluronan-based native and sulfated hydrogels, *Carbohydr. Res.*, **341**, 1848–1858.

37. Barbucci, R., Rappuoli, R., Borzacchiello, A., Ambrosio, L. (2000) Synthesis, chemical and rheological characterization of new hyaluronic acid-based hydrogels, *J. Biomater. Sci. Polym. Ed.*, **11**, 383–399.

38. Barbucci, R, Magnani, A., Lamponi, S, Rappuoli, R., Consumi, M. (2000) Immobilisation of sulphated hyaluronan for improved biocompatibility, *Inorg. Biochem.*, **79**, 119–125.

39. Girotto, D., Urbani, S., Brun, P., Renier, D., Barbucci, R., Abatangelo, G. (2003) Tissue-specific gene expression in chondrocytes grown on three-dimensional hyaluronic acid scaffolds, *Biomaterials*, **24**, 3265–3275.

40. Zhong, S. P., Capoccia, D., Doherty, P. J., Williams, R. L. Benedetti, L., Williams, D. F. (1994) Biodegradation of hyaluronic acid derivatives by hyaluronidase, *Biomaterials*, **15**, 359–365.

41. Barbucci, R., Leone, G., Monici, M., Pantalone, D., Fini, M., Giardino, R. (2005) The effect of amidic moieties on polysaccharides: Evaluation of the physico-chemical and biological properties of amidic carboxymethylcellulose (CMCA) in the form of linear polymer and hydrogel, *J. Mater. Chem.*, **22**, 2234–2241.

42. Leone, G., Delfini, M., Di Cocco, M. E., Borioni, A., Barbucci, R. (2008) The applicability of an amidated polysaccharide hydrogel as a cartilage substitute: Structural and rheological characterization, *Carbohydr. Res.*, **343**, 317–327.

43. Leone, G., Torricelli, P., Chiumiento, A., Facchini, A., Barbucci, R. (2008) Amidic alginate hydrogel for nucleus pulposus replacement, *J. Biomed. Mater. Res.*, **84A**, 391–401.

44. Leone, G., Fini, M., Torricelli, P., Giardino, R., Barbucci, R. (2008) An amidated carboxymethylcellulose hydrogel for cartilage regeneration, *J. Mater. Sci. Mater. Med.*, **19**, 2873–2880.

45. Barbucci, R., Arturoni, A., Panariello, G., Di Canio, C., Lamponi, S. (2010) A new amido phosphonate derivative of carboxymethylcellulose with an osteogenic activity and which is capable of interacting with any Ti surface, *J. Biomed. Mater. Res.*, **95A**, 58–67.

46. Pasqui, D., Danieli, E., Leone, G., Torricelli, P., Barbucci, R.(2008) *Macromol. Symp.* (Biohydrogel special issue), **266**, 23–29.

47. Pasqui, D, Rossi, A., Di Cintio, F., Barbucci, R. (2007) Functionalized titanium oxide surfaces with phosphated carboxymethyl cellulose: Characterization and bonelike cell behavior, *Biomacromolecules*, **8**, 3965–3972.

48. Barbucci, R., Magnani, A., Metal-ion complexes in the angiogenetic effect (2000) *Macromol. Symp.*, **156**, 239–251.

49. Barbucci, R., Magnani, A., Lamponi, S., Mitola, S., Ziche, M., Morbidelli, L., Bussolino, F. (2001) Cu(II) and Zn(II) complexes with hyaluronic acid and its sulphated derivative: Effect on the motility of vascular endothelial cells, *J. Inorg. Biochem.*, **81**, 229–237.

50. Barbucci, R., Lamponi, S., Magnani, A., Peluso, G., Petillo, O. (2001) Metal complexes with linear and crosslinked polysaccharides as mediators of angiogenesis, *Polym. Adv. Techn.*, **12**, 1–8.

51. Barbucci, R., Leone, G., Magnani, A., Montanaro, L., Arciola, C. R., Peluso, G., Petillo, O. (2002) $Cu^{2+}$- and $Ag^{+}$-complexes with a hyaluronane-based hydrogel, *J. Mater. Chem.*, **12**, 3084–3092.

52. Barbucci, R., Leone, G., Lamponi, S. (2006) Thixotrophy property of hydrogels to evaluate the cell growing on the inside of the material bulk (Amber effect), *J. Biomed. Mater. Res. B Appl. Biomater.*, **76B**, 33–40.

53. Leone, G., Barbucci, R., Borzacchiello, A., Ambrosio, L., Netti, P. A., Migliaresi, C. (2004) Preparation and physico-chemical characterisation of microporous polysaccharidic hydrogels, *J. Mater. Sci. Mater. Med.*, **15**, 463–467.

54. Barbucci, R., Leone, G., Vecchiullo, A. (2004) Novel carboxymethylcellulose-based microporous hydrogels suitable for drug delivery, *J. Biomater. Sci. Polym. Ed.*, **15**, 607–609.

55. Barbucci, R., Lamponi, S., Borzacchiello, A., Ambrosio, L., Fini, M., Torricelli, P., Giardino, R. (2002) Hyaluronic acid hydrogel in the treatment of osteoarthritis, *Biomaterials*, **23**, 4503–4513.

56. Barbucci, R., Fini, M., Giavaresi, G., Torricelli, P., Giardino, R., Lamponi, S., Leone, G. (2005) Hyaluronic acid hydrogel added with ibuprofen-lysine for the local treatment of chondral lesions in the knee: *In vitro* and *in vivo* investigations, *J. Biomed. Mater. Res. B Appl. Biomater.*, **75**, 42–48.

57. Fantozzi, F., Arturoni, E., Barbucci, R. (2010) The effects of the electric fields on hydrogels to achieve antitumoral drug release, *Bioelectrochemistry*, **78**, 191–195.

58. Giavaresi, G., Torricelli, P., Fornasari, P. M., Giardini, R., Barbucci, R., Leone, G. (2005) Blood vessel formation after soft-tissue implantation of hyaluronan-based hydrogel supplemented with copper ions, *Biomaterials*, **26**, 3001–3008.

# Protein-Based Hydrogels

Proteins constitute another source of hydrogels for tissue engineering scaffolds. Gels of proteins, such as collagen, fibroin, or fibrin have been widely proposed and also used to the tissue engineering of cartilage, intervertebral disks, skin, bone, and other tissues. Proteins network gels generally occur via hydrophobic interactions, hydrogen bonds, or covalent bonds such as disulfide bonds. Gelation can be induced by heat, pH lowering that promotes electrostatic interactions, shear stresses, pressure, or ion addition.

## 1. Collagen

Collagen is the main component of the mammalian tissues, about 25% of their weight, where it exploits mainly mechanical functions while being a template for the ECM and tissue organization. It is mainly produced by fibroblasts. Collagen is not only abundant in mammals, but is also present in many organisms and making up almost the total body of marine organisms, such as jelly fishes, or sea cucumbers, or others. Native collagen has a fibrous structure that is dominated by triple-helices conformation and stabilized by intramolecular bonds. As such the molecule is not water-soluble. After denaturation and hydrolysis, the quaternary and tertiary structure is lost and collagen, now named gelatin, becomes water-soluble. Slightly crosslinked collagens (e.g., immature collagen that is present in some organisms) can also be solubilized in acidic environments.

Gelatin has an incredibly large number of uses in different fields. In biomedicine, it is used as coating of cell culture plates, and also in diagnostics, in immunohistoichemistry.

Most of the gelatin originates from mammals, mainly pig skin. In recent years, gelatin from fishes has attracted increasing attention and jelly fishes, sea cucumber, or fish scale-derived gelatins have been investigated, also for tissue engineering applications.[1,2]

As well discussed,[3] among the 29 different types of collagen that are present in mammals, collagen type I is the most suitable for implantation since only few persons possess specific immunity against it, that can be easily revealed by serologic allergenic tests. Collagen is degraded by collagenase, an enzyme present in our body, which breaks the peptide bonds in collagen.

Collagen gels can be easily prepared by heating collagen type I solutions. Heating generates assembling of the collagen monomers into fibrils that further aggregate in bundles or fibers, with fibrils morphology and aggregation controlled by pH, ionic strength, and temperature.[4]

Composites of poly(lactic acid-co-ε-caprolactone) (PLAC) and collagen gels have been used as scaffolds for the regeneration of the bladder wall.[5] Collagen gels contract when seeded with fibroblasts and other connective tissue cells (Fig. 1).[6] This ability of the cells is the basis of the modeling and remodeling of tissues. The ability of tissues to remodel should be always considered in the tissue regeneration pathway.

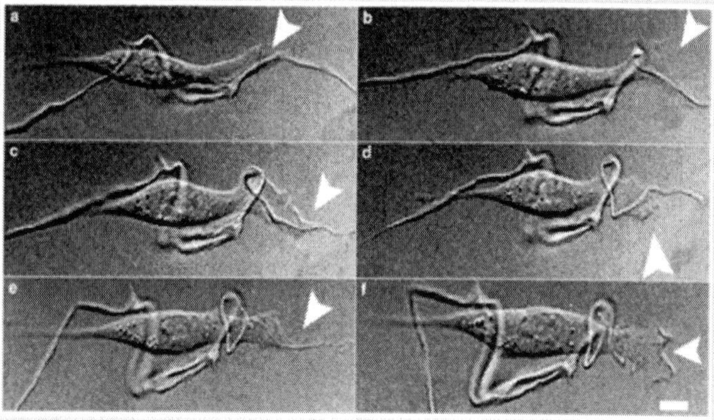

**Figure 1** Sequence showing how a fibroblast cell extents and contracts over a collagen fiber, orienting itself along the long axis of the fiber and extends lamellipodia along the fiber scale bar, 5 μm. Reprinted by permission from Macmillan Publishers Ltd: *Nature Cell Biology* (Ref. 6), copyright (2005).

## 2. Fibroin

As reported in the chapter on natural polymers, fibroin from silkworm silk acquires progressive interest for applications in tissue engineering. Fibroin gels can be easily prepared with different methods, such as sonication, acidification, shear stresses application, or supercritical $CO_2$. The physical characteristics of fibroin gel depend on the concentration of the starting water solution and on the method used for gelation. For instance, acidification of water solutions form flowable gels, while more compact and stiffer gels can be obtained by supercritical $CO_2$ treatments[7] or sonicating higher fibroin concentration water solutions[8]. Gelation occurs via $\beta$-sheets and intermolecular bonds formation. Injectable fibroin gels have been used to induce bone repair in critical size defects in rabbits[9] or to encapsulate cells.[8]

For instance, Fini *et al.*[8] (Fig. 2), have shown that fibroin gels can induce critical size bone defect regeneration after four weeks, with the formation of smaller bone trabeculae, well integrated with the existing bone. A full remodeling process occurring within three months reconstituted the normal anatomy of the newly generated bone.

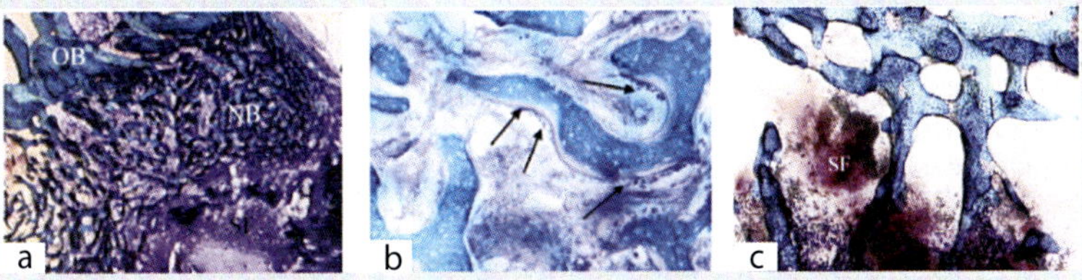


**Figure 2** A critical size defect in the femur condyle of a rabbit filled with a fibroin gel. (a) The hole populated by small bone trabeculae after four weeks; OB, original bone; NB, ew bone. (b) Mineralizing new bone with osteoblasts lining the trabecular surface (arrows) near the center of the defect. (c) Remodeling after 12 weeks and the recovery of the morphology of the original bone. Reprinted from Ref. 9, Copyright (2005), with permission from Elsevier.

## 3. Fibrin

The proteolithic cleavage by thrombin of fibrinopeptides transforms fibrinogen in fibrin, a fibrous protein involved in the coagulation of the blood.[10] The resulting network of fibers forms hydrogels even at very low volume fraction of polymer. The characteristics (rheology and stability) of the gels can be modulated by changing fibrinogen and thrombin concentration.

Unlike synthetic hydrogels, fibrin can directly interact with cells being provided with two pairs of RGD sites. Fibrin gels have been long used for many applications. Examples are drug release, and in the field of tissue engineering, the regeneration of adipose tissue, bone, cartilage, liver, nervous tissue, skin, tendons, ligaments, and cardiovascular tissue engineering.[11–14]

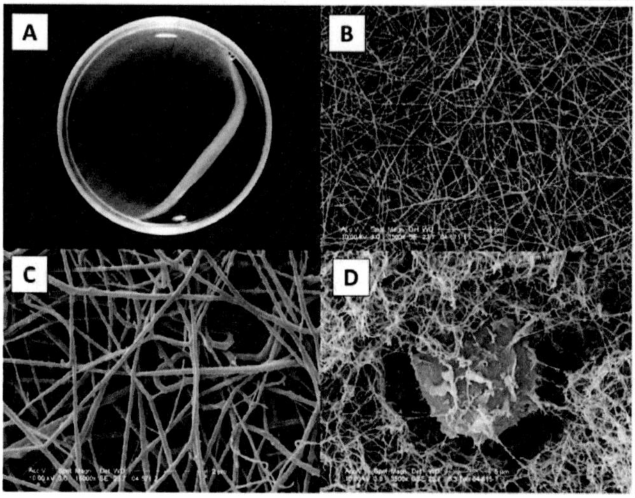


**Figure 3** (A) Fibrin gel after polymerization. (B–D) SEM images of the nanofiber network structure of the fibrin, which enables the gentle embedding of cells, with a vascular smooth muscle cell (SMC) shown in (D) within the fibrin network (Image courtesy: InTech).[13]

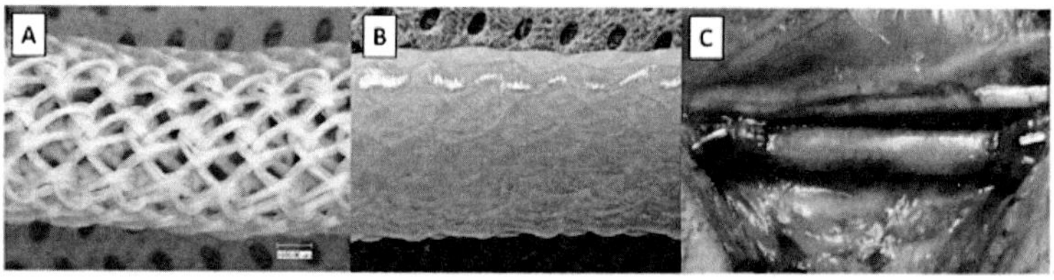


**Figure 4** A vascular graft made of (A) macroporous mesh of poly(*D,L*) lactic acid, (B) embedded into a fibrin pre-seeded with SMC, and (C) implanted in a ovine carotid model (Image courtesy: InTech).[13]

## 4. Other Systems

Uriel *et al.*[15] prepared gel matrices starting from solutions of proteins and glycoproteins from soft tissues, showing that the gel matrices supported tissue-specific, integrin-mediated cell growth, and EC network formation. Ritzoulis *et al.* investigated milk protein-derived gels for bone tissue engineering.[16] Santin *et al.* also investigated the use of soybean gels and demonstrated that soy flour is an inexpensive source for the preparation of gels with different physicochemical properties, nontoxic behavior, and intrinsic bioactivity on cells.[17]

## 5. Remarks

Tissue engineering scaffolds are usually designed and fabricated to match as closely as possible the native tissue properties and architecture. A general opinion is that the more a scaffold resembles the ECM, the more successful the scaffold will be. It is generally true; however, the fact that a scaffold should always accurately match properties or architecture of the target tissue can be controversial.

An example, reported in this book, is the ability of some gels to promote complete and physiologically consistent bone regeneration after their insertion or injection in the defect. Spontaneous regeneration of natural tissues occurs via the remodeling of the ECM matrix. The process of regeneration always involves remodeling, and in favorable conditions the newly formed tissue remodels and assumes the target tissue morphology, structure, composition, and properties. Moreover, the implanted scaffold undergoes modification after implantation and its structure, composition, and properties change.

The regeneration of a tissue is a developmental process of constructive tissue remodeling, and scaffolds should promote it and be compatible with it.

*Importantly, our concept of a scaffold includes both the 3-dimensional traditional geometrically defined construct and the newer injectable material, which does not provide a distinct macroscopic architecture but still provides a controlled microenvironment for the cells.*[18]

## References


1. Jeong, S. I., Kim, S. Y., Cho, S. K., Chong, M. S., Kim, K. S., Kim, H., Lee, S. B., and Lee, Y. M. (2007) Tissue-engineered vascular grafts composed of marine collagen and PLGA fibers using pulsatile perfusion bioreactors, *Biomaterials*, **28**(6), 1115–1122.
2. Krishnan, S., Sekar, S., Katheem, M. F., Krishnakumar, S., and Sastry, T. P. (2012) Fish scale collagen: A novel material for corneal tissue engineering, *Artif. Organs*, **36**(9), 829–835.
3. Parenteau-Bareil, R., Gauvin, R., and Berthod, F. (2010) Collagen-based biomaterials for tissue engineering applications, *Materials*, **3**, 1863–1887.
4. Achilli, M. and Mantovani, D. (2010) Tailoring mechanical properties of collagen-based scaffolds for vascular tissue engineering: The effects of pH, temperature and ionic strength on gelation, *Polymers*, **2**, 664–680.
5. Engelhardt, E.-V., Micol, L. A., Houis, S., Wurm, F. M., Hilborn, J., Hubbell, J. A., and Frey, P. (2011) A collagen-poly(lactic acid-co-ε-caprolactone) hybrid scaffold for bladder tissue regeneration, *Biomaterials*, **32**(16), 3969–3976.
6. Meshel, A. S., Wei, O., Adelstein, R. S., and Sheetz, M. P. (2005) Basic mechanism of three-dimensional collagen fibre transport by fibroblasts, *Nat. Cell Biol.*, **7**, 157–164.
7. Floren, M., Spilimbergo, S., Motta, A., and Migliaresi, C. (2012) Carbon Dioxide induced silk protein gelation for biomedical applications, *Biomacromolecules*, **13**(7), 2060–2072.
8. Wang, X., Kluge, J., Leisk, G. G., and Kaplan, D. L. (2008) Sonication-induced gelation of silk fibroin for cell encapsulation, *Biomaterials*, **29**(8), 1054–1064.
9. Fini, M., Motta, A., Torricelli, P., Giavaresi, G., Nicoli Aldini, N., Tschon, M., Giardino, R., and Migliaresi, C. (2005) The healing of confined critical size cancellous defects in the presence of silk fibroin hydrogels, *Biomaterials*, **26**(17), 3527–3536.
10. Janmey, P. A., Winer, J. P., and Weisel, J. W. (2009) Fibrin gels and their clinical and bioengineering applications, *J. R. Soc. Interface*, **6**(30), 1–10.
11. Eyrich, D., Brandl, F., Appel, B., Maier, G., Wenzel, M., Staudenmaier, R., Goepferich, A., and Blunk, T. (2007) Long-term stable fibrin gels for cartilage engineering, *Biomaterial*, **28**(1), 55–65.

12. Ahmed, T. A., Dare, E.V., and Hinke, M. (2008) Fibrin: A versatile scaffold for tissue engineering applications, *Tissue Eng. Part B Rev.*, **14**(2), 199–215.

13. Jockenhoevel, S., and Flanagan, T. C. (2011). Cardiovascular tissue engineering based on fibrin-gel-scaffolds. In: *Tissue Engineering for Tissue and Organ Regeneration*. Eberli, D. (Ed.). http://www.intechopen.com/books/tissue-engineering-for-tissue-and-organ-regeneration/cardiovascular-tissue-engineering-based-on-fibrin-gel-scaffolds.

14. Tschoeke, B., Flanagan, T. C., Koch, S., Harwoko, M. S., Deichmann, T., Ella, V., Sachweh J. S., Kellomaki, M., Gries, T., Schmitz-Rode, T., and Jockenhoevel, S. (2009) Tissue-engineered small-caliber vascular graft based on a novel biodegradable composite fibrin-polylactide scaffold, *Tissue Eng. Part A*, **15**(8), 1909–1918.

15. Uriel, S., Labay, E., Francis-Sedlak, M., Moya, M. L., Weichselbaum, R. R., Ervin, N., Cankova, Z., and Brey, E. M. (2009) Extraction and assembly of tissue-derived gels for cell culture and tissue engineering, *Tissue Engin. Part C Methods*, **15**(3), 309–321.

16. Ritzoulis, C., Scoutaris, N., Papademetriou, K., Stavroulias, S., and Panayiotou, C. (2005) Milk protein-based emulsion gels for bone tissue engineering, *Food Hydrocoll.*, **19**(3), 575–581.

17. Meikle, S. T., Standen, G., Salvage, J., De Santis, R., Nicolais, L., Ambrosio, L., and Santin, M. (2012) Synthesis and characterization of soybean-based hydrogels with an intrinsic activity on cell differentiation, *Tissue Engin. Part A*, **18**(17-18), 1932–1939.

18. Shoichet, M. S. (2010) Polymer scaffolds for biomaterials applications, *Macromolecules*, **43**, 581–591.

## Chapter 11

# Fabrication Methods of Tissue Engineering Scaffolds

**Vitor M. Correlo,**[a,b] **Joaquim M. Oliveira,**[a,b] **Ana R. C. Duarte,**[a,b] **Albino Martins,**[a,b] **Ana L. Oliveira,**[a,b,c] **Joana Silva-Correia,**[a,b] **Rui L. Reis,**[a,b] **and Nuno M. Neves**[a,b]

[a]*3B's Research Group—Biomaterials, Biodegradables and Biomimetics, University of Minho, Headquarters of the European Institute of Excellence on Tissue Engineering and Regenerative Medicine, AvePark, 4806-909 Taipas, Guimarães, Portugal*

[b]*ICVS/3B's—PT Government Associate Laboratory, Braga/Guimarães, Portugal*

[c]*CBQF—Center for Biotechnology and Fine Chemistry, School of Biotechnology, Portuguese Catholic University, Porto, 4200-072, Portugal*

nuno@dep.uminho.pt

Tissue engineering and regenerative medicine aims at developing new strategies for the regeneration of tissues lost by disease or trauma. The most frequent approaches involve the use of porous scaffolds, eventually in combination with bioactive molecules and/or communities of cells. Those hybrid implants are intended to lead to the desired regeneration of the tissues upon implantation at the defect site.

*Scaffolds for Tissue Engineering: Biological Design, Materials, and Fabrication*
Edited by Claudio Migliaresi and Antonella Motta

ISBN 978-981-4463-20-1 (Hardcover), 978-981-4463-21-8 (eBook)
www.panstanford.com

This chapter covers a large range of available processing methods that may be used to produce porous biodegradable scaffolds from natural or synthetic polymers. Each processing method is discussed starting by an in-depth review of the literature analyzing the main advances in the technology and the main challenges requiring future developments to enable obtaining highly functional scaffolds. The limitations and specific characteristics of each technology are critically discussed. The chapter also provides a detailed analysis of the structure, porous morphology and the consequent mechanical properties of the scaffolds obtained by each method for developing successful tissue engineering strategies.

## 11.1 Introduction

Tissue engineering and regenerative medicine (TERM) is an interdisciplinary area that aims at developing new interdisciplinary strategies to provide new and more effective solutions for unmet clinical needs. The typical components required to obtain a functional regeneration of tissues may include temporary scaffolds, bioactive molecules and different communities of cells (either autologous or heterologous). Those components can be implanted either isolated or in combinations both after substantial manipulation of cells and maturation *in vitro* or minimizing this stage. From a regulatory viewpoint, an acellular scaffold may be considered as a medical device if the mode of action is based on the recruitment and assistance of endogenous cells in the regeneration process. If the scaffold is intended for the delivery of bioactive agents it may be considered as a new drug and follow the regulatory procedures that are common for pharmacological drugs. If the scaffold is intended to act as a delivery device for cells and bioactive molecules, it is considered as an Advanced Therapy Medicinal Product and follows the most stringent rules in the regulatory process.[1] Thus, the mode of action of the strategy being implemented has fundamental implications in the development process.

Biodegradable scaffolds are the cornerstones of modern strategies of TERM.[2] The scaffolds, either synthetic or biologically derived, are aimed at providing specific extracellular environments that help the cells in the regeneration process and in the homeostasis of tissues. Indeed, recent reports in the literature involving decellularization of organs followed by repopulation with patient-specific cells show the efficacy of the strategy to obtain nonimmunogenic organs to overcome the shortage in organ donors for transplantation.[3] This strategy, however, still relies on obtaining donor organs that are frequently too scarce to satisfy the clinical demand.

Synthetic scaffolds for use in TERM have a number of critical requisites that include the biodegradability to obtain a temporary structure, the cytocompatibility and nontoxicity both of the scaffold and of its products of degradation, to ensure device safety.[4] Another requisite is the level and dimensions of the porosity to obtain enough surface area for attachment and growth of sufficiently large communities of cells required to fill a defined volumetric space. The interconnectivity of that porosity is a critical issue in ensuring opportunities for migration of the cells and further colonization of the scaffold surface and also to maintain the viability of the cells by enabling diffusion in of the required nutrients and diffusion out of metabolic residues. The bioactivity of the surface is important to provide the right cues for the cells to secrete and mature extracellular matrix (ECM) and to remodel the construct. Another important aspect contributing the efficacy of the scaffold in mimicking the functionality of the ECM is the balance between the level of porosity and the mechanical properties. The aim is to reach a good compromise between the level of porosity and a sufficient mechanical stability for the cell-populated construct to sustain the stresses applied upon implantation. Ideally, the mechanical properties should provide a good and stable integration with the surrounding host tissues. Furthermore, they should provide the required cues for the cells to promote the intended functional tissue regeneration. Indeed, it was shown in the literature that the differentiation of stem cells is highly sensitive to the stiffness of the culturing substrate.[5] It is acceptable that other supporting devices may assist in the mechanical stabilization of the cell/scaffold hybrid construct.

The criteria of biodegradability of the scaffold define precise boundaries in the range of candidate biomaterials for the development of scaffolds that includes biologically derived materials, synthetic polymers and some ceramics. By defining the material of interest, it also defined the range of processing methods that can be used to process the scaffolds.[6]

This chapter will cover a large range of available processing methods that may be used to produce porous biodegradable scaffolds from natural and synthetic polymers. For each processing method, we will provide a review the literature and analyze the main advances in the technology and the main challenges still in need of further developments to obtain highly functional scaffolds enabling developing successful tissue engineering (TE) strategies.

## 11.2 Introduction to Scaffold Fabrication Methods

The methods enabling the fabrication of biodegradable porous scaffolds may be classified as solvent-based or melt-based.[7] Both types of methods have strengths and limitations in developing devices for biomedical applications, in particular for porous scaffolds. The solvent-based methods require very careful processes and verifications ensuring that no harmful residual solvents are retained in the scaffolds after processing. Conversely, the melt-based processing, although not involving the use of solvents, requires the use of mild temperatures ensuring that no thermal degradation is involved during the processing of sensitive biomaterials.

Porous scaffolds can be achieved by very established and relatively simple methods such as solvent casting with different sacrificial porogen agents, freeze-drying or compounding with chemical or physical blowing agents.[8] Most of those techniques do enable the processing of biomedical products, although offering limited control in tailoring the porous microstructure of the scaffolds. The porogen-based methods involve the need to subject the scaffolds to immersion in liquids aimed at removing the porogen agents by selective dissolution. Those postprocessing steps may limit the functionality of the developed scaffolds. Other techniques were also adapted to process porous scaffolds from other areas such as textile technology,[9] supercritical fluid (SCF) processing[10] or polymer processing.[8]

The ideal scaffold processing technology enabling obtaining very reproducible predefined porous morphologies and architectures in a sufficiently large range of available biomaterials is yet to be developed. However, the advent of rapid prototyping or solid free-form fabrication enabled obtaining a sufficiently higher degree of control in the scaffold morphology and of the final shape of an implantable device.[11] Indeed, those technologies also enable designing patient-specific implants that will fit precisely the tissue defect, by obtaining three-dimensional (3D) data from tomography scans or by magnetic resonance imaging.

### 11.2.1 Critical Issues in the Control of the Porosity and Interconnectivity of Scaffolds

A common feature of the most frequently used methods for the fabrication of porous scaffolds is the low level of reproducibility of the porous morphology. This issue poses evident difficulties and constraints when a particular scaffolds aims at being validated for clinical applications.

Most processing methods do not enable specifying the size, shape or the spatial distribution of the porosity. Frequently, the methods enable a tight control of the raw materials and may enable obtaining stable processing routines that result in scaffolds having similar levels of porosity and of the macroscopic properties. Most methods may also allow obtaining some control of the level of interconnectivity of the porosity. However, the eventual inhomogeneous distribution of the interconnectivity implies that some parts of the scaffold will not be able to be populated by cells. Furthermore, the lower interconnectivity causes difficulty in the local diffusion of the aqueous phase and hinders the hydrolysis of the biomaterial, resulting in uneven degradation kinetics.

The amount, size and level of interconnectivity of the porosity are all interrelated, but they have different impacts on the quality of the scaffolds processed. The amount of the porosity is inversely related to the mechanical properties. High levels of porosity are desirable by maximizing the opportunities for cell colonization and for faster tissue development. However, those high levels of porosity imply lower mechanical stability of the scaffold. Thus, an optimized balance between those conflicting requirements needs to be obtained for each specific application.[12]

The size of the porosity is a critical issue in conditioning the performance of the scaffolds with respect to the cell migration. If the pores are too small, they may hinder the colonization of the internal structure by the cells. Indeed, this is a common problem in electrospun meshes that may be counteracted by some strategies proposed in the literature.[13]

The pore morphology is another aspect that may influence the performance of the scaffolds. The porosity of trabecular bone is concave and shows no sharp edges or flat surfaces.[14] Thus, when considering the development of a porous scaffold, it may be relevant to select the processing method that enables obtaining pore morphology and architecture with higher affinity with native tissues, mimicking as closely as possible the porous structure of the ECM.

In some TE applications, it may be interesting to obtain functionally graded porous scaffolds, having controlled variation of properties.[15] Another powerful concept in increasing the level of sophistication and detail of the scaffold structure reports the production of hierarchically organized scaffolds, mimicking closely the architecture found in organs and tissues.[16] Such hierarchical scaffolds provide enhanced opportunities to develop ECM-biomimetic structures that may further improve the functionality of the scaffolds.

In summary, all aspects related with the porosity of the scaffolds may contribute to obtain enhanced performance or may lead to inconsistency and anisotropy of the properties of the scaffolds. Thus, it is highly desirable to select the most appropriate biomaterial and production method and to further validate in detailed experiments with relevant and specific cells an optimized pore structure to maximize the performance of the scaffolds.

## 11.3 Melt-Based Processing

Melt-based processing methods are often used for producing TE scaffolds. In melt-based processing methods, temperature is used to decrease polymer viscosity allowing to obtain complex structures and shapes according to the desired application, avoiding the use of solvents. Our group has developed extensive research efforts concerning the various processing methods available to produce porous scaffolds for TE, such as (i) compression molding combined with particulate leaching; (ii) extrusion or injection molding combined with blowing agents; (iii) fiber bonding and (iv) rapid prototyping. Those methods will be reviewed in the following sections.

(i) *Compression molding combined with particulate leaching*

Methods based on the leaching of soluble particulates are among the most commonly used in the fabrication of 3D porous scaffolds. Compression molding combined with particulate leaching consists in obtaining a solid mixing of a polymer or a polymeric blend with leachable particles. This mixture is further loaded into a mould and heated above the melting temperature of the polymer and pressure is applied, for a certain period of time, aiming to maximize packaging. After cooling, the molded polymer-porogen composite is immersed in a solvent for the selective dissolution of the porogen agent fraction. The porosity and pore size of the scaffold are both controlled, respectively, by the amount and particle size of the porogen used.[12]

This methodology was applied by Gomes *et al.*[17] to developed starch-based scaffolds exhibiting pore sizes between 10 μm and 500 μm and porosity of about 50%. In a different study, Correlo *et al.*[12] produced scaffolds from chitosan/polyester blends and composites using melt-based compression molding followed by salt leaching. The scaffolds were prepared using two different granulometric salt size ranges and with different levels of NaCl content. The produced scaffolds are characterized by having distinct properties concerning porosity, pore size and mechanical performance depending on the average particle size and the amount of NaCl used. Studies conducted with these scaffolds seeded with bovine articular chondrocytes demonstrated that pore size and geometry of the pore had an effect on cell proliferation and ECM production.[18] *In vitro* studies confirmed that the developed chitosan/polyester-based scaffolds support adhesion, viability/proliferation and osteogenic or chondrogenic differentiation of a mouse mesenchymal stem cell line (BMC-9).[19,20] Recent studies have shown that chitosan-based scaffolds prepared by compression molding combined with particulate leaching, besides supporting *in vitro* proliferation and osteogenic differentiation of hBMSCs, induced bone formation *in vivo*.[21] Therefore, these results indicate that those scaffolds can be further developed for the delivery of cells for bone TE strategies.

(ii) *Extrusion or injection molding combined with blowing agents*

By exploring the thermoplastic behavior of polymers it is possible to produce 3D porous scaffolds using traditional melt-based technologies, such as extrusion or injection molding combined with blowing agents. In both processes, the blowing agent is mixed with a polymer. During the melting process, the temperature activates the blowing agent to react, releasing gases. Thus, porosity results from gases released by the thermal decomposition of the blowing agent that is dissolved in the polymer melt.

Gomes *et al.*[17] produced scaffolds from a blend of starch with cellulose acetate (SCA) by extrusion with blowing agents (BA). By using different types and amounts of BA it was possible to obtain scaffolds with different porosities in the range of 50–500 μm. The same approach was used to produce scaffolds from a blend of corn starch/ethylene-vinyl alcohol (SEVA-C).[22] In this case, the developed porous structures had 60% porosity with pore sizes between 200 and 900 μm with an acceptable degree of interconnectivity. *In vitro* tests confirmed that these scaffolds are cytocompatible, supporting the growth, both at the surface and in the inner pore structure, of human osteoblast-like cells. The limitation of this method is the limited level of control of the pore size and interconnectivity between the pores. A similar strategy, based on injection molding, has been applied for the production of TE scaffolds.[23,24] This method allows obtaining scaffolds with the precise shape of the mould designed for specific applications. However, injection molding causes the formation of a compact outer layer on processed samples, which needs to be subsequently removed.

(iii) *Fiber bonding*

Fiber-based scaffolds were among the first porous structures proposed for TE.[25] Fiber meshes scaffolds can be developed by a fiber bonding process consisting on applying a thermal treatment to bond random fibers bundles. By controlling the compression of the bundles, the amount and size of fibers, it is possible to obtain scaffolds with different morphologies regarding porosity and pore size.[26] This approach was followed to develop starch-based scaffolds from melt spun starch-polycaprolactone (SPCL) and also starch-polylactic acid (SPLA) fibers.[27] The two types of scaffolds produced by this method exhibited a typical fiber-mesh structure, with highly interconnected pores and a porosity of approximately 75%.

Oliveira *et al.*[28] tested fiber mesh SPCL scaffolds and polyglycolic acid (PGA) nonwoven scaffolds with bovine articular chondrocytes. The results have showed that SPCL scaffolds can support bovine articular chondrocytes adhesion, proliferation, contrarily to PGA scaffolds that presented a central area of cells depletion.

It has also been demonstrated that the use of starch-based scaffolds, in conjunction with the fluid flow obtained in bioreactor cultures, enhanced the differentiation of marrow stromal cells, leading to the development of bone-like mineralized tissue.[29] Aiming the development of functional vasculature in bone TE applications, Santos *et al.*[30] demonstrated that starch-based fiber mesh scaffolds are an excellent substrate for the growth of human endothelial cells (ECs) required for the vascularization process.

Chitosan-based fiber mesh scaffolds were produced using microextruded chitosan-polyester fibres.[31] Several sequential studies have shown that these scaffolds obtained by fiber bonding processes have adequate structure for bone[32] and cartilage[33,34] TE applications.

(iv) *Rapid prototyping*

Recently, rapid prototyping technologies emerged as powerful methods to prepare highly organized and reproducible 3D porous scaffolds. Rapid prototyping is a general name used to define a family of techniques between others, fused deposition modeling (FDM), precise extruding deposition and 3D plotting. A rapid prototyping method uses a computer-controlled four-axis machine with a multiple-dispenser head. One of the main advantages of these processing methods includes the possibility of using computer-based medical imaging technologies (e.g., CT scans or magnetic resonance imaging) to design and fabricate customized and anatomically adapted scaffolds.[35]

Hoque *et al.*[36] reported the production of 3D scaffolds, from PCL and PCL-based copolymers, with complete interconnected and highly regular honeycomb-like pore morphology using a desktop robot-based rapid prototyping technique. This study also demonstrated that the processing parameters directly influenced the pore size, porosity and, consequently, the mechanical properties of the scaffolds. Our group has been developing TE scaffolds based on SPCL using 3D bioplotting.[37,38] Nevertheless, a limitation of these scaffolds is the low cell-seeding efficiency since pore size is relatively large as compared to cell dimensions.[16] SPCL tubular structure prepared by means of 3D bioplotting has been used to assure mechanical stability on TE strategies to recover spinal cord injuries[36, 37].

## 11.4 Solvent-Based Fabrication

A number of different methods have been described in the literature for the production of 3D porous structures to be used in TERM. Solvent-based fabrication is among the most commonly employed techniques for the production of such structures. To obtain a desired scaffold first, the polymer needs to be dissolved in a suitable solvent.

This type of processing methods often involves the use of organic solvents which can present mutagenic and toxicological properties. Thus, an appropriate method is required to completely remove the solvent prior to scaffolds further use in TERM approaches. Within the scope of this section, we highlight the most relevant processing methods involving the use of solvents that attracted the interest of the research community. Herein, the tunable properties of scaffolds that can be obtained are also described.

(i) *Emulsification/Freeze-drying*

The freeze-drying technique is an effective and bioclean procedure that has been used in a wide range of pharmaceutical, medical and industrial applications.[39–41] It is also a multi-step and energy-consuming methodology, where scaffolds can be obtained by means of freezing a polymer solution and removal of solvent through sublimation in the freeze-dryer equipment.[42] Since this procedure requires several days for sample drying, this constitutes the main drawback of the technique.[43] Despite being time-consuming, the freeze-drying is a versatile method since it allows varying the different experimental parameters, and thus to tailor the final scaffolds properties. In practice, pores are formed after complete evaporation of the frozen solvent and by this mean a 3D porous structure can be produced. Therefore, the polymer should be first dissolved into a suitable solvent that can be frozen at temperatures typically comprised between −20°C and −80°C.[44] To the polymeric solution (not aqueous) may be added water, and the two liquids are then mixed in order to obtain an emulsion, in a process known as emulsification.[45,46] The final scaffold properties can be tuned during the first step, i.e., by means of adjusting the type of polymer and its concentration in the formulations[47,48] and the water content. This way it is possible to tailor its resorption rate, mechanical properties and porous architecture.[49,50] During this step the loading of drugs and other molecules or even other materials[51-53] is possible, and thus the functional customization can also be obtained. In the second step, the polymer solution should be transferred into a mould and kept frozen during several hours. This step is sensitive and in the case of emulsions should be as fast as possible since the separation of the two phases should be avoided. Besides varying the freezing temperature and the cooling kinetics,[54–57] other solvents and surfactants can be added to the polymer solution prior to freezing.

By this way, it is possible to control the solvent crystal growth and the final porous architecture of the scaffold namely the porosity, interconnectivity, pore size and shape, and geometry.[58,59] The final procedure is the drying step, which is performed as aforementioned in the freeze-dryer equipment. This is a critical step, as we need to assure that the sample is completely frozen prior to drying, in order to prevent the final structure to collapse or originate an undesired dense external skin-like layer.[43] Moreover, the sublimation of the frozen solvent should be performed under the same conditions inside the freeze-dryer in order to obtain reproducibility of the scaffolds structure batch after batch. Parameters of the process include the pressure of the vacuum chamber, surface temperature, plate temperature and temperature of the ice condenser. All those parameters need to be optimized and well controlled and also kept constant during all process.[60,61] To obtain high-quality scaffolds, our group has been reporting that by means of using this technique, it is possible to produce suitable scaffolds for different TERM applications. Interestingly, several foam-like (Fig. 11.1) and bilayered structures can be obtained by means of freeze-drying alone or in combination with other techniques.[62–65] These have shown to be adequate scaffolds for bone, cartilage and osteochondral applications. The scaffolds present adequate mechanical properties, support *in vitro* cell adhesion, proliferation and differentiation.

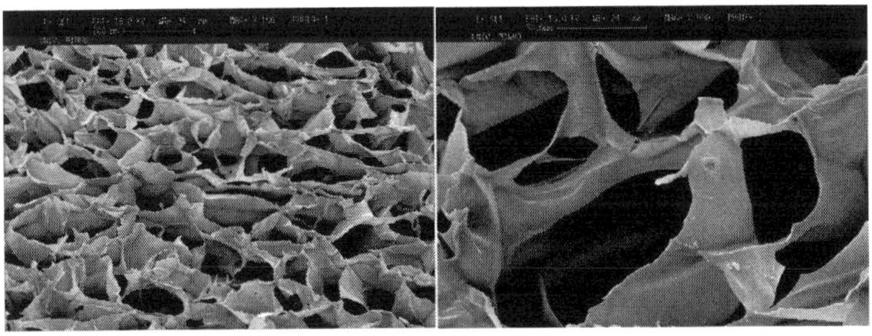

**Figure 11.1** Chitosan scaffold (3 wt.%) obtained by means of freeze-drying. The solvent used is an acetic acid (3%, v/v) aqueous solution. Liquid nitrogen (–196°C) was used to freeze the polymer solution, prior to freeze-drying.

Kang *et al.*[66] proposed using the freeze-drying method to produce gelatin porous hydrogels. Hydrogels possessing different porosity were obtained by varying the freezing and cooling rate.

They have shown that as the freezing temperature was lowered, the hydrogel structure gradually became more two-dimensionally ordered. Wongpanit *et al.*[67] produced silk fibroin/chitin whiskers sponges by using a freeze-drying technique. These nanocomposites possess improved mechanical properties and cell adhesion ability as compared to neat silk fibroin sponges. Recently, Oliveira *et al.*[68] also reported silk-based scaffolds resembling bone lamellar structure using a freeze-drying technique. Methanol, water annealing and steam sterilization post-treatments were performed to induce water stability, and tailor porous architecture and mechanical properties. Moreover, the lamellar structures supported the osteogenic differentiation of human bone marrow–derived mesenchymal stem cells (MSCs), thus producing a multilamellar hybrid construct *in vitro*.

(ii) *Electrospinning*

The electrospinning technique consists of the ejection of a polymer solution, controllably dispensed by a syringe pump, from a spinneret. The solution undergoes extensional flow and deposits into a collector by the application of an external electrostatic field. Basically, when a pendant droplet of the polymer solution in the tip of the spinneret is subjected to an electric field strong enough to overcome its surface tension, a tiny jet is ejected in the direction of the collector. Before reaching the collector, the solvent mostly evaporates leading to the deposition of long and very thin fibers.[69] Important parameters in the electrospinning process are not only the ones related with the polymer and solution properties, but also the processing and the environmental conditions. Up to date, more than one hundred different polymers (i.e., from synthetic- and natural-origin) have been successfully spun into ultrafine fibers using the electrospinning technique.[70] Electrospinning has attracted tremendous interest in the research community as a simple and versatile technique to produce synthetic polymeric ultrafine fibers with diameters ranging from a few micrometers to tens of nanometers. Owing to their very small diameter, polymeric nanofibers exhibit unusual properties such as high specific surface area and flexibility in the surface functionalities.[13] Typically, the electrospun fibers are randomly organized, but some alignment can be also achieved, in a mesh-like structure characterized by a high interconnectivity and porosity, and pores size in the micrometer range. In addition, electrospun nonwoven meshes could physically mimic the ECM structure of native tissues.[71] These remarkable properties render electrospun nanofibers that are useful in many applications, particularly those related to the field of biomedical engineering such as scaffolding[69,71] and drug release.[72] So far, the production of electrospun nanofiber meshes (NFM) with complex ordered topographies and patterns remains a challenge.[73,74] Additionally, the inherent planar structure,[16] a pore size in the micron range[75] and limited mechanical properties may constitute limitations to their application as scaffolds for TERM.[71]

Different approaches based on electrospun fiber have been recently developed for cartilage and bone TE. In an initial study, electrospun PCL and starch-PCL (i.e., SPCL) NFMs were used to seed and culture bovine articular chondrocytes to evaluate their potential for cartilage TE.[76] Biological results point out that PCL and SPCL NFMs are suitable as supports for cartilaginous ECM formation. Human mesenchymal stem cells (hMSCs) isolated from human bone marrow aspirates were also seeded onto PCL NFMs and cultured in a multi-chamber flow perfusion bioreactor to determine their ability to produce cartilage-like ECM.[77] Results indicate that the dynamic culture conditions increased the chondrogenic differentiation of hMSCs and stimulated ECM deposition. Bone ECM is composed mainly of hydroxyapatite crystals formed at the surface of collagen type I nanofibrous networks and bundles. Thus, many systems developed for bone TE involve combinations of polymer with calcium phosphate in composites aimed at the fabrication of scaffolds. A very innovative embodiment,[78] described the electrospun NFM coated with a biomimetic calcium phosphate (BCP) layer, by means of a surface modification performed at physiological-compatible conditions, without disturbing the initial microporous structure. The biological assays demonstrated that the polycaprolactone NFM coated with a BCP layer facilitate the adhesion and enhance the proliferation of human osteoblastic cells for long culture periods. Moreover, those BCP-coated NFMs were cultured with osteoblastic cells (MG63 cell line), within a rotating bioreactor that simulates microgravity conditions.[79] Results showed that, under dynamic culture conditions, the production of proteins associated with the ECM of bone (i.e., type I collagen and osteopontin) was higher on BCP-NFM constructs than in the PCL NFM. These evidences suggested that combining BCP-

NFM constructs with dynamic culture conditions, which provides an efficient mass transfer, homogeneous cell distribution, and minimal damaging turbulence enabled the production of a highly efficient model with potential applications in bone TE.

(iii) *Other methods*

Thermally induced phase separation (TIPS) also involves the use of a solvent with a low melting point that is easy to sublime.[80–83] Processing parameters such as quenching temperature, polymer concentration, solvent composition and molecular weight have been shown to influence the scaffold morphology.[82,84–86] Hua *et al.*[84] reported on ternary systems to produce porous scaffolds. This comprises the use of dioxane to dissolve poly(D,L-lactic acid-co-glycolic acid) (PLGA) and then phase separation is induced through the addition of water. A polymer-rich and a polymer-poor phase are formed. By means of cooling below the solvent melting point and vacuum-drying to obtain sublimation of the solvent, regular and highly interconnected macroporous PLGA scaffolds are obtained. When preparing scaffolds by means of liquid–liquid phase separation, it is necessary to optimize the set of parameters aforementioned for emulsification/freeze-drying technique.

Solvent casting followed by particulate leaching is another popular and simple technique that requires the dissolution of the polymer in an appropriate solvent and then the solution is cast into a mould filled with porogen particles.[87,88] Yan *et al.*[63] recently reported the production of silk fibroin scaffolds derived from high concentration aqueous silk fibroin solutions following this process. The silk fibroin scaffolds were prepared with different initial silk concentrations [8%, 10%, 12%, and 16% (wt.%)]. These scaffolds revealed a trabecular structure and possess a homogeneous porosity. By increasing the silk fibroin concentration from 8% to 16%, the mean porosity decreased from 90.8 ± 0.9% to 79.8 ± 0.3%, and the mean interconnectivity decreased from 97.4 ± 0.5% to 92.3 ± 1.3%. Based on the physicochemical properties of the silk scaffolds, these were proposed for applications in meniscus and cartilage TE scaffolding.

Other solvent-based fabrication methods were analyzed in detail elsewhere.[89–94] Details of other available fabrication methods for obtaining natural-based polymeric and ceramic scaffolds may be found elsewhere.[41,42]

## 11.5 Supercritical Processing of Porous Scaffolds

For nearly two decades, SCF processing has advanced toward the discovery of new processes that offer various possibilities to prepare materials in different morphologies or formulations.[95,96] SCF technology offers a clean and environmentally friendly, single-step operation, which provides an attractive platform technology able to meet the demands of many industries. Carbon dioxide is the most commonly used SCF. Its low critical parameters (pressure 71 bar and temperature 34°C), specially its low critical temperature, make it very attractive for the processing of thermosensitive active compounds, such as proteins and/or pharmaceuticals. It presents other important advantages, since it is environmentally benign, nontoxic, nonflammable, noncorrosive, readily available and inexpensive. Its elimination and the recovery of final products are easier (no residue is left and a dry solid product is easily obtained, just by manipulating the pressure), leading to processes with low energy consumption.

SCF technology comprises several processes, leading to very different matrices in terms of size, shape and porous morphology.[97,98] In this chapter, the production of 3D porous structures using SCF technology is reviewed. 3D polymeric structures can be prepared by processes based in the use of either SCF foaming or supercritical-assisted phase inversion.

SCF foaming takes advantage of the plasticizing properties of carbon dioxide.[99] It is known that the compression of a pure substance with a gas alters its phase equilibria. When a polymer is compressed with carbon dioxide, its glass transition temperature ($T_g$) is lowered, in some cases significantly, and a glass-to-rubber transition is observed. The extent of this effect depends greatly on the interactions between carbon dioxide and the polymer. Indeed, stronger interactions will enhance the extent of $T_g$ depression. A schematic representation of the SCF foaming process is presented in Fig. 11.2.

This process consists of the pressurization of a polymer with carbon dioxide, in a high pressure vessel. The solubility of carbon dioxide in the polymer will lead to the depression of the glass transition temperature of the polymer and upon venting, the depressurization of the system leads to a thermodynamic instability which causes the super saturation of carbon dioxide dissolved in the polymer and hence the nucleation of pores occurs.

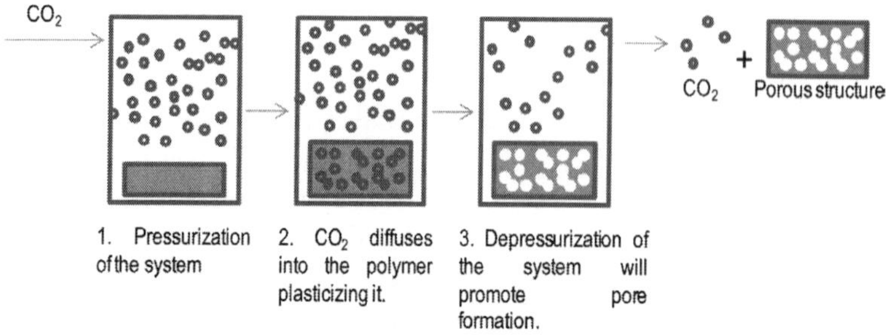


**Figure 11.2** Schematic representation of the supercritical gas foaming process.

**Table 11.1** Advantages and disadvantages of gas foaming technique

| Advantages | Disadvantages |
|---|---|
| • No use of organic solvents<br>• High production capacity<br>• Low pressure operation<br>• Low $CO_2$ consumption<br>• Easy separation of gas and polymeric matrix | • Limited by the solubility of $CO_2$ in the polymer<br>• Not applicable to natural-derived polymers<br>• Batch process |

The production of porous matrixes for TE purposes using the SCF foaming process was first described by Mooney and coworkers.[100] In their work, they showed the possibility to foam PLGA and to produce structures with a porosity of nearly 93%. Since then many studies have been reported and these include mostly processing of homopolymers and copolymers of lactic and glycolic acids, as well as caprolactone.[101–105] This technique has also been described to prepare complex systems combining one or two different materials, particularly ceramics,[106–108] pharmaceutical compounds, proteins,[109–110] and/or mammalian cells.[111]

The preparation of a bioactive system able to promote cell signaling and differentiation has recently been proposed by Santo *et al.*[112,113] In this work, it reported the successful incorporation of growth factors (GFs) within a 3D matrix of PDLLA, without the loss of protein activity. Figure 11.3 shows a scanning electron micrograph of the PDLLA scaffolds prepared showing the high level of porosity obtained.

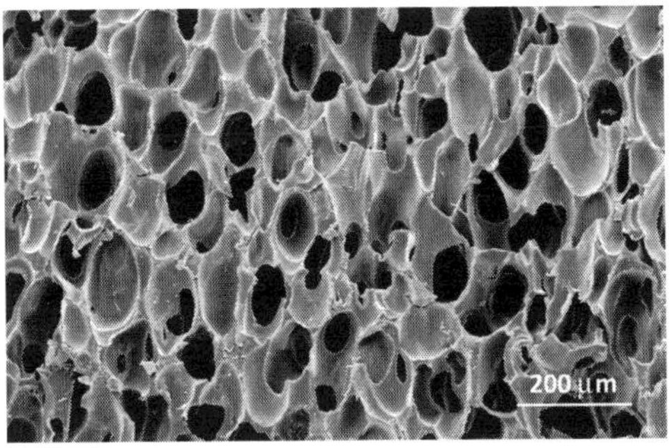


**Figure 11.3** Scanning electron microscopy image of a PDLLA scaffold produced by supercritical fluid foaming.

In general, the supercritical gas foaming is a rather difficult process to control, in the sense that it is the interplay of different variables which determines the final pore size and morphology structure. Nonetheless, it was observed that lower pressures, higher temperatures and slower depressurization rates, independently, may favor the formation of matrixes with larger pores (Table 11.1).[114]

The supercritical-assisted phase inversion method involves casting of a polymer solution onto an inert support followed by immersion of the support with the cast film into a bath filled with a nonsolvent for the polymer that can be an SCF. The contact between the solvent and nonsolvent will promote the precipitation of the polymer as it causes the phase-separation of the solution.[115] The main advantages of the use of carbon dioxide as a nonsolvent are the fact that it is possible to tailor the final structure of the matrix by changing the operating conditions of the process, namely, pressure, temperature and polymer concentration. Furthermore, when using an SCF no liquid–vapor interface exists, therefore, the matrix can be dried without the collapse of its porous structure (Table 11.2). A schematic representation of the process is presented in Fig. 11.4.

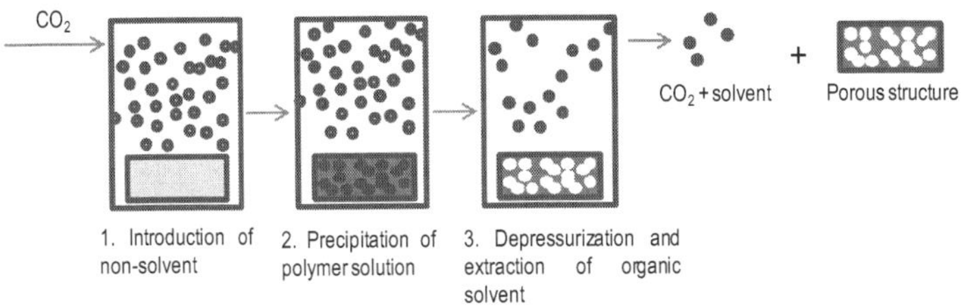


**Figure 11.4** Schematic representation of the supercritical-assisted phase inversion process.

**Table 11.2** Advantages and disadvantages of supercritical-assisted phase inversion

| Advantages | Disadvantages |
|---|---|
| • Applicability to a wide range of materials<br>• Dry product is obtained<br>• Processing can be conducted at near ambient temperatures<br>• Morphology of the structures obtained can be fine-tuned changing the operating parameters | • Use of organic solvents<br>• Difficult scale-up of the process<br>• Residual solvent might be present in the product<br>• Separation of gas and solvent may be required at industrial level |

This technique has been widely explored in the preparation of 3D porous scaffolds for TERM purposes and either natural or synthetic polymers have been processed. Using this technique, it is also possible to prepare bioactive matrixes that combine polymer and bioactive glasses or hydroxyapatite, or a pharmaceutical compound.[116–122] Figure 11.5 shows a cross section of a scaffold of a polymeric mixture of chitosan:PLLA (50:50) processed at 100 bar and 60°C.

**Figure 11.5** 3D scaffold of chitosan:PLLA (50:50) prepared by supercritical-assisted phase inversion.

The successful production of highly porous and interconnected scaffolds using this technique depends to a great extent on the ternary phase diagram between polymer-solvent-nonsolvent.[123] In this case, some thermodynamic insights may help the understanding of the matrix formation, as well as the knowledge of solubility parameter and the interaction coefficients nonsolvent/solvent, nonsolvent/polymer and solvent/polymer. Different operating conditions also influence supercritical-assisted phase inversion process and hereafter the morphology of the structures obtained, namely pressure, temperature and polymer concentration.

The choice of the best processing technique will be determined by the characteristics of the polymers to be processed and on the desired final application. It is the principle inherent to each technique, which will ultimately limit their applicability; nonetheless, they present interesting advantages over conventional polymer processing methodologies.

## 11.6 Hierarchical Scaffolds

Generally, the biomaterial scaffolds are engineered to mimic some architectural features of the macroscopic physical structure of tissues. Frequently, it is also aimed at simultaneously recreating the natural hierarchical architecture of tissues at the cellular and sub-cellular levels.[124] The fibrous nature of the natural ECM has inspired many researchers to focus on the development of fiber-based hierarchical scaffolds. Current attempts to replicate the complexity and hierarchical organization of natural ECM are mostly restricted to dispensing nanofibers in previously developed implants and biomedical devices. Indeed, electrospun nonwoven meshes have the distinctive feature of physically mimic the ECM structure of many human tissues. Therefore, the ability to engineer hierarchical scaffolds comprising some morphological features of the native ECM may be embodied through the combination of electrospun nanofibers with well-established techniques for biomaterials processing.[71] In this section, four alternative strategies are presented allowing the development of complex fibrous structures that enable mimicking the typical hierarchical organization of tissues and, consequently, to increase the biological performance of those 3D scaffolds.

The integration of electrospun nanofibers with macroscale fibers, processed by other polymer processing techniques (i.e., wet-spinning, rapid prototyping and fiber extrusion), was explored by our research group and will be herein described. In a first and simplest approach, electrospun nanofibers were directly deposited over a pre-fabricated wet-spun microfiber scaffold (Fig. 11.6A).[125,126]

This combined structure was obtained by a two-step methodology and structurally consist of a nano-network incorporated on a macrofibrous support. Its biological functionality was demonstrated by the culturing of human osteoblast-like cells, bone marrow stromal cells and ECs (i.e., human umbilical vein ECs and microvascular ECs).[30,126,127] This hierarchical structure was developed to mimic the highly organized fibrous structure of bone tissue, not forgetting the vascular network, which is identified as the main pitfall in bone TE and the major hurdle for the clinical application of the engineered constructs.

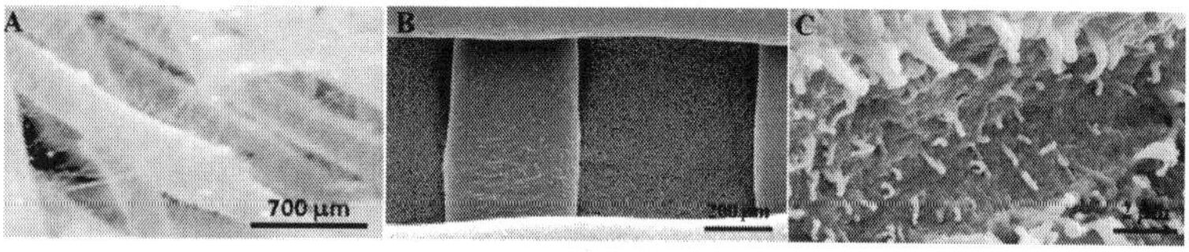


**Figure 11.6** Hierarchical fibrous scaffolds comprising electrospun nanofibers associated with macro- and microfibers produced by wet-spinning (A), rapid prototyping (B) and melt extrusion (C) techniques.

The second 3D micro/nanofibrous scaffold involves the intercalation of electrospun nanofibers, pre-fabricated or deposited on time, in between rapid prototyped microfibers (Fig. 11.6B).[16,128–131] The integrated electrospun nanofibers provide topological cues at the ECM level, while the micro-3D fibrous structure ensures the mechanical stability of the engineered hierarchical scaffold. So far, most part of these biomaterial structures were developed envisioning their application in the regeneration of cartilage tissue. However, the hierarchical organization of collagen fibers is more prominent on the natural bone ECM. Cell culture experiments demonstrated the preferential adhesion of bovine or porcine primary chondrocytes to the electrospun nanofiber matrices, as well as a statistically significant increment of cell proliferation on the micro/nanofibrous scaffolds.[128–131] Similar biological results were achieved when human osteoblastic cells were dynamically seeded on these hierarchical fibrous scaffolds, showing significantly higher proliferation and maturation.[16] Particularly, SEM micrographs demonstrated that the osteoblastic cells preferentially adhered and spread on the electrospun NFMs, constituting an innovative strategy to enhance cell seeding efficiency/cell adhesion into the microfibrous scaffolds and, simultaneously, obtaining a scaffold with mechanical stability.

Submicron electrospun polymer fibers are also good candidates as reinforcing agents in the development of advanced nanocomposites due to their continuity, orientation, inherent flexibility and potential compatibility with polymeric matrices. Consequently, another type of hierarchical fibrous structure is composed of randomly distributed microfibers reinforced by electrospun chitosan nanofibers (Fig. 11.6C), processed by melt extrusion and assembled by fiber bonding.[132] Those composite scaffolds showed improved mechanical (i.e., tensile modulus) and degradation properties (i.e., water uptake). However, only a limited number of composites reinforced with electrospun nanofibers have been developed, and mainly for providing some outstanding physical characteristics (i.e., mechanical properties).[132–139] An extensive physical characterization of electrospun nanofiber-reinforced composites was conducted in dental restorative composites resins,[133–140] but their biological functionality still needs to be explored. Recent *in vitro* results with human bone marrow MSCs cultured on nanofibers-reinforced microfibrous composite scaffolds, under osteogenic induction conditions, showed a sustained ECM deposition and mineralization, as demonstrated by the increased amount of calcium phosphate produced.[141] The data reported about the biological performance of the structures undoubtedly demonstrate the superior performance of electrospun NFMs over the cellular adhesion, proliferation and also differentiation of MSCs.

Aiming at replicating the multiscale hierarchical architecture and the function of an intervertebral disc (IVD), novel disc ECM analogues were developed using agarose to replicate the morphology of the nucleus pulposus (NP) tissue. Oriented electrospun nanofibrous scaffolds were also proposed to mimic the structure of annulus fibrosus (AF).[142] Mechanical testing in compression and torsion revealed equilibrium and time-dependent behaviors that are qualitatively similar to those of the IVD, although lower in magnitude. Furthermore, bovine MSCs seeded into the both regions (i.e., NP and AF) adopted distinctive morphologies of the disc tissue. Similar cellular orientation was also reported by Yang and co-workers[143] in a novel highly organized scaffold. The scaffold is produced through a fabrication methodology based on the incorporation of electrospun NFMs into 3D hydrogels via a bottom-up, layer-by-layer assembly process. Three-dimensional micropatterned fibrous scaffolds, consisting of a polystyrene (PS) fiber matrices and polyethylene glycol (PEG) hydrogel microstructures, were fabricated using electrospinning and photolithography, respectively.[144] The micropatterned scaffold was obtained as a free-standing and bidirectional-porous sheet, which may be stacked in multiple layers or rolled into tubes to build up more complicated constructs. Because of the nonadhesiveness of proteins and cells to the PEG hydrogel, the cells selectively adhered and remained viable within the PS fiber region, creating cellular micropatterns. Altogether, these nanofiber-hydrogel composites demonstrate the huge potential for mimicking highly organized and complex architectures of soft tissues, such as the AF and the corneal tissue.

Summarizing, to create a tissue-engineered scaffold capable of regenerating a fully functional tissue, it is believed that it should mimic both the fibrous morphology and the complex ordered structure/functionality of the native ECM. This ECM analogue should have topographical features and geometry at the macro-, micro- and even nanoscale levels, as each level may influence passively the cell response to the scaffold. Therefore, all the presented strategies are focused on the development of multiscale polymeric fibrous-based scaffolds, trying to mimic the hierarchical organization of the native ECM. In this way, the integration of electrospun nanofiber meshes into macrostructured scaffolds provides a strong strategy to obtain unique and improved biological functionality for bone and cartilage tissue-engineering strategies.

## 11.7 Advanced Textile Technologies

In the biomedical field, TE represents a specialized niche in which textile technologies can have distinctive contributions. Maximization of tissue attachment to materials requires a highly organized porous structure for tissue integration and a template for cell assembly, combined with structural properties analogous of the living tissue. Scaffolds developed for TE need to facilitate and promote cellular proliferation and tissue regeneration. The intra-architectural scaffold geometry, as well as the scaffold material and its surface properties, plays an important role in this process. Most of the conventional fabrication techniques available for scaffold production are not yet enabled to obtain the desired scaffold properties, as many of the processing routes still depend heavily on manual intervention, which can lead to inconsistent and processing routines and low reproducibility. In this regard, textile-based technologies are considered as potential routes for the production of complex scaffolds for TE applications, as they can present superior control over the design, manufacturing precision and reproducibility.

The textile field offers a collection of state-of-the-art technologies that can markedly contribute to produce innovative complex fiber-based scaffolds for TE applications.[145,146] In fact, over the last decades, advances made on the development of superior textile products allowed for the upgrading of the processing technologies resulting in a diversity of 3D textured materials with highly controlled mechanical properties and porosities. Moreover, being computer assisted, these technologies have potential for the production of predesigned architectures with predictable properties based on the patient's own needs. Therefore, bringing textile technologies to the TE field represents a breakthrough still far from being fully explored.

The idea of bridging textiles with the biomedical field is not new. As more research has been completed, textiles have found their way into a variety of medical applications.[147] This offers a competitive opportunity for the textile industry since the traditional textile market is declining worldwide. Therefore, there is a strong need and push for seeking the niche applications for more functional, customized and high-tech textiles, which leads to the biotextiles paradigm. The term "biotextiles" can be defined as a "structure composed of textile fibers and designed for use in a specific biological environment (e.g., surgical implants), where its performance depends on its interactions with cells and biological fluids as measured in terms of its biocompatibility and biostability.[147] Textile fabrics have been used in a variety of biomedical applications such as the augmentation and reconstruction of knee[148] and shoulder[149] ligaments, tendon[150] and IVD replacement.[151] In the therapy of cardiovascular diseases, fabrics have been utilized for cardiac support devices,[152] tissue regenerative vascular grafts[153] and prosthetic heart valves for percutaneous implantation.[154] In the repair and regeneration of the peripheral nerves, woven tubular shaped guides were used.[155] Skin, being a very compliant tissue, constitutes one of the most obvious applications for biotextiles, as in the case of wound dressings or TE products.[146]

The optimal design of such textile implants requires a multi- and interdisciplinary combination of skills. Despite the fact that some of the traditional textiles have fulfilled primary quality requirements such as biocompatibility, flexibility or strength, there is a need for further developing of these systems to meet more demanding and specialized functions. It is evident that the disciplines of polymer chemistry, fiber science, textile technology and engineering have unique roles to play when combined with molecular biology, biochemistry and biotechnology to design and develop novel and high performance biotextile scaffolds for TE applications.[145–147]

Scaffolds shaped with biodegradable polymers such as PGA, polylactic acid (PLA), polycaprolactone (PCL), starch or silk fibers were investigated for cell transplantation and regeneration of various tissues such as nerve, skin, ligament, bladder, cartilage and bone.[156,157] Several methods were already developed and proposed to prepare porous 3D biodegradable scaffolds for TE, including gas foaming, fiber extrusion and bonding, 3D printing, phase separation, emulsion freeze-drying, and porogen leaching or rapid prototyping.[157,158] Most of these techniques have been extensively studied using different biodegradable polymers.[157,159] Among those, fiber-based structures represent a wide range of morphological and geometric possibilities that can be tailored for each specific TE application (as previously discussed in Sections 11.4 and 11.5).[160] Interconnected fiber networks of SPCL produced either by rapid prototyping or by using the so-called fiber bonding technique are particularly interesting in promoting cell attachment and proliferation.[29,38,161] Nevertheless, the later reported process is obviously not the most suitable for a fine control of the morphology since the fibers are not individually manipulated and a random distribution is obtained.

Presently, several efforts are being done in order to apply precise and reproducible textile technologies to develop original 3D architectures for the regeneration of different tissues such as skin and bone. In Fig. 11.7 several examples of combined different textile technologies for tuning the textile constructs to specific needs are presented.[162,163] The processing control is taken to the fiber level by using state-of-the-art technology for extrusion enabling to obtain fibers with complex cross-section shape or to produce of multicomponent fibers. The co-extrusion of polymers controlling different physical and chemical properties allows for the development of novel multifunctional scaffolds. By using this technique, it is possible to produce tailored fibers in terms of degradation, mechanical behavior or the chemical and surface properties. The synergistic effect of combining those properties will then maximize the resulting biological performance. The control can be taken to the nano-level by investigating new formulations, changing the surface properties through functionalization or using bioactive agents and nanoparticles dispersed in the polymer matrix.

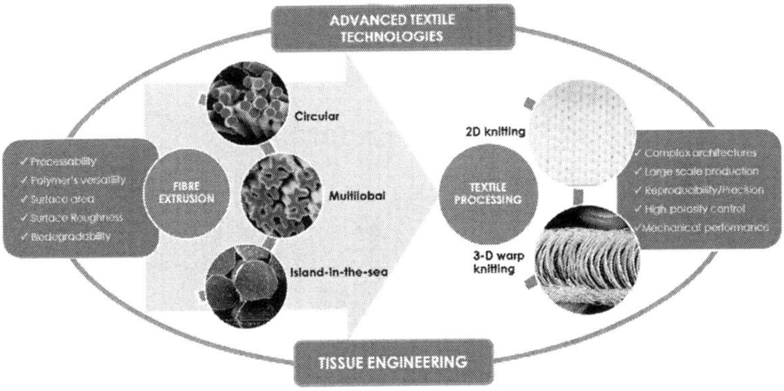


**Figure 11.7** Examples on the use of advanced textile technologies for potential tissue engineering applications currently under development.[162,163]

Recently, natural and synthetic biodegradable polymers such as silk, polybutylene succinate (PBS) and polyethylene terephthalate (PET) were processed into multifilament yarns and knitted into different structures such as Jersey, Rib and Piqué (Fig. 11.8) and 3D spacer (Fig. 11.9). In the case of synthetic polymers, the extrusion process was optimized in order to produce fibers with the mechanical properties necessary to be knitted. Fibers with circular and multilobal cross sections were developed as a way of achieving different surface area/roughness. Additionally, bicomponent fibers were also proposed using polyethylene oxide (PEO). Since PEO is a water soluble polymer, the strategy is to dissolve the Island in order to achieved multi-hollow fibers for further incorporation of GFs, antibiotics, etc. The cross-section configuration selected was 64 Islands-in-the-Sea. Furthermore, different surface modifications were performed (e.g., acid/alkaline treatment, UV radiation and plasma) for increasing cell adhesion and proliferation and also for further surface immobilization of biomolecules of interest.

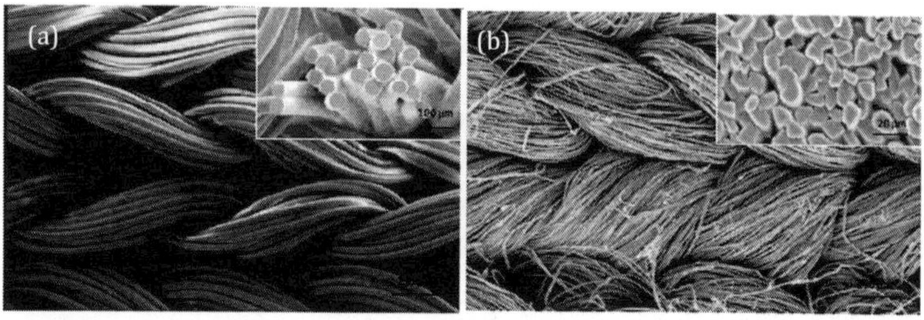


**Figure 11.8** SEM micrographs of (a) PBS and (b) silk fibroin knitted structures revealing a very reproducible intra-architectural scaffold geometry, with high surface area.[162]

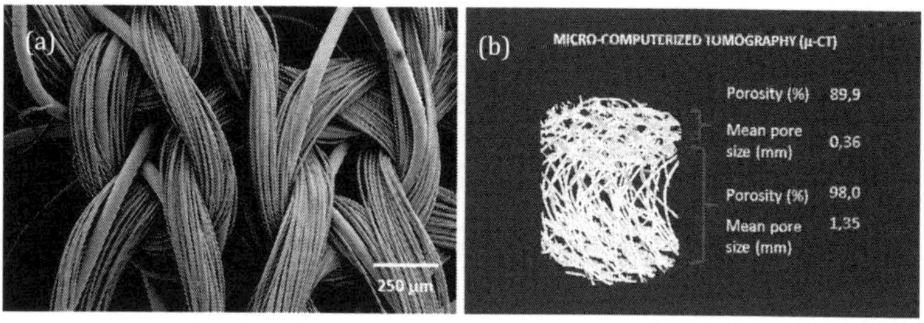


**Figure 11.9** PET warp knitted structures: (a) SEM micrographs revealing the intra-architectural scaffold geometry and (b) calculated values for porosity and mean pore size of each region of the scaffold by micro-computerized tomography.[162]

The obtained constructs present very reproducible intra-architectural scaffold geometry with high surface area and exhibiting a wide range of porosities. With the above-mentioned techniques, it was possible to validate the effectiveness of the proposed treatments in modifying the surface of the materials. The immobilization of different bioactive molecules on the surface of modified materials is presently being investigated. Preliminary cellular tests with L929 mouse fibroblasts were seeded on the different materials up to 14 days and have shown that the proposed materials can support cell adhesion and proliferation. Silk and PBS, in particular, seem promising materials for developing biotextiles to be applied in TE.

## 11.8 Hydrogel-Based Technologies

The objective of this section is to briefly describe some relevant methods reported in the literature for processing hydrogels and to provide some examples of their application in acellular and cellular TERM strategies and drug delivery. Hydrogels have been attracting great deal of interest due to their tunable physical, chemical and biological properties. They possess interesting features to be used in TERM applications due to their ability for swelling and retaining high amounts of water, and mimic the ECM environment. These versatile systems enable using different methods for gelation, while by means of varying the type of polymer and formulations, degree of crosslinking, chemistry or functionalization allow fine tuning the final properties of the hydrogel. When choosing a hydrogel for a particular application, different parameters have to be taken in consideration, namely, mechanical properties, degradation rate, permeability, rheology, noncytotoxicity and molecular response.[164] Hydrogels may present different sol–gel transition mechanisms. In order to be suitable for TERM purposes, it must be liquid to facilitate homogeneous cell distribution and drug loading,[165] while being capable of undergoing gelation after or during implantation. Examples of processing techniques used for inducing gelation and obtaining hydrogel scaffolds and particles are summarized in Table 11.3.

Electrospinning combined with hydrogel lithography[175] and templating water-in-oil emulsion and subsequent *in situ* gelation are some of the methods that have been developed to produce hydrogels with interesting features.[166] The first method allows producing micropatterned fiber scaffolds that are capable of controlling the spatial positioning of mammalian cells. Templating water-in-oil emulsion/*in situ* gelation enables to prepare hybrid alginate hydrogel beads with shells of porous $CaCO_3$ microparticles (MPs), which showed great promise for applications as delivery carriers for drugs or cosmetics, food supplements or cells. Emulsification/ *in situ* thermal gelation was described as a mean to promote kappa-carrageenan gelation, for example.[176] The group of Kaplan[168] has shown that ultrasonication can induce the gelation of silk, and thus it is a useful method to initiate rapid sol–gel transitions. Recently, Lu *et al.*[169] described a method for obtaining silk fibroin gel using direct application of voltage, through electrogelation. This work reported that a silk fibroin gel may be reversibly converted to a liquid by applying reverse voltage. Yan *et al.*[170] described an enzymatic method to produce silk-fibroin hydrogels that supports the encapsulation of cells and drugs to be applied in several settings of a TE approach. These bioadhesive hydrogels are obtained by means of peroxidase-mediated crosslinking in the presence of peroxide or other oxidizers at mild temperatures and within the physiological pH range. Interestingly, one of the embodiments comprises shape-memory properties of the peroxidase-mediated crosslinked of silk fibroin hydrogels.

**Table 11.3** Different methodologies for producing hydrogels aimed at find applications in TERM and drug delivery

| Processing methodology | Composition | References |
|---|---|---|
| Water-in-oil emulsion followed by *in situ* gelation | Hybrid alginate/$CaCO_3$ beads | 166 |
| Electrostatic interaction | Collagen-chitosan | 167 |
| Electrogelation, ultrasonication and enzymatic gelation | Silk | 168–170 |
| Ionic-crosslinking (cations) | Gellan gum and methacrylated gellan gum | 171–173 |
| Photopolymerization (UV light) | Methacrylated gellan gum | 173 |
| Reductive alkylation crosslinking | Starch | 174 |
| Electrospinning followed by lithography | Polycaprolactone | 175 |
| Emulsification/thermal gelation | Kappa-carrageenan Beads | 176 |

Baran *et al.*[174] reported on the preparation of starch-chitosan hydrogels by reductive alkylation crosslinking, which possess adequate properties for cartilage regeneration. Stimuli-responsive hydrogels based on natural origin polymers as chitosan-*β*-glycerophosphate (CGP) and chitosan-*β*-glycerophosphate-1% starch (CST) can be implanted in a minimally invasive technique that was proposed by Sa-Lima *et al.*[177] These hydrogel systems represent a great promise as thermosensitive templates or matrices for cartilage TE. The study demonstrated that the CGP and CST hydrogels supported the chondrogenic differentiation of encapsulated adipose-derived stromal cells. The CST hydrogels were shown to have improved biological performance as compared to CGP. Pereira *et al.*[165,178] described carrageenan/fibrin/hyaluronic acid-based hydrogels with *in situ* gelation ability for cartilage TE applications. These studies demonstrated that hydrogels seeded with bovine cells were able to regenerate and repair the damaged articular cartilage in a bovine model. Similarly, promising injectable gellan gum hydrogels were proposed by Oliveira *et al.*[171,179–181]

Bone TE approaches would benefit by the development of materials that can promote vascularization. Santo *et al.*[182] used kappa-carrageenan to develop hydrogel beads with the ability to incorporate GFs for promoting angiogenesis. Platelet-derived growth factor (PDGF) was loaded into the beads. Results demonstrated that the system is able to control the release profile of PDGF at levels similar to those required for the development of a fully functional vascular network and, thus, has great potential for being used in bone TE applications. Similarly, Chiu *et al.*[167] reported a collagen-chitosan hydrogel for the loading of thymosin $\beta$4, a 43-amino acid peptide, that is both angiogenic and cardioprotective. The study demonstrated that the hydrogels allowed obtaining a controlled release of peptide over 28 days to elicit prolonged effects and promoted angiogenesis *in vivo*, thus showing great promise for cardiac repair.

Some TERM applications may require using hydrogels possessing anti-angiogenic properties. This is the case of hydrogels for use as acellular and cellular NP substitutes. Silva-Correia *et al.*[173,183] have been addressing this important problem. Recently, they have reported on methacrylated gellan gum-based (GG-MA) hydrogels, which are able to be ionic and photo-crosslinked for application in IVD regeneration. Figure 11.10 illustrates the several steps for the preparation of ionic- and photo-crosslinked hydrogels.

The study revealed that photo-crosslinked GG-MA hydrogels present higher stability and mechanical properties as compared to ionic-crosslinked hydrogels. Regarding their biological performance, the results[184] showed that both hydrogels are cytocompatible since it supported the encapsulation and viability of mouse lung fibroblast (L929) cells during 21 days. We have demonstrated that the developed hydrogels also supported the viability of human IVD cells (Fig. 11.11).

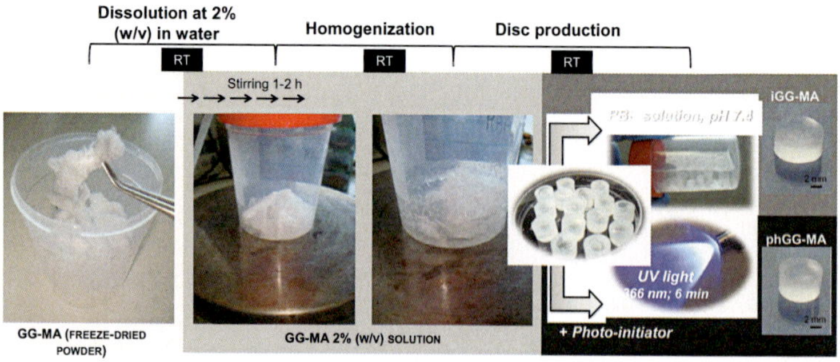


**Figure 11.10** Schematic representation for the preparation of the GG-MA hydrogels by means of ionic- and photo-crosslinking.

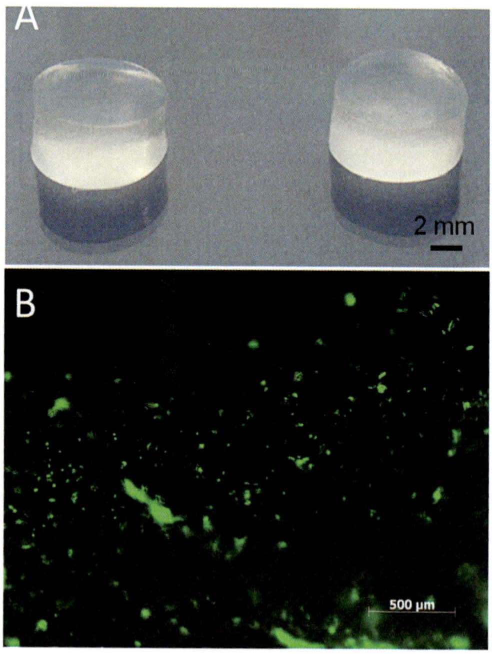


**Figure 11.11** Photograph of the ionic-crosslinked GG-MA hydrogel (A). Fluorescence microscopy image of the ionic-crosslinked GG-MA hydrogel with encapsulated human IVD cells after 21 days of culturing; Calcein-AM assay—green color shows live cells (B).

The *in vivo* studies in chicken chorioallantoic membrane (CAM assay) revealed that the ionic- and photo-crosslinked GG-MA hydrogels do not allow the ingrowth of chick ECs and do not elicit any acute inflammatory response. The results also showed that both GG-MA hydrogels present different permeability to cells and prevented vascular invasion and possess a nonangiogenic properties.[185] These hydrogels present promising and tunable properties for being used as NP substitutes in the treatment of degenerative IVD. Moreover, the hydrogels were shown to be well tolerated and nontoxic when subcutaneously implanted in rat. Pereira *et al.*[172] also proposed a substitute to regenerate NP by means of using gellan gum hydrogel matrices formulations reinforced with cell-loaded MPs. The *in vitro* results demonstrated that the hydrogels do not have a deleterious effect over L929 cells, as the cells encapsulated in the MPs remained viable over a period of 72 hours. Thus, the gelation mechanisms of gellan gum-based hydrogels allow processing these hydrogels in the form of scaffolds, MPs, fibers or 3D structure with open interconnected pores.

Briefly, by means of using the above-mentioned gelation methods, it is possible to develop different hydrogels that support the encapsulation, viability, and differentiation of cells for the several TERM applications. The loading of different bioactive molecules/GFs within the hydrogels is also possible. Silk- and gellan gum–based hydrogels are the most promising hydrogels to be applied in the repair and regeneration of cartilage and IVD.

## 11.9 Final Remarks

Despite significant advances in TE research, there are a number of challenges, such as regeneration of complex tissue structures, especially vascularized tissues like bone, that still remain unsolved until now. These barriers include the ability to produce 3D porous structures that simultaneously supports large dimension tissue regeneration and network vascularization. For a tissue to grow beyond 100–200 μm (the diffusion limit of oxygen), the existence of a vascular system is required for delivering oxygen and nutrients supply to the cells.[186] Moreover, scaffolds architecture and design, namely pore size and pore interconnectivity, have a high effect on the rate of vascularization after implantation.[187,188] It is also recognized that micro- and nanotopography can affect cellular responses from initial attachment and migration to differentiation and production of new tissue.[189,190]

The traditional technologies to produce scaffolds are limited to randomly control of individual pore geometry, size, interconnectivity and morphology. Recent technologies such as 3D printing and solid free-form are also limited to hundreds of microns of spatial resolution.[191] As result of these limitations, microfabrication and bio-microelectromechanical system (bio-MEMS) technologies have immerged as versatile and powerful approaches for generating precisely engineered scaffolds for TE as they allow the control of features at scales of relevance for cells and molecules.[192] These technologies can be used to control features at length scales from <1 μm to >1 cm offering the possibility of building structures and devices in dimensional scales smaller to those of cells[193] and expand functionality of a scaffold. However, microfabrication techniques are essentially based on planar patterning being mainly used to produce micro/nanofeatures on 2D polymeric substrates for cells contact guidance. Combining micro- and nanofabrication technologies, e.g., fiber bonding or rapid prototyping with electrospinning, can be the key for the successful development of highly organized scaffolds for complex TE applications.

## References


1. Schneider, C.K., *et al.* (2010). Challenges with advanced therapy medicinal products and how to meet them, *Nature Rev. Drug Discov.*, **9**(3), 195–201.
2. Fisher, O.Z., *et al.* (2010). Bioinspired Materials for controlling stem cell fate, *Accounts Chem. Res.*, **43**(3), 419–428.
3. Ott, H.C., *et al.* (2010). Regeneration and orthotopic transplantation of a bioartificial lung, *Nat. Med.*, **16**(8), 927–933.
4. Freyman, T.M., Yannas, I.V., and Gibson, L.J. (2001). Cellular materials as porous scaffolds for tissue engineering, *Prog. Mater. Sci.*, **46**(3/4), 273–282.
5. Engler, A.J., *et al.* (2011). Matrix elasticity directs stem cell lineage specification, *Cell*, **126**(4), 677–689.
6. Kim, M.S., *et al.* (2011). Polymeric scaffolds for regenerative medicine, *Polym. Rev.*, **51**(1), 23–52.
7. Liu Tsang, V. and Bhatia, S.N. (2004). Three-dimensional tissue fabrication, *Adv. Drug Del. Rev.*, **56**(11), 1635–1647.
8. Neves, N.M., Kouyumdzhiev, A., and Reis, R.L. (2005). The morphology, mechanical properties and ageing behavior of porous injection molded

starch-based blends for tissue engineering scaffolding, *Mat. Sci. Eng. C-Bio. S.*, **25**(2), 195–200.

9. Hemmrich, K., *et al.* (2008). Three-dimensional nonwoven scaffolds from a novel biodegradable poly(ester amide) for tissue engineering applications, *J. Mater. Sci-Mater. M.*, **19**(1), 257–267.
10. Hu, X.B. and Lesser, A.J. (2006). Solid-state processing of polymer in the presence of supercritical carbon dioxide, *J. Cell. Plast.*, **42**(6), 517–527.
11. Yeong, W.-Y., *et al.* (2004). Rapid prototyping in tissue engineering: Challenges and potential, *Trends Biotechnol.*, **22**(12), 643–652.
12. Correlo, V.M., *et al.* (2009). Melt-based compression-molded scaffolds from chitosan-polyester blends and composites: Morphology and mechanical properties, *J. Biomed. Mater. Res - A*, **91A**(2), 489–504.
13. Martins, A., *et al.* (2009). Surface modification of electrospun polycaprolactone nanofiber meshes by plasma treatment to enhance biological performance, *Small*, **5**(10), 1195–1206.
14. Ding, M., *et al.* (2002). Age-related variations in the microstructure of human tibial cancellous bone, *J. Orthop. Res.*, **20**(3), 615–621.
15. Leong, K., *et al.* (2008). Engineering functionally graded tissue engineering scaffolds, *J. Mech. Behav. Biomed. Mater.*, **1**(2), 140–152.
16. Martins, A., *et al.* (2009). Hierarchical starch-based fibrous scaffold for bone tissue engineering applications, *J. Tissue Eng. Regen. Med.*, **3**(1), 37–42.
17. Gomes, M.E., *et al.* (2002). Alternative tissue engineering scaffolds based on starch: Processing methodologies, morphology, degradation and mechanical properties, *Mater. Sci. Eng. C*, **20**(1/2), 19–26.
18. da Silva, M.L.A., *et al.* (2010). Chitosan/polyester-based scaffolds for cartilage tissue engineering: Assessment of extracellular matrix formation, *Acta Biomater.*, **6**(3), 1149–1157.
19. Costa-Pinto, A.R., *et al.* (2008). Adhesion, proliferation, and osteogenic differentiation of a mouse mesenchymal stem cell line (BMC9) seeded on novel melt-based chitosan/polyester 3D porous scaffolds, *Tissue Eng. Part A*, **14**(6), 1049–1057.
20. Oliveira, J.T., *et al.* (2008). Assessment of the suitability of chitosan/polybutylene succinate scaffolds seeded with mouse mesenchymal progenitor cells for a cartilage tissue engineering approach, *Tissue Eng. Part A*, **14**(10), 1651–1661.

21. Costa-Pinto, A.R., *et al.* (2012). Chitosan-poly(butylene succinate) scaffolds and human bone marrow stromal cells induce bone repair in a mouse calvaria model, *J. Tissue Eng. Regen. Med.*, **6**(1), 21–28.

22. Salgado, A.J., Coutinho, O.P., and Reis, R.L. (2004). Novel starch-based scaffolds for bone tissue engineering: Cytotoxicity, cell culture, and protein expression, *Tissue Eng.*, **10**(3/4), 465–474.

23. Gomes, M.E., *et al.* (2001). A new approach based on injection moulding to produce biodegradable starch-based polymeric scaffolds: Morphology, mechanical and degradation behavior, *Biomaterials*, **22**(9), 883–889.

24. Neves, N.M., Kouyumdzhiev, A., and Reis, R.L. (2005). The morphology, mechanical properties and ageing behavior of porous injection molded starch-based blends for tissue engineering scaffolding, *Mater. Sci. Eng.-C*, **25**(2), 195–200.

25. Mikos, A.G., *et al.* (1993). Preparation of poly(Glycolic Acid) bonded fiber structures for cell attachment and transplantation, *J. Biomed. Mater. Res.*, **27**(2), 183–189.

26. Chung, S., *et al.* (2008). Dependence of cell proliferation on the morphology of starch based scaffolds for tissue engineering, *Tissue Eng. Part A*, **14**(5), 718.

27. Gomes, M.E., *et al.* (2008). Starch-poly(epsilon-caprolactone) and starch-poly(lactic acid) fiber-mesh scaffolds for bone tissue engineering applications: Structure, mechanical properties and degradation behavior, *J. Tissue Eng. Regen. Med.*, **2**(5), 243–252.

28. Oliveira, J.T., *et al.* (2007). A cartilage tissue engineering approach combining starch-polycaprolactone fiber mesh scaffolds with bovine articular chondrocytes, *J. Mater. Sci-Mater. M.*, **18**(2), 295–302.

29. Gomes, M.E., *et al.* (2006). Influence of the porosity of starch-based fiber mesh scaffolds on the proliferation and osteogenic differentiation of bone marrow stromal cells cultured in a flow perfusion bioreactor, *Tissue Eng.*, **212**(4), 801–809.

30. Santos, M.I., *et al.* (2007). Response of micro- and macrovascular endothelial cells to starch-based fiber meshes for bone tissue engineering, *Biomaterials*, **28**(2), 240–248.

31. Correlo, V.M., *et al.* (2010). Melt processing of chitosan-based fibers and fiber-mesh scaffolds for the engineering of connective tissues, *Macromol. Biosci.*, **10**(12), 1495–1504.

32. Costa-Pinto, A.R., *et al.* (2009). Osteogenic differentiation of human bone marrow mesenchymal stem cells seeded on melt based chitosan

scaffolds for bone tissue engineering applications, *Biomacromolecules*, **10**(8), 2067–2073.

33. da Silva, M.L.A., *et al.* (2011). Chondrogenic differentiation of human bone marrow mesenchymal stem cells in chitosan-based scaffolds using a flow-perfusion bioreactor, *J. Tissue Eng. Regen. Med.*, **5**(9), 722–732.

34. Oliveira, J.T., *et al.* (2011). Novel melt-processable chitosan-polybutylene succinate fiber scaffolds for cartilage tissue engineering. *J. Biomat. Sci-Polym. E.*, **22**(4/6), 773–788.

35. Hutmacher, D.W. (2000). Scaffolds in tissue engineering bone and cartilage, *Biomaterials*, **21**(24), 2529–2543.

36. Hoque, M.E., *et al.* (2009). Processing of polycaprolactone and polycaprolactone-based copolymers into 3D scaffolds, and their cellular responses, *Tissue Eng. Part A*, **15**(10), 3013–3024.

37. Silva, N.A., *et al.* (2008). Starch/gellan gum hybrid 3D guidance systems for spinal cord injury regeneration: Scaffolds processing, characterization and biological evaluation, *Tissue Eng. Part A*, **14**(5), 779.

38. Silva, N.A., *et al.* (2010). Development and characterization of a novel hybrid tissue engineering-based scaffold for spinal cord injury repair, *Tissue Eng. Part A*, **16**(1), 45–54.

39. Mladenov, D.A., Tsvetkov, T.D., and Vulchanov, N.L. (1993). Freeze drying of biomaterials for the medical practice, *Cryobiology*, **30**(3), 335–348.

40. Felix, F., (1998). Freeze-drying of bioproducts: Putting principles into practice, *Eur. J. Pharm. Biopharm.*, **45**(3), 221–229.

41. Hottot, A., Vessot, S., and Andrieu, J. (2007). Freeze drying of pharmaceuticals in vials: Influence of freezing protocol and sample configuration on ice morphology and freeze-dried cake texture, *Chem. Eng. Process.*, **46**(7), 666–674.

42. Daoussi, R., *et al.* (2009). Sublimation kinetics and sublimation end-point times during freeze-drying of pharmaceutical active principle with organic co-solvent formulations, *Chem. Eng. Res. Des.*, **87**(7), 899–907.

43. Correlo, V.M., *et al.* (2007). Tissue engineering using natural polymers, in *Biomedical Polymers* (ed., Jenkins, M.). Cambridge: Woodhead Publishing Limited, 197–217.

44. Mandal, B.B. and Kundu, S.C. (2009). Cell proliferation and migration in silk fibroin 3D scaffolds, *Biomaterials*, **30**(15), 2956–2965.

45. Grant, N. and Zhang, H. (2011). Poorly water-soluble drug nanoparticles via an emulsion-freeze-drying approach, *J. Coll. Interf. Sci.*, **356**(2), 573–578.

46. Whang, K., Goldstick, T.K., and Healy, K.E. (2000). A biodegradable polymer scaffold for delivery of osteotropic factors, *Biomaterials*, **21**(24), 2545–2551.

47. Davidenko, N., *et al.* (2010). Collagen–hyaluronic acid scaffolds for adipose tissue engineering, *Acta Biomater.*, **6**(10), 3957–3968.

48. She, Z., *et al.* (2008). Preparation and *in vitro* degradation of porous three-dimensional silk fibroin/chitosan scaffold, *Polym. Degrad. Stabil.*, **93**(7), 1316–1322.

49. Autissier, A., *et al.* (2010). Fabrication of porous polysaccharide-based scaffolds using a combined freeze-drying/cross-linking process, *Acta Biomater.*, **6**(9), 3640–3648.

50. Kim, J.-W., *et al.* (2008). Preparation of poly(L-lactic acid) honeycomb monolith structure by unidirectional freezing and freeze-drying, *Chem. Eng. Sci.*, **63**(15), 3858–3863.

51. Sachlos, E., *et al.* (2008). The impact of critical point drying with liquid carbon dioxide on collagen–hydroxyapatite composite scaffolds, *Acta Biomater.*, **4**(5), 1322–1331.

52. Luetzow, K., *et al.* (2007). Formation of poly(ε-caprolactone) scaffolds loaded with small molecules by integrated processes, *J. Biomech.*, **40, Suppl. 1**(0), S80–S88.

53. Prabaharan, M. and Jayakumar, R. (2009). Chitosan-graft-*β*-cyclodextrin scaffolds with controlled drug release capability for tissue engineering applications, *Int. J. Biol. Macromol.*, **44**(4), 320–325.

54. Fergal J. O' Brien, *et al.* (2004). Influence of freezing rate on pore structure in freeze-dried collagen-GAG scaffolds, *Biomaterials*, **25**(6), 1077.

55. Yuan, N.-Y., *et al.* (2009). Effects of the cooling mode on the structure and strength of porous scaffolds made of chitosan, alginate, and carboxymethyl cellulose by the freeze-gelation method, *Carbohyd. Polym.*, **78**(2), 349–356.

56. Landi, E., Valentini, F., and Tampieri, A. (2008). Porous hydroxyapatite/gelatine scaffolds with ice-designed channel-like porosity for biomedical applications, *Acta Biomater.*, 2**4**(6), 1620–1626.

57. Nakagawa, K., *et al.* (2006). Influence of controlled nucleation by ultrasounds on ice morphology of frozen formulations for pharmaceutical proteins freeze-drying, *Chem. Eng. Process.*, **45**(9), 783–791.

58. Wu, X., *et al.* (2010). Preparation of aligned porous gelatin scaffolds by unidirectional freeze-drying method, *Acta Biomater.*, **6**(3), 1167–1177.

59. Zhang, F., *et al.* (2011). Fabrication of gelatin–hyaluronic acid hybrid scaffolds with tunable porous structures for soft tissue engineering, *Int. J. Biol. Macromol.*, **48**(3), 474–481.

60. Gan, K.H., *et al.* (2005). Freeze-drying of pharmaceuticals in vials on trays: Effects of drying chamber wall temperature and tray side on lyophilization performance, *Int. J. Heat Mass Tran.*, **48**(9), 1675–1687.

61. Velardi, S.A. and Barresi, A.A. (2008). Development of simplified models for the freeze-drying process and investigation of the optimal operating conditions, *Chem. Eng. Res. Des.*, **86**(1), 9–22.

62. Oliveira, J.M., *et al.* (2006). Novel hydroxyapatite/chitosan bilayered scaffold for osteochondral tissue-engineering applications: Scaffold design and its performance when seeded with goat bone marrow stromal cells, *Biomaterials*, **27**(36), 6123–6137.

63. Yan, L.-P., *et al.* (2012). Macro/microporous silk fibroin scaffolds with potential for articular cartilage and meniscus tissue engineering applications, *Acta Biomater.*, **8**(1), 289–301.

64. Silva, S.S., *et al.* (2008). Novel Genipin-cross-linked chitosan/silk fibroin sponges for cartilage engineering strategies, *Biomacromolecules*, **9**(10), 2764–2774.

65. Yan, L.P., *et al.* (2010). Genipin-cross-linked collagen/chitosan biomimetic scaffolds for articular cartilage tissue engineering applications, *J. Biomed. Mater. Res. A*, **95A**(2), 465–475.

66. Kang, H.-W., Tabata, Y., and Ikada, Y. (1999). Fabrication of porous gelatin scaffolds for tissue engineering, *Biomaterials*, **20**(14), 1339–1344.

67. Wongpanit, P., *et al.* (2007). Preparation and characterization of chitin whisker-reinforced silk fibroin nanocomposite sponges, *Eur. Polym. J.*, **43**(10), 4123–4135.

68. Oliveira, A.L., *et al.* (2012). Aligned silk-based 3-D architectures for contact guidance in tissue engineering. *Acta Biomater.*, **8** (4), 1530–1542.

69. Martins, A., Reis, R.L., and Neves, N.M. (2008). Electrospinning: Processing technique for tissue engineering scaffolding, *Int. Mater. Rev.*, **53**(5), 257–274.

70. Ashammakhi, N., *et al.* (2007). Biodegradable nanomats produced by electrospinning: Expanding multifunctionality and potential for tissue engineering, *J. Nanosci. Nanotechnol.*, **7**(3), 862–882.

71. Martins, A., *et al.* (2007). Electrospun nanostructured scaffolds for tissue engineering applications, *Nanomedicine (Lond.)*, **2**(6), 929–942.

72. Martins, A., *et al.* (2010). Osteogenic induction of hBMSCs by electrospun scaffolds with dexamethasone release functionality, *Biomaterials*, **31**(22), 5875–5885.

73. Martins, A., *et al.* (2011). The influence of patterned nanofiber meshes on human mesenchymal stem cell osteogenesis, *Macromol. Biosci.*, **11**(7), 978–987.

74. Neves, N.M., *et al.* (2007). Patterning of polymer nanofiber meshes by electrospinning for biomedical applications, *Int. J. Nanomedicine*, **2**(3), 433–448.

75. Guimaraes, A., *et al.* (2010). Solving cell infiltration limitations of electrospun nanofiber meshes for tissue engineering applications, *Nanomedicine (Lond.)*, **5**(4), 539–554.

76. da Silva, M.A., *et al.* (2009). Evaluation of extracellular matrix formation in polycaprolactone and starch-compounded polycaprolactone nanofiber meshes when seeded with bovine articular chondrocytes, *Tissue Eng. Part A*, **15**(2), 377–385.

77. Alves da Silva, M.L., *et al.* (2010). Cartilage tissue engineering using electrospun PCL nanofiber meshes and MSCs, *Biomacromolecules*, **11**(12), 3228–3236.

78. Araujo, J.V., *et al.* (2008). Surface controlled biomimetic coating of polycaprolactone nanofiber meshes to be used as bone extracellular matrix analogues, *J. Biomat. Sci-Polym. E.*, **19**(10), 1261–1278.

79. Araujo, J.V., *et al.* (2010). Dynamic culture of osteogenic cells in biomimetically coated poly(caprolactone) nanofibre mesh constructs, *Tissue Eng. Part A*, **16**(2), 557–563.

80. Chen, J.-S., Tu, S.-L., and Tsay, R.-Y. (2010). A morphological study of porous polylactide scaffolds prepared by thermally induced phase separation, *J. Taiwan Inst. Chem. E.*, **41**(2), 229–238.

81. Kim, H.D., *et al.* (2004). Effect of PEG–PLLA diblock copolymer on macroporous PLLA scaffolds by thermally induced phase separation, *Biomaterials*, **25**(12), 2319–2329.

82. Martel-Estrada, S.A., *et al.* (2010). Synthesis and thermo-physical properties of chitosan/poly(dl-lactide-co-glycolide) composites prepared by thermally induced phase separation, *Carbohyd. Polym.*, **81**(4), 775–783.

83. Rowlands, A.S., *et al.* (2007). Polyurethane/poly(lactic-co-glycolic) acid composite scaffolds fabricated by thermally induced phase separation, *Biomaterials*, **28**(12), 2109–2121.

84. Hua, F.J., Park, T.G., and Lee, D.S. (2003). A facile preparation of highly interconnected macroporous poly(d,l-lactic acid-co-glycolic acid) (PLGA) scaffolds by liquid–liquid phase separation of a PLGA–dioxane–water ternary system, *Polymer*, **44**(6), 1911–1920.

85. Liu, X. and Ma, P.X. (2009). Phase separation, pore structure, and properties of nanofibrous gelatin scaffolds, *Biomaterials*, **30**(25), 4094–4103.

86. He, L., *et al.* (2009). Fabrication and characterization of poly(l-lactic acid) 3D nanofibrous scaffolds with controlled architecture by liquid–liquid phase separation from a ternary polymer–solvent system, *Polymer*, **50**(16), 4128–4138.

87. Cai, Q., *et al.* (2002). A novel porous cells scaffold made of polylactide–dextran blend by combining phase-separation and particle-leaching techniques, *Biomaterials*, **23**(23), 4483–4492.

88. Antunes, J.C., *et al.* (2010). Novel poly(L-lactic acid)/hyaluronic acid macroporous hybrid scaffolds: Characterization and assessment of cytotoxicity, *J. Biomed. Mater. Res. A*, **94A**(3), 856–869.

89. Li, J., *et al.* (2010). A one-step method to fabricate PLLA scaffolds with deposition of bioactive hydroxyapatite and collagen using ice-based microporogens, *Acta Biomater.*, **6**(6), 2013–2019.

90. Parks, W.M. and Guo, Y.B. (2008). A casting based process to fabricate 3D alginate scaffolds and to investigate the influence of heat transfer on pore architecture during fabrication, *Mater. Sci. Eng., C*, **28**(8), 1435–1440.

91. Ryu, W., *et al.* (2007). The construction of three-dimensional micro-fluidic scaffolds of biodegradable polymers by solvent vapor based bonding of micro-molded layers, *Biomaterials*, **28**(6), 1174–1184.

92. Liu, C., Xia, Z., and Czernuszka, J.T. (2007). Design and development of three-dimensional scaffolds for tissue engineering, *Chem. Eng. Res. Des.*, **85**(7), 1051–1064.

93. Rezwan, K., *et al.* (2006). Biodegradable and bioactive porous polymer/inorganic composite scaffolds for bone tissue engineering, *Biomaterials*, **27**(18), 3413–3431.

94. Zhang, X., Reagan, M.R., and Kaplan, D.L. (2009). Electrospun silk biomaterial scaffolds for regenerative medicine, *Adv. Drug Delivery Rev.*, **61**(12), 988–1006.

95. Knez, Z., *et al.* Processing polymeric biomaterials using supercritical CO(2), *Chem. Ing. Tech.*, **83**(9), 1371–1380.

96. Barry, J.J.A., *et al.* (2006). Supercritical carbon dioxide: Putting the fizz into biomaterials, *Philos. T. Roy. Soc. A*, **364**(1838), 249–261.

97. Duarte, A.R.C., Mano, J.F., and Reis, R.L. (2009). Perspectives on: Supercritical fluid technology for 3D tissue engineering scaffold applications, *J. Bioact. Compat. Pol.*, **24**(4), 385–400.

98. Duarte, A.R.C., Mano, J.F., and Reis, R.L. (2009). Supercritical fluids in biomedical and tissue engineering applications: A review, *Int. Mater. Rev.*, **54**(4), 214–222.

99. Dehghani, F. and Annabi, N. (2011). Engineering porous scaffolds using gas-based techniques, *Curr. Opin. Biotechnol.*, **22**(5), 661–666.

100. Mooney, D.J., *et al.* (1996). Novel approach to fabricate porous sponges of poly(D,L-lactic-co-glycolic acid) without the use of organic solvents, *Biomaterials*, **17**(14), 1417–1422.

101. Singh, L., Kumar, V., and Ratner, B.D. (2004). Generation of porous microcellular 85/15 poly ((DL)-lactide-co-glycolide) foams for biomedical applications, *Biomaterials*, **25**(13), 2611–2617.

102. White, L.J., *et al.* The effect of processing variables on morphological and mechanical properties of supercritical CO(2) foamed scaffolds for tissue engineering, *Acta Biomater.*, **8**(1), 61–71.

103. Zhu, X.H., *et al.* (2008). Characterization of porous poly(D,L-Lactic-co-glycolic acid) sponges fabricated by supercritical CO2 gas-foaming method as a scaffold for three-dimensional growth of hep3B cells, *Biotechnol. Bioeng.*, **100**(5), 998–1009.

104. Annabi, N., *et al.* (2011) Fabrication of porous PCL/elastin composite scaffolds for tissue engineering applications, *J. Supercrit. Fluid.*, **59**, 157–167.

105. Intranuovo, F., *et al.* Uniform cell colonization of porous 3-D scaffolds achieved using radial control of surface chemistry, *Acta Biomater.*, **7**(9), 3336–3344.

106. Montjovent, M.-O., *et al.* (2008). Human fetal bone cells associated with ceramic reinforced PLA scaffolds for tissue engineering, *Bone*, **42**(3), 554–564.

107. Mou, Z.-L., *et al.* Preparation of porous PLGA/HA/collagen scaffolds with supercritical CO(2) and application in osteoblast cell culture, *J. Supercrit. Fluid.*, **58**(3), 398–406.

108. Salerno, A., *et al.* Design of bimodal PCL and PCL-HA nanocomposite scaffolds by two step depressurization during solid-state supercritical CO(2) foaming, *Macromol. Rapid Comm.*, **32**(15), 1150–1156.

109. Hile, D.D., *et al.* (2000). Active growth factor delivery from poly(D,L-lactide-co-glycolide) foams prepared in supercritical CO2, *J. Control. Release*, **66**(2/3), 177–185.

110. Jang, J.H. and Shea, L.D. (2003). Controllable delivery of non-viral DNA from porous scaffolds, *J. Control. Release*, **86**(1), 157–168.

111. Ginty, P.J., *et al.* (2008). A supercritical CO2 injection system for the production of polymer/mammalian cell composites, *J. Supercrit. Fluid.*, **43**(3), 535–541.

112. Santo, V.E., *et al.* (2010) Hybrid 3D structure of poly(D,L-lactic acid) loaded with chitosan/chondroitin sulfate nanoparticles to be used as carriers for biomacromolecules in tissue engineering, *J. Supercrit. Fluid.*, **54**(3), 320–327.

113. Santo, V.E., *et al.* (2012). Enhancement of osteogenic differentiation of human adipose derived stem cells by the controlled release of platelet lysates in a combined 3D construct, *J Control Release*, **162**(1), 19–27.

114. Tai, H.Y., *et al.* (2007). Control of pore size and structure of tissue engineering scaffolds produced by supercritical fluid processing, *Eur. Cell. Mater.*, **14**, 64–76.

115. vandeWitte, P., *et al.* (1996). Phase separation processes in polymer solutions in relation to membrane formation, *J. Membrane Sci.*, **117**(1/2), 1–31.

116. Duarte, A.R.C., Mano, J.F., and Reis, R.L. Enzymatic degradation of 3D scaffolds of starch-poly-(epsilon-caprolactone) prepared by supercritical fluid technology, *Polym. Degrad. Stabil.*, **95**(10), 2110–2117.

117. Duarte, A.R.C., Mano, J.F., and Reis, R.L. Novel 3D scaffolds of chitosan-PLLA blends for tissue engineering applications: Preparation and characterization, *J. Supercrit. Fluid.*, **54**(3), 282–289.

118. Duarte, A.R.C., *et al.* (2009). Processing of novel bioactive polymeric matrixes for tissue engineering using supercritical fluid technology, *Mater. Sci. Eng. C Mater. Biol. Appl.*, **29**(7), 2110–2115.

119. Duarte, A.R.C., Mano, J.F., and Reis R.L. Supercritical phase inversion of starch-poly(epsilon-caprolactone) for tissue engineering applications, *J. Mater. Sci. Mater. Med.*, **21**(2), 533–540.

120. Duarte, A.R.C., Mano, J.F., and Reis, R.L. (2009). Dexamethasone-loaded scaffolds prepared by supercritical-assisted phase inversion, *Acta Biomater.*, **5**(6), 2054–2062.

121. Duarte, A.R.C., Mano, J.F., and Reis, R.L. (2009). Preparation of starch-based scaffolds for tissue engineering by supercritical immersion precipitation, *J. Supercrit. Fluid.*, **49**(2), 279–285.

122. Temtem, M., *et al.* (2009). Supercritical CO(2) generating chitosan devices with controlled morphology. Potential application for drug delivery and mesenchymal stem cell culture, *J. Supercrit. Fluid.*, **48**(3), 269–277.

123. Duarte, A.R.C., Mano, J.F., and Reis, R.L. (2012). The role of organic solvent on the preparation of chitosan scaffolds by supercritical assisted phase inversion, *J. Supercrit. Fluid.*, **72**, 326–332.

124. Stevens, M.M. and George, J.H. (2005). Exploring and engineering the cell surface interface, *Science*, **310**(5751), 1135–1138.

125. Tuzlakoglu, K., *et al.* (2005). Nano- and micro-fiber combined scaffolds: A new architecture for bone tissue engineering, *J. Mater. Sci. Mater. Med.*, **16**(12), 1099–1104.

126. Tuzlakoglu, K., *et al.* (2011). Design of nano- and microfiber combined scaffolds by electrospinning of collagen onto starch-based fiber meshes: A man-made equivalent of natural extracellular matrix, *Tissue Eng. Pt. A*, **17**(3/4), 463–473.

127. Santos, M.I., *et al.* (2008). Endothelial cell colonization and angiogenic potential of combined nano- and micro-fibrous scaffolds for bone tissue engineering, *Biomaterials*, **29**(32), 4306–4013.

128. Kim, G., *et al.* (2008). Hybrid process for fabricating 3D hierarchical scaffolds combining rapid prototyping and electrospinning, *Macromol. Rapid Commun.*, **29**(19), 1577–1581.

129. Moroni, L., *et al.* (2008). 3D fiber-deposited electrospun integrated scaffolds enhance cartilage tissue formation, *Adv. Funct. Mater.*, **18**, 53–60.

130. Park, S.H., *et al.* (2008). Development of dual scale scaffolds via direct polymer melt deposition and electrospinning for applications in tissue regeneration, *Acta Biomater.*, **4**(5), 1198–1207.

131. Thorvaldsson, A., *et al.* (2008). Electrospinning of highly porous scaffolds for cartilage regeneration, *Biomacromolecules*, **9**(3), 1044–1049.

132. Pinho, E.D., *et al.* (2009). Degradable particulate composite reinforced with nanofibres for biomedical applications, *Acta Biomater.*, **5**, 1104–1114.

133. Bergshoef, M.M. and G.J. Vancso (1999). Transparent nanocomposites with ultrathin, electrospun nylon-4,6 fiber reinforcement, *Adv. Mater.*, **11**(16), 1362–1365.

134. Chen, G. and Liu, H. (2008). Electrospun cellulose nanofiber reinforced soybean protein isolate composite film, *J. Appl. Polym. Sci.*, **110**(2), 641–646.

135. Dodiuk-Kenig, H., *et al.* (2008). Performance enhancement of dental composites using electrospun nanofibers, *J. Nanomater.* DOI: 10.1155/2008/840254

136. Fong, H. (2004). Electrospun nylon 6 nanofiber reinforced BIS-GMA/TEGDMA dental restorative composite resins, *Polymer*, **45**(7), 2427–2432.

137. Gao, Y., *et al.* (2008). Electrospun nano-scaled glass fiber reinforcement of bis-GMA/TEGDMA dental composites, *J. Appl. Polym. Sci.*, **110**(4), 2063–2070.

138. Lin, S., *et al.* (2008). Electrospun nanofiber reinforced and toughened composites through *in situ* nano-interface formation, *Compos. Sci. Technol.*, **68**(15/16), 3322–3329.

139. Tian, M., *et al.* (2007). Bis-GMA/TEGDMA dental composites reinforced with electrospun nylon 6 nanocomposite nanofibers containing highly aligned fibrillar silicate single crystals, *Polymer*, **48**(9), 2720–2728.

140. Tian, M., *et al.* (2008). Fabrication and evaluation of Bis-GMA/TEGDMA dental resins/composites containing nano fibrillar silicate, *Dent. Mater.*, **24**(2), 235–243.

141. Martins, A., *et al.* (2010). Biodegradable nanofibers-reinforced microfibrous composite scaffolds for bone tissue engineering, *Tissue Eng. Part A*, **16**(12), 3599–3609.

142. Nerurkar, N.L., *et al.* (2010). Engineered disc-like angle-ply structures for intervertebral disc replacement, *Spine*, **35**(8), 867–873.

143. Yang, Y., Wimpenny, I., and Ahearne, M. (2011). Portable nanofiber meshes dictate cell orientation throughout three-dimensional hydrogels, *Nanomed. Nanotechnol.*, **7**(2), 131–136.

144. Lee, H.J., *et al.* (2010). Micropatterned fibrous scaffolds fabricated using electrospinning and hydrogel lithography: New platforms to create cellular micropatterns, *Sens. Actuators B*, **148**(2), 504–510.

145. Mather, R.R. (2006). Developments in textiles, *Med. Device Technol.*, **17**(8), 32–35.

146. Wollina, U., *et al.* (2003). Functional textiles in prevention of chronic wounds, wound healing and tissue engineering, *Curr. Probl. Dermatol.*, **31**, 82–97.

147. Sumanasinghe, R.D. and King, W. (2003). New trends in biotextiles—The challenge of tissue engineering, *J Text. Apparel Tech. Manag.*, **3**(2), 1–13.

148. Liu, Z.T., *et al.* (2010). Four-strand hamstring tendon autograft versus LARS artificial ligament for anterior cruciate ligament reconstruction, *Int. Orthop.*, **34**(1), 45–49.

149. Burns, J.P. and Snyder, S.J. (2009). Biologic patches for management of irreparable rotator cuff tears, *Tech Shoulder Elbow Surg*, **10**(1), 11–21. DOI: 10.1097/BTE.0b013e3181938499.

150. Inui, A., *et al.* (2010). Potency of double-layered poly L-lactic acid scaffold in tissue engineering of tendon tissue, *Int. Orthop.*, **34**(8), 1327–1332.

151. Shikinami, Y., *et al.* (2010). A biomimetic artificial intervertebral disc system composed of a cubic three-dimensional fabric, *Spine J.*, **10**(2), 141–152.

152. Konertz, W.F., *et al.* (2001). Passive containment and reverse remodeling by a novel textile cardiac support device, *Circulation*, **104**(12), I270–I275.

153. Riepe, G., *et al.* (2002). What can we learn from explanted endovascular devices?, *Eur. J. Vasc. Endovasc. Surg.*, **24**(2), 117–122.

154. Heim, F., Durand, B., and Chakfe, N. (2010). Textile for heart valve prostheses: Fabric long-term durability testing, *J. Biomed. Mater. Res. B*, **92B**(1), 68–77.

155. Inada, Y., *et al.* (2004). Regeneration of peripheral nerve gaps with a polyglycolic acid-collagen tube, *Neurosurgery*, **55**(3), 640–646.

156. Burg, K.J.L., Porter, S., and Kellam, J.F. (2000). Biomaterial developments for bone tissue engineering, *Biomaterials*, **21**(23), 2347–2359.

157. Salgado, A.J., Coutinho, O.P., and Reis, R.L. (2004). Bone tissue engineering: State of the art and future trends, *Macromol. Biosci.*, **4**(8), 743–765.

158. Hutmacher, D.W. (2001). Scaffold design and fabrication technologies for engineering tissues—State of the art and future perspectives, *J. Biomat. Sci-Polym. Ed.*, **12**(1), 107–124.

159. Gomes, M., *et al.* (2008). Natural polymers in tissue engineering applications, in *Tissue Eng.* Burlington: Academic Press, 145–192.

160. Tuzlakoglu, K. and Reis, R.L. (2009). Biodegradable polymeric fiber structures in tissue engineering, *Tissue Eng. Part B-Rev.*, **15**(1), 17–27.

161. Oliveira, A.L., *et al.* (2007). Micro-computed tomography (mu-CT) as a potential tool to assess the effect of dynamic coating routes on the

formation of biomimetic apatite layers on 3D-plotted biodegradable polymeric scaffolds, *J. Mater. Sci. Mater. Med.*, **18**(2), 211–223.

162. Almeida, L.R., *et al.* (2011). Innovative 2D/3D biotextiles for potential bone tissue engineering applications, *Proc. 4th ESAO-IFAO Joint Congress.*

163. Ribeiro, A.S., *et al.* (2011). Innovative bio-based multifilament fibers and textiles structures for bone regeneration, *Proc. 50th Dornbirn Man Made Fibres Congress.*

164. Pereira, D.R., *et al.* (2011). Hydrogels in acellular and cellular strategies for intervertebral disc regeneration, *J. Tissue Eng. Regen. Med.*, **7**(2), 85–98.

165. Pereira, R.C., *et al.* (2011). Encapsulation of human articular chondrocytes into 3D hydrogel: Phenotype and genotype characterization, *Methods Mol. Biol.*, **695**, 167–181.

166. Liu, H., *et al.* (2008). Fabrication of novel core-shell hybrid alginate hydrogel beads, *Int. J. Pharm.*, **351**(1/2), 104–112.

167. Chiu, L.L.Y. and Radisic, M. (2011). Controlled release of thymosin $\beta 4$ using collagen–chitosan composite hydrogels promotes epicardial cell migration and angiogenesis, *J. Control. Release*, **155**(3), 376–385.

168. Wang, X., *et al.* (2008). Sonication-induced gelation of silk fibroin for cell encapsulation, *Biomaterials*, **29**(8), 1054–1064.

169. Lu, Q., *et al.* (2011). Silk fibroin electrogelation mechanisms, *Acta Biomater.*, **7**(6), 2394–2400.

170. Yan, L., *et al.*, (2011). Hydrogels derived from silk fibroin: Methods and uses thereof, in *PT Provisional Patent*, INPI, Portugal.

171. Oliveira, J.T., *et al.* (2009). Performance of new gellan gum hydrogels combined with human articular chondrocytes for cartilage regeneration when subcutaneously implanted in nude mice, *J. Tissue Eng. Regen. Med.*, **3**(7), 493–500.

172. Pereira, D.R., *et al.* (2011). Development of gellan-gum based microparticles/hydrogel matrices for application in the intervertebral disc regeneration, *Tissue Eng. Part C-Meth.*, **17**(10), 961–972.

173. Silva-Correia, J., *et al.* (2011). Gellan gum-based hydrogels for intervertebral disc tissue-engineering applications, *J. Tissue Eng. Regen. Med.*, **5**(6), e97–e107.

174. Baran, E.T., Mano, J.F., and Reis, R.L. (2004). Starch-chitosan hydrogels prepared by reductive alkylation cross-linking, *J. Mater. Sci-Mater Med.*, **15**(7), 759–765.

175. Lee, H.J., *et al.* (2010). Micropatterned fibrous scaffolds fabricated using electrospinning and hydrogel lithography: New platforms to create cellular micropatterns, *Sens. Actuators, B*, **148**(2), 504–510.

176. Belyaeva, E., *et al.* (2004). New approach to the formulation of hydrogel beads by emulsification/thermal gelation using a static mixer, *Chem. Eng. Sci.*, **59**(14), 2913–2920.

177. Sa-Lima, H., *et al.* (2010). Stimuli-responsive chitosan-starch injectable hydrogels combined with encapsulated adipose-derived stromal cells for articular cartilage regeneration, *Soft Matter*, **6**(20), 5184–5195.

178. Pereira, R.C., *et al.* (2009). Novel injectable gel (system) as a vehicle for human articular chondrocytes in cartilage tissue regeneration, *J. Tissue Eng. Regen. Med.*, **3**(2), 97–106.

179. Oliveira, J.T., *et al.* (2010). Injectable gellan gum hydrogels with autologous cells for the treatment of rabbit articular cartilage defects, *J. Orthop. Res.*, **28**(9), 1193–1199.

180. Oliveira, J.T., *et al.* (2010). Gellan gum: A new biomaterial for cartilage tissue engineering applications, *J. Biomed. Mater. Res. A*, **93A**(3), 852–863.

181. Oliveira, J.T., *et al.* (2010). Gellan gum injectable hydrogels for cartilage tissue engineering applications: *In vitro* studies and preliminary *in vivo* evaluation, *Tissue Eng. Part A*, **16**(1), 343–353.

182. Santo, V.E., *et al.* (2009). Carrageenan-based hydrogels for the controlled delivery of PDGF-BB in bone tissue engineering applications, *Biomacromolecules*, **10**(6), 1392–1401.

183. Silva-Correia, J., *et al.*, (2011). *Photo-crosslinked gellan gum-based hydrogels: Methods and uses thereof*, INPI, Portugal.

184. Silva-Correia, J., *et al.* (2012). Biocompatibility evaluation of ionic- and photo-crosslinked methacrylated gellan gum hydrogels: *In vitro* and *in vivo* study, *Adv. Healthcare Mater*, doi: 10.1002/adhm.201200256.

185. Silva-Correia, J., *et al.* (2012). Angiogenic potential of Gellan gum-based hydrogels for application in nucleus pulposus regeneration: *In vivo* study, *Tissue Eng. Part A*, **18**(11/12), 1203–1212.

186. Volkmer, E., *et al.* (2008). Hypoxia in static and dynamic 3D culture systems for tissue engineering of bone, *Tissue Eng. Part A*, **14**(8), 1331–1340.

187. Karageorgiou, V. and Kaplan, D. (2005). Porosity of 3D biomaterial scaffolds and osteogenesis, *Biomaterials*, **26**(27), 5474–5491.

188. Druecke, D., *et al.* (2004). Neovascularization of poly(ether ester) block-copolymer scaffolds *in vivo*: Long-term investigations using intravital fluorescent microscopy, *J. Biomed. Mater. Res. A*, **68A**(1), 10–18.

189. Dalby, M.J., *et al.* (2002). *In vitro* reaction of endothelial cells to polymer demixed nanotopography, *Biomaterials*, **23**(14), 2945–2954.

190. Miller, D.C., *et al.* (2004). Endothelial and vascular smooth muscle cell function on poly(lactic-co-glycolic acid) with nano-structured surface features, *Biomaterials*, **25**(1), 53–61.

191. Zhang, H., *et al.* (2005). Microrobotics and MEMS-based fabrication techniques for scaffold-based tissue engineering, *Macromol. Biosci.*, **5**(6), 477–489.

192. Bettinger, C.J., *et al.* (2006). Three-dimensional microfluidic tissue-engineering scaffolds using a flexible biodegradable polymer, *Adv. Mater.*, **18**(2), 165–169.

193. Whitesides, G.M., *et al.* (2001). Soft lithography in biology and biochemistry, *Annu. Rev. Biomed. Eng.*, **3**, 335–373.

Chapter 12

# Engineered Scaffolds: Materials and Microstructure from Nanostructures to Macrostructures for Tissue Engineering

**Venu G. Varanasi,[a] Panayiotis S. Shiakolas,[b] and Pranesh B. Aswath[c]**

[a]*Biomedical Sciences, Texas A&M Health Science Center, Baylor College of Dentistry, 3302 Gaston Avenue, Dallas, TX 75246, USA*

[b]*Mechanical and Aerospace Engineering Department, University of Texas at Arlington, Arlington, TX 76019, USA*

[c]*College of Engineering and Professor, Materials Science and Engineering Department, University of Texas at Arlington, Arlington, TX 76019, USA*

Aswath@uta.edu

Scaffolds for tissue engineering and drug delivery application have made significant progress since the first engineered scaffolds were developed more than 20 years ago. Scaffolds are used for reconstructive and regenerative applications of both soft and hard tissues. Materials used for scaffold range from biodegradable polymers, bioactive ceramics, amorphous and glass ceramics, and combinations of the polymers and ceramics. The mechanism of bioactive ceramics and glass ceramics is reviewed with special attention on the combinatorial effect of ion release on gene expression

*Scaffolds for Tissue Engineering: Biological Design, Materials, and Fabrication*
Edited by Claudio Migliaresi and Antonella Motta

ISBN 978-981-4463-20-1 (Hardcover), 978-981-4463-21-8 (eBook)
www.panstanford.com

and mineralization process. The scaffolds developed should have an open architecture to promote vascularization while retaining the structural integrity during the integration to surrounding tissue. Latest developments in scaffold production including transfer printing replication, stereolithography, selective laser sintering, fused deposition modeling, 3D printing, robocasting, two photon polymerization, and inkjet printing are reviewed.

## 12.1 Introduction

Tissues are 3D self-assembled constructs developed to serve very specific needs based on their location. Typically the cells in a tissue have a common embryonic origin and similar structure and tend to perform a specific function. The organization of the cells in a tissue is based on function and may be classified into four basic types: (i) epithelial, (ii) connective, (iii) muscle, and (iv) nervous. The epithelial tissue covers the body and all internal cavities with the major function of protection, secretion, absorption, and filtration. The connective tissue is widely dispersed in the human body and serves the function of support and protection and includes loose connective tissues such as fat, dense connective tissue such as cartilage and bone, and vascular tissues such as erythrocytes, leucocytes, and platelets. Muscle tissue includes smooth, skeletal, and cardiac and can be broadly classified as voluntary or involuntary muscle tissue. Smooth muscle tissue and cardiac muscle tissue are involuntary while skeletal tissue is considered to be voluntary tissues. Nervous tissue is made up of specialized cells that constitute the central nervous system including the brain and spinal cord.

The seminal work by Langer and Vacanti[1] established the importance of scaffolds and still forms the basis of many of the developments today.[2–40] Scaffolds for tissue engineering would involve the development of a construct that can incorporate living cells together with bioactive molecules such as growth factors, enzymes, and proteins that can be used to repair, regenerate, or reconstruct damaged or diseased tissue. The basic principles of scaffold design include the following: (i) The scaffold material must be biocompatible and allow for cells to attach to the surface, differentiate, proliferate, and migrate into the scaffold and be able to lay down new matrix. (ii) The porous structure must allow for vascularization and integration with the surrounding tissue. One of the limiting factors in scaffold design that limit their application is improper vascularization which limits the removal of degradation products resulting in an inflammatory response in the surrounding tissue. (iii) The scaffold must have mechanical integrity to maintain form and function while being integrated with surrounding tissue and at the same time provide the signaling required to help in cell proliferation and tissue regeneration to integrate the scaffold with surrounding tissue. (iv) The scaffold must be biodegradable at a predictable rate that does not compromise the integrity of the scaffold and the surrounding tissue.

Application of scaffolds for hard tissue reconstruction and regeneration requires a structure with the morphology and chemistry that favors continuous cell and tissue remodeling together with stable vascularization of the construct. In particular, scaffolds for bone defects must be able to stimulate new bone growth and promote remodeling. In addition, the constructs must be able to provide the mechanical strength to compensate for the loss in strength of the diseased or damaged tissue and retain strength as new bone is deposited on the scaffold until such a time when the newly mineralized tissue completely replaces the scaffold.

Much of the early work has involved the use of autologous scaffolds for repair of skeletal defects with the iliac crest being the location of choice to harvest the bone.[41] The success of autologous bone grafts stems from the fact that these grafts already contain the important signaling molecules that assist in initial and subsequent integration with surrounding tissue, they are osteoconductive, osteoinductive, and osteogenic. Even in the case where autologous bone is used, failure rates approaching 50% are not uncommon. Moreover, the need for a second surgery, longer recovery times, and recurrent pain are issues of concern. In cases where sufficient bone or bone of suitable quality is not available in the patient, bone allografts have been used with as many as one-third of the bone grafts currently being allografts. However, the long-term clinical success of allografts is significantly lower than autologous grafts. In some cases, xenografts have been used with limited success as they suffer from the same issues as allografts that include rejection, inflammation, and infection. There is a need for engineered scaffolds to bridge the gap to augment or supplement autologous grafts and allografts.

## 12.2 Design Requirements for Scaffolds for Hard Tissue Regeneration

A basic principle of tissue engineering is that the scaffold used must provide a biomimetic environment where cells can differentiate and proliferate concomitant with the vascularization of the scaffold. This generally creates a set of design requirements that are in conflict, as the requirements for cell migration are usually an open and porous microstructure that facilitates the easy permeability and transport of cells and nutrients, while the design requirements for support and stiffness generally require a scaffold to have a denser microstructure and fewer pores in them.

Bone has a complex 3D hierarchical structure from the nanoscale to the macro scale and any scaffold that is used to replace diseased or damaged tissue should replicate this structure if possible. Natural bone derives it toughness by virtue of the multiple interfaces at all length scales that provide resistance to crack growth. Figure 12.1 illustrates the 3D structure of bone that shows collagen protein molecules making up the collagen fibrils at the nanoscale with nanocrystallites of hydroxyapatite crystals attached to them.[42] The fibrils self-assemble in the form of bundles to form microstructures, which at the next size scale integrate with osteocytes and canaliculi to form the building blocks known as Haversian canals. However, engineering such a complex structure is quite complex and a good theoretical foundation is essential to engineer a scaffold from ground up.

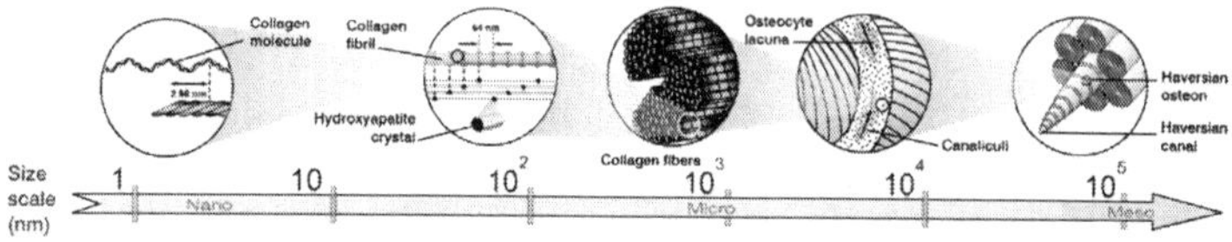


**Figure 12.1** Hierarchical structure of bone from the nano to the macro size scales. Reprinted from Ref. 42, Copyright 2013, with permission from Elsevier.

Newer developments in μCT and computer tomography have made it possible to develop 3D models of defect sites that need scaffolds for tissue engineering.[30] Using computational topographic design and solid free-form fabrication, Hollister[19] has shown that it is possible to reproduce a scaffold that can be perfect fit for bone regeneration. They used a series of basic unit cells with different levels of porosity to develop a hierarchical structure that mimics the anatomical structure of bone. They then used homogenization theory that takes a heterogonous problem and approximates it with a homogenous one. The homogenization theory may be applied across multiple length scales to arrive at the relationship between the smallest and largest features on the overall scaffold properties. Figure 12.2 illustrates schematically how an image file such as a computer-aided design (CAD)/computed tomography (CT) scan can be used as a template to develop a 3D model which then can be used to develop an anatomically correct 3D scaffold which can be manufactured by a rapid prototyping process.[19]

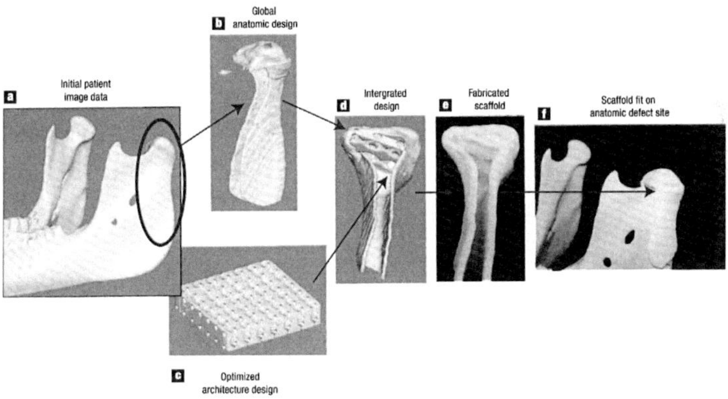


**Figure 12.2** Image-based procedure for integrating designed microstructure with anatomic shape. (a) A CT (as shown here) or an MRI scan serves as starting point for designing scaffold exterior. (b) The scaffold exterior shape is created with additional features for surgical fixation. (c) Architecture image design is created using computational topographic design (CTD). (d) Global anatomic and architecture design are integrated using boolean image techniques. (e) Solid freeform fabrication (SFF) is used to fabricate design from degradable biomaterial, in this case selective laser sintering (SLS) was used to fabricate a polycaprolactone (PCL) scaffold (fabricated scaffold created by Suman Das). (f) Final fabricated scaffold fits well on the intended anatomic reconstruction site. Reprinted by permission from Macmillan Publishers Ltd: Nature Materials19, copyright 2005.

Once the basic design of the scaffold has been developed, one has to choose the right material to achieve the functions for replacement and regeneration. The matrix material has to be osteoconductive and must have incorporated into it the osteoconductive signaling material and lastly have osteogenic precursor cells, all of which are supplied with an adequate blood supply to help in the regenerative process leading to the formation of osteoblasts that lay down new bone.[42] Figure 12.3 schematically illustrates this approach for the regeneration of bone.

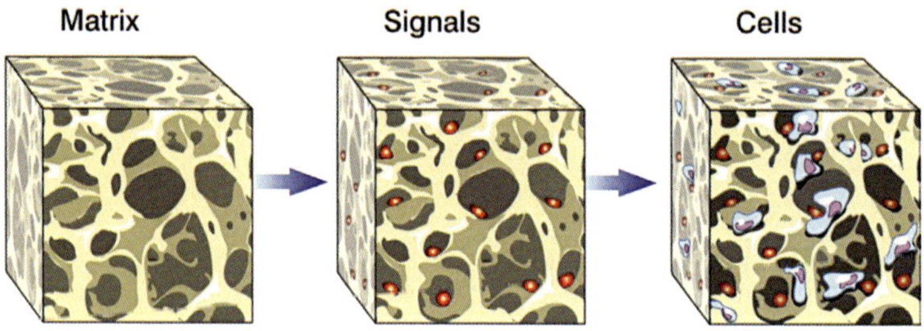


**Figure 12.3** A schematic figure that illustrates the steps in developing a scaffold, incorporating signaling molecules and osteogenic precursor cells to help in the development of fresh tissue on the scaffold. Reprinted from Ref. 42, Copyright 2013, with permission from Elsevier.

There are several materials that have been used in the development of scaffolds and they range from natural biomaterials such as autologous scaffolds, allografts, and xenografts, all of which have different limitations. Several natural and synthetic materials are currently explored for the development of scaffold matrix materials. These include synthetic polymeric materials, natural biomaterials, and crystalline and amorphous ceramic materials. The next section explores the different classes of materials, their properties, and their potential uses as scaffold materials.

## 12.3 Materials Used for Tissue Engineering and Drug Delivery Applications

### 12.3.1 Biopolymers

Natural polymers, synthetic polymers, natural and synthetic bioceramics, and their composites are the major classes of materials used in the development of scaffolds for hard tissue applications.

Earlier chapters have dealt in detail with natural and synthetic polymeric materials for scaffold applications and a brief summary of the materials used will be reviewed here. A more detailed focus will be placed on the role of bioceramics and polymer matrix composites in scaffold construction. Table 12.1 lists the most common synthetic biodegradable polymers used as scaffold materials. Synthetic polymers are hydrolytically degradable while natural polymers are enzymatically degradable.[24] Synthetic polymers are used more extensively as they have more predictable properties and fewer variations as compared to natural polymers.

#### 12.3.1.1 Polylactides

Polylactides are the most popular biodegradable biopolymers used in tissue engineering applications.[2,3,44–46] Lactides are chiral molecules that can exist as L-lactide and D-lactide and polymerization of the monomers using stannous octoate as a catalyst results in semicrystalline polymers. Polymerization of mixtures of D and L lactides usually results in DL-lactides (PDLLA) that are amorphous in nature. The most common structure that forms among lactides is Poly-L lactic acid (PLLA) which is a slow degrading polymer. PLLA together with PDLLA have been used extensively as scaffold materials and can be made into porous structures with as much as 98% open porosity by a variety of methods including salt leaching, sub-critical sublimation, etc. PDLLA by virtue of its amorphous structure typically has lower strength in comparison to PLLA and tends to lose it strength in 1–2 months and undergoes complete hydrolytic degradation over a period of 12 months. The hydrolytic degradation products of PLLA/PDLLA are typically just metabolic products such as water and carbon dioxide. More details of scaffolds from polylactides are provided in the earlier chapters.

#### 12.3.1.2 Polyglycolides

Polyglycolides have high crystallinity and has higher stiffness in comparison to PLLA and has low solubility in organic solvents.[2,24] Polyglycolic acid (PGA) degrades faster than PLLA and was first used as sutures. It generally degrades by scission of the ester linkages in the structure. Typically PGA loses its strength in the first two months and fully degrades in about 12 months. The high acidic by-products of degradation of PGA limit its use as a scaffold material where it is not possible to elute the by-products of the degradation. Applications with glycolide functional units typically are incorporated with other structures such as lactides to modulate the degradation process.

#### 12.3.1.3 Poly(lactide-co-glycolide) (PLGA)

Copolymers of glycolide and lactide have been made with ratios ranging from 10% to 90% and are amorphous in structure.[2,6,21,24,46–50] The copolymers are generally more unstable in comparison to the homopolymers with the copolymers with larger amounts of lactide being more stable. The degradation of the copolymers proceeds by hydrolytic degradation with the rate of degradation dependent on the ratio of lactide to glycolide as well as on the porous structure of the scaffold.

#### 12.3.1.4 Polycaprolactone (PCL)

They are semicrystalline polymers that are easily processable and soluble in a number of organic solvents making them suitable materials for biological applications. Moreover their low $T_g$ makes them attractive as a ductile scaffold for applications ranging from drug delivery to tissue engineering applications. However, their very long degradation times have resulted in the development of several copolymers that incorporate their functionality in their structure yielding a product that has the benefits of both the constituent elements.[24,26,51–54]

### 12.3.2 Bioceramics

A number of natural and synthetic bioceramics have been used in hard tissue repair and regeneration.[55–65] The bioceramics can be classified as bioinert, bioactive, and bioresorbable, the classification based on the reaction of the surface of the scaffold with the surrounding tissue. These bioceramics can be made in an open porous or dense microstructure based on whether the scaffold is used for tissue engineering or implant application and in some cases with a graded structure where they satisfy the needs of both structural requirements and tissue engineering applications. In addition, bioceramics may be broadly classified based on whether they are crystalline or amorphous in nature. Crystalline bioceramics include a number of phosphates that range from the bioresorbable tricalcium phosphate [$Ca_3(PO_4)_2$] to the bioactive hydroxyapatite [$Ca_{10}(PO_4)_6(OH)_2$ or HAp] to the bioinert $Al_2O_3$. Amorphous

**Table 12.1** Typical physical and mechanical properties of some common synthetic biopolymers[24]

| Polymer | Structure | $\rho$(g/cm$^3$) | $T_m$ (°C) | $T_g$ (°C) | Degradation time (months) | Tensile strength (MPa) | Strain to failure (%) | Modulus (GPa) |
|---|---|---|---|---|---|---|---|---|
| PLLA | 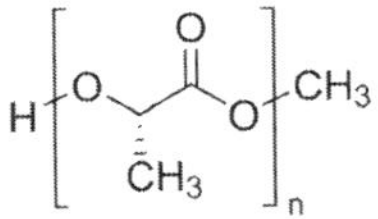 | 1.24–1.3 | 173–178 | 60–65 | >24 | 15–150 | 2.5–6 | 2.7–4.14 |
| PGA | 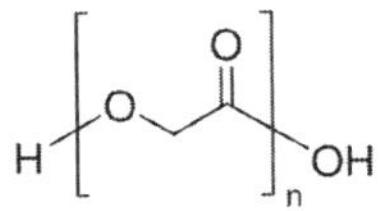 | 1.5–1.7 | 225–230 | 35–40 | 6–12 | 60–100 | 1.5–20 | 6–7 |
| PDLLA | 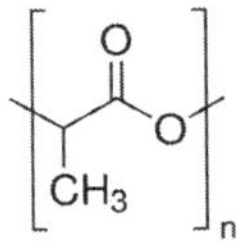 | 1.25–1.27 | Amorphous | 55–60 | 12–16 | 27–50 | 2–10 | 1–3.5 |
| PLGA | 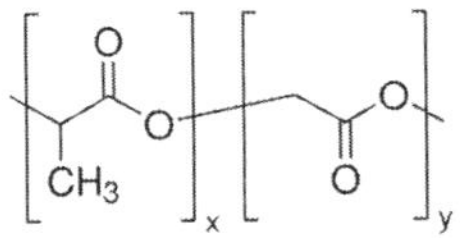 | 1.3–1.4 | Amorphous | 45–55 | 1–12 | 41–55 | 2–10 | 1–4.3 |
| PCL | 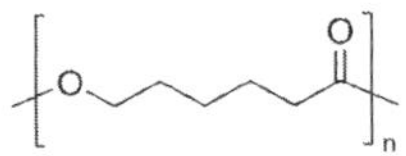 | 1.1– 1.15 | 58 | –72 | >24 | 20–42 | 300–000 | 0.2–0.4 |

*Source:* Adapted from Ref. 43.

bioceramics include customized chemistries that include Bioglass® 45S5 comprising 45 wt.% $SiO_2$, 24.5 wt.% $Na_2O$, 24.5 wt.% CaO, and 6 wt.% $P_2O_5$ that is the most bioactive of a series of glass ceramics developed by Hench. In addition, it has been shown that amorphous silica is significantly more bioactive in comparison to crystalline silica by virtue of the elution of $Si^{4+}$ ions that enhances the biomineralization process by activating the deposition and crosslinking of collagen and mineralized tissue.

#### 12.3.2.1 Crystalline bioceramics

HAp, $\beta$-TCP, and $\alpha$-TCP are some of the most common bioceramics used in scaffolds and coatings for implants and tissue regeneration applications. Table 12.2 shows some of the characteristic properties of these ceramics that make them attractive candidates for these applications.[3,29,58,66–76]

##### 12.3.2.1.1 *Hydroxyapatite (HAp)*

HAp is a widely used bioactive ceramic in scaffolds for tissue regeneration and repair as well as coatings of implants for prosthetic applications. HAp is one of the most stable of the phosphates of calcium under physiological conditions and is most similar in structure to natural biological apatite. In addition, due to the similarity in structure between natural apatite and synthetic Hap, there is a quick integration between the scaffold and the surrounding tissue. However, the mechanical properties of synthetic HAp are much poorer in comparison to natural cortical bone. Recent studies have indicated that by managing the initial and final grain size of HAp to be in the nanocrystalline size scale it is possible to improve the properties of the scaffold.

##### 12.3.2.1.2 *Tricalcium phosphate (TCP)*

$Ca_3(PO_4)_2$ (TCP) exists in two allotropes: $\alpha$-TCP and $\beta$-TCP, both exist at room temperature in metastable states. Both $\alpha$-TCP and $\beta$-TCP are bioresorbable with $\alpha$-TCP being more bioactive compared to $\beta$-TCP, the resorbed material is physiologically redeposited as HAp/Carbonated HAp in mineralized tissue. The slower degradation rate of $\beta$-TCP makes it a much more attractive candidate for bone scaffold engineering where the degradation rate may match up the ingrowth of new bone into the scaffold.

**Table 12.2** Typical properties of common crystalline bioceramics

| Chemistry (phases) | Ca/P ratio | Density ($g/cm^3$) | Solubility at 25°C, $-\log(K_{sp})$ | pH stability in aqueous solution at 25°C | Tensile strength (MPa) | Compressive strength (MPa) | Young's modulus (GPa) |
|---|---|---|---|---|---|---|---|
| Hydroxyapatite $Ca_{10}(PO_4)_6(OH)_2$ | 1.66 | 3.16 | 116.8 | 9.5–12 | 40–100[a] | 9–120[a] | 40–100[a] |
| $\alpha$-Tricalcium Phosphate $\alpha$-$Ca_3(PO_4)_2$ | 1.5 | 2.86 | 28.9 | Cannot be precipitated from aqueous solution | | | |
| $\beta$-Tricalcium Phosphate $\beta$-$Ca_3(PO_4)_2$ | 1.5 | 3.07 | 25.5 | Cannot be precipitated from aqueous solution | | | |
| Cortical Bone | 1.65–1.75[b] | 1.6–2.1[b] | — | — | Longitudinal 130–150[b]<br>Transverse 30–60[b] | Longitudinal 180–200[b]<br>Transverse 120–140[b] | Longitudinal 17–25[b]<br>Transverse 10–12[b] |
| Trabecular Bone | — | — | — | — | 1–5[b] | 4–12[b] | Longitudinal 0.2–1.0[b]<br>Transverse 0.01–0.4[b] |

*Source:* Adapted from Refs. 29 and 77.

[a]Depends on density and pore distribution in sintered product

[b]Depends on location and species of mammal chosen

#### 12.3.2.2 Bioactive glass ceramics

Glass ceramics have been developed in the 1960s with several different chemical combinations developed with varying degrees of bioactivity and bioresorbability. Some of the glass ceramics are entirely amorphous while some have crystalline phases embedded within an amorphous matrix.

##### 12.3.2.2.1 *Synthesis of bioactive glasses*

The first compositions of bioactive glasses were melt-derived.[62,65,78–88] This involves the use of ionic salts of the cationic and anionic sources that comprise the glass chemistry and combusting them at high temperature to form the powdered bioactive glass chemistry. These powders are solvated in ethanol, spread into a Pt crucible, and then melted at 1400–1500°C to form the melt, which is poured into a graphite mold and allowed to anneal at 500–600°C in air. This is the first FDA-approved bioactive glasses composition made in this fashion as 45S5. Since this time, many other glass compositions have been made. The advantage of this method is the myriad of chemistries that can be made with this method. The disadvantage of this method is the use of high temperatures and lack of flexibility of this method to be used with other materials, such as polymers, for scaffold fabrication. Other issues with this method of fabrication include a lack of control over properties of the glass matrix such as porosity and design that are needed for ensuring proper ingress of tissues occurs.

Sol–gel processing has also been adopted for bioactive glass fabrication.[65,78–80] Sol–gel involves the use of a solution of reagents (typically alkoxide precursors of cations) that are homogenized and subjected to a series of hydrolysis and polymerization reactions. A dense gel is formed after the precipitation of a wet gel that is then heated and dried. A porous gel/aerogel is formed by the extraction of the gel under super critical conditions. Certain types of glasses made from this method are FDA approved for amorphous silica-based compositions.

Bioactive glass nanoparticles have become increasingly used in various bone regeneration applications.[45,89–91] They can be fabricated by various milling processes after fabrication by melt-derived methods and sol–gel. Typically this involves the milling (ball milling, jet milling) and sieving fabricated samples to the desired particle size. Another method that has shown better control over the shape and chemistry of nanoparticle fabrication is flame-spray deposition.[90] In this method, liquid reagent sources of metals comprising the glass are fed into a flame reactor. The flame creates a "fly ash" that oxidizes rapidly in the high temperature oxygen environment. The nanopowder that results is then collected and used in its nanoparticulate form.[92] Table 12.3 reviews the chemistry and characteristics of some of the common bioactive/bioresorbable glass ceramics, all of which have the common theme of having fairly high $SiO_2$ content.

Single phase amorphous glasses are generally not structurally viable and when used as scaffolds it has been shown that when crystallized the glass ceramics have much better structural viability without sacrificing the bioactivity much. Crystallization of the amorphous glass is achieved by a high temperature treatment. All the compositions shown in Table 12.3 exhibit bioactivity and form a bond with bone. Several studies have indicated that in comparison to crystalline phosphate-based ceramics, amorphous and partly crystalline silicate-based glasses exhibit significantly higher levels of bioactivity. In addition, in studies with HAp with and without Si as a dopant it was demonstrated that in the presence of Si bone grew much more rapidly. All of these results indicate that the presence of soluble Si plays an important role in the tissue repair and regeneration process. Hence, a detailed understanding of the role of soluble Si on mineralization is warranted.

#### 12.3.2.3 Amorphous silica

Amorphous silica ($SiO_2$) is a material used in a number of applications, from glasses for optoelectronics industry to biomedical devices. Amorphous silica consists of four tetrahedrally bound O atoms that surround each Si atom with the bond direction toward the more electronegative O atom. The $SiO_2$ constituent is considered a network former by forming the basic glass matrix with an –Si–O–Si– repeating bridged structure (Fig. 12.4a).[94]

In this structure, there is no long-range ordering, and oxygen acts as a bridge between adjacent Si atoms. Use of amorphous silica in biomedical applications results in the formation of a hydration layer interface between the base material and tissues (Fig. 12.4c), which is also called a silica-gel layer. This layer forms on amorphous

**Table 12.3** Typical chemistry of Bioactive Glasses

| | Bioglass® | | | | Glass Ceramics | | | |
|---|---|---|---|---|---|---|---|---|
| **Parameter** | **45S5** | **S45PZ** | **45S5.4F** | **45B15S5** | **KG Cera Ceravital®** | **A/W Glass Ceramic®** | **Mina 13 Ceravital®** | **Bioverit®** |
| $Na_2O$ | 24.5 | 24 | 24.5 | 24.5 | 5–10 | 0 | 0 | |
| $K_2O$ | 0 | 0 | 0 | 0 | 0.5–3.0 | 0.4 | 0 | 5–9 |
| MgO | 0 | 0 | 0 | 0 | 2.5-5 | 4.6 | 5 | 6–28 |
| CaO | 24.5 | 22 | 14.7 | 24.5 | 20 | 44.9 | 33 | 13–28 |
| $Al_2O_3$ | 0 | 0 | 0 | 0 | 0 | 0 | 0 | 0 |
| $SiO_2$ | 45 | 45 | 45 | 30 | 46.2 | 34.2 | 46 | 29.5–50 |
| $P_2O_5$ | 6 | 7 | 6 | 6 | 0 | 16.3 | 0 | 8–18 |
| CaF | 0 | 0 | 9.8 | 0 | 0 | 0 | 0 | 2.5–7 (F) |
| $B_2O_3$ | 0 | 2 | 0 | 15 | 0 | 0 | 0 | 0 |
| $Ca_3(PO_4)_2$ | 0 | 0 | 0 | 0 | 25.5 | 0 | 16 | 0 |

*Source:* Adapted from Refs. 61 and 93.

silica-based materials when protons ($H^+$) from the physiological environment invade the glass network and bond with O molecules within the glass network to form hydroxyl groups ($OH^-$). When the density of protons is sufficiently high, the Si atom is surrounded by 4 hydroxyl groups, which forms silanols ($Si(OH)_4$, Eq. 12.1[95]).

$$SiO_x + xH_2O \rightarrow Si(OH)_{2x}. \tag{12.1}$$

$$\sum_{2}^{n} nSi\ (OH)_{2x} \rightarrow OH - [Si(OH)_x - O]_n - H + (n-1)H_2O \tag{12.2}$$

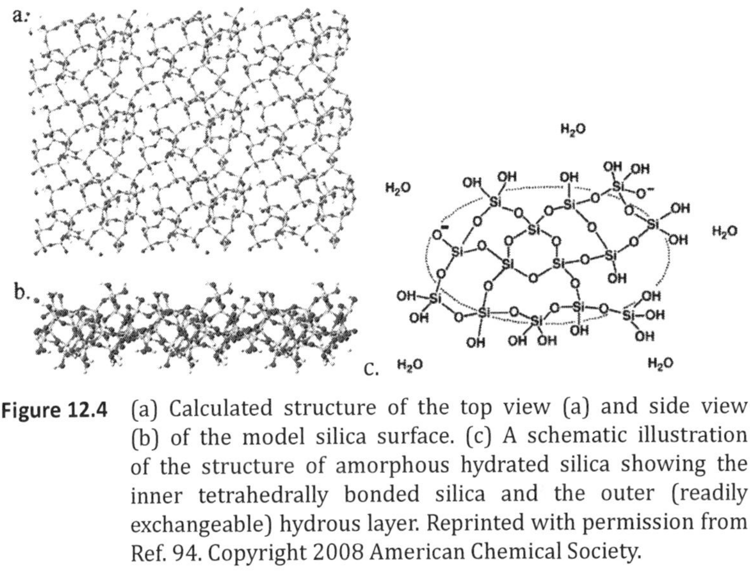


**Figure 12.4** (a) Calculated structure of the top view (a) and side view (b) of the model silica surface. (c) A schematic illustration of the structure of amorphous hydrated silica showing the inner tetrahedrally bonded silica and the outer (readily exchangeable) hydrous layer. Reprinted with permission from Ref. 94. Copyright 2008 American Chemical Society.

In this generalized model, it is assumed that the amorphous silica ($SiO_x$) is nearly stoichiometric ($x = 2$). This reaction first starts at the surface and results in silanol desorption. The sites left vacant are then filled by matrix silanols that migrate through the glass matrix, occupy the residual sites, and then release physiologically. As the number of silanols ($n$, Eq. 12.2) increases in the near surface layer, they spontaneously polymerize to form a silica-gel network on the amorphous silica surface, thus forming the hydration layer observed in Fig. 12.4b. This hydration layer is considered a nucleation site for tissue attachment.

### 12.3.2.4 Role of dopants in amorphous silica

Although dissolution does occur in amorphous silica networks, the process is very slow. The slow degradation is due to the relatively high Si–O bond energy. For this reason, minor constituents, known as dopants, can be added into the network to increase the degradation of amorphous silica. These dopants can come from four basic elemental categories when used in biological applications: Group I, Group II, Non-metal, and Transition metal elements.

Molecular dynamics and *ab initio* calculations of NMR, FTIR, and Raman spectra have been used in recent years to describe the effects that dopants have on the structure and migration of these dopants in amorphous silica-based biomaterials. In general, these dopants either substitute O atoms or create new coordination centers within the glass matrix. Bridging remains centered on Si–O–Si bonds. Cations substituted within the glass matrix create a polyhedral coordinated structure with the existing Si and O atoms. These new sites are created with a lower relaxation energy that facilitates cationic movement within glasses. The energy associated with such movement is lower for cations to migrate through the glass via "like sites." Relaxation energies for cation transport to unlike sites are greater since the site energy and coordination with adjacent O atoms must accommodate for the change in charge distribution. If an element substitutes for O, bridging between cations may or may not occur. In some cases, endpoints occur and ionic bonding, instead of covalent bonding, can occur.

**Group I dopants:** Typically comprise Na and K incorporation. These dopants are typically added in melt-derived processes. When these dopants are added, their network modification results in single alkali cation that is arranged in a polyhedron with coordinated O atoms. Cation migration occurs via discontinuous, thermally activated like site to like site movement. At higher concentrations of cations, ion channels within the glass matrix form that act as conduits for cation migration. Energetically, cation movement occurs most favorably through like site to like site nearest neighbors.[96]

Multicomponent alkali incorporation leads to reduced dissolution as compared to single alkali incorporation. Multi-cation incorporation results in a reduced density of like sites and an increased density of unlike sites. For example, a glass containing either K or Na is fully composed of ions that occupy like sites, whereas the multicomponent glass composition has reduced like sites and more unlike sites. The migration is thus reduced because it takes more energy for ions to transport between unlike sites. This is primarily due to local relaxation of the site once occupied and these are different for different cations. Moreover, these factors reduce ion diffusivity within the glass matrix and thus reduce the ion release from the glass.[96]

**Group II dopants:** Cationic migration is similar to that for group I with some exceptions. First, cations tend to migrate slower.[96] Single and multicomponent incorporation allows for chelated structures to form within glasses. $Ca^{2+}$ has also shown an ability to interrupt $Na^{+}$ diffusion within ion channels in the glass matrix by occupying sites within the diffusion path. All of these factors thus slow water ingress and formation of a hydration layer on the surface and within the glass matrix. [97]

**Transition metal dopants:** Typically, these consist of transition metal oxides that become incorporated in the glass through high temperature processing. Transition metal substitution follows similar trends as borate substitution and Group II substitutions. The exception in this case is that transition metals can have multiple valence states. This allows them to coordinate with O in such a way to more balance charge distribution in the glass. Just like Group II elements, these elements can block Group I ion channels within the glass matrix. Some examples include Ti, Fe, Co, Ni, Cu, and Zn. [98]

**Nonmetals:** Substitution of nonmetals into the glass matrix has a mixed effect. On one hand, elements such as N or F can substitute for O for direct bonding and coordination with Si. In the case of N, such substitution results in a change in the coordination of the glass matrix in which four N atoms are coordinated with three Si atoms.[99] Partial substitution of N for O, thus making silicon oxynitride, can also be accomplished to introduce more complex coordination chemistry. Typically, these materials are made using nanotechnology-based methods, such as chemical vapor deposition. In the case of F, there is some controversy on its bonding within glasses. It has been determined that such coordination is mostly oriented toward alkali or alkali earth cation substituted components within the glass matrix.[100] Only a small percentage of these atoms are actually bound directly to Si. Moreover, when F substitutes for O, the F atom terminates the bond and no bridging between adjacent Si atoms occur.

#### 12.3.2.5 Bioactivity of bioactive glasses

Bioactive glasses have been used in a myriad of biomedical applications. Since their inception by Hench and colleagues,[83, 84, 95] they have been adopted, modified, and used clinically to repair small bone fractures. The base composition most notably used is 45S5 (45 wt.% $SiO_2$, 24.5 wt.% CaO, 24.5 wt.% $Na_2O$, 5 wt.% $P_2O_5$). Since this time, number of compositions have been developed that incorporate dopants that we have described above. Before introducing other glass compositions, the basic concepts in the reaction of these glasses under physiological conditions are presented.

These materials are considered bioactive because they biodegrade. When these materials degrade they release ions to cells in their local physiological environment. During this degradation process, these materials form a surface hydroxyapatite layer that has the chemical, morphological, and mechanical properties of bone-like HA, a summary of the bioactive and bioinert range of glass compositions is given (Fig. 12.5). The bioactive range is considered to be below 59 wt.% $SiO_2$. When the composition remains above 40 wt.% $SiO_2$, these glasses partially degrade until their surfaces are completely covered with HA. When the composition of $SiO_2$ is at 25 wt.%, these glasses completely dissolve in aqueous environments (known as water glass). Thus, these materials have a range of compositions that can be used to design for the desired degradation properties.

Bioactive glass surface HA formation has been well-established.[42,65,88] Upon immersion in physiological conditions, an initial rapid ion exchange of alkali ions ($Na^+$) with protons ($H^+$) occurs. This facilitates invasion of these protons in the vacated sites that remain after $Na^+$ release. This is followed by a slower release of divalent cations ($Ca^{2+}$) and phosphate ions ($PO_4^{3-}$). The pH of the local environment increases as a result. Concurrently, the proton ingress into the glass surface and matrix bonds directly to bridging O atoms and as such ends the bridge and form an OH group around the Si atom. This forms a silanol that is subsequently released from the surface and is spontaneously polymerized to form a silica-gel network. This also creates a chemical potential for Ca and $PO_4^{3-}$ migration within the glass to the surface and from the physiological environment to the surface. This causes the adsorption, migration, and nucleation of Ca–P aggregates on the surface that eventually crystallizes to form hydroxyapatite. As we show in Fig. 12.6, the HA

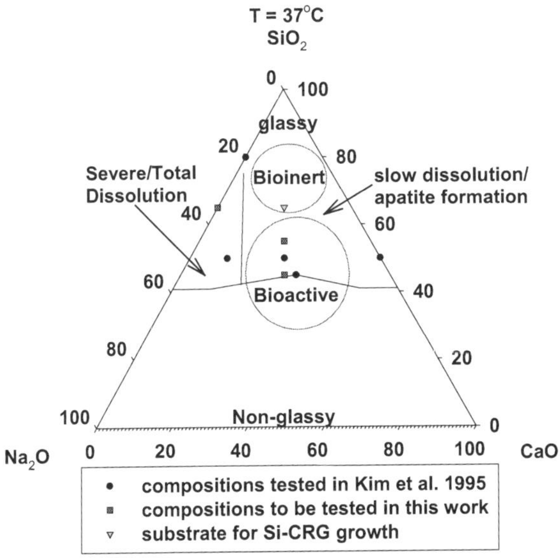


**Figure 12.5** Schematic phase diagram of bioactive glasses used for biomedical research. Data for this ternary phase diagram obtained from Kim, HM, Miyaji, F, *et al.* Bioactivity of $Na_2O$-CaO-$SiO_2$ glasses. *J. Am. Ceram. Soc.* 1995; 78(9):2405–2411.

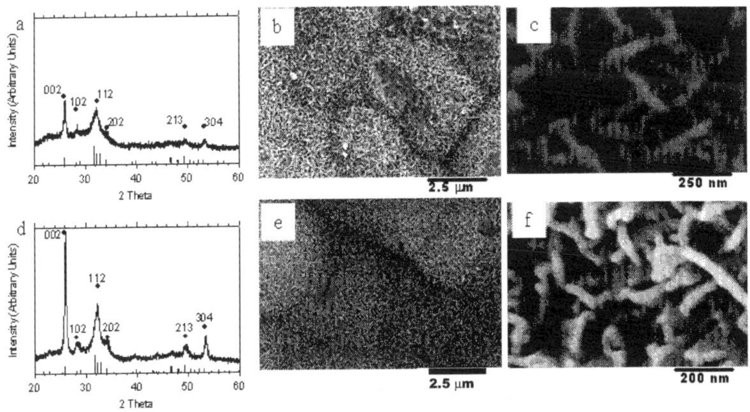


**Figure 12.6** Characterization of HA formation on bioactive glass surface: (a) XRD pattern for HA on 6P55 surface, (b) microstructure of HA on 6P55, (c) randomly oriented crystals of HA on 6P55, (d) XRD result of HA on 45S5, (e) HA microstructure on 45S5, (f) randomly oriented HA crystals of HA on 45S5.

layer has a needle-like shape with crystals sizes between 200 and 250 nm. On these surfaces, bulk grains have been observed to have an (0 0 2) preferred orientation after 4 weeks of *in vitro* degradation. In contrast, glass compositions that include multiple monovalent and divalent cations degrade slower and thus do not form as evolved HA structures. This particular crystal structure and orientation is amenable to osteoblast attachment and growth.[84]

#### 12.3.2.6 Response of osteoblasts to bioactive glass ions

The ions released from these materials were long thought to be used by cells as raw materials for mineralized tissue production. These glasses were primarily used as Ca and P delivery devices for osteoblast biomineral production. However, over the past decade, these ions have been found to play more of an inductive role in cell behavior.

Several cell types are considered as progenitors of osteogenesis including human periodontal ligament fibroblasts (hPDLF), MC3T3-E1 osteoblast precursors, human bone marrow stromal cells, and others. For example, hPDLF osteogenesis follows a timeline associated with its differentiation into a mineralizing phenotype that is similar to osteoblast differentiation. Starting from senescence (10–12 days), hPDLF express collagen type 1 (0–2 days after senescence), followed by alkaline phosphatase expression (2–8 days after senescence), and then osteocalcin expression and mineralized tissue synthesis (6–10 days after senescence).[101–104] Other cell types may exhibit a variance on this timeline, but the sequence of events is similar.[103,105] A summary of the osteogenic pathway affected by these ions is given in Fig. 12.7a. These markers are impacted by upstream expression of membranous receptors and signaling molecules that upregulate osteoblast-specific transcription factors that denote the osteoblast phenotype. Thus, for osteoproduction, bioactive glasses must target the expression of these markers.

Several studies have shown that all of these cellular processes are affected by these ions. The combined effect of the released ions enhanced collagen gene and matrix production, enzymes-associated collagen matrix development, and noncollagenous proteins involved in biomineral incorporation.[62,104,106–108] Moreover, transcription factors and membranous receptors were implicated in the regulation of such osteogenic effects. For example, growth factors (e.g., insulin-like growth factor, BMP-2)[109] and transcription factors (e.g., osterix, runx2, atf4, msx1, and SMADS) are tightly regulated osteogenic markers that themselves may only have 2- to 3-fold increased expression, but they have a much larger effect on downstream collagenous and noncollagenous genes and protein expression.[110,111] But, no conceptualized theory has been proposed as to how these ions affect osteogenesis, and no mechanism underlying these effects has been uncovered.

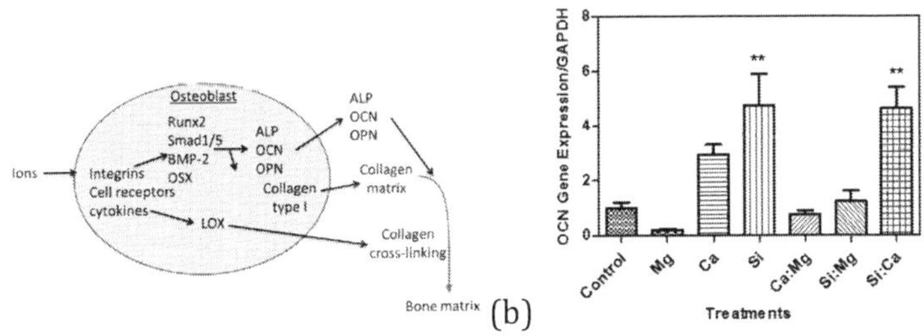


(a) (b)

**Figure 12.7** (a) Schematic of generalized osteogenic pathway for osteoblasts in the presence of bioactive glass ionic products. (b) Combinatorial effect of $Si^{4+}$, $Ca^{2+}$, and $Mg^{2+}$ on osteoblast differentiation after 6 days. OCN expression was enhanced by $Si^{4+}$ and $Ca^{2+}$ both individually and combinatorially; however, $Mg^{2+}$ individually downregulated osteocalcin (OCN) expression while it combinatorially decreased the enhanced effects of $Si^{4+}$ and $Ca^{2+}$.

More recently, a new concept in the effect these ions play has emerged. This role is involved in the combinatorial effect on osteogenic gene expression. It is well known that gene regulatory proteins (GRPs) are involved in the control over cellular gene expression in that each GRP is involved in the control of the expression of many genes while multiple GRPs are involved in the control of each expressed gene. It was found that an analogous role occurred with several of these bioactive ions (Fig. 12.7b).[112] It was found that Si and Ca combinatorially upregulate OCN expression,[106] while Si and Mg combinatorially regulate OCN expression. Individually,[109] Si and Ca upregulate many osteogenic markers,[113] while Mg downregulates these same markers.[114] This concept could be used in the design of many new biomaterials designs for future bioactive glass use.

#### 12.3.2.7 Clinical applications of bioactive glasses

Although numerous patients have been treated with bioactive glass, 45S5 is currently the only FDA-approved composition being widely used in biomedical applications. It is currently being used as a microparticulate known as Perioglass™. Other types of bioactive glasses that have been derived from 45S5 are being used as dental materials in the form of toothpaste and bone fillers (Fig. 12.8). These materials have shown some clinical success in stimulating bone formation in nonload bearing noncritical size defects.

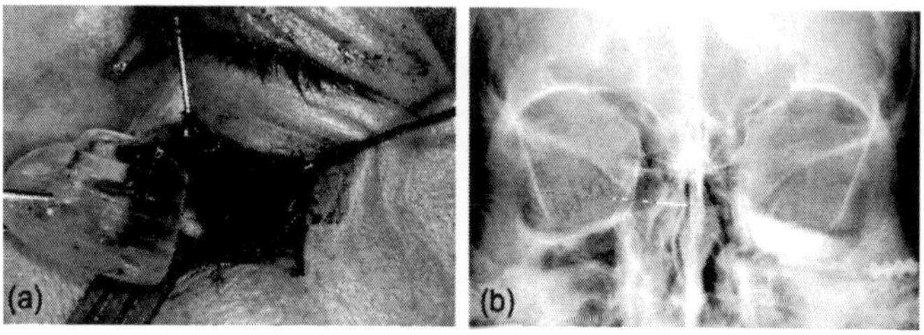


**Figure 12.8** Using cast bioactive glass monoliths for repair of orbital floors. (a) Inserting the glass implant beneath the eye. (b) Post-operative X-ray, showing that the bioactive glass implant has repaired the orbital floor and the eyes are now the same height.

Despite these clinical successes, several limitations have prevented further use. First, some evidence has shown that these materials can be micro-encapsulated into the connective tissues, even after more than a decade postimplantation, or the presence of glass granules.[115,116] Other than being used as particles, these materials are not easily formed into other shapes for clinical use in order to adapt to irregular defects. Finally, these materials are too brittle to be used in larger defects or defects that require strength in supporting the surrounding bone. Ideally, the scaffold should temporarily hold the defect stable until the body is ready to resorb it and put natural tissue in its place. The chemical release of ions and mechanical support of the defect can enhance new bone formation. Therefore, many new approaches have shifted to the use of hybrid scaffolds to address these issues.

## 12.4 Engineering Scaffolds for Drug Delivery and Tissue Engineering Applications

Several composites involving bioactive glasses and polymers have been attempted. Polymers such as polylactide (PLA), PCL, poly-

L-glutamic acid (PLGA), collagens, and gelatin have been used. In general, the incorporation of bioactive glasses into these materials is done to make more "ECM-like" scaffolds for bone regeneration. As a result, the polymer-bioactive glass composite has a higher elastic modulus than the polymer alone. In some cases, an increased compressive strength is also observed. The polymeric phase takes on the fibrous characteristics of the scaffold while the bioactive glass takes on more mineralized components of the scaffold. Up to this point, these composites have been fabricated by nonsynthetic routes and were simply combined together through solvent mixing.

More recently, attempts have been made to address the issues related to inadequate compressive strength by addressing the issue of covalent bonding between the glass and polymer phases. Two methods were attempted: (1) synthetic nanofibrous gelatin-bioactive glass scaffolds and (2) collagen nanofiber-amorphous silica-scaffolds. In the case of gelatin-bioactive glass composites, bioactive glass particles mimic HA in bone. It was found through the process of the porous architecture formation, a siloxane-nanofiber covalent bond was formed. In the case of collagen-amorphous silica scaffolds, Si was impregnated into the intrafibrillar collagen nanofibers (Fig. 12.9).[117] This was done to prevent the rapid degradation and collapsing of collagen that occurs with collagen sponges. The penetration of Si induced the formation of intrafibrillar mineral, which help slow the degradation of collagen. As compared to the relatively fast degradation of collagen scaffolds (within a few days after implantation), the collagen-bioactive glass scaffolds degrades in 6 weeks. During the degradation process, high levels of capillary angiogenesis occurred.

Although these scaffolds have potential to regenerate bone, there are still limitations to their use. First, it is unclear if the covalent bonding that occurs between the siloxane and polymeric chains will impart improved compressive strength to the scaffolds. This can be particularly challenging in large bone defects where significant bone volume is needed to support the surrounding bone as the scaffold promotes healing. Second, the degradation period is still too short. The degradation period must last longer than 1 month to properly sustain the defect while the body naturally resorbs and replaces the scaffold. This must be addressed by improving the similarity in degradation rates between the polymeric and glass materials. Therefore, these issues must be addressed before these scaffolds can be used for bone regeneration.

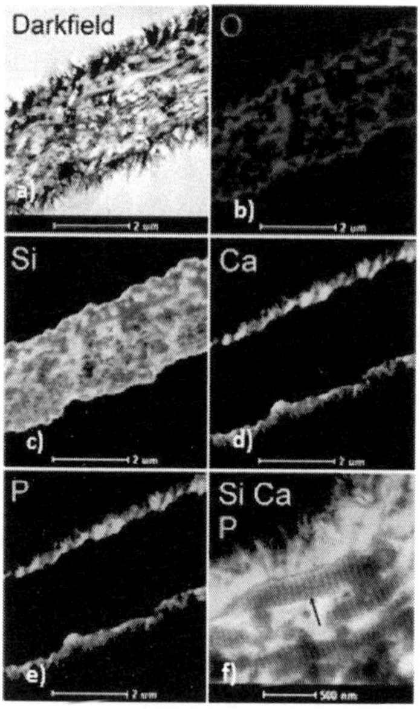


**Figure 12.9** STEM-EDX images/analysis of silicified collagen scaffold leaflet after immersion *in vitro*. (a) Dark field image of a collagen leaflet (consisting of multiple collagen fibrils) within the scaffold. Needle-shaped crystals were deposited along the periphery of the collagen leaflet. Distribution of O (b), Si (c), Ca (d), and P (e) within and around the collagen leaflet. (f) Bright magnification, merged distributions of Ca, P, and Si. Banded appearance of the collagen fibrils was reproduced by the hierarchical distribution of intrafibrillar silicon (arrow).

### 12.4.1 Rapid Prototyping

Rapid Prototyping (RP) is a loosely used term that refers to the use of various approaches and technologies for the quick fabrication of controlled geometry and net shape components of functional devices under computer control. Rapid prototyping is an additive fabrication technique where net 3D shapes are produced one layer at a time. These fabrication techniques include among others stereolithography (SLA), selective laser sintering (SLS), 3D printing, fused deposition modeling (FDM), two photon polymerization (TPP or 2PP), inkjet-based bioimprinting, and electrohydrodynamic (EHD) hot-jet plotting. Usually the name is associated with the type of materials used during the fabrication phase. A comparison of various RP techniques based on certain characteristics is shown in Table 12.4.

**Table 12.4** Comparison of different rapid prototyping (RP) methods on the basis of their achievable resolution, advantages, and disadvantages

| RP Technique | Resolution (μm) | Advantages | Disadvantages |
|---|---|---|---|
| SLA | 70–250 | Easy to remove support materials, easy to achieve small features | Limited choice of photopolymerizable and biocompatible liquid polymer materials |
| 3D printing | 200 | Low costs, no toxic components, fast processing | Weak bonding between powder particles rough surface postprocessing |
| SLS | 500 | High accuracy, good mechanical strength, broad range of materials | High temperatures during process, trapped powder is difficult to be removed |
| FDM | 250 | Low costs | Elevated temperatures during process, small range of materials |
| 3D plotting | 250 | Broad range of materials and conditions, incorporation of cells, proteins and fillers | Slow process, limited resolution |
| 2PP | 0.1 | Simple, fast, true 3D process, high resolution | Requires a femtosecond laser, material has to be photosensitive |

*Source:* Adapted from Ref. 118.

The common thread between the RP approaches is that a net shape component is fabricated based on a virtual component. The virtual component is drawn in a 3D solid modeling CAD software and exported in the standard SLA format. The SLA model is a discretized form of the virtual object. The SLA file is processed for each fabrication approach and a number of cross-sectional layers are developed. The thickness of each layer is based on the capabilities of the actual fabrication approach to be used, capabilities of equipment, and properties of the raw material used for fabrication.

Even though RP techniques are in general slow and usually fabricate one component at a time, they have a major advantage over other replication process. Their main advantage is the ability to fabricate personalized or customized scaffolds based on actual scanned data from scanning devices such a CT, MRI, or 3D echocardiography for the actual location where the scaffold will be embedded. Even though scaffolds with complex internal architectures could be fabricated with RP, one must assure that scaffold interconnectivity is preserved. The limitations of the scaffold features (feature size, porosity and pore size, interconnectivity, and surface roughness) depend on the actual RP technique, materials, and hardware used for fabrication.[119–121]

Many processes have been proposed and demonstrated in the open literature for the fabrication of scaffolds for tissue engineering and drug delivery applications. These processes have mainly been demonstrated in research environments. In clinical settings one could argue that various implants are scaffold-type materials designed to perform one or more functions. In this review, the focus will be on fabrication processes employed for the fabrication of polymeric and polymer matrix composite scaffolds as well as scaffolds that include bioactive glass ceramics. An overview of the fabrication process will be presented, and a discussion on the advantages and disadvantages, limitation on materials processing, and challenges will be addressed.

The type of application for a scaffold will dictate the scaffold properties including scaffold material, scaffold architecture, geometry, pore size, porosity and interconnectivity, short- and long-term structural strength and integrity, biodegradability rate, desired feature size for cell seeding and proliferation, etc. The scaffold properties or characteristics must be considered in deciding a particular fabrication process. The fabrication processes are characterized by the size of features they could produce at the micro to nano level, the properties of the materials employed such as glass transition, viscosity, thermal stability, adhesiveness and elastic/plastic strength, the ability to fabricate amorphous or controlled net shape 3D scaffolds, and the ability to fabricate one of a kind (personalized scaffold), multiple but limited number or transition to mass production for clinical settings.

Polymeric-based scaffolds in general could be fabricated using either subtractive or additive processes or a combination of the two. Processes employed to fabricate 2D, 2½D, or 3D, amorphous and/or controlled net shape features, and scaffolds are examined. These fabrication processes include salt leaching and solvent extraction, transfer printing replication processes such as hot embossing and nanoimprint lithography (NIL), rapid prototyping-based processes such as fused deposition, inkjet printing, and TPP. Secondary processes could be performed for further processing of a fabricated scaffold for surface texturing or subtractive material removal using femtosecond lasers.

#### 12.4.1.1 Transfer printing replication

Transfer printing replication processes examined are hot embossing and NIL. These processes are well suited to produce 2D and 2 ½D features with sub-micron or nano-feature size and not direct net shape 3D scaffolds. These are replication processes and as such they require the fabrication of a transfer mold to be replicated. The features to be replicated on the transfer mold at best could have straight walls but for ease removing the mold, a release angle must be considered in the design of the mold. The final replication quality cannot be better than the quality of the mold. As such, a high-quality mold is desired and it is not unusual for the mold to degrade as a function of the number of replications and processing conditions. As the desired size of the features to be replicated decreases in size, the fabrication of the transfer mold might require the use of more specialized equipment such as LIGA, chemical etching, or femtosecond laser micromachining. The primary mold degradation issue could be addressed through a two stage process where the primary mold is not the negative but rather the actual desired structure.[122]

The versatility of these replication processes lie in the fact that at least in theory, thermoplastic polymers heated at and held above their glass transition temperature could replicate the negative of the transfer mold. The replication success, as mentioned, depends on the quality of the transfer mold, the correct definition of the replication process parameters such as replication temperature, temperature control, and stability, the applied pressure between the highly viscous polymer and transfer mold, and the time of replication. Proper control of the process parameters will improve the outcome of the process; for example, keeping the temperature above the glass transition level while maintaining the correct pressure will provide enough time for the viscous polymer to flow and fill the voids of the transfer mold. Even though these replication processes could be employed to fabricate high aspect ratio structures, the user must be extremely careful during the separation of the mold and polymeric substrate as the size of the replicated features are relatively small and can be easily distorted during mold separation. Figure 12.10 illustrates some of the hot embossed features.

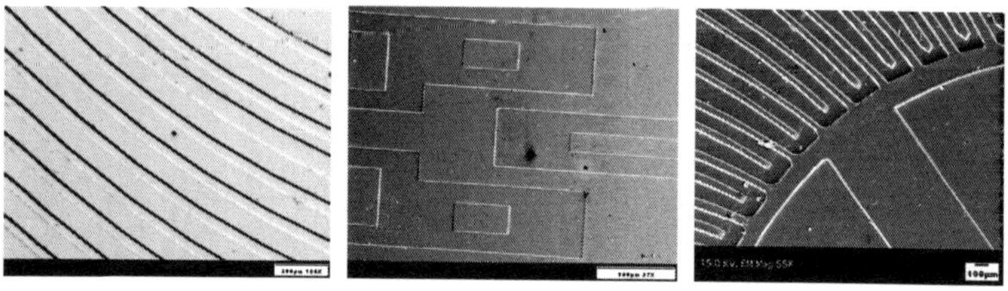

**Figure 12.10** Hot embossing replicated features.

These replication processes are traditionally used to fabricate 2½D polymer sheets with many features of varying size and geometry and not net shape 3D scaffolds. In order to fabricate a 3D scaffold using replication processes a number of secondary steps are required. Three-dimensional structures may be fabricated by sequentially applying a transfer mold on the previously fabricated layer.[20]

The challenges in using transfer printing processes for 3D scaffold fabrication through repeated replication relate to the control of the process parameters to avoid damaging the micro-nano features on the existing layers. Materials used for the thermal transfer processes, such as PMMA, PC, and PDMS, varied. The application of thermal transfer processes to biocompatible and biodegradable polymers PLLA (poly-L-lactic acid) and PLGA [poly(lactic-co-glycolic acid)] has been successfully demonstrated.[21,123,124]

#### 12.4.1.2 Stereolithography

This process involves the use of a laser system coupled with a computer aided three or four axis stage system that allows for a laser to be used to cure a photocurable monomer that in turn may or may not contain ceramic powder.[30,125] A laser beam scans the vat in a controlled path and causes the monomer to cure and polymerize. The polymerized polymer now acts as a binding agent holding the ceramic powder together. Once a layer is polymerized, the stage holding the newly polymerized layer is lowered causing the polymerized layer to be covered by a new layer of monomer. The laser beam is directed to scan this new layer according to the instructions from the SLA layered file. The process is repeated until the scaffold is fabricated (see Fig. 12.11). Scaffolds matching the elastic properties and porosity of a human trabecular bone have been fabricated with intricate internal scaffold architecture using SLA.[30,125]

#### 12.4.1.3 Selective laser sintering (SLS)

This is an additive process and operates on similar principles as SLA (see Fig. 12.11) but without the need of a photocurable monomer vat, thus eliminating the possibility of foreign or toxic material entrapped in a complex geometry scaffold.[53,126,127] The process lays a thin layer of a powder (~100–200 μm thick) material using a cylindrical roller to guarantee layer uniformity. Then, through computer control as the laser traverses a predefined path, the laser beam sinters the powder locally, thus creating the part geometry for that particular layer and fused on the previous layer. Subsequently another layer of powder is laid and the process repeats itself. The nonsinter powder serves as a support structure during the fabrication process. Any powder material that will not decompose when irradiated with the laser beam but rather meld and fuse is a candidate for SLS. Even though this process could be employed and has been demonstrated in fabricating scaffolds with complex geometries, care must be taken during scaffold design to assure that pathways for removing the nonsintered support powder are provided once scaffold fabrication has been completed. The SLS method has been extensively used to fabricate 3D porous bioresorbable PCL scaffolds as well as scaffolds that incorporate biodegradable polymer along with a bioactive/ bioresorbable glass ceramic.[53,126–130]

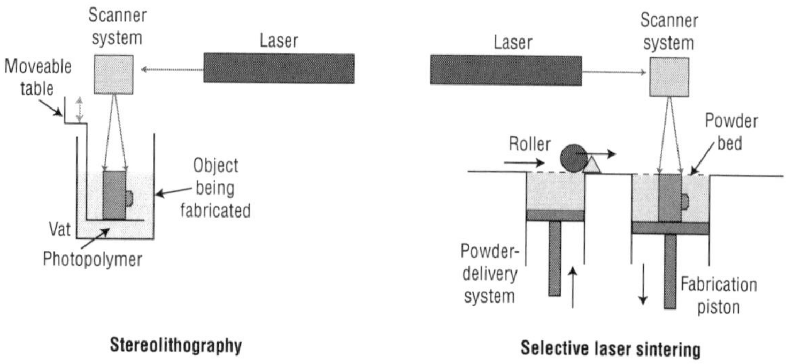


**Figure 12.11** Schematics of rapid prototyping systems showing the mechanism of SLA and selective laser sintering. Reprinted from Ref. 42, Copyright 2013, with permission from Elsevier.

### 12.4.1.4 Fused deposition modeling

This is an additive fabrication process that has seen exponential growth in the last few years not only in biomedical applications but also in general manufacturing and rapid prototyping for hobbyists. This revolution is due to the low prices for home devices that come with all the required software. The working principle of FDM is different from SLA and SLS.[30,131] In the FDM process, the additive polymeric material is extruded—pushed through a controlled temperature heated nozzle. As the polymeric material is pushed through the nozzle, the nozzle moves in a layer-based predefined path according to the SLA file instructions and deposits the highly viscous polymer causing it to adhere to the previous layer. Once the whole layer is completed, the whole built assembly is lowered by one layer thickness and the procedure is repeated until the component is fabricated. The size of the deposited polymer and eventual features depends on the size of the nozzle used and the solidification properties of the polymer. In addition, some FDM manufacturers require a controlled temperature environment for quality fabrication. One of the issues that FDM equipment with soluble supports easily addresses is that of the fabrication of complex internal architecture—geometry scaffolds. Once the scaffold is fabricated, the support structure is chemically removed. Thus, one must assure that the whole support structure has been removed and that no toxic chemicals entered the scaffold and possibly adhere to it during the cleaning process. Thus, the FDM process might prove more versatile for the fabrication of 3D scaffolds as compared to the SLS process especially in removing the support structure.

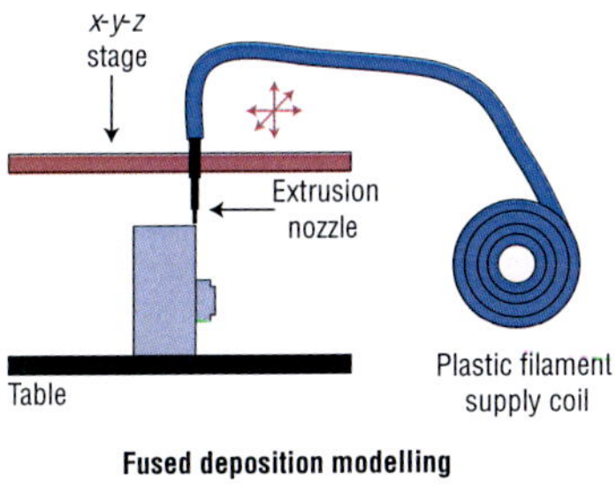


(a)

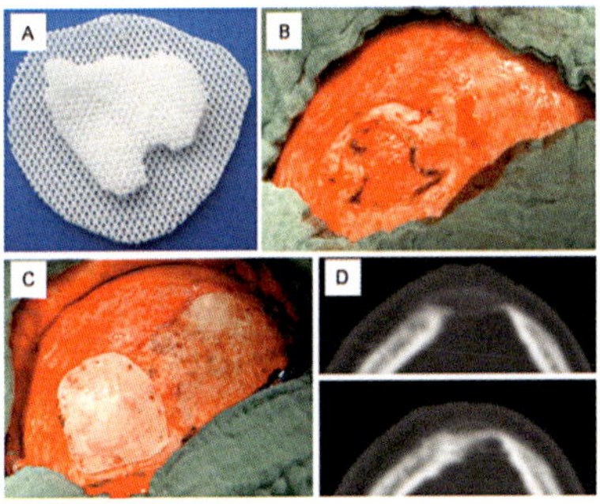


(b)

**Figure 12.12** (a) Schematics of rapid prototyping systems showing the mechanism of FDM. Reprinted from Ref. 42, Copyright 2013, with permission from Elsevier. (b) Example of cell-free clinical application of tissue engineering: calvarial reconstruction using PCL-calcium phosphate scaffolds. (A) Scaffold designed from medical CT imaging data and fabricated by FDM. (B) Calvarial defect. (C) Defect after implantation of scaffold. (D) CT images showing beginning of bony consolidation of the defect after 6 months. Reprinted from Ref. 148, Copyright 2012, with permission from Elsevier.

The FDM process is based on the controlled extrusion of polymer material through a nozzle. As such, this process is suitable for experimentation with various polymers to assess their suitability as potential scaffold material. Also, the FDM process could be employed to fabricate scaffolds using multiple materials depending on the desired scaffold properties and is well suited for applications requiring blends of biocompatible and biodegradable polymers. The FDM process has been used to fabricate 3D scaffolds of poly(D,L-lactide-co-glycolide) (PLGA), where it was observed that the average molecular weight of PLGA decreased to about 60% of the original value after the process.[131]

#### 12.4.1.5 3D printing process

The 3D printing process first starts with deriving a CT scan of the patient and/or the region that needs to be reconstructed with a scaffold. Using the appropriate conversion, the CT scan is converted using an engineering design language such as a CAD program to create a 3D model of the region to be reconstructed. The 3D model is now recreated in physical space using an additive manufacturing process where layers of material are laid down in stages. A printer not unlike an inkjet printer is used to lay down the different constituents on a surface, and by using multiple passes it is able to build up a 3D object. The resolution of the features in the 3D printing process is an engineering limitation, and current technology is able to develop features with size ranging from 20 μm to 100 μm (see Fig. 12.13). The advantage of this process lies in the ability to customize the scaffold for individual patients from the CT scan of their particular defect. Typical times to create a scaffold range from a few minutes to a few hours depending on the complexity and size of the scaffold that needs to be reconstructed. It is possible to construct composites constituted of biodegradable and bioactive polymer matrix with a bioactive and bioresorbable ceramic or glass ceramic particles dispersed in the matrix.

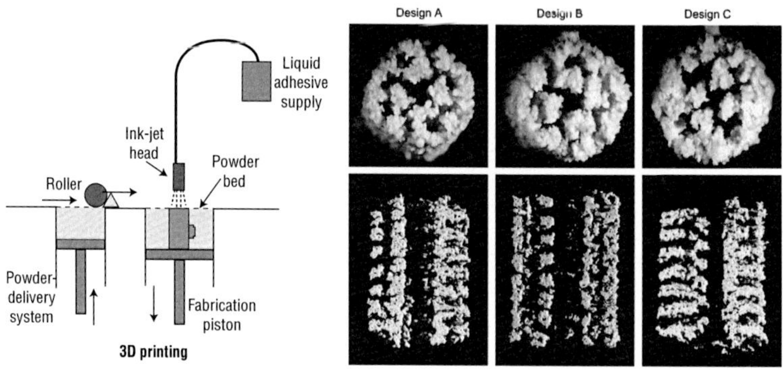


(a) (b)

**Figure 12.13** (a) Schematics of rapid prototyping systems showing the mechanism of FDM and 3D printing.[30] Reprinted from Ref. 42, Copyright 2013, with permission from Elsevier. (b) The photographs in the first row provide a top view of the opaque 3D-printed HA scaffolds with the central channel. The tomographic data can be used to produce any desired virtual cut of the different scaffolds, for example visualizing the internal morphology by the axial cut presented in the second row. The diameters of the scaffolds range from 3.9 to 4.2 mm and the minimal diameters of the central channel from 0.70 mm to 0.87 mm for designs A, B, and C, respectively. Reprinted from Ref. 132, Copyright 2008, with permission from Elsevier.

#### 12.4.1.6 Robocasting

In this method, the synthesized bioactive glass is suspended in a solvent and immediately withdrawn into a syringe. The syringe is mounted into a plunger-depression assembly. The motorized printing device, such as a robocast-assisted deposition system, then extrudes the suspended materials into porous scaffolds. The solvents used typically have high boiling points (40–50°C). Scaffold designs can be seen in Fig. 12.14. These materials can also be hybridized with organic materials (e.g., biopolymers) as long as the boiling points of both organic and inorganic precursors are very close to each other.

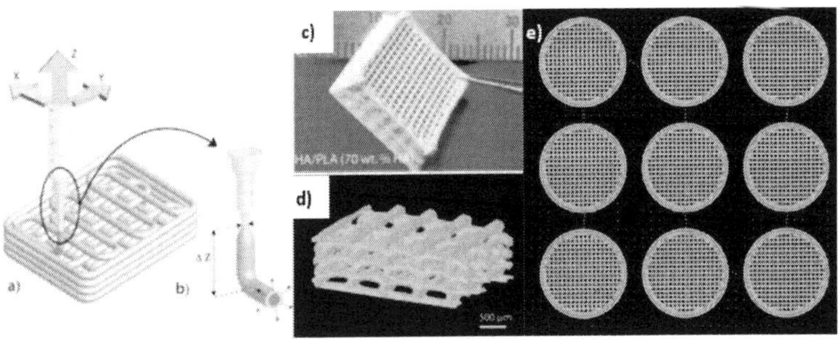

**Figure 12.14** (a) Schematic representation of the RAD process. The syringe displacement and the liquid flow through the nozzle are controlled by the computer. (b) During the printing, the inks swell after leaving the capillaries and the evaporation of the solvent creates a solid skin favoring the consolidation of the printed line. (c) Low-magnification image showing one of the scaffolds fabricates (ceramic/polymer, 70 wt.% ceramic scaffolds with 17 layers). (d) Three-dimensional reconstructed image of a ceramic/polymer (70 wt.% HA) grid obtained by synchrotron X-ray computed tomography. It can be observed that the material does not exhibit large defects. (e) Layout of multiple scaffolds fabricated in one session of robocasting. Reprinted from Ref. 42, Copyright 2013, with permission from Elsevier.

#### 12.4.1.7 Two photon polymerization

Two photon polymerization (TPP or 2PP) is similar in concept to other prototyping techniques where a device or component is available in CAD software and then exported as SLA.[133,134] The exported file is processed and machine instruction codes are generated for each layer of the component. The component is then fabricated on a brick-by-brick layer-by-layer process by scanning a femtosecond laser pulsing at high repetition rates (~80 MHz) in a transparent liquid vat of monomer and photoinitiator solution. The high repetition rate is required to deliver two photons at the same molecule almost instantaneously to initiate the polymerization solidification process. The 2PP process has been employed to produce components and features in the micro- and nanoscales. A range of monomer and photoinitiators have been used in the process. The concentration of photoinitiator present must be carefully controlled to avoid exceeding threshold limits that are considered to be toxic.

The interaction of the tightly focused femtosecond laser pulses with the monomer/photoinitiator solution is via two-photon absorption process. This initiates the photopolymerization process and generates a solidified voxel (volumetric pixel). A component is prototyped by scanning the laser to generate a series of voxels on a layer by layer. The 2PP process has the ability to fabricate multiscale structures with features from the sub-100 nm level to dimensions comparable with the vat holding the liquid solution. The smaller feature size to be fabricated with 2PP is the size of the voxel. The size of the voxel is a function of the material properties of the monomer and photoinitiator and the laser process parameters such as pulse repetition rate, laser power level, laser scanning speed, and focusing lens. Recent theoretical and experimental work indicated that the size of the voxel could be predicted and employed in defining process parameters. Experimentally defined process parameters yielded a voxel with an error of 5% between the estimated and experimental ones.[133,134] Even though 2PP process is a single step, it is a linear process and is slow. Recently, work has been reported where multiple identical components have been fabricated by generating multiple foci in the solution or exhibited higher scanning rates to reduce fabrication time.[135] The 2PP process has been extensively investigated in the fabrication of 3D scaffolds for tissue engineering with various organic–inorganic resists and photoinitiators. The 2PP process is similar to SLA but at a smaller scale and due to the fact that SLA is an established RP fabrication process, a range of biocompatible materials is available.

The list of materials is reproduced here and the interested reader could refer to Gittard *et al.* for the detailed references where these materials were first discussed and used in the fabrication of biocompatible scaffolds (see Fig. 12.14).[136] The material list includes Ormocer® (organically modified ceramics), SU-8, polyethylene glycol diacrylate, biodegradable PCL-based polymers, biodegradable methacrylated polylactide, Zr-based sol–gel composites, Ti-based sol–gel composites, metal ion–doped acrylates, commercial acrylates (e-Shell 300), bovine serum albumin, fibrinogen, gelatin, and collagen. Ormocers are materials containing both organic (polymer) and inorganic (ceramic) functional groups. Other materials used to fabricate biocompatible 3D scaffolds to assess cell-structure interaction include hybrid organic-inorganic sol–gel (SZ2080) and 4,40-bis(diethylamino)benzophenone at 1% concentration.[137] Three-dimensional scaffolds using Ormocore b59 and Ormoclear and sol–gels based on zirconium and silicon oxides SZ2080 have been successfully fabricated using 2PP and seeded with cells. A biodegradable gelatin material, Methacrylamide-Modified Gelatin (GelMod), was modified to control its degradation behavior and used in the fabrication of 3D scaffolds using 2PP and Irgacure 2959 photoinitiator.[138,139] The use of acrylate-based resins dissolved in different water concentrations was tested for 2PP structuring and toxicity for *in vivo* applications. A water-soluble molecule 1,4-bis(4-(*N,N*-bis(6-(*N,N,N*-trimethylammonium) hexyl)amino)-styryl)-2,5-dimethoxybenzene tetraiodide (WSPI) was used as photoinitiator with polyethylene glycol (PEG) spacer with acrylate reactive end groups.[140]

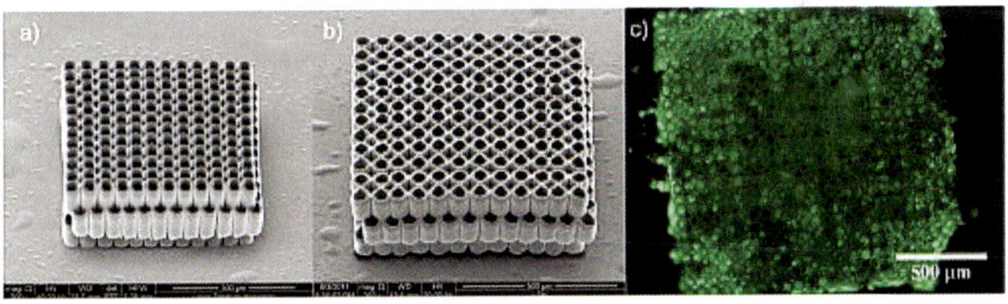


**Figure 12.15** In this image, a multilayer tissue scaffold created with single-focus two-photon polymerization ("2PP") (a) is shown beside one created with multifocal two-photon polymerization (b). The two scaffolds are structurally similar, but the one created with multifocal 2PP was completed in approximately one-fourth the time. In image c, cells from the inner lining of bovine blood vessels are shown growing on the multifocal-created scaffold. Reprinted from Ref. 136, Copyright 2011, with permission from The Optical Society.

The 2PP was also applied to the fabrication of 3D scaffolds using photosensitive hydrogels with whole organisms present in the solution during fabrication instead of being introduced after the scaffold is fabricated. In their work they used water-based soluble initiator and photopolymer and verified biocompatibility with the living organisms.[140]

#### 12.4.1.8 Inkjet printing

The technological advancements in the various RP processes and the investigations in improving each individual process as it relates to tissue engineering are at a stage where more than one process could be combined toward development of a true customized bioimprinting environment.[28,127,141–146] Each process does best what is designed to do but the combination of multiple processes affords the researcher more control on the overall outcome. Bioimprinting (bioprinting) is a technique based on inkjet printer technology where the colors are replaced with bioinks: biological or functional fluids and entities such as cells, enzymes, and proteins. The ability to deliver controllably multiple bioinks in a single environment could find applications in biosensors, drug delivery, and tissue engineering through the fabrication of biomimetic scaffolds and patterns.[140]

The bioimprinting technology has been used to produce *in vitro* 3D biological scaffolds composed of both living cells and hydrogels with micrometer resolution and in arrangements similar to biological tissues.[142–144] The versatility of the bioprinting approach has been employed on a modified inkjet printer to both deliver chondrocytes and photopolymerize the surrounded biomaterial scaffold [poly(ethylene glycol) dimethacrylate (PEGDMA)]with application to 3D cartilage scaffolds. The success of the inkjet-based bioimprinting approach, especially the ability to control the type and location of delivered cells, is laying the foundations for new research areas based on scaffold free self-assembly of bioink particles. It is postulated that no matter how complex and sophisticated the man-fabricated 3D scaffolds are, problems might still be caused since the scaffolds could biodegrade or react in different ways than the expected ones or due to *a priori* unforeseen complications not considered during the design phase.[127,146,147]

## 12.5 Final Remarks

Significant progress has been made in the development of materials and methods to develop scaffolds for tissue engineering with the ultimate goal of repairing and reconstructing damaged tissue and organs (see Fig. 12.16). Development of an understanding of the interrelationship between chemistry, structure, and morphology has been an important step in furthering this knowledge and helping in the development of scaffolds for both hard and soft tissue repair and reconstruction. However, more work needs to be done before a truly biomimetic scaffold is developed that can be translated from the bench to a patient in a short time frame.

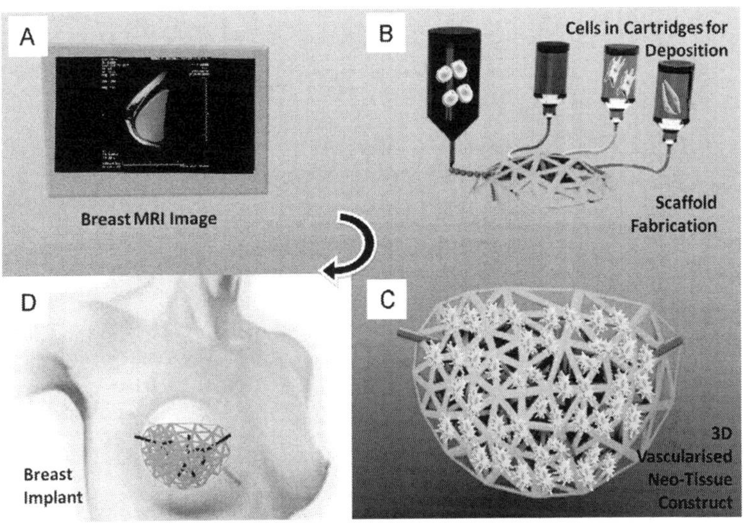


**Figure 12.16** Schematic elucidating the principle of additive tissue manufacturing using inkjet printing. (A) Imaging of an organ to obtain 3D digital blueprint. (B) Concurrent additive manufacturing of scaffolding structure (biodegradable thermoplastic) and cells suspended in gels: pre-adipocytes in adipose-mimetic ECM gel and smooth muscle cells in gel mimicking their native ECM. (C) Manufactured 3D neo-tissue construct. (D) Implantation after mastectomy. Reprinted from Ref. 148, Copyright 2012, with permission from Elsevier.

## References


1. Langer R, Vacanti J. Tissue engineering. *Science* 1993; 260(5110):920–960.
2. Boccaccini AR, Maquet V. Bioresorbable and bioactive polymer/bioglass® composites with tailored pore structure for tissue engineering applications. *Composites Sci Technol* 2003; 63:2417–2429.
3. Blaker JJ, Gough JE, Maquet V, Notingher I, Boccaccini AR. *In vitro* evaluation of novel bioactive composites based on bioglass®-filled polylactide foams for bone tissue engineering scaffolds. *J Biomed Mater Res* 2003; 67:1401–1411.
4. Wintermantel E, Mayer J, Blum J, Eckert K, Luscher P, Mathey M. Tissue engineering scaffolds using superstructures. *Biomaterials* 1996; 17:83–91.
5. Agrawal CM, Athanasiou KA, Heckman JD. Biodegradable PLA-PGA polymers for tissue engineering in orthopaedics. *Mater Sci Forum* 1997; 250:115–128.
6. Jun JY, Park TG. Degradation behaviors of biodegradable macroporous scaffolds prepared by gas foaming of effervescent salts. *J Biomed Mater Res* 2001; 55:401–408.
7. Hollister SJ, Maddox RD, Taboas JM. Optimal design and fabrication of scaffolds to mimic tissue properties and satisfy biological constraints. *Biomaterials* 2002; 23:4095–4103.
8. Landers R, Pfister A, Hubner U, John H, Schmelzeisen R, Mulhaupt R. Fabrication of soft tissue engineering scaffolds by means of rapid prototyping techniques. *J Mater Sci* 2002; 37:3107–3116.
9. Trantolo DJ, Sonis ST, Thompson MJ. Evaluation of a porous, biodegradable biopolymer scaffold for mandibular reconstruction. *Int J Oral Maxillofac Implants* 2003; 18:182–188.
10. Uemura T, Dong J, Wang Y, *et al.* Transplantation of cultured bone cells using combinations of scaffolds and culture techniques. *Biomaterials* 2003; 24:2277–2286.
11. Maquet V, Boccaccini AR, Pravata L, Notingher I, Jerome R. Preparation, characterization, and *in vitro* degradation of bioresorbable and bioactive composites based on bioglass®-filled polylactide foams. *J Biomedic Mater Res* 2003; 66:335–346.
12. Leong KF, Cheah CM, Chua CK. Solid freeform fabrication of three-dimensional scaffolds for engineering replacement tissues and organs. *Biomaterials* 2003; 24:2363–2378.

13. Lin ASP, Barrows TH, Cartmell SH, Guldberg RE. Microarchitectural and mechanical characterization of oriented porous polymer scaffolds. *Biomaterials* 2003; 24:481–489.

14. Almirall A, Larrecq G, Delgado JA, Martinez S, Planell JA, Ginebra MP. Fabrication of low temperature macroporous hydroxyapatite scaffolds by foaming and hydrolysis of an alpha-TCP paste. *Biomaterials* 2004; 25:3671–3680.

15. Woodfield TBF, Malda J, Wijn JD, Peters F, Riesle J, van Blitterswijk CA. Design of porous scaffolds for cartilage tissue engineering using a three-dimensional fiber-deposition technique. *Biomaterials* 2004; 25:4149–4161.

16. Meinel L, Karageorgiou V, Hofmann S, *et al.* Engineering bone-like tissue *in vitro* using human bone marrow stem cells and silk scaffolds. *J Biomedic Mater Res* 2004; 71:25–34.

17. Kim H, Knowles JC, Kim H. Development of hydroxyapatite bone scaffold for controlled drug release via poly(ε-caprolactone) and hydroxyapatite hybrid coatings. *J Biomed Mater Res* 2004; 70:240–249.

18. Nazarov R, Jin H, Kaplan DL. Porous 3-D scaffolds from regenerated silk fibroin. *Biomacromolecules* 2004; 5:718–726.

19. Hollister SJ. Porous scaffold design for tissue engineering. *Nat Mater* 2005; 4:518–524.

20. Hu W, Yim EKF, Reano RM, Leong KW, Pang SW. Effects of nanoimprinted patterns in tissue-culture polystyrene on cell behavior. *J Vac Sci Tech* 2005; 23:2984.

21. Yang Y, Basu S, Tomasko DL, Lee LJ, Yang S. Fabrication of well-defined PLGA scaffolds using novel microembossing and carbon dioxide bonding. *Biomaterials* 2005; 26:2585–2594.

22. Adachi T, Osako Y, Tanaka M, Hojo M, Hollister SJ. Framework for optimal design of porous scaffold microstructure by computational simulation of bone regeneration. *Biomaterials* 2006; 27:3964–3972.

23. Zhang Y, Xu HHK, Takagi S, Chow LC. *In situ* hardening hydroxyapatite-based scaffold for bone repair. *J Mater Sci Mater Med* 2006; 17:437–445.

24. Nair LS, Laurencin CT. Biodegradable polymers as biomaterials. *Prog Polym Sci* 2006; 32:762–798.

25. Murugan R, Ramakrishna S, Panduranga Rao K. Nanoporous hydroxy-carbonate apatite scaffold made of natural bone. *Mater Lett* 2006; 60:2844–2847.

26. Mondrinos M, Dembzynski R, Lu L, *et al.* Porogen-based solid freeform fabrication of polycaprolactone—Calcium phosphate scaffolds for tissue engineering. *Biomaterials* 2006; 27:4399–4408.
27. Best SM, Porter AE, Thian ES, Huang J. Bioceramics: Past, present and for the future. *J Eur Ceram Soc* 2007; 28:1319–1327.
28. Campbell PG, Weiss LE. Tissue engineering with the aid of inkjet printers. *Exp Opin Biol Ther* 2007; 7:1123–1127.
29. Kalita SJ, Bhardwaj A, Bhatt HA. Nanocrystalline calcium phosphate ceramics in biomedical engineering. *Mater Sci Eng* 2007; 27:441–449.
30. Hollister SJ, Lin CY. Computational design of tissue engineering scaffolds. *Comput Methods Appl Mech Eng* 2007; 196:2991–2998.
31. Kretlow JD, Mikos AG. Review: Mineralization of synthetic polymer scaffolds for bone tissue engineering. *Tissue Eng* 2007; 13:927–938.
32. Uttarwar M, Aswath PB. Fabrication of porous, drug releasing, biodegradable, polymer scaffolds for sustained drug release. *J Biomed Mater Res* 2008; 87: 121–131.
33. Jones AC, Arns CH, Hutmacher DW, Milthorpe BK, Sheppard AP, Knackstedt MA. The correlation of pore morphology, interconnectivity and physical properties of 3D ceramic scaffolds with bone ingrowth. *Biomaterials* 2009; 30:1440–1451.
34. Boccaccini AR, Gorustovich AA, Roether JA. Effect of bioactive glasses on angiogenesis: A review of *in vitro* and *in vivo* evidences. *Tissue Eng* 2010; 16:199–207.
35. O'Brien FJ. Biomaterials & scaffolds for tissue engineering. *Mater Today* 2011; 14:88–95.
36. Levengood SKL, Polak SJ, Wheeler MB, *et al.* Multiscale osteointegration as a new paradigm for the design of calcium phosphate scaffolds for bone regeneration. *Biomaterials* 2010; 31:3552–3563.
37. Bose S, Roy M, Bandyopadhyay A. Recent advances in bone tissue engineering scaffolds. *Trends Biotechnol* 2012; 30:546–554.
38. Danilevicius P, Rekstyte S, Balciunas E, *et al.* Micro-structured polymer scaffolds fabricated by direct laser writing for tissue engineering. *J Biomed Opt* 2012; 17:081405.
39. Willie BM, Petersen A, Schmidt-Bleek K, *et al.* Designing biomimetic scaffolds for bone regeneration: Why aim for a copy of mature tissue properties if nature uses a different approach? *Soft Matter* 2010; 6:4976.
40. Schroeder JE, Mosheiff R. Tissue engineering approaches for bone repair: Concepts and evidence. *Injury* 2011; 42:609–613.

41. Reichert JC, Saifzadeh S, Wullschleger ME, *et al.* The challenge of establishing preclinical models for segmental bone defect research. *Biomaterials* 2009; 30:2149–2163.

42. Saiz E, Zimmermann EA, Lee JS, Wegst UGK, Tomsia AP. Perspectives on the role of nanotechnology in bone tissue engineering. *Dental Mater* 2013; 29:103–115.

43. Rezwan K, Chen QZ, Blaker JJ, Boccaccini AR. Biodegradable and bioactive porous polymer/inorganic composite scaffolds for bone tissue engineering. *Biomaterials* 2005; 27:3413–3431.

44. Lima KM, Rodrigues JM. Poly-DL-lactide-co-glycolide microspheres as a controlled release antigen delivery system. *Braz J Med Biol Res* 1999; 32:171–180.

45. Liu AX, Hong ZK, Zhuang XL, *et al.* Surface modification of bioactive glass nanoparticles and the mechanical and biological properties of poly(L-lactide) composites. *Acta Biomater* 2008; 4:1005–1015.

46. Patterson J, Stayton PS, Li X. *In situ* characterization of the degradation of PLGA microspheres in hyaluronic acid hydrogels by optical coherence tomography. *IEEE Trans Med Imaging* 2009; 28:74–81.

47. Lu L, Garcia CA, Mikos AG. *In vitro* degradation of thin poly(DL-lactic-co-glycolic acid) films. *J Biomed Mater Res* 1999; 46:236–244.

48. Ko YK, Kim SH, Ha HJ, *et al.* Hybrid type of tissue engineered biodisc using annulus fibrosus seeded PLGA scaffold and nucleus pulposus seeded MPEG-PCL hydrogel: Preliminary study. *Key Eng Mat* 2007; 342/343:173–176.

49. Hu X, Zhou J, Zhang N, Tan H, Gao C. Preparation and properties of an injectable scaffold of poly(lactic-co-glycolic acid) microparticles/chitosan hydrogel. *J Mech Behav Biomed Mater* 2008; 1:352–359.

50. Kawazoe N, Lin X, Tateishi T, Chen G. Three-dimensional cultures of rat pancreatic RIN-5F cells in porous PLGA-collagen hybrid scaffolds. *J Bioact Compatible Polym* 2009; 24:25–42.

51. Calandrelli L, Immirzi B, Malinconico M, Volpe MG, Oliva A, Della Ragione F. Preparation and characterization of composites based on biodegradable polymers for "*in vivo*" application. *Polymer* 2000; 41:8027–8033.

52. Wang X, Lu X, Jiang F, Yang X, Deng C, Weng J. The formation of bone-like apatite on the surfaces of the bioceramics and biocomposites. *Gongneng Cailiao/J Funct Mater* 2008; 39:647–650.

53. Yeong WY, Sudarmadji N, Yu HY, *et al.* Porous polycaprolactone scaffold for cardiac tissue engineering fabricated by selective laser sintering. *Acta Biomater* 2010; 6:2028–2034.

54. Wei J, Chen F, Shin J, *et al.* Preparation and characterization of bioactive mesoporous wollastonite—polycaprolactone composite scaffold. *Biomaterials* 2009; 30:1080–1088.

55. Lemons JE. Ceramics: Past, present, and future. *Bone* 1996; 19:S121–S128.

56. Ohgushi H, Caplan AI. Stem cell technology and bioceramics: From cell to gene engineering. *J Biomed Mater Res* 1999; 48:913–927.

57. Oonishi H, Hench L, Wilson J, *et al.* Comparative bone growth behavior in granules of bioceramic materials of various sizes. *J Biomed Mater Res* 1998; 44:31–43.

58. Langstaff S, Sayer M, Smith TJN, Pugh SM, Hesp SAM, Thompson WT. Resorbable bioceramics based on stabilized calcium phosphates. Part I: Rational design, sample preparation and material characterization. *Biomaterials* 1999; 20:1727–1741.

59. Langstaff S, Sayer M, Smith TJN, Pugh SM. Resorbable bioceramics based on stabilized calcium phosphates. Part II: Evaluation of biological response. *Biomaterials* 2001; 22:135–150.

60. Beletskii BI, Gaidak TI. Prediction of the structure of sintered bioactive composite materials for dental implants. *Glass Ceram* (English translation of Steklo i Keramika). 2003; 60:379–381.

61. Cao W, Hench LL. Bioactive materials. *Ceram Int* 1996; 22:493–507.

62. Xynos ID, Hukkanen MVJ, Batten JJ, Buttery LD, Hench LL, Polak JM. Bioglass (R) 45S5 stimulates osteoblast turnover and enhances bone formation *in vitro*: Implications and applications for bone tissue engineering. *Calcified Tissue Int Calcified Tissue Int* 2000; 67:321–329.

63. Oonishi H, Hench LL, Wilson J, *et al.* Quantitative comparison of bone growth behavior in granules of bioglass, A-W glass-ceramic, and hydroxyapatite. *J Biomed Mater Res* 2000; 51:37–46.

64. Peitl O, Dutra Zanotto E, Hench LL. Highly bioactive P2O5-Na2O-CaO-SiO2 glass-ceramics. *J Non Cryst Solids* 2001; 292:115–126.

65. Jones J. Review of bioactive glass: From hench to hybrids. *Acta Biomater* 2013; 9:4457–4486.

66. Akao M, Aoki H, Kato K. Mechanical properties of sintered hydroxyapatite for prosthetic applications. *J Mater Sci* 1981; 16:809–812.

67. Moore DC, Chapman MW, Manske D. Evaluation of biphasic calcium phosphate ceramic for use in grafting long-bone diaphyseal defects. *J Orthop Res* 1987; 5:356–365.

68. Neo M, Kotani S, Nakamura T, *et al.* Comparative study of ultrastructures of the interfaces between four kinds of surface-active ceramic and bone. *J Biomed Mater Res* 1992; 26:1419–1432.

69. Vrouwenvelder WCA, Groot CG, de Groot K. Histological and biochemical evaluation of osteoblasts cultured on bioactive glass, hydroxylapatite, titanium alloy, and stainless steel. *J Biomed Mater Res* 1993; 27:465–475.

70. Lin F, Liao C, Chen K, Sun J, Lin C. Preparation of beta TCP/HAP biphasic ceramics with natural bone structure by heating bovine cancellous bone with the addition of (NH4)(2)HPO4. *J Biomed Mater Res* 2000; 51:157–163.

71. Balas F, Pérez-Pariente J, Vallet-Regí M. *In vitro* bioactivity of silicon-substituted hydroxyapatites. *J Biomed Mater Res* 2003; 66:364–375.

72. Pena J, Vallet-Regi M. Hydroxyapatite, tricalcium phosphate and biphasic materials prepared by a liquid mix technique. *J Eur Ceram Soc* 2003; 23:1687–1696.

73. Wang Q, Ge S, Zhang D. Highly bioactive nano-hydroxyapatite/partially stabilized zirconia ceramics. *J Bionics Eng* 2004; 1:215–220.

74. Patel N, Best SM, Bonfield W. Characterisation of hydroxyapatite and substituted-hydroxyapatites for bone grafting. *J Australasian Ceram Soc* 2005; 41:1–22.

75. Patel N, Brooks RA, Clarke MT, *et al. In vivo* assessment of hydroxyapatite and silicate-substituted hydroxyapatite granules using an ovine defect model. *J Mater Sci Mater Med* 2005; 16:429–440.

76. Webster TJ, Ergun C, Doremus RH, Siegel RW, Bizios R. Specific proteins mediate enhanced osteoblast adhesion on nanophase ceramics. *J Biomed Mater Res* 2000; 51:475–483.

77. Hing K. Bone repair in the twenty-first century: Biology, chemistry or engineering? *Phil Trans R Soc Lond* 2003; 362:2821–2850.

78. Jones JR, Sepulveda P, Hench LL. Concentration dependence of bioactive glass dissolution *in vitro*. *Bioceramics* 14. 2002; 218–2:303–306.

79. Jones JR, Hench LL. Factors affecting the structure and properties of bioactive foam scaffolds for tissue engineering. *J Biomed Mater Res* 2004; 68B:36–44.

80. Bielby RC, Christodoulou IS, Pryce RS, Radford WJP, Hench LL, Polak JM. Time- and concentration-dependent effects of dissolution products of 58S sol-gel bioactive glass on proliferation and differentiation of murine and human osteoblasts. *Tissue Eng* 2004; 10:1018–1026.

81. Brauer DS, Karpukhina N, O'Donnell MD, Law RV, Hill RG. Fluoride-containing bioactive glasses: Effect of glass design and structure on degradation, pH and apatite formation in simulated body fluid. *Acta Biomater* 2010; 6:3275–3282.

82. Foppiano S, Marshall SJ, Marshall GW, Saiz E, Tomsia AP. The influence of novel bioactive glasses on *in vitro* osteoblast behavior. *J Biomed Mater Res* 2004; 71A:242–249.

83. Foppiano S, Marshall SJ, Saiz E, Tomsia AP, Marshall GW. Functionally graded bioactive coatings: Reproducibility and stability of the coating under cell culture conditions. *Acta Biomater* 2006; 2:133–142.

84. Foppiano S, Marshall SJ, Marshall GW, Saiz E, Tomsia AP. Bioactive glass coatings affect the behavior of osteoblast-like cells. *Acta Biomater* 2007; 3:765–771.

85. Jones JR, Sepulveda P, Hench LL. Dose-dependent behavior of bioactive glass dissolution. *J Biomed Mater Res* 2001; 58:720–726.

86. Sepulveda P, Jones JR, Hench LL. *In vitro* dissolution of melt-derived 45S5 and sol-gel derived 58S bioactive glasses. *J Biomed Mater Res* 2002; 61:301–311.

87. Sepulveda P, Jones JR, Hench LL. Drug-releasing scaffolds fabricated from drug-loaded microspheres. *J Biomed Mater Res* 2002; 59:349–356.

88. Sepulveda P, Jones JR, Hench LL. Characterization of melt-derived 45S5 and sol-gel-derived 58S bioactive glasses. *J Biomed Mater Res* 2001; 58:734–740.

89. Oku T, Suganuma K, Wallenberg LR, Tomsia AP, Gomez-Vega JM, Saiz E. Structural characterization of the metal/glass interface in bioactive glass coatings on ti-6Al-4V. *J Mater Sci* 2001; 12:413–417.

90. Vollenweider M, Brunner TJ, Knecht S, *et al.* Remineralization of human dentin using ultrafine bioactive glass particles. *Acta Biomater* 2007; 3:936–943.

91. Jayalekshmi AC, Victor SP, Sharma CP. Magnetic and degradable polymer/bioactive glass composite nanoparticles for biomedical applications. *Colloids Surf B Biointerfaces* 2013; 101:196–204.

92. Brunner TJ, Grass RN, Stark WJ. Glass and bioglass nanopowders by flame synthesis. *Chem Commun* 2006:1384–1386.

93. Dubok V. Bioceramics—Yesterday, today, tomorrow. *Powder Metall Metal Ceram* 2000; 39:381–394.

94. Tielens F, Gervais C, Lambert JF, Mauri F, Costa D. *Ab initio* study of the hydroxylated surface of amorphous silica: A representative model. *Chem Mater* 2008; 20:3336–3344.

95. Hench L. Bioceramics: From concept to clinic. *J Am Chem Soc* 1991; 74:1487–1510.

96. Tilocca A. Sodium migration pathways in multicomponent silicate glasses: Car-parrinello molecular dynamics simulations. *J Chem Phys* 2010; 133.

97. Gao H, Tan T, Wang D. Effect of composition on the release kinetics of phosphate controlled release glasses in aqueous medium. *J Controlled Release* 2004; 96:21–28.

98. Azooz MA, Aiad THMA, ElBatal FH, ElTabii G, Abdollahi M. Characterization of bioactivity in transition metal doped-borosilicate glasses by infrared reflection and dielectric studies. *Ind J Pure Appl Phys* 2008; 46:880–888.

99. Bachar A, Mercier C, Tricoteaux A, *et al.* Effects of addition of nitrogen on bioglass properties and structure. *J Non Cryst Solids* 2012; 358:693–701.

100. Pedone A, Charpentier T, Menziani MC. The structure of fluoride-containing bioactive glasses: New insights from first-principles calculations and solid state NMR spectroscopy. *J Mater Chem* 2012; 22:12599–12608.

101. Inanc B, Elcin AE, Elcin YM. Osteogenic induction of human periodontal ligament fibroblasts under two- and three-dimensional culture conditions. *Tissue Eng* 2006; 12:257–266.

102. Nagata S, Nozaki T, Ohura K, Daito M. Analysis of gene expression during differentiation induction of normal human periodontal ligament cells into an osteoblast lineage. *J Oral Tissue Eng* 2008; 6:88–96.

103. Arceo N, Sauk JJ, Moehring J, Foster RA, Somerman MJ. Human periodontal cells initiate mineral-like nodules *in vitro*. *J Periodontol* 1991; 62:499–503.

104. Varanasi VG, Owyoung JB, Saiz E, Marshall SJ, Marshall GW, Loomer PM. The ionic products of bioactive glass particle dissolution enhance periodontal ligament fibroblast osteocalcin expression and enhance early mineralized tissue development. *J Biomed Mater Res* 2011; 98A:177–184.

105. Zhou Y, Hutmacher DW, Sae-Lim V, Zhou Z, Woodruff M, Lim TM. Osteogenic and adipogenic induction potential of human periodontal cells. *J Periodontol* 2008; 79:525–534.

106. Varanasi VG, Leong KK, Jue SM, Dominia LM, Loomer PM, Marshall GW. Si and Ca combinatorially target and enhance early MC3T3-E1 osteoblast expression of osteocalcin. *J Oral Implantol* 2012; Accepted.

107. Varanasi VG, Saiz E, Loomer PM, *et al.* Enhanced osteocalcin expression by osteoblast-like cells (MC3T3-E1) exposed to bioactive coating glass ($SiO_2$–CaO–$P_2O_5$–MgO–$K_2O$–$Na_2O$ system) ions. *Acta Biomater* 2009; 5:3536–3547.

108. Xynos ID, Edgar AJ, Buttery LDK, Hench LL, Polak JM. Ionic products of bioactive glass dissolution increase proliferation of human osteoblasts and induce insulin-like growth factor II mRNA expression and protein synthesis. *Biochem Bioph Res Co* 2000; 276:461–465.

109. Tousi NS, Velten M, Bishop TJ, *et al.* Combinatorial effect of $Si^{4+}$, $Ca^{2+}$, and $Mg^{2+}$ released from bioactive glasses on osteoblast osteocalcin expression and biomineralization. *Mater Sci Eng* 2013; Accepted.

110. Xiao G, Jiang D, Ge C, *et al.* Cooperative interactions between ATF4 and Runx2/Cbfa1 stimulate osteoblast-specific osteocalcin gene expression. *J Bone Miner Res* 2005; 20:S356–S356.

111. Xiao GZ, Jiang D, Ge CX, *et al.* Cooperative interactions between activating transcription factor 4 and Runx2/Cbfa1 stimulate osteoblast-specific osteocalcin gene expression. *J Biol Chem* 2005; 280:30689–30696.

112. Alberts B. *Molecular Biology of the Cell, 4th ed.* New York: Garland Science, 2002.

113. Maeno S, Niki Y, Matsumoto H, *et al.* The effect of calcium ion concentration on osteoblast viability, proliferation and differentiation in monolayer and 3D culture. *Biomaterials* 2005; 26:4847–4855.

114. Abed E, Moreau R. Importance of melastatin-like transient receptor potential 7 and cations (magnesium, calcium) in human osteoblast-like cell proliferation. *Cell Proliferat* 2007; 40:849–865.

115. Barboza EP, Caula AL, Caula FD, *et al.* Effect of recombinant human bone morphogenetic protein-2 in an absorbable collagen sponge with space-providing biomaterials on the augmentation of chronic alveolar ridge defects. *J Periodontol* 2004; 75:702–708.

116. Lindfors NC, Koski I, Heikkila JT, Mattila K, Aho AJ. A prospective randomized 14-year follow-up study of bioactive glass and autogenous bone as bone graft substitutes in benign bone tumors. *J Biomed Mater Res* 2010; 94:157–164.

117. Niu LN, Jiao K, Qi YP, *et al.* Intrafibrillar silicification of collagen scaffolds for sustained release of stem cell homing chemokine in hard tissue regeneration. *FASEB J* 2012; 26:4517–4529.

118. Peltola SM, Melchels FPW, Grijpma, Dirk W. and Kellomaki,Minna. A review of rapid prototyping techniques for tissue engineering purposes. *Ann Med* 2008; 40:268–280.

119. Kim MS, Hansgen AR, Wink O, Quaife RA, Carroll JD. Rapid prototyping: A new tool in understanding and treating structural heart disease. *Nature* 2008; 117:2388–2394.

120. Binder TM, Moertl D, Mundigler G, *et al.* Stereo lithographic biomodeling to create tangible hard copies of cardiac structures from echocardiographic data. *J Am Coll Cardiol* 2000; 35:230–237.

121. Gilon D, Cape EG, Handschumacher MD, *et al.* Effect of three-dimensional valve shape on the hemodynamics of aortic stenosis: Three-dimensional echocardiographic stereolithography and patient studies. *J Am Coll Cardiol* 2002; 40:1479–1486.

122. Belligundu S, Shiakolas PS. Study on two-stage hot embossing microreplication: Silicon to polymer to polymer. *J Microlith, Microfabrication Microsyst* 2006; 5:021103.

123. Belligundu S, Shiakolas PS, Pandey, A, Aswath, PB. A systemic approach toward optimization of the hot embossing of poly-L-lactic acid for biomedical applications. *J Biomed Mater Res* 2008; 85:469–477.

124. Wang Z, Hu H, Wang Y, Wang Y, Wu QaL, Litian, Chen G. Fabrication of poly(3-hydroxybutyrate-co-3-hydroxyhexanoate) (PHBHH*x*) microstructures using soft lithography for scaffold applications. *Biomaterials* 2006; 27:2550–2557.

125. Lin CY, Kikuchi N, Hollister SJ. A novel method for biomaterial scaffold internal architecture design to match bone elastic properties with desired porosity. *J Biomech* 2004; 37:623–636.

126. Coelho P, Flanagan C, Hollister S, Fernandes P, Rodrigues H. Design and structural evaluation of scaffolds prototypes obtained by selective laser sintering. *J Biomech* 2012; 45, Supplement 1:S343.

127. Marga F, Jakab K, Khatiwala C, *et al.* Toward engineering functional organ modules by additive manufacturing. *Biofabrication* 2012; 4:022001.

128. Williams JM, Adewunmi A, Schek RM, *et al.* Bone tissue engineering using polycaprolactone scaffolds fabricated via selective laser sintering. *Biomaterials* 2005; 26:4817–4827.

129. Eshraghi S, Das S. Mechanical and microstructural properties of polycaprolactone scaffolds with 1-D, 2-D, and 3-D orthogonally oriented porous architectures produced by selective laser sintering. *Acta Biomater* 2010; 6:2467.

130. Shuai C, Mao Z, Lu H, Nie Y, Hu H, Peng S. Fabrication of porous polyvinyl alcohol scaffold for bone tissue engineering via selective laser sintering. *Biofabrication* 2013; 5:015014.

131. Yen H, Tseng C, Hsu S, Tsai C. Evaluation of chondrocyte growth in the highly porous scaffolds made by fused deposition manufacturing (FDM) filled with type II collagen. *Biomed Microdev* 2009; 11:615–624.

132. Fierz FC, Beckmann F, Huser M, *et al.* The morphology of anisotropic 3D-printed hydroxyapatite scaffolds. *Biomaterials* 2008; 29:3799–3806.

133. Uppal N, Shiakolas PS. Process sensitivity analysis and resolution prediction for the two photon polymerization of micro/nano structures. *J Manufact Sci Eng* 2009; 131:051018.

134. Uppal N, Shiakolas PS. Modeling of temperature-dependent diffusion and polymerization kinetics and their effects on two-photon polymerization dynamics. *J Micro/Nanolith MEMS MOEMS* 2008; 7:043002.

135. Hsieh TM, Benjamin Ng CW, Narayanan K, Wan, ACA, Ying JY. Three-dimensional microstructured tissue scaffolds fabricated by two-photon laser scanning photolithography. *Biomaterials* 2010; 31:7648–7652.

136. Gittard SD, Nguyen A, Obata K, Koroleva A, Narayan RJ, Chichkov BN. Fabrication of microscale medical devices by two-photon polymerization with multiple foci via a spatial light modulator. *Biomed Opt Express* 2011; 2:3167–3178.

137. Aprile V, Eaton SM, Lagana M, Cerullo G, Raimondi MT, Osellame R. Femtosecond laser two-photon polymerization of three-dimensional scaffolds for tissue engineering and regenerative medicine applications. *Proc. SPIE 8247.*

138. Ovsianikov A, Deiwick A, van Vlierberghe S, Dubruel P, Moller L, Drager G, Chichkov B. Laser fabrication of three-dimensional CAD scaffolds from photosensitive gelatin for applications in tissue engineering. *Biomacromolecules* 2011; 12:851–858.

139. Ovsianikov A, Deiwick A, van Vlierberghe S, Pflaum M, Wilhelmi M, Dubruel P, Chichkov B. Laser fabrication of 3D gelatin scaffolds for the generation of bioartificial tissues. *Materials* 2011; 4:288–299.

140. Torgersen J, Baudrimont A, Pucher N, Stadlmann K, Cicha K, Heller C, Liska R, Stampfl J. *In vivo Writing using Two-Photon-Polymerization*, Stuttgart and Germany, 2010.

141. Boland T, Xu T, Damon B, Cui X. Application of inkjet printing to tissue engineering. *Biotech J* 2006; 1:910–917.

142. Nakamura M, Nishiyama HC. 3D micro-fabrication by inkjet 3D biofabrication for 3D tissue engineering. *International Symposium on Micro-NanoMechatronics and Human Science, MHS 2008*:451–456.

143. Nakamura M, Kobayashi A, Takagi F, *et al.* Biocompatible inkjet printing technique for designed seeding of individual living cells. *Tissue Eng* 2005; 11:1658–1666.

144. Nishiyama Y, Nakamura M, Henmi C, *et al.* Development of a three-dimensional bioprinter: Construction of cell supporting structures using hydrogel and state-of-the-art inkjet technology. *J Biomech Eng* 2009; 131:035001.

145. Phillippi JA, Miller E, Weiss L, Huard JaW, Alan, Campbell P. Microenvironments engineered by inkjet bioprinting spatially direct adult stem cells toward muscle- and bone-like subpopulations. *Stem Cells* 2008; 26:127–134.

146. Jakab K, Norotte C, Marga F, Murphy K, Vunjak-Novakovic G, Forgacs G. Tissue engineering by self-assembly and bio-printing of living cells. *Biofabrication* 2010; 2:022001.

147. Rajaram A, Chen X, Schreyer DJ. Strategic design and recent fabrication techniques for bioengineered tissue scaffolds to improve peripheral nerve regeneration. *Tissue Eng* 2012; 18:454.

148. Melchels FPW, Domingos MAN, Klein TJ, Malda J, Bartolo PJ, Hutmacher DW. Additive manufacturing of tissues and organs. *Prog Polym Sci* 2012; 37:1079–1104.

# Decellularized Tissues as Scaffolds

Cells produce the extra cellular matrix (ECM), which in turn makes up cartilage, bone, ligament, and all other tissues. At the same time, the ECM constitutes the scaffold for cells, provides structural support to them, provides binding sites for cells, and is a reservoir of nutrients or biological molecules that are used in the continuous remodeling process of the tissues or in their fabrication.

Making scaffolds means mimicking the role and the functions of ECM. However, even if man-made scaffolds can get close to the morphology and mechanical properties of ECM, mimicking its functional properties is not easy. On the role of ECM as a scaffold readers are invited to consult, among others, the paper of Badylak.[1]

Supposedly, decellularized ECM (Fig. 1) could be the ideal scaffold for tissue regeneration, provided that the signaling sites are preserved, that is, they provide sites for the biomolecular recognition. The decellularization process should remove only cells, lipids, and all protein except structural proteins (e.g., collagen or elastin), without affecting their integrity. However, during decellularization, besides the cells being removed, generally the biochemical structure, composition, physical properties remodeling response, and functional outcome of the resulting scaffold[2,3] also gets affected by the mechanical, chemical, and enzymatic processes that are used for the decellularization.

In addition, ineffective or incomplete decellularization can influence host response to implantation, macrophage response, and remodeling outcome.[3]

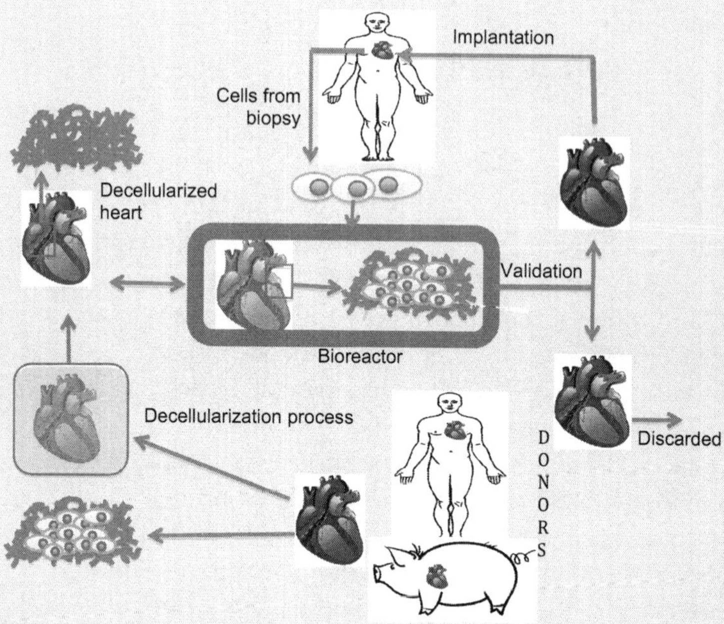


**Figure 1** Schematic diagram of an organ decellularization–application pathway. Adapted from Ref. 4.

In addition to the methods reported in Fig. 2, supercritical $CO_2$ extraction has been developed that does not require the use of surfactants.[6] The application of supercritical $CO_2$ provides a clean technology to obtain acellular ECM with a good guarantee of absence of the pathogens and proteins, avoiding the use of solvents that could be retained in the ECM after the drying process.

However, the application of supercritical technologies has some limitations, the most important being the relatively limited dimensions of the chamber or reactor for the application of supercritical conditions. The application of this process is very attractive if water is absent. Otherwise, the presence of small quantities of water produces the formation of carbonic acid from $CO_2$ and therefore, the elimination of water by washing with organic solvents previous to the treatment with $CO_2$ is necessary.

Nevertheless, decellularized tissues and organs forms have been successfully used for clinical needs in humans, such as for trachea regeneration by using a decellularized human trachea,[7] tested in animals for blood vessels,[8,9] and even approved for commercialization (e.g., for burn wound covering or heart valves) (Table 1). For the sake of clarity, pig skin for burn treatments and porcine heart valves are an example of wide and successful use of decellularized tissue, however, they are not scaffolds for engineering new tissue.

Recenlty, a decellularized pig kidney has been proven to be effective for the regeneration of the kidney in pigs.[10]

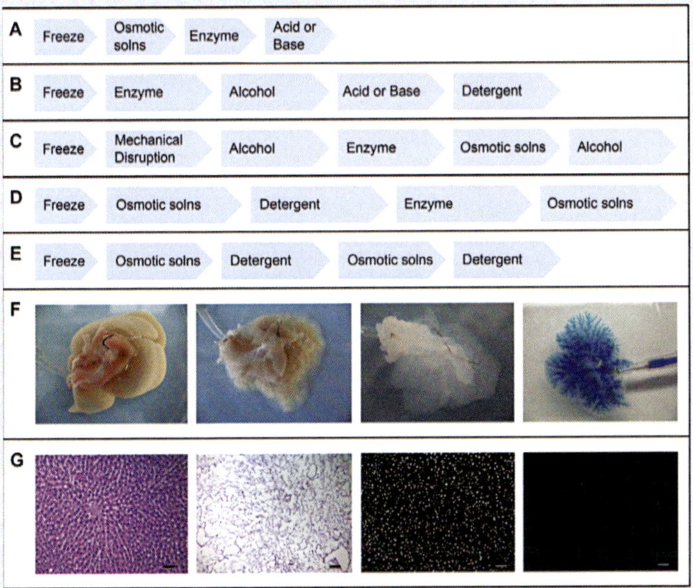


**Figure 2** Example decellularization protocols for (A) thin laminates such as pericardium, (B) thicker laminates such as dermis, (C) fatty, amorphous tissues such as adipose, (D) composite tissues or whole simple organs such as trachea, and (E) whole vital organs such as liver. Arrow lengths are proportional to exposure times. (F) Appearance of rat liver subjected to decellularization: (left to right) before, during, and after decellularization; decellularized liver perfused with blue dye. (G) Photomicrographs after nuclear staining reveal the removal of cells: (left to right) native rat liver, HE stain; decellularized liver-ECM, HE stain; native rat liver DAPI stain; liver-ECM, DAPI stain. Scale bars: 50 mm. Reprinted from Ref. 5, Copyright (2011), with permission from Elsevier.

**Table 1** Some commercial decellularized materials

| Product | Company | Material | Form |
|---|---|---|---|
| Alloderm® | LifeCell™ | Human skin | Natural |
| Axis™ dermis | Mentor | Human dermis | Natural |
| Bard Dermal Allograft | C R Bard | Cadaveric human dermis | Natural |
| CuffPatch™ | Biomet Sports Medicine | Porcine small intestinal submucosa (SIS) | Crosslinked |
| DuraADAPT™ | Pegasus Biologicals | Horse pericardium | Crosslinked |
| Dura-Guard® | Synovis Surgical | Bovine pericardium | Crosslinked |
| Durasis | Cook® Medical | Porcine small intestinal submucosa (SIS) | Natural |
| Durepair® | TEI Biosciences/ Medtronic | Fetal bovine skin | Natural |
| FasLata® | CRBard | Cadaveric fascia lata | Natural |
| Graft Jacket® | Wright Medical Tech | Human skin | Natural |
| Oasis® | Cook® Biotech/ Healthpoint | Porcine small intestinal submucosa (SIS) | Natural |
| OrthADAPT™ | Pegasus Biologicals | Horse pericardium | Crosslinked |
| Pelvicol™ | CRBard | Porcine dermis | Crosslinked |
| Peri-Guard® | Synovis® Surgical Innovations | Bovine pericardium | Crosslinked |
| Permacol™ | Covidien | Porcine skin | Crosslinked |
| PriMatrix™ | TEI Biosciences | Fetal bovine skin | Natural |

| Product | Company | Material | Form |
|---|---|---|---|
| Restore® | DePuy | Porcine small intestinal submucosa (SIS) | Natural |
| SurgiMend® | TEI Biosciences | Fetal bovine skin | Natural |
| Surgisis® | Cook® Medical | Porcine small intestinal submucosa (SIS) | Natural |
| Suspend™ | Mentor | Human Fascia lata | Natural |
| TissueMend® | TEI Biosciences | Fetal bovine skin | Natural |
| Veritas® | Synovis® Surgical Innovations | Bovine pericardium | Crosslinked |
| Xenform® | TEI Biosciences/ Boston Scientific | Fetal bovine skin | Natural |

*Source:* Reprinted from Ref. 3, Copyright (2012), with permission from Elsevier.

## References


1. Badylak, S. F. (2002) The extracellular matrix as a scaffold for tissue reconstruction, *Cell Dev. Biol.*, **13**, 377–383.
2. Gilbert, T. W., Sellaro, T. L., and Badylak, S. F. (2006) Decellularization of tissues and organs, *Biomaterials*, **27**(19), 3675–3683.
3. Keane, T. J., Londono, R., Turner, N. J., and Badylak, S. F. (2012) Consequences of ineffective decellularization of biologic scaffolds on the host response, *Biomaterials*, **33**(6), 1771–1781.
4. He, M. and Callanan, A. (2013) Comparison of methods for whole-organ decellularization in tissue engineering of bioartificial organs, *Tissue Engin. B*, **19**(3), 1–15.
5. Crapo, P. M., Gilbert, T. W., and Badylak S. F. (2011) An overview of tissue and whole organ decellularization processes, *Biomaterials*, **32**, 3233–3242.
6. Sawada, K., Terada, D., Yamaoka, T., Kitamura, S., and Fujisato, T. (2008) Cell removal with supercritical carbon dioxide for acellular artificial tissue, *J. Chem. Technol. Biotechnol.*, **83**(6), 943–949.

7. Macchiarini, P., Jungebluth, P., Go, T., Asnaghi, M. A., Rees, L. E., Cogan, T. A., Dodson, A., Martorell, J., Bellini, S., Parnigotto, P. P., Dickinson, S. C., Hollander, A. P., Mantero, S., Conconi, M. T., and Birchall, M. A. (2008) Clinical transplantation of a tissue-engineered airway, *Lancet*, **372**(9655), 2023–2030.
8. Zhao, Y., Zhang, S., Zhou, J., Wang, J., Zhen, M., Liu, Y., Chen, J., and Qi, Z. (2010) The development of a tissue-engineered artery using decellularized scaffold and autologous ovine mesenchymal stem cells, *Biomaterials*, **31**(2), 296–307.
9. Quint, C., Kondo, Y., Manson, R. J., Lawson, J. H., Dardik, A., and Laura E. Niklason, L. E. (2011) Decellularized tissue-engineered blood vessel as an arterial conduit, *Proc. Natl. Acad. Sci. USA*, **108**(22), 9214–9219.
10. Orlando, G., Farney, A. C., Iskandar, S. S., Mirmalek-Sani, S. H., Sullivan, D. C., Moran, E., AbouShwareb, T., De Coppi, P., Wood, K. J., Stratta, R. J., Atala, A., Yoo, J. J., and Soker, S. (2012) Production and implantation of renal extracellular matrix scaffolds from porcine kidneys as a platform for renal bioengineering investigations, *Ann. Surg.*, **256**(2), 363–370.

# Chapter 13

# Organ Printing and Cell Encapsulation

**Savas Tasoglu,[a] Umut Atakan Gurkan,[a] Sinan Guven,[a] and Utkan Demirci[a,b]**
[a]*Bio-Acoustic-MEMS in Medicine (BAMM) Laboratory, Division of Biomedical Engineering and Infectious Diseases, Department of Medicine, Brigham and Women's Hospital, Harvard Medical School, Boston, MA, USA*
[b]*Harvard-MIT Health Sciences and Technology, Cambridge, MA, USA*
udemirci@rics.bwh.harvard.edu

Cells, extracellular matrix components, and neighboring biological agents in tissues and organs are intricately organized in a three-dimensional (3D) microenvironment. Tissue engineering aims to create 3D microarchitecture of cells *in vitro* in a reproducible and precise manner. The major challenges of tissue engineering are: (i) achieving complex 3D cellular architecture and organization that facilitates vital cell–cell and cell–ECM interactions, (ii) control over cellular proximity and microscale resolution, and (iii) enhancing

Savas Tasoglu and Umut Atakan Gurkan have contributed equally to this chapter. Dr. Umut Atakan Gurkan is now assistant professor at Case Biomanufacturing and Microfabrication Laboratory, Mechanical and Aerospace Engineering Department, Case Western Reserve University, Cleveland, OH; Department of Orthopaedics, Case Western Reserve University, Cleveland, OH; Advanced Platform Technology Center, Louis Stokes Cleveland Veterans Affairs Medical Center, Cleveland, OH.

*Scaffolds for Tissue Engineering: Biological Design, Materials, and Fabrication*
Edited by Claudio Migliaresi and Antonella Motta

ISBN 978-981-4463-20-1 (Hardcover), 978-981-4463-21-8 (eBook)
www.panstanford.com

transport through scaffold porosity mimicking vascular network *in vivo*. Bioprinting is an emerging bottom-up approach to provide spatial microcontrol over placement of living cells and other biological components, and to bio-engineer tissues and organs that mimic physiological systems. In this chapter, we will survey several emerging technologies and applications of bioprinting in creating spatially well-organized heterogeneous tissue constructs. There is also growing interest in developing computational and mathematical models to develop a better understanding of underlying mechanisms of cell encapsulation and bioprinting, such as detachment of cell-encapsulating droplets from ejector tip or landing of droplets onto receiving substrate. These approaches will enable faster simultaneous investigations of several governing parameters, such as material properties of surface (hydrophobicity/hydrophilicity), material properties of encapsulating fluid and cell, and ejection speed. Following the discussion of cell encapsulation and bioprinting, both experimentally and theoretically, we illustrate an emerging set of applications. Among these emerging applications of bioprinting and cell encapsulation are (i) cardiomyocyte encapsulation and printing, (ii) stem cell encapsulation and printing, (iii) cancer cell encapsulation and printing, (iv) cryoprinting: cryobiology applications of cell encapsulation and bioprinting, (v) bioprinting blood vessels, (vi) bioprinting in bone tissue engineering, (vii) cell printing as a platform for genomic analyses, (viii) generating porous bioprinted hydrogels, and (ix) *in vivo* applications of bioprinting.

## 13.1 Introduction

Tissue engineering holds promise to enable alternative therapies for diseases such as diabetes, kidney, liver, and heart failure. A common approach in tissue engineering is seeding cells in biodegradable scaffolds (top-down approach), which brings cells together to mimic native tissues. The classical paradigm seeds cells on scaffolds either statically or with perfusion systems.[1] A crucial aspect of tissue engineering is to reproduce the body's architectural and geometric intricacies that would facilitate the vital cell–cell, and cell–ECM interactions creating a physiological environment.[2–7] These scaffolds are expected to degrade and be replaced by cellular growth and extracellular matrix deposition by cells over time. Cells cultured in 2D monolayers have demonstrated major differences in their gene expression as well as their functionality compared to cells in native tissues and in 3D culture conditions.[8,9] These differences can be attributed to elimination of the interactions between cells and their microenvironments.[9] The need for 3D systems in tissue engineering is well known since 2D systems do not mimick the complex tissue microenvironment.[8,10]

Current engineered 3D tissues are designed to culture cells within natural or synthetic scaffolds.[11–15] These engineered scaffolds function as a 3D structure on which cells grow, interact with the ECM, and communicate with other cells in their proximity.[14–16] Printing biocompatible hydrogels accompanying cells and biologically active molecules with high precision is an emerging strategy in tissue engineering.[17,18] Organ printing can be defined as biofabrication of 3D living human organs from raw materials including cells, extracellular matrices, nutrients, therapeutic drugs, growth factors, and other biomaterials. Computer-aided design of a target organ/tissue is initially developed to generate the blueprint that enables the robotic bio-dispenser to print the designed structure layer-by-layer. Bioprinting techniques enable researchers to generate geometrically well-defined scaffolds using polymers or blends of polymers and other cell inductive particles (e.g., hydroxy apatite) providing support and stimulation for seeded cells.[17–19] Tissue engineering scaffolds are expected to enhance and promote cell adhesion, proliferation, extracellular matrix (ECM) production, and cell motility. Bioprinting offers scalable and automated fabrication of tissue engineered products.

Bioprinting is an emerging technology, which can provide control over placement of living cells, extracellular matrix components, biochemical factors, therapeutic drugs, and biomaterials to form tissues and organs that mimic physiological systems.[20–26] Examples of tissues and organs that can be fabricated by bioprinting include heart, bone, skin, muscle, and cancer tissue models. Bioprinting of cell-based assays can create platforms for drug discovery, drug toxicity and efficacy testing, and cancer research. Bioprinted patient-specific tumor assays can be used as predictive models. For example, ovarian cancer models have been created by bioprinting cancer cells, which can provide *in vitro* models to improve our understanding of cancer pathology.[27] Therefore, bioprinting of functional tissue models can provide alternatives to animal testing and potentially reduce the use of animals in biomedical research.

Current 3D tissue scaffolding methods can potentially benefit from spatial and temporal control over cell seeding and extracellular matrix composition.[13,14,28] Challenges in current tissue engineering approaches are: (i) achieving complex 3D cellular architecture and organization,[13,29] and (ii) control over cellular proximity and microscale resolution.[14,30] To address these challenges, various bioprinting and patterning methods have been developed to control cell positioning in 2D and 3D settings, such as acoustic cell encapsulation, inkjet based, ejector based, and laserprinting.[23–26,31–39] Commercially available desktop inkjet printers have been modified and used as cell printers.[40] Since Mironov *et al.* demonstrated the bioink and bioprinting process,[41] this field has been growing with emerging technologies and applications. In these systems, Bioink, which is composed of cells in a suspension, is placed in a cartridge of a printer and the printing pattern is controlled through a computer. Another technique to generate cell encapsulating hydrogel droplets is the valve-based method. In this method, cell containing hydrogel droplet is ejected onto a surface. Size and number of cells in a single droplet and amount of droplets are controlled with valve opening time and frequency.[42] Another way to pattern cells is to use a laser guided technique.[43] In laser based bioprinting, due to the different refractive index of cells in media, photons from laser beam traps and guides cells for patterning.[43] To obtain reproducible results for the placement of cell encapsulating droplets, individual printing of cells has to be performed with high precision (comparable to the cell size <10 μm).[33,44,45] Yet, exposure of excessive thermal energy via laser light onto the cell biolayer and overheating of the cell can be one of the challenges.[20,33,41,46–49] Although microfluidic approaches offer deterministic control over the cell encapsulation process,[50–52] they require complex setups involving hydrodynamic focusing and flow control for tracking multiple cells in these systems.[53–56]

In this chapter, we will survey various technologies and applications of bioprinting in generating spatially organized multicellular tissues and platforms for drug screening and clinical diagnostics. We will also describe the theoretical understanding of cell encapsulation and bioprinting.

## 13.2 Cell Encapsulation

Cells in tissues and organs are intricately organized in a 3D architecture. Even though there are original approaches to engineer the 3D microenvironment of cells *in vitro* (e.g., scaffold-free, scaffold-based, and printing-based),[9,14,34–39,57,58,118] the capability to repeatedly and precisely position individual cells in a 3D configuration remains a challenge. High cell densities of native tissues pose significant challenges for tissue engineering. Current methods for generating 3D tissue constructs offer a limited control over cell density, which is well below the physiologically relevant values.[59] For example, cardiac tissue is composed of high density myocytes ($\sim 10^8$ cells/cm$^3$),[60] which introduces additional challenges in developing engineered cardiac tissue. The ability to generate droplets encapsulating single cells from a suspension of cells is integral to achieve higher cell densities in patterning. To create intricate 3D tissue constructs, acoustic-based nozzleless cell encapsulation and patterning technology has been developed.[20] Emerging technologies in biopatterning and bioprinting specifically address the following challenges: (i) limited control over the number of encapsulated cells in droplets, (ii) frequent nozzle blockage, (iii) unpredictability regarding cell encapsulating droplets *vs.* cell-free droplets, (iv) excessive thermal energy, (v) high shear forces on cells, and (vi) absence of reproducibility and low throughput function. The ability to encapsulate and pattern single cells in a 3D architecture would allow replication and engineering of complex tissue structures.[34–39,118] Such technology would benefit broad range of fields including biological research, biomedical sciences, and tissue regeneration technologies. In this section, we will present the fundamentals of cell encapsulation in picoliter droplets, and applications in bioprinting.

### 13.2.1 Cell Encapsulation in Picoliter Droplets

A picoliter droplet generation and cell encapsulation technology via acoustics has been developed (Fig. 13.1).[20,48,61–63] This technology does not apply excessive thermal energy, pressure, or shear on cells encapsulated in droplets. With this approach, well defined hydrogel droplets in micrometer precision can be generated on demand and patterned by using acoustic waves.[20] Among the various advantages that this technology presents, some are: (i) nozzle is not required since droplets are created from a liquid reservoir, which circumvents complications related to shear or clogging; (ii) this system allows straightforward reliable process; (iii) acoustic waves are harmless to cells because their wavelength is larger than the diameter of

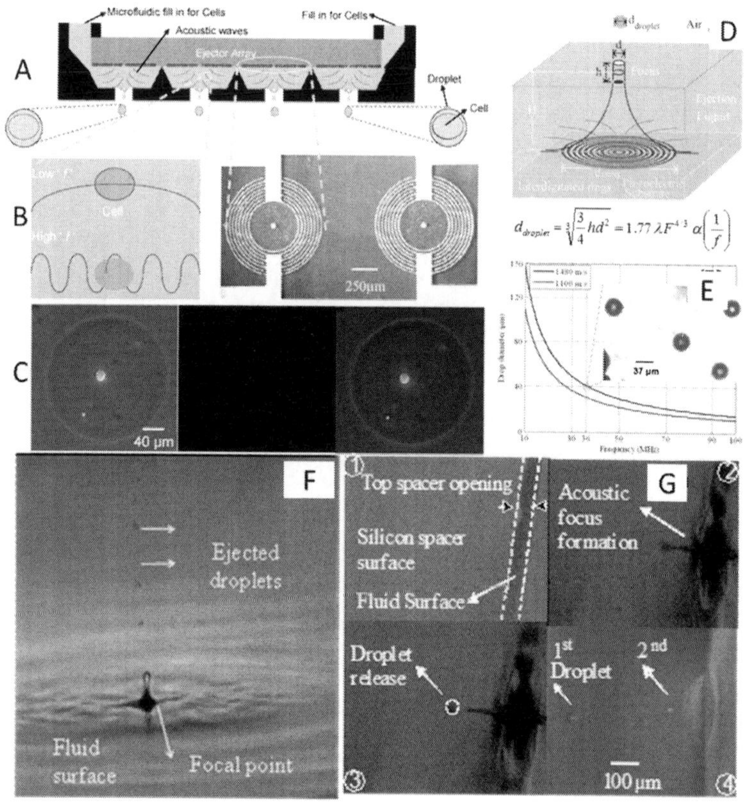


**Figure 13.1** Schematic for acoustic nozzles droplet generation technology. (A) Schematic of multiple ejectors and ejected cell encapsulating droplets from different reservoirs. (B) (left) Comparison of acoustic wavelength with the diameter of the cell; (right) The interdigitated circular micromachined device. Scale bar is 250 μm. (C) Computerized printing of single cell encapsulating droplets via x-y-z stage control. (D) Cylindrical acoustic focus at fluid surface. (E) Theoretical droplet size as a function of frequency, plotted for two fluids. Ejected droplets at 36.0 MHz were measured to be 37.0 ± 0.1 μm in diameter. The simple analytical theory predicts the droplet size to be 40.0 μm for such droplets. (F, G) 28 μm droplets ejected upward from an open pool. Droplets ejected downward from a 100 μm wide microfluidic channel spacer opening. The ejector generated single droplets drop-on-demand without satellites. Reproduced from Refs. 20 and 62 by permission of the Royal Society of Chemistry.

cell; (iv) multiple acoustic ejectors can be potentially combined uniformly to amplify the rate and still it would act as a single ejector; and (v) capability to eject several biomaterials such as living cells, extracellular matrices, nutrients, therapeutic drugs, and growth factors simultaneously from the same platform. In this system, separate microfluidic channels can feed different ejectors with microscale precision and droplet directionality.[20,48,61–63]

### 13.2.2 Theory Behind Picoliter Acoustic Cell Encapsulation

This section demonstrates analytical,[61] finite element,[62] and experimental validation of acoustic picoliter droplet cell encapsulation technology.[20] The key points for design specifically targeting bioprinting applications are described. The unit cell of an acoustic picoliter droplet generator is an interdigitated circular transducer located on a piezoelectric surface (Fig. 13.1A). Periodic placement of unit cell forms 2D arrays. The droplet diameter can be modeled by a fluid cylinder breaking from a fluid surface (Fig. 13.1D). The acoustic beam equations are $d = 1.02\lambda F$ and $h = 7.1\lambda F^2$ for limited diffraction. Here, $F$ is the focal number. The droplet size scales inversely with frequency (Fig. 13.1E). This model estimates droplet diameter to be 40 μm at 36 MHz, which agrees with the experimental 37 μm acoustic droplet diameter. Using finite element methods, the wave distribution was simulated in an ejection reservoir (using ANSYS 5.7) and the key parameters of the acoustic droplet generation mechanism were investigated. This analysis determines location, width, and height of the focal point, as well as the resonant frequency. It also determines the minimum amount of acoustic cycles required for droplet ejection. Finally, it determines the evolution of an acoustic wave as it travels from the device surface to the focal point. This ensures gentle cell handling, facilitating design of devices and ejection reservoir. Viscosity and surface tension of the fluid are crucial to determine the energy required (acoustic cycle length) for droplet generation. The resonance frequency is inherent to device geometry and substrate. As a sinusoidal signal (1V, 32.8–38.5 MHz) is applied at varying frequencies to interdigitated rings, the wave distribution in the fluid medium from substrate surface into the fluid along a vertical path, was monitored as a function of frequency. The simulations predicted that at 34.7 MHz the focal point would be $H$ = 575 μm above the piezoelectric substrate surface. Experimental and simulated wave results agreed. Theoretical and experimental results for lateral wave distribution at the focal point of 575 μm were at 34.7 MHz coincide. Hence, the wave distribution that the cells experience as they are being placed into droplets was controlled. In addition, the lateral wave distribution was determined in 3D, demonstrating the wave distribution at every point in reservoir at various heights from the channel bottom.

The analysis of the wave distribution in the cell suspension fluid is significant, since it predicts the forces that cells experience in the fluid reservoir during printing. The acoustic wave distribution in the reservoir is determined by the acoustic wavelength (~50 μm at 36 MHz) and reservoir geometry. In general, cells (~10–20 μm diameter) are significantly smaller than the wavelength at the frequencies utilized in the instrument (Fig. 13.1). Therefore, the cells are minimally affected from the travelling acoustic waves in the medium and cells do not experience a sharp pressure gradient at frequencies, where the acoustic wavelength is larger than the cell size. Instead, cells move around in the fluid reservoir as in mixing.

The capability to print functional and viable cells encapsulated in hydrogels has been shown, which is critical for cell printing approach. Single to few cells were encapsulated with micrometer precision, high cell viability/functionality, with well-controlled directionality using acoustic picoliter droplets.[20,62] The device does not have a nozzle and prints from an open-pool (Fig. 13.1F). Hence, adverse nozzle effects were eliminated on cell viability and functionality. Fluids of various viscosity and surface tension have been tested and validated for ejection with this method, such as: water (1 Centistoke at 20°C); Dulbecco's modified essential media (DMEM) with cells; solvents (isopropanol, ethyl alcohol, acetone); ethylene glycol (18 Centistokes at 20°C). Cell-hydrogel suspensions of interest have been successfully utilized and ejected in experiments, of which viscosity falls in the ejectable range.

A micro-electromechanical system (MEMS) based fabricated droplet generator is shown in Fig. 13.1A. Cell deposition onto a surface was performed with a stage placed under a droplet generator ~1 cm away and parallel to the surface (Fig. 13.1C). This distance has to be much smaller (<0.1 cm) for the inkjet due to lack of directionality. This device can generate droplets from an open pool (Fig. 13.1F) drop-on-demand. Fluids are kept reliably without leaking in open-pool fluidic channels by the surface tension even if the ejector is facing down.

## 13.3 Computational Modeling of Cell Encapsulation and Bioprinting

Computational approaches offer developing a deeper understanding of experimental steps, and faster investigations of effects of governing parameters, such as material properties of surface (hydrophobicity/hydrophilicity), material properties of encapsulating fluid and cell, and ejection speed. These are vital to enable integration of cell printing technology with tissue engineering and regenerative medicine. In this section, (i) a computational approach to cell printing process, and (ii) a statistical method to predict single target cell encapsulation are described.

### 13.3.1 Statistical Modeling of Cell Encapsulation

Statistical approaches can help to estimate the parameters that govern cell encapsulation in droplets and develop a better understanding of the process. Statistical methods can eventually provide a reliable and repeatable control over the cell encapsulation process.

A statistical model for cell encapsulation process of droplets was developed.[64] In this study, for several target cell concentrations and types of cell loading, encapsulation process was modeled and experimentally verified (Fig. 13.2A,B). To enable encapsulation of a single target cell from a heterogeneous cell solution, cell encapsulation probability functions, $P(X_t)$ (Eq. 3.6 in Moon *et al.*),[64,65] were plotted in Figs. 13.2C–E for heterogeneous cell mixture including several cell loading concentrations. The experimental results determined the shape of Poisson distributions. For several concentrations of cell loading, binomial distributions and Poisson distribution were shown in both continuous random variable space and the discrete space. As shown in Fig. 13.2C–E, while the percentage of target cells and homogeneity decreased in solutions, each $P(X_t)$ approached a Poisson function.

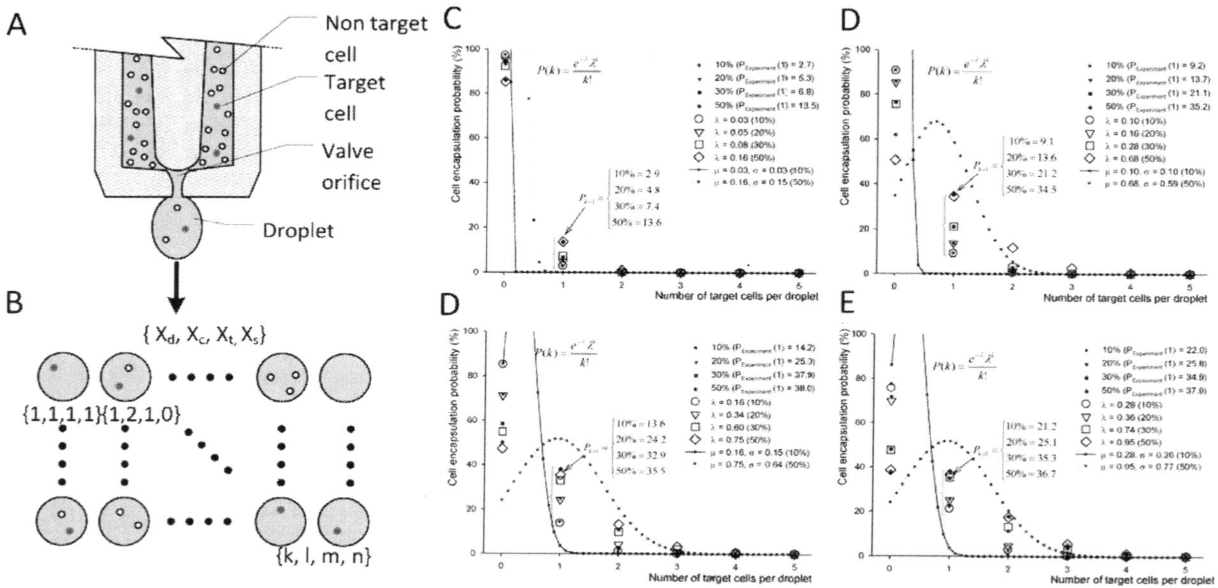


**Figure 13.2** Schematic for statistical analysis of target cell encapsulation. (A) A droplet ejector was filled with heterogeneous mixture including target and non-target cells for random cell encapsulation process. (B) ($X_d$) the number of droplets that contain cells, ($X_c$) number of cells per droplet, ($X_t$) number of target cells, and ($X_s$) droplets encapsulating a single target cell, were mapped onto a matrix of cell encapsulating droplets. Cell encapsulation probabilities, $P(X_t)$, from (C–E) for a range of cell concentrations from $0.5 \times 10^5$ cells/ml to $2.0 \times 10^5$ cells/ml. Reprinted from Ref. 64.

### 13.3.2 Computational Modeling of Cell Printing

Cells could be damaged during various steps of the bioprinting process including: (i) ejection and (ii) impact of cell encapsulating droplets. During ejection and impact, shear stresses applied on cells are high. These stresses can be manipulated by tuning ejection speed or encapsulating cells with fluids that have different material properties, such as viscosity, density, and surface tension. Here, "impact" refers to landing of cell-encapsulating droplets onto a receiving flat substrate, onto which the cell deposition is performed. Cell viability has been correlated with cell deformation.[66] Computational models can be used to predict and reduce cell deformation that occurs during impact of cell-encapsulating droplets onto a receiving surface. Therefore, a better understanding of deposition of viscous compound droplets mimicking encapsulated cells has to be developed.

A computational model of impact and spreading of a cell-encapsulating droplet on a horizontal surface was developed employing a finite-difference/front-tracking technique.[67] The cell, the outer droplet encapsulating the cell, and the ambient air were all assumed to be Newtonian fluid with different interfacial and material properties. The inner droplet was assumed to be a highly viscous fluid, mimicking the cell. Also, it was assumed that the outer droplet partially wetted the surface whereas the cell was non-wetting (not sticking to the substrate). This computational work[67] employed moving contact line model and previously confirmed for the simple droplet collision.[68] The dynamic contact angle was calculated at each time step,[69] and contact line was moved to set the evaluated dynamic contact angle. It was also assumed that the rate of deformation was significant in predicting viability. Therefore, the circumstances that lead to the minimum cell deformation and the rate of deformation were identified. For this purpose, simulations were performed for a range of governing dimensionless parameters, that is, Reynolds number, Weber number, viscosity ratio, surface tension ratio, diameter ratio, and equilibrium contact angle. The effects of dimensionless numbers on the cell deformation and deformation rate were investigated. The Reynolds number (Re) and the Weber number (We) are non-dimensional numbers extensively used in fluid dynamics field.[70–72] While the former provides the ratio of inertial forces and viscous forces, the latter shows the relative importance of fluid's inertia compared to its surface tension.[73,74]

A mathematical approach to cell printing process was reported.[67] Figure 13.3 shows the velocity vectors (left side) and shear stress contours (right side). During the initial collision, shear stress reached its maximum value in the vicinity of the triple point. Triple point is the point where encapsulating droplet, wall, and ambient air were in contact. Shear stress consistently decreased as the cell encapsulating droplet spread. During the initial phase of impact, the shear stress was 50% of the maximum hydrodynamic pressure (10 kPA).[67] Maximum shear stress was observed at the interface of encapsulating droplet and surrounding air. This was due to the non-zero ejection velocity of all fluid particles inside the cell encapsulating droplet, and the quiescent surrounding air at the beginning. During the ongoing collision, spreading and relaxation of droplet, maximum pressure changed its location. For instance, pressure amplified in the vicinity of the triple point at the beginning of collision and spreading phase. In addition, the maximum pressure was positioned in the vicinity of the contact line just before recoil. Yet, the maximum pressure started to move toward the distal end from the substrate, where it stayed there until the recoil stage.

The success of cell printing technology primarily relies on the relation between cell and flow dynamics during the spreading of compound droplet, as well as the other complications regarding ejection itself, such as shear. Simulations were presented for a range of non-dimensional numbers to explore the cell rate of deformation and cell deformation itself. In this study, gross deformation and rate of deformation were linked to cell damage. A gross deformation of cell was defined as, $D = (W_b - H_b)/(W_b + H_b)$.

The results showed that the deformation of the cell consistently increased: (i) as the inertial forces over viscous forces, that is, Reynolds number, increased; (ii) as the diameter ratio of the outer droplet to the inner droplet (cell) decreased; (iii) as the ratio of surface tensions of the air-encapsulating fluid interface to the encapsulating fluid–inner droplet interface increased; (iv) as the viscosity ratio of the inner droplet to outer droplet decreased; or (v) as the equilibrium contact angle decreased. A local minimum of maximum deformation was observed at We = 2 after investigating the effects of Weber number.

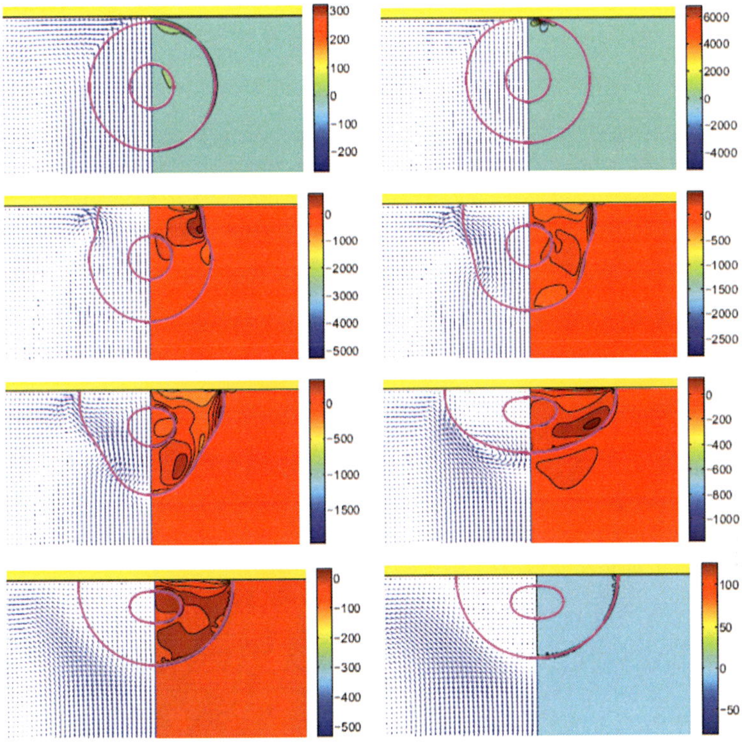


**Figure 13.3** As cell encapsulating droplet impacted on a flat surface, velocity vectors and shear contours were plotted at the left half and the right half, respectively. Time increases from left to right and from top to bottom. Images were taken at times $t^* = 2.69 \times 10^{-4}$, $5.41 \times 10^{-2}$, 0.1351, 0.2162, 0.2703, 0.5135, 1.0270, and 3.8432. Other non-dimensional numbers are We = 0.5, Re = 30, $d_o/d_i$ = 2.85, $\sigma_o/\sigma_i$ = 2541, $\mu_c/\mu_d$ = 10. Reprinted with permission from Ref. 67. Copyright 1994, American Institute of Physics.

The cell viabilities were predicted by linking them to cell deformation. This was done by using an experimental correlation of compression of cells between parallel plates.[66] The compression of the inner droplet during the spreading process was also plotted to show the goodness of the assumption, that is, almost symmetric compression (Fig. 13.4). The cell viabilities based on the computational model were given in Table 13.1. Based on the experimental results and requirements for computational convergence and stability, the set of the non-dimensional parameters were selected: We = 0.5, Re = 30,

$d_o/d_i = 2.85$, $\sigma_o/\sigma_i = 2541$, $\mu_c/\mu_d = 10$, $\mu_d/\mu_o = 40$, $\rho_c/\rho_d = 1$, $\rho_d/\rho_o = 20$, and $\theta_e = 90°$ as the base case. Only single parameter was changed in each row of Table 13.1 while all other parameters were set as the same values as those in the base case. As $\theta_e$ and $\mu_c/\mu_d$ were reduced, the cell viability rapidly decreased. Note that even though the minimum static contact angle in this computational work, i.e., 30°, was still larger than the value in the experimental work, i.e., 10°, cell viability was declined to an intolerable value such as 26%. However, this was attributed to relatively small value of $\mu_c/\mu_d$. This ratio was set to 10 in varying $\theta_e$ case. In reality, cell viscosity is much higher than that of encapsulating fluid and higher $\mu_c/\mu_d$ should cause to larger cell viability. Therefore, the mutual effect of smaller $\theta_e$ and larger $\mu_c/\mu_d$ on cell viability is still unknown.

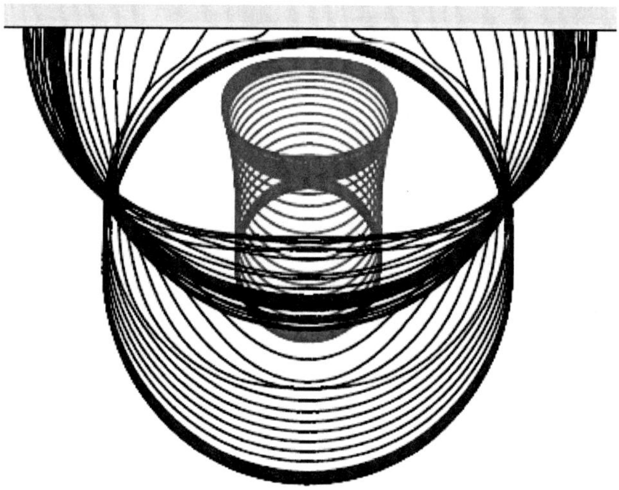

**Figure 13.4** Sequential impact images of cell encapsulating droplet for We = 0.5, Re = 30, $d_o/d_i = 2.85$, $\sigma_o/\sigma_i = 2541$, and $\mu_c/\mu_d = 10$. Reprinted with permission from Ref. 67. Copyright 1994, American Institute of Physics.

There is an unmet need for computational studies to model cell printing process including non-Newtonian features of fluids, lower contact angles, microstructured models for cells, and multi-impact of cell encapsulating droplets. The results described above provided initial vision into identifying conditions for utmost cell viability during cell encapsulating droplet printing process.

**Table 13.1** Cell viabilities

| Re | 15 | 20 | 30 | 40 | 45 | | |
|---|---|---|---|---|---|---|---|
| $\eta$ | 100 | 100 | 100 | 90.38 | 86.31 | | |
| We | 0.25 | 0.5 | 1 | 2 | 5 | 10 | |
| $\eta$ | 88.62 | 100 | 100 | 100 | 100 | 98.67 | |
| $d_o/d_i$ | 1.5 | 2 | 2.5 | 2.85 | 3 | 3.5 | |
| $\eta$ | 88.05 | 92.97 | 97.9 | 100 | 100 | 100 | |
| $\sigma_o/\sigma_i$ | 10 | 20 | 50 | 500 | 2541 | 5000 | |
| $\eta$ | 100 | 100 | 100 | 100 | 100 | 100 | |
| $\mu_c/\mu_d$ | 2 | 5 | 10 | 20 | 40 | | |
| $\eta$ | 44.07 | 88.43 | 100 | 100 | 100 | | |
| $\theta_e$ | 30° | 45° | 60° | 75° | 90° | 105° | 120° |
| $\eta$ | 26.15 | 54.81 | 80.77 | 91.84 | 100 | 100 | 100 |

*Source:* Reprinted with permission from Ref. 67. Copyright 1994, American Institute of Physics.

## 13.4 Applications of Cell Encapsulation and Bioprinting

### 13.4.1 Cardiomyocyte Encapsulation and Printing

Cardiovascular diseases affect 70 million Americans, resulting in an economic burden of $150 billion and accounting for nearly 40% of all deaths in the United States.[5,76] Conventional treatments of cardiac injury cannot achieve myocardial regeneration. Therefore, it has been suggested that *in vitro* engineered cardiac tissue grafts can be surgically placed into the defect site to facilitate cardiac regeneration.[75] The current challenges faced in cardiac tissue engineering can point to the importance of bioprinting and biopatterning of cells in 3D. The unique challenges of engineering the cellular architecture and organization of cardiac tissue include (i) spatial organization of multiple cell types (i.e., cardiomyocytes and cardiofibroblasts), (ii) orientation and directionality of cardiomyocytes, (iii) presence of endothelial cells forming microvasculature among the densely packed cardiac cells, and (iv) high cell density (~$10^8$ cells/$cm^3$).[60] Cardiac muscle (myocardium) is mainly composed of myocytes (~60–70% of total cell population).[77] Myocytes are connected to each other by intercalated disks (end to end) and each myocyte is in contact with several others to form a 3D network.[77] Structural hierarchy of cardiac tissue is composed of: (i) interconnections between myocytes, (ii) connection between myocytes and capillaries, and (iii) a collagen weave surrounding groups of myocytes. Cardiomyocytes have unique contractile properties due to their layered composition within the cardiac tissue. Cardiac fibroblasts are responsible for maintaining the intracellular collagen network and ECM,[77] therefore, they are an integral part of cardiac tissue structure.

The encapsulation and ejection of single to few cells with acoustic droplets using various fluids (e.g., PBS, media and hydrogels) have been demonstrated. Various cell types were encapsulated and bioprinted (NIH-3T3 fibroblasts, human RAJI cells, mouse embryonic stem cells, HL-1 cardiomyocytes, AML-12 hepatocytes) in ~37 ± 0.1 μm diameter droplets at rates varying from 1–1,000 droplets per second with relative viabilities exceeding 89.8%.[20] Live/dead assay of cell printing experiments of ejected HL-1 cardiomyocytes in a 0.5% agarose hydrogel is shown in Fig. 13.5. Bright field images indicate cell locations whereas green images indicate live cells (Fig. 13.5A,B). The initial cell printing experiments indicate repeatability and reliability across multiple repetitions. A thin hydrogel coating of a cell is achieved by droplet sizes comparable to the cell size (Fig. 13.5B). Cells that are encapsulated and deposited this way can be successfully used to pattern the unit cells of tissues in a layer-by-layer approach to form 3D structures. It should be noted that each cell encapsulating hydrogel droplet is a microscale construct, a unit cell of a larger 3D construct. We have investigated how to control the patterning periodicity to create fully coated surfaces with polymers drop-by-drop. We have shown that the patterning distances are controllable, drop sizes are uniform and cells can be encapsulated in hydrogel droplets.

After performing live/dead tests and beyond morphological observations, the bioprinted AML-12 cells were cultured for 12 days to observe cell functionality (Fig. 13.5C–E). Printed AML-12 and HL-1 cells adhered to the bottom of the flasks after ejection as these cells do in culture. Printed AML-12 cells at the 3rd, 6th, and 12th day in culture are healthy, and they divide and proliferate regularly. They were similar in morphology to those in a control flask that were not ejected (Fig. 13.5). Moreover, a cell metabolic assay was performed on the ejected AML-12 cells. We used nonfluorescent resazurin, which can be reduced by viable cells to red-fluorescent resorufin, to detect metabolic activity of ejected AML-12 cells (Fig. 13.5F).

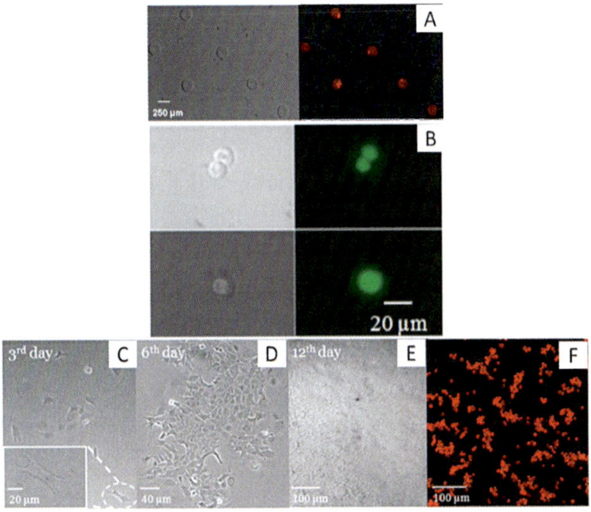


**Figure 13.5** (A) Cells in acoustic droplets were printed on a glass surface. (B) We ejected HL-1 cardiomyocytes in 0.5% agarose hydrogel in live/dead assay, where cells were green indicating they were alive. Gel layer around the cell is thinner (~5 µm) than cell diameter. (C–E) Printed AML-12 cells in cell culture after 3, 6, and 12 days, respectively. The morphology and growth of ejected cells are comparable to those of non-printed AML-12 cells. Cells adhered to the culture flask surface after ejection and became confluent in 12 days. (F) Red-fluorescent resorufin, shows metabolic activity of printed AML-12 cells. Reproduced from Ref. 20 by permission of the Royal Society of Chemistry.

### 13.4.2 Stem Cell Encapsulation and Printing

Inkjet printing technology has been employed for stem cell research.[26] Embryonic stem cells (ESCs) have the potential to differentiate into almost all types of cell. Hence, these cells offer potential for regenerative medicine and cell therapy. Initial phase of *in vitro* differentiation of ESCs is embryoid body (EB) formation. EB has a central role in *in vitro* differentiation of embryonic stem cell, and forms in a way similar to the initial phases of embryogenesis. The size and uniformity of EB are significant parameters that have direct effects on the phenotype expression of ESCs. Several approaches have been proposed to create EBs involving natural cell aggregation. Yet, there are still challenges in generating uniform EBs with smooth shapes in a repeatable manner. The present conventional hanging-drop technologies are time consuming and labor intensive and not capable of generating evenly sized droplets. An alternative technology was demonstrated to create EBs with uniform sizes by incorporating with bioprinting technologies.[26] The method was straightforward, fast, and robust (Fig. 13.6). Compared to the manual hanging-drop technology, uniform sizes of EBs were obtained. Microenvironments of stem cells were also engineered via inkjet technology for immobilized growth factor patterning.[78] A laser direct write method was used to print mouse ESCs onto defined arrays.[79]

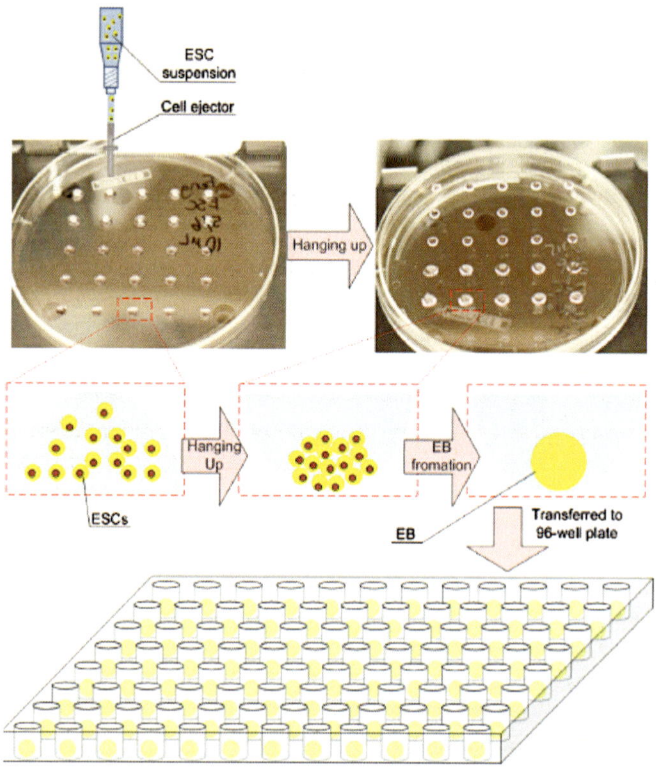


**Figure 13.6** Schematic of the embryoid body (EB) formation method integrated with bioprinting technology. To allow for EB aggregation, we bioprinted cell-medium suspension droplets onto a petri dish and hung up them for 24 hours. Then, EBs were transferred into a 96-well plate for extra culture up to 96 hours. Reprinted with permission from Ref. 26. Copyright 2011, American Institute of Physics.

### 13.4.3 Cancer Cell Encapsulation and Printing

To develop a better understanding of cancer pathology and to advance cancer therapies, 3D *in vitro* models offering a more precise illustration of disease *in vivo* are needed.[80] Yet, there are limitations in development of 3D manual models. Among these limitations are poor reproducibility, low level of control over cell density, low throughput, and, for co-culture systems, lack of repeatable control over placing different cell types, such as stromal and cancer cells. In a recently proposed method, a fast computerized cell printing system was introduced to bioprint human ovarian cancer (OVCAR-5) cells onto Matrigel.[81] Fastly and repeatedly, patterning of two cell types was performed within a spatially controlled microenvironment (e.g., cell density, cell-cell distance). After patterning, both cell types stayed live while printing process continued. By using this method, miniaturization of a conventional 3D macro-scale culture system was attained. This technology would also enable methodical examination into the several unknown interactions between stromal cells and tumor; and offer an instrument for rapid drug screening. The details of the cell biopatterning system were described in Ref. 23,24. The setup was composed of a computerized micrometer-resolution x-y-z stage, dispensing valves, and a pulse generator to control dispensing in nanometer scale (Fig. 13.7a). The complete experimental setup was placed inside a sterile hood enabling long-term (up to 3 weeks) cultures for patterned ovarian cancer model constructs (Fig. 13.7b).

### 13.4.4 Cryoprinting: Cryobiology Applications of Cell Encapsulation and Bioprinting

Bioprinting has also enabled new methods in cryobiology.[82] Cell-encapsulating droplets are ejected into a liquid reservoir via this approach, rather than being deposited onto a solid substrate or thin film. For instance, an ejection technology was developed to vitrify nanoliter droplets of oocytes.[83] This system was used to cryopreserve oocytes inside droplets in a continuous manner.[83] After each step of vitrification, parthenogenetic development of oocytes, their morphologies, and survival rates were evaluated compared to fresh oocytes. Following loading/unloading of cryoprotectant agents, oocytes were recovered, and comparable survival rates were absorbed between the results of nanoliter droplet encapsulation

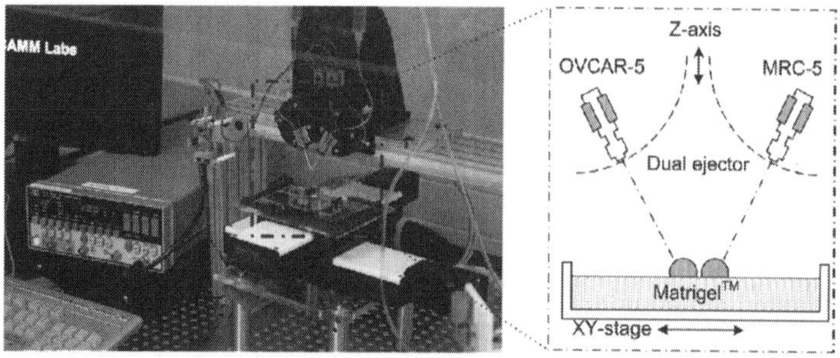


(a)

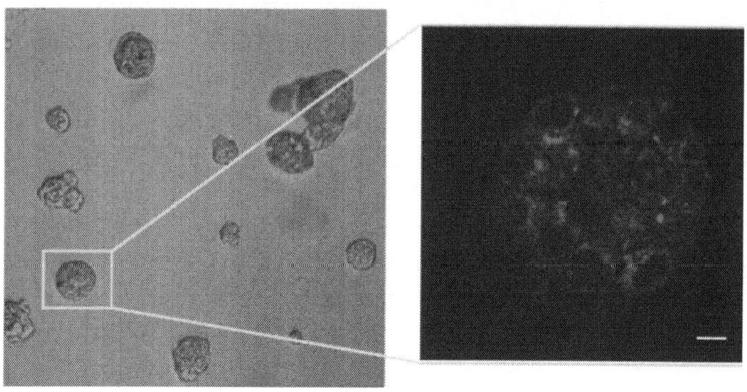

(b)

**Figure 13.7** Schematic for experimental setup composed of two ejectors and an automated stage. Different cell types, that is, cancer cells (OVCAR-5) and fibroblasts (MRC-5) were ejected simultaneously by dual ejector heads. (a) To prevent contamination, the setup was mounted in a sterile hood and HEPA filters were utilized. The setup is composed of a computerized x-y-z stage, nanoliter valves, and a pulse generator to control dispensing. The programmed control mechanisms synchronized the location of substrate and droplet generator. (b) Three-dimensional acini were formed inside GFR Matrigel™. Cells (OVCAR5) were ejected onto Matrigel™ substrate and grown in culture medium. After seven days, formation of 3D acini from ovarian cancer cells was observed (see two-photon auto-fluorescence images). Scale bar is 20 μm. Reproduced from Ref. 81 with permission by Wiley-VCH Verlag GmbH & Co. KGaA, Weinheim.

and fresh oocytes after 24 hours in culture. Similar morphologies were reported as well between oocytes retrieved after vitrification/

thawing and fresh oocytes. Another similar parameter between nanoliter droplet encapsulation and fresh oocytes was the rate of oocyte parthenogenetic activation. This technology presented a great potential to advance oocyte cryopreservation technology by allowing the vitrification of oocytes with smaller amount of cryoprotectants at larger warming and cooling rates.

Another bioprinting application of cryopreservation has been presented for blood banking,[84] which has potential to improve public health. In this application, red blood cell (RBC) encapsulating micrometer-sized (i.e., <100 μm)[24] droplets were vitrified in a high-throughput manner. This technique can allow mass-vitrification of blood in a very short amount of time. In this work,[83] the effects of three governing parameters: (i) the nitrogen flow rate (3.2–4.8 l/min), (ii) the RBCs (loaded with CPA) flow rate (160–220 ml/min), and (iii) the distance between ejector tip and droplet receiving film (60–90 mm), were investigated (Fig. 13.8). To enable low CPA level vitrification, small droplets (<100 μm) were generated by tuning nitrogen flow rate. This approach enabled also reducing potential negative osmotic and toxic effects.[84,85] In this context, the theory of cryoprinting[24] and microfluidic approaches to control loading and unloading cryoprotectants[86] were also studied.

## 13.4.5 Bioprinting Blood Vessels

Vascularization of tissue engineered constructs still remains a challenge in regenerative medicine. Generation of viable large scale tissue engineered grafts is limited due to the lack of vascular network. Cells within native tissues reside approximately within 100–200 μm distance to the closest capillary providing vital nutrients and oxygenation.[87] Many scaffold based approaches has been developed to engineer blood vessels that can mimic native vascular tree.[88,89] Scaffold free techniques target to create vasculature using living engineered cell sheet to increase engrafting process, flexibility, and to avoid any foreign body reactions.[90–92]

Endothelial capillary formation is the basis of angiogenesis. Endothelial cells isolated from umbilical vein, skin, or cord blood have shown to form endothelial capillary networks within gel systems. Co-culture of endothelial cells with adipose tissue derived stromal cell gives rise to robust and stable vascular network.[93] Laser-

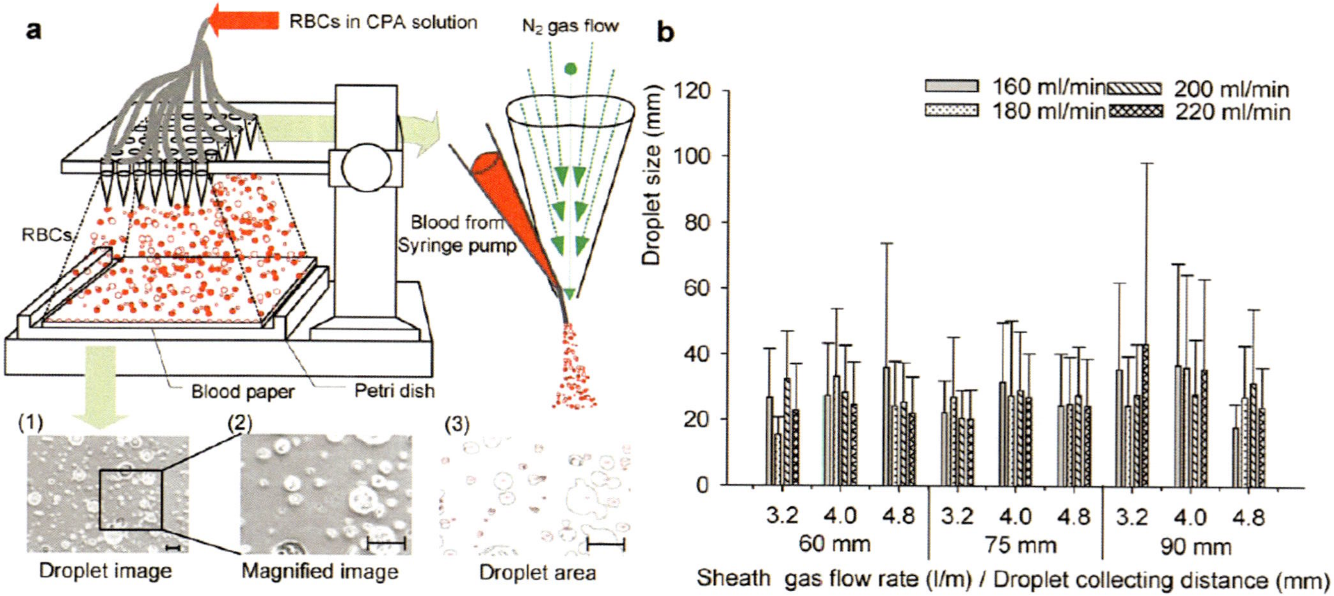


**Figure 13.8** A schematic of the blood cryopreservation platform. (a) Ejection and deposition of RBC encapsulating droplets on receiving paper (top) and droplet pictures (bottom). Scale bar is 500 μm. (b) Average size of ejected droplets are plotted for a range of sheath gas flow rate (3.2, 4.0, and 4.8 l/m), blood flow rate (160, 180, 200, and 220 ml/min), and droplet collecting distance (60, 75, and 90 mm).[83]

assisted bioprinting system was used to generate 3D multicellular arrays to study cell to cell interactions, particularly endothelial cell migration toward VEGF secreting adipose tissue derived stromal cells.[94] The study showed that laser-assisted bioprinting did not significantly affect the viability or functionality of endothelial cells, which are among the most vulnerable cells that can be damaged or lose functionality. Bioprinting can be used to generate computer aligned multi-cellular arrays to study angiogenesis.[40,95]

Advances in bioprinting can also be applied directly to build microcapillaries.[40,96–98] It has been demonstrated that microtubular structures with 500 μm inner diameter can serve as an intermediate vascular component, which can connect bioprinted arteries and capillary network (Fig. 13.9).[97] Work has been performed to fabricate artificial arteries using bioprinting approach.[99] Cell sheet based technique to generate functional human blood vessels has been demonstrated extremely promising results in the clinic.[100] Building blood vessels with bioprinting approach is on the way to reach the success of cell sheet based methods, overcoming the drawbacks of 2D culture.

### 13.4.6 Bioprinting in Bone Tissue Engineering

Bioprinting approaches have been also applied in generation of osteogenic grafts (Fig. 13.10).[101] Osteogenic grafts usually need good mechanical properties. At this point combining solid free form (SFF) scaffold fabrication with bioprinting can provide valuable solutions. Solid supporting scaffold was fabricated for bioprinted hydrogel using multi head deposition system.[102] Bone tissue engineering can benefit from such hybrid graft manufacturing. There are already some steps taken toward hybrid constructs. Hybrid multifunctional bioprinted grafts were fabricated for an osteochondral defect where the both osteogenic and chondrogenic functionality are needed.[103] Bioink is based on alginate gel where there are two compartments one seeded with mesenchymal stromal cells (MSC) and supplemented with osteoinductive biphasic ceramic particles to support osteo-differentiation and the second compartment is seeded with chondrocytes. The bioprinted graft has been cultured *in vitro* with differentiation media containing both osteogenic and chondrogenic differentiation factors. After ectopic *in vivo*

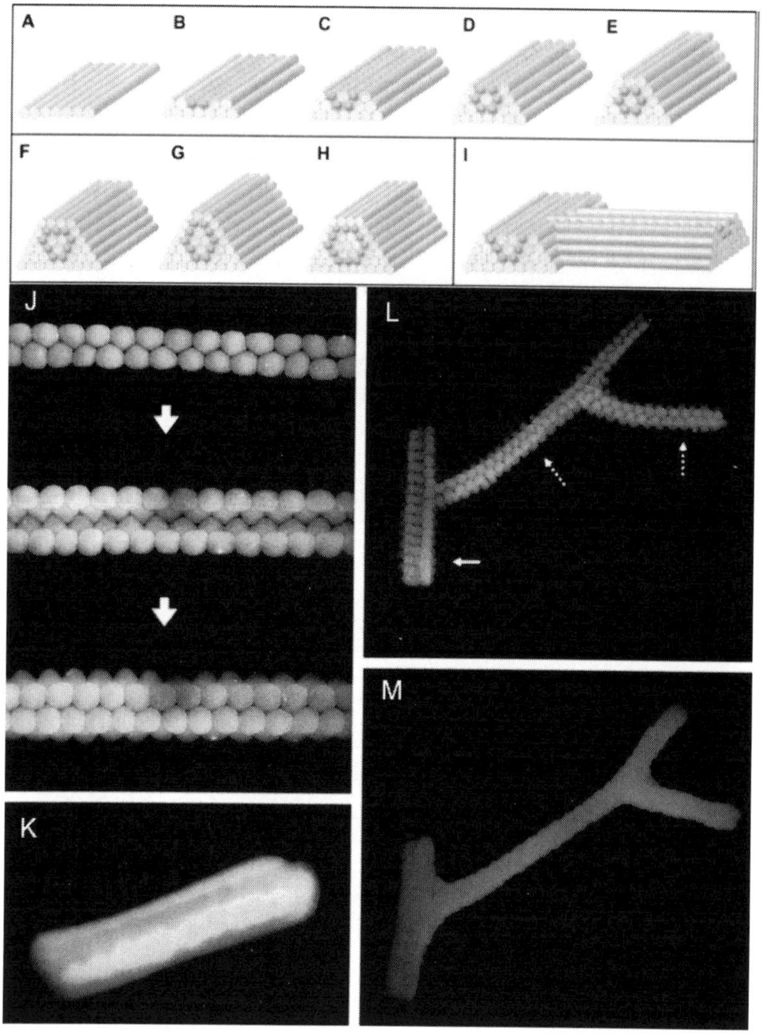


**Figure 13.9** Schematic representation of blood vessel design (A–E). Microspherical bioink is deposited on rods of biopaper. (F–H) Small and large diameter blood vessels can be supported with inert hydrogel to be removed after maturation. (I) Design of branched blood vessel. Fusion of multicellular spheroids integrated into tubular structures. (J) Spheroids assembled. (K) 7 days post tissue assembly of labeled sequences of seeded spheroids. (L) Branched structure built of 300 μm spheroids with branches of 1.2 mm (solid arrow) and 0.9 mm (broken arrows). (M) 6 days post fiber deposition. Reprinted from Ref. 97, with permission from Elsevier.

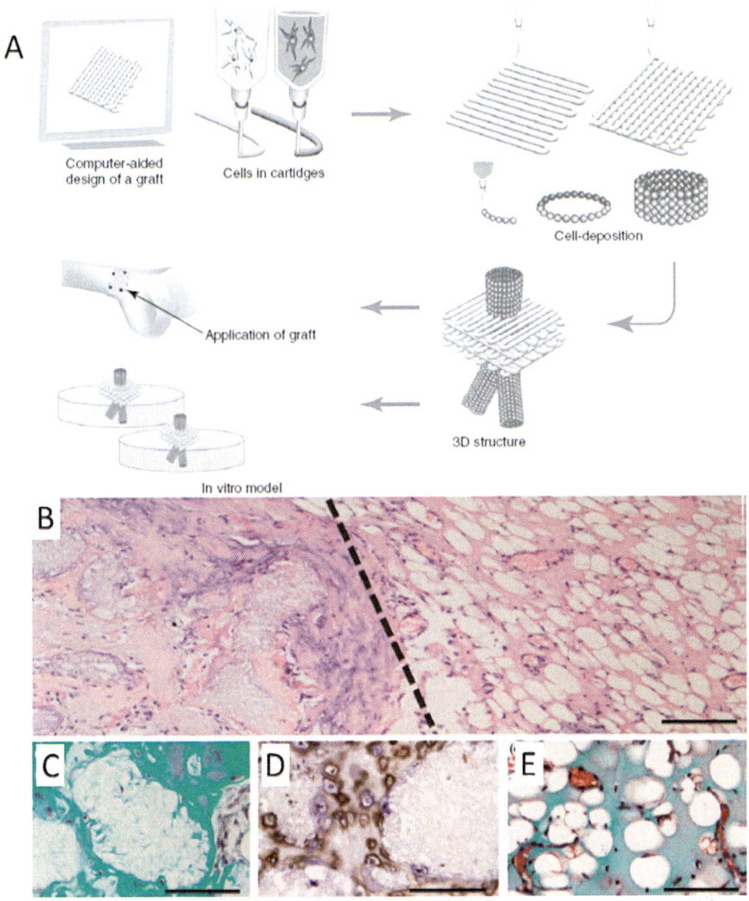


**Figure 13.10** (A) Schematic for layer-by-layer production of cell encapsulating hydrogel architectures with the goal of providing both *in vitro* and *in vivo* models. Images of heterogeneous tissue formation.[103] (B) Dashed line between images shows the transition. At the left side, multipotent stromal cells (MSCs) in Matrigel supplemented with BCP granules were printed. At the right side, EPCs in Matrigel were shown. (C) MSCs in Matrigel complemented with BCP (cell stained blue, collagen stained green, and BCP is gray). (D) Collagen type II in brown. (E) Endothelial progenitors cells in Matrigel (Matrigel is green and erythrocytes are red). (C) and (D) were stained with Goldner's trichrome. Scale bar is 200 μm (top image) and 50 μm (bottom images). Reprinted from Ref. 101, Copyright (2011), with permission from Elsevier.

implantation printed cells preserved their phenotype and produced considerable levels of endogenous osteogenic and chondrogenic ECM. Introducing vascular compartment hosting endothelial cells adjacent to the osteogenic part might boost bone tissue formation and increase engraftment.[101] Hybrid multifunctional bioprinted grafts are promising for mimicking native tissue.

### 13.4.7 Cell Printing as a Platform for Genomic Analyses

Cells that are present either in homogenous or heterogeneous populations, communicate with each other via complicated signaling pathways. Though having same phenotype, metabolic activities or responses to specific signals may vary from one cell to another.[104] Stem cell, tissue engineering, system biology and many more disciplines require single cell level analyses for better understanding of biological features and thereby characterize cells in more reliable ways. Cell printing techniques can provide extremely powerful platforms to obtain single cells in patterned ways. A fast and efficient high-throughput system has been shown to encapsulate single stem cells and provide platform for subsequent genomic analyses.[42] Single cells have been isolated from heterogeneous cell population using "drop-on-demand" cell printing technology with direct access to patterned cells with high viability (98.0 ± 0.8%). The system is simple, fast and has been validated with both theoretical and experimental applications. Genomic functional analyses of ejected cells based on mRNA expression levels shows that the printing procedure preserves properties of encapsulated stem cells and can be used as single cell isolation technique. Eleven stem cell markers including Kit and Notch1 were expressed in both the printed cells and non-printed control groups. Hence, genomic analysis performed on bioprinted cells indicating that the cells are not genomically affected from the printing process compared to the controls.[42] Therefore, bioprinting can be used to isolate and pattern cells for genomic and clonal studies without altering their baseline genetic properties.

### 13.4.8 Generating Porous Bioprinted Hydrogels

Porosity of scaffolds used in tissue engineering is vital for achieving efficient cell seeding and cell viability. Due to deficiency of vascularization in early phases, oxygen, nutrient supply and disposal metabolic wastes are performed through interconnected porous network within engineered scaffold. Hydrogels are widely used for tissue engineering. A novel approach using living bacterial sacrificing porogens to create interconnected pores within bioprinted hydrogels has been introduced.[25] *E. Coli* within hydrogel have been printed in specific pattern and incubated *in vitro* to achieve proliferation and followed with decellularization. After decellularization porogens formed interconnected pores that merged over a period of culture time and formed microchannels through the printing pattern. After subsequent washing procedures biocompatibility of decellularized hydrogel has been achieved by seeding of mammalian cells. Growth and viability of seeded 3T3 cells showed no adverse effect of bacteria culture and decellularization of porous hydrogel. Organ decellularization procedures are well established and provide preorganized natural extracellular matrices for tissue engineering approaches.[105] This novel method allows controlling pore size and interconnectivity enabling tuning mechanical and diffusion characteristics of hydrogels.

## 13.4.9 *In vivo* Applications of Bioprinting

Post-printing handling of bioprinted tissues is a challenge due to fragility of the printed graft and difficulties to provide fully effective bioreactor systems for bioprinting applications. Studies on *in vitro* maturation of printed grafts for clinical applications are needed. Highly developed bioprinting technologies can control and provide precision in bioprinting. An *in vivo* bioprinting application was demonstrated on critical size bone defect in mice calvarias.[107] A workstation based on high-throughput biological laser printing has been developed and hydroxy apatite based slurry has been applied on open calvaria defect of a mouse. Prior the laser printing the adverse effects of laser irradiation has shown inflammation in a range of magnetic resonance imaging. System has been well set in terms of stabilization of mouse and x-y-z orientation of automated platform. This work shows satisfying results of bone formation in critical size calvaria defects and opens discussions to use host body as ultimate bioreactor for bioprinted tissues.

## 13.5 Current Challenges and Future Perspectives in Bioprinting

The ability to encapsulate and pattern single cells in a complex 3D architecture at physiologically relevant cell densities would allow engineering of complex tissue structures *in vitro*,[59,75] benefiting regenerative medicine and biomedical research.[75] It is crucial to control the geometrical shape of the fabricated constructs. Generation of new synthetic hydrogels with enhanced biological and biochemical properties[116,117] or bioreactor based maturation strategies of printed tissues aim to overcome existing limitations.[22] Development of optical, biomechanical, physical, and biochemical monitoring systems and mechanisms are going to be critical tools to create new generations of bioprinting systems. Further, implementation of structural and functional biomarkers is also crucial to understand the maturation level of the fabricated tissues.[21] Another promising strategy to improve stability of printed tissue constructs is to combine rapid prototyped solid scaffold fabrication with bioprinting and bioreactors. Studies applying these approaches to generate hybrid grafts are promising[102] to meet both biological and mechanical design parameters. Developing validated, sterile and closed bioprinting systems for target tissues is the ultimate goal.

*In vivo* bioprinting can be potentially combined in the future with an intraoperative approach where cells (for instance from adipose tissue) can be *in situ* isolated with an automated system in an operation theatre, introduced in a hydrogel, and directly printed into a defect zone.[119] Bioprinting technologies, research and applications are exciting, since they have the potential to provide custom therapies for patients, and have a positive impact on costly and demanding *in vitro* cell culture and tissue maturation phases and regulations. Clinical applications of bioprinting will benefit from further studies as the field continues to grow rapidly. Shorter-term bioprinting applications are also promising including testing systems for drug screening, diagnostic kits for diseases, cryopreservation and models for *in vitro* organ systems.

## Acknowledgments

We would like to acknowledge R21 NIH grant 1R21HL112114-01 and NIH grant R21HL095960. UD is a founder of, and has an equity interest in, DxNow Inc., a company that is developing microfluidic and imaging technologies for point-of-care diagnostic solutions. UD's interests were reviewed and are managed by the Brigham and Women's Hospital and Partners HealthCare in accordance with their conflict of interest policies.

## References


1. Martin, I., Smith, T. and Wendt, D. (2009) Bioreactor-based roadmap for the translation of tissue engineering strategies into clinical products, *Trends Biotechnol.*, **27**(9), 495–502.
2. Cho, T. J., Gerstenfeld, L. C. and Einhorn T. A. (2002) Differential temporal expression of members of the transforming growth factor beta superfamily during murine fracture healing, *J. Bone Miner. Res.*, **17**(3), 513–520.
3. Le, A. X., *et al.* (2001) Molecular aspects of healing in stabilized and non-stabilized fractures, *J. Orthop. Res.*, **19**(1), 78–84.
4. Tatsuyama, K., *et al.* (2000) Expression of various growth factors for cell proliferation and cytodifferentiation during fracture repair of bone, *Eur. J. Histochem.*, **44**(3), 269–278.
5. Muraglia, A., *et al.* (1998) A nude mouse model for human bone formation in unloaded conditions, *Bone*, **22**(5), 131s–134s.
6. Histing, T., *et al.* (2010) An internal locking plate to study intramembranous bone healing in a mouse femur fracture model, *J. Orthop. Res.*, **28**(3), 397–402.
7. Thompson, Z., *et al.* (2002) A model for intramembranous ossification during fracture healing, *J. Orthop. Res.*, **20**(5), 1091–1098.
8. Birgersdotter, A., Sandberg, R. and Ernberg, I. (2005) Gene expression perturbation *in vitro*—A growing case for three-dimensional (3D) culture systems, *Semin. Cancer Biol.*, **15**(5), 405–412.
9. Gurkan, U., *et al.* (2011) A scaffold-free multicellular three-dimensional *in vitro* model of osteogenesis, *Calcified Tissue Int.*, **88**(5), 388–401.
10. Tibbitt, M. W. and Anseth, K. S. (2009) Hydrogels as extracellular matrix mmics for 3D cell culture, *Biotechnol. Bioeng.*, **103**(4), 655–663.
11. Li, R. K., *et al.* (2000) Construction of a bioengineered cardiac graft, *J. Thorac. Cardiovasc. Surg.*, **119**(2), 368–375.
12. Masters, K. S., *et al.* (2005) Crosslinked hyaluronan scaffolds as a biologically active carrier for valvular interstitial cells, *Biomaterials*, **26**(15), 2517–2525.

13. Geckil, H., *et al.* (2010) Engineering hydrogels as extracellular matrix mimics, *Nanomedicine*, **5**(3), 469–484.
14. Khademhosseini, A., *et al.* (2006) Microscale technologies for tissue engineering and biology, *P. Natl. Acad. Sci. USA*, **103**(8), 2480–2487.
15. Lee, K. Y. and Mooney, D. J. (2001) Hydrogels for tissue engineering. *Chem. Rev.*, 101(7), 1869–1879.
16. Griffith, L. G. and Swartz, M. A. (2006) Capturing complex 3D tissue physiology *in vitro*, *Nat. Rev. Mol. Cell Biol.*, **7**(3), 211–224.
17. Tasoglu, S. and Demirci, U. (2013) Bioprinting for stem cell research, *Trends Biotechnol.*, 31(1), 10–19.
18. Durmus, N.G., Tasoglu, S., and Demirci, U. (2013) Functional droplet networks, *Nat. Mater.*, 12(6), 478–79.
19. Hamid, Q., *et al.* (2011) Fabrication of three-dimensional scaffolds using precision extrusion deposition with an assisted cooling device, *Biofabrication*, **3**(3), 034109.
20. Demirci, U. and Montesano, G. (2007) Single cell epitaxy by acoustic picolitre droplets, *Lab on a Chip*, **7**(9), 1139–1145.
21. Mironov, V., *et al.* (2008) Organ printing: promises and challenges, *Regen. Med.*, **3**(1), 93–103.
22. Mironov, V., V. Kasyanov, and Markwald, R. R. (2011) Organ printing: From bioprinter to organ biofabrication line, *Curr. Opin. Biotechnol.*, **22**(5), 667–673.
23. Moon, S., *et al.* (2010) Layer by layer three-dimensional tissue epitaxy by cell-laden hydrogel droplets,*Tissue Eng. Part C Methods*, **16**(1), 157–166.
24. Xu, F., *et al.* (2010) A droplet-based building block approach for bladder smooth muscle cell (SMC) proliferation, *Biofabrication*, **2**(1), 014105. doi:10.1088/1758-5082/2/1/014105.
25. Xu, F., *et al.* (2011) Living bacterial sacrificial porogens to engineer decellularized porous scaffolds, *PLoS One*, **6**(4), e19344.
26. Xu, F., *et al.* (2011) Embryonic stem cell bioprinting for uniform and controlled size embryoid body formation, *Biomicrofluidics*, **5**(2), 22207.
27. Xu, F., *et al.* (2011) A three-dimensional *in vitro* ovarian cancer coculture model using a high-throughput cell patterning platform, *Biotechnol. J.*, **6**(2), 204–212.
28. Calvert, P. (2007) Materials science: Printing cells, *Science*, **318**(5848), 208–209.

29. Mikos, A. G., *et al.* (2006) Engineering complex tissues, *Tissue Eng.*, **12**(12), 3307–3339.

30. Langer, R. (2007) Tissue engineering: perspectives, challenges, and future directions, *Tissue Eng.*, **13**(1), 1–2.

31. Nakamura, M., *et al.* (2005) Biocompatible inkjet printing technique for designed seeding of individual living cells, *Tissue Eng.*, **11**(11–12), 1658–1666.

32. Boland, T., *et al.* (2006) Application of inkjet printing to tissue engineering, *Biotechnol. J.*, **1**(9), 910–917.

33. Barron, J.A., *et al.* (2004) Application of laser printing to mammalian cells, *Thin Solid Films*, **453**, 383–387.

34. Gurkan, U. A., *et al.* (2013) Simple precision creation of digitally specified, spatially heterogeneous, engineered tissue architectures, *Adv. Mater.*, 25(8), 1192–1198.

35. Asghar, W., *et al.* (2013) *In vitro* three-dimensional cancer culture models, in *Cancer Targeted Drug Delivery* (eds, Park, K., Bae, Y. H., and Mrsny, R.), 635–665.

36. Tasoglu, S., *et al.* (2013) Manipulating biological agents and cells in micro-scale volumes for applications in medicine, *Chem. Soc. Rev.*, 42(13), 5788–5808.

37. Gurkan, U. A., *et al.* (2012) Bioprinting anisotropic stem cell microenvironment, *J. Tissue Eng. Regen. Med.*, 6(SI), 366–366.

38. Tasoglu, S., *et al.* (2012) Magnetic 3D assembly of microgels for tissue engineering and regenerative medicine, *J. Tissue Eng. Regen. Med.*, 6, 224–224.

39. Gurkan, U. A., *et al.* (2012). Emerging technologies for assembly of microscale hydrogels, *Adv. Healthcare Mater.*, 1(2), 149–158.

40. Cui, X. and Boland, T. (2009) Human microvasculature fabrication using thermal inkjet printing technology, *Biomaterials*, **30**(31), 6221–6227.

41. Mironov, V., *et al.* (2003) Organ printing: computer-aided jet-based 3D tissue engineering, *Trends Biotechnol.*, **21**(4), 157–161.

42. Moon, S., *et al.* (2011) Drop-on-demand single cell isolation and total RNA analysis, *PLoS One*, **6**(3), e17455.

43. Nahmias, Y. and Odde, D. J. (2006) Micropatterning of living cells by laser-guided direct writing: Application to fabrication of hepatic-endothelial sinusoid-like structures, *Nat. Protoc.*, **1**(5), 2288–2296.

44. Ringeisen, B. R., *et al.* (2002) Generation of mesoscopic patterns of viable Escherichia coli by ambient laser transfer, *Biomaterials*, **23**(1), 161–166.

45. Wu, P. K., *et al.* (2001) The deposition, structure, pattern deposition, and activity of biomaterial thin-films by matrix-assisted pulsed-laser evaporation (MAPLE) and MAPLE direct write, *Thin Solid Films*, **398**, 607–614.

46. Langer, R. S. and *Vacanti*, J. P. (1999) Tissue engineering: The challenges ahead, *Sci. Am.*, **280**(4), 86–89.

47. Demirci, U. and Montesano, G. (2007) Cell encapsulating droplet vitrification, *Lab Chip*, **7**(11), 1428–1433.

48. Demirci, U. and Toner, M. (2006) Direct etch method for microfludic channel and nanoheight post-fabrication by picoliter droplets, *Appl. Phys. Lett.*, **88**(5), 053117.

49. Chang R., Nam J., and Sun, W. (2008) Direct cell writing of 3D microorgan for *in vitro* pharmacokinetic model, *Tissue Eng. Part C*, **14**(2), 157–166.

50. Utada, A. S., *et al.* (2005) Monodisperse double emulsions generated from a microcapillary device, *Science*, **308**(5721), 537–541.

51. Hong, J. W. and Quake, S. R. (2003) Integrated nanoliter systems, *Nat. Biotechnol.*, **21**(10), 1179–1183.

52. AHA (2006) *Heart and Stroke Statistical Update*, American Heart Association (AHA): New York.

53. Baret, J.-C., *et al.* (2009) Fluorescence-activated droplet sorting (FADS): Efficient microfluidic cell sorting based on enzymatic activity, *Lab on a Chip*, **9**(13), 1850–1858.

54. Abate, A.R., *et al.* (2009) Beating Poisson encapsulation statistics using close-packed ordering, *Lab on a Chip*, **9**(18), 2628–2631.

55. Abate, A.R., *et al.* (2009) Impact of inlet channel geometry on microfluidic drop formation, *Phys. Rev. E*, **80**(2), 026310.

56. Mazutis, L., *et al.* (2009) Multi-step microfluidic droplet processing: Kinetic analysis of an *in vitro* translated enzyme, *Lab on a Chip*, **9**(20), 2902–2908.

57. Jakab, K., *et al.* (2010) Tissue engineering by self-assembly and bio-printing of living cells, *Biofabrication*, **2**(2), 022001.

58. Xu, F., *et al.* (2011) Microengineering methods for cell-based microarrays and high-throughput drug-screening applications, *Biofabrication*, **3**(3), 034101.

59. Guillotin, B., *et al.* (2010) Laser assisted bioprinting of engineered tissue with high cell density and microscale organization, *Biomaterials*, **31**(28), 7250–7256.

60. Radisic, M. and Vunjak-Novakovic, G. (2005) Cardiac tissue engineering, *J. Serb. Chem. Soc.*, **70**(3), 541–556.

61. Demirci, U. (2005) Picoliter droplets for spinless photoresist deposition, *Rev. Sci.Instrum.*, **76**(6), 065103.

62. Demirci, U. (2006) Acoustic picoliter droplets for emerging applications in semiconductor industry and biotechnology, *J. Microelectromech. S.*, **15**(4), 957–966.

63. Demirci, U., *et al.* (2004) Forward-viewing CMUT arrays for medical imaging, *IEEE Trans. Ultrason. Ferroelectr. Freq. Control*, **51**(7), 887–895.

64. Moon, S., *et al.* (2011) Statistical modeling of single target cell encapsulation, *PLoS One*, **6**(7), e21580.

65. Ceyhan, E., Xu, F., Gurkan, U. A., Emre, A. E., Turali, E. S., El Assal, R., Acikgenc, A., Wu, C.-a.M., and Demirci, U. (2012). Prediction and control of number of cells in microdroplets by stochastic modeling. *Lab Chip*, 12(22), 4884–4893; doi:10.1039/c2lc40523g

66. Takamatsu, H. and Rubinsky, B. (1999) Viability of deformed cells., *Cryobiology*, **39**(3), 243–251.

67. Tasoglu, S., *et al.* (2010) Impact of a compound droplet on a flat surface: A model for single cell epitaxy, *Phys. Fluids*, **22**(8), 082103.

68. Muradoglu, M. and Tasoglu, S. (2010) A front-tracking method for computational modeling of impact and spreading of viscous droplets on solid walls, *Comput. Fluids*, **39**(4), 615–625.

69. Kistler, S. F. (1993) *Wettability*, (ed. Berg, J. C.) Dekker, New York.

70. Tasoglu, S., Demirci, U., and Muradoglu, M. (2008) The effect of soluble surfactant on the transient motion of a buoyancy-driven bubble, *Phys. Fluids*, 20, 040805.

71. Tasoglu, S., *et al.* (2011) The effects of inhomogeneous boundary dilution on the coating flow of an anti-HIV microbicide vehicle, *Phys. Fluids*, 23, 093101.

72. Tasoglu, S., *et al.* (2013) Transient swelling, spreading and drug delivery by a dissolved anti-HIV microbicide-bearing film, *Phys. Fluids*, 25(3), 031901.

73. Tasoglu, S., *et al.* (2011) The consequences of yield stress on deployment of a non-Newtonian anti-HIV microbicide gel, *J. Non-Newtonian Fluid Mech.*, 166, 1116–1122.

74. Tasoglu, S., Katz, D. F., and Szeri, A. J. (2012) Transient spreading and swelling behavior of a gel deploying an anti-HIV microbicide, *J. Non-Newtonian Fluid Mech.*, 187–188, 36–42.

75. Lanza, R. P., Langer, R. S., and Vacanti, J. (eds) (2007) *Principles of Tissue Engineering*, 3rd ed., Elsevier Academic Press, Amsterdam, Boston, xxvii, p. 1307.

76. Trogdon, J. G., *et al.* (2007) The economic burden of chronic cardiovascular disease for major insurers, *Health Promot. Pract.*, 8, 234, DOI: 10.1177/1524839907303794.

77. LeGrice, I., Pope, A., and Smaill, B. (2005) The architecture of the heart: Myocyte organization and the cardiac extracellular matrix, in *Interstitial Fibrosis in Heart Failure* (ed., F.J. Villarreal), Springer: New York.

78. Phillippi, J.A., *et al.* (2008) Microenvironments engineered by inkjet bioprinting spatially direct adult stem cells toward muscle- and bone-like subpopulations, *Stem Cells*, **26**(1), 127–134.

79. Raof, N.A., *et al.* (2011) The maintenance of pluripotency following laser direct-write of mouse embryonic stem cells, *Biomaterials*, **32**(7), 1802–1808.

80. Rizvi, I., *et al.* (2011) Biologically relevant 3D tumor arrays: Treatment response and the importance of stromal partners, *SPIE*, doi:10.1117/12.875892, 788609.

81. Xu, F., *et al.* (2011) A three-dimensional *in vitro* ovarian cancer coculture model using a high-throughput cell patterning platform, *Biotechnol. J.*, **6**(2), 204–212.

82. Zhang, X., *et al.* (2012) Nanoliter droplet vitrification for oocyte cryopreservation, *Nanomedicine* (Lond), **7**(4), 553–564.

83. Samot, J., *et al.* (2011) Blood banking in living droplets, *PLoS One*, **6**(3), e17530.

84. Meryman, H.T. and Hornblower, M. (1972) A method for freezing and washing red blood cells using a high glycerol concentration, *Transfusion*, **12**(3), 145–156.

85. Tullis, J.L., *et al.* (1958) Studies on the *in vivo* survival of glycerolized and frozen human red blood cells, *J. Am. Med. Assoc.*, **168**(4), 399–404.

86. Song, Y.S., *et al.* (2009) Microfluidics for cryopreservation, *Lab Chip*, **9**(13), 1874–1881.

87. Carmeliet, P. and Jain, R. K. (2000) Angiogenesis in cancer and other diseases, *Nature*, **407**(6801), 249–257.

88. Peck, M., *et al.* (2012) The evolution of vascular tissue engineering and current state of the art, *Cells Tissues Organs*, **195**(1–2), 144–158.

89. da Graca, B. and Filardo, G. (2011) Vascular bioprinting, *Am. J. Cardiol.*, **107**(1), 141–142.

90. L'Heureux, N., *et al.* (1998) A completely biological tissue-engineered human blood vessel, *FASEB J.*, **12**(1), 47–56.

91. L'Heureux, N., *et al.* (2006) Human tissue-engineered blood vessels for adult arterial revascularization, *Nat. Med.*, **12**(3), 361–365.

92. Elloumi-Hannachi, I., Yamato, M. and Okano, T. (2010) Cell sheet engineering: A unique nanotechnology for scaffold-free tissue reconstruction with clinical applications in regenerative medicine, *J. Intern. Med.*, **267**(1), 54–70.

93. Traktuev, D. O., *et al.* (2009) Robust functional vascular network formation *in vivo* by cooperation of adipose progenitor and endothelial cells, *Circ. Res.*, **104**(12), 1410–1420.

94. Gruene, M., *et al.* (2011) Laser printing of three-dimensional multicellular arrays for studies of cell-cell and cell-environment interactions, *Tissue Eng. Part C Methods*, **17**(10), 973–982.

95. Chang, C. C., *et al.* (2012) Determinants of microvascular network topologies in implanted neovasculatures, *Arterioscler. Thromb. Vasc. Biol.*, **32**(1), 5–14.

96. Skardal, A., Zhang, J. and Prestwich, G. D. (2010) Bioprinting vessel-like constructs using hyaluronan hydrogels crosslinked with tetrahedral polyethylene glycol tetracrylates, *Biomaterials*, **31**(24), 6173–6181.

97. Norotte, C., *et al.* (2009) Scaffold-free vascular tissue engineering using bioprinting, *Biomaterials*, **30**(30), 5910–5917.

98. Visconti, R. P., *et al.* (2010) Towards organ printing: engineering an intra-organ branched vascular tree, *Expert Opin. Biol. Ther.*, **10**(3), 409–420.

99. Mironov, V., Kasyanov, V. and Markwald, R. R. (2008) Nanotechnology in vascular tissue engineering: from nanoscaffolding towards rapid vessel biofabrication, *Trends Biotechnol.*, **26**(6), 338–344.

100. L'Heureux, N., McAllister, T. N., and de la Fuente, L. M. (2007) Tissue-engineered blood vessel for adult arterial revascularization, *N.Engl. J. Med.*, **357**(14), 1451–1453.

101. Fedorovich, N. E., *et al.* (2011) Organ printing: The future of bone regeneration?, *Trends Biotechnol.*, **29**(12), 601–606.

102. Shim, J. H., *et al.* (2011) Development of a hybrid scaffold with synthetic biomaterials and hydrogel using solid freeform fabrication technology, *Biofabrication*, **3**(3), 034102.

103. Fedorovich, N. E., *et al.* (2012) Biofabrication of osteochondral tissue equivalents by printing topologically defined, cell-laden hydrogel scaffolds, *Tissue Eng. Part C Methods*, **18**(1), 33–44.

104. Tay, S., *et al.* (2010) Single-cell NF-kappaB dynamics reveal digital activation and analogue information processing, *Nature*, **466**(7303), 267–271.

105. Uygun, B. E., *et al.* (2010) Organ reengineering through development of a transplantable recellularized liver graft using decellularized liver matrix, *Nat. Med.*, **16**(7), 814–820.

106. Fedorovich, N. E., *et al.* (2011) Distinct tissue formation by heterogeneous printing of osteo- and endothelial progenitor cells, *Tissue Eng. Part A*, **17**(15–16), 2113–2121.

107. Keriquel, V., *et al.* (2010) *In vivo* bioprinting for computer- and robotic-assisted medical intervention: Preliminary study in mice, *Biofabrication*, **2**(1), 014101.

108. Gruene, M., *et al.* (2011) Adipogenic differentiation of laser-printed 3D tissue grafts consisting of human adipose-derived stem cells, *Biofabrication*, **3**(1), 015005.

109. Lee, W., *et al.* (2009) Three-dimensional bioprinting of rat embryonic neural cells, *Neuroreport*, **20**(8), 798–803.

110. Lee, Y. B., *et al.* (2010) Bio-printing of collagen and VEGF-releasing fibrin gel scaffolds for neural stem cell culture, *Exp. Neurol.*, **223**(2), 645–652.

111. Snyder, J. E., *et al.* (2011) Bioprinting cell-laden matrigel for radioprotection study of liver by pro-drug conversion in a dual-tissue microfluidic chip, *Biofabrication*, **3**(3), 034112.

112. Wang, X., *et al.* (2006) Generation of three-dimensional hepatocyte/gelatin structures with rapid prototyping system, *Tissue Eng.*, **12**(1), 83–90.

113. Mironov, V., Drake, C. and Wen, X. (2006) Research project: Charleston Bioengineered Kidney Project, *Biotechnol. J.*, **1**(9), 903–905.

114. Xu, T., *et al.* (2009) Fabrication and characterization of bio-engineered cardiac pseudo tissues, *Biofabrication*, **1**(3), 035001.

115. Gaetani, R., *et al.* (2012) Cardiac tissue engineering using tissue printing technology and human cardiac progenitor cells, *Biomaterials*, **33**(6), 1782–1790.

116. Lutolf, M. P. and Blau, H. M. (2009) Artificial stem cell niches, *Adv. Mater.*, **21**(32–33), 3255–3268.

117. Van Vlierberghe, S., Dubruel, P. and Schacht, E. (2011) Biopolymer-based hydrogels as scaffolds for tissue engineering applications: A review, *Biomacromolecules*, **12**(5), 1387–1408.

118. Tasoglu, S. *et al.* (2013) Untethered micro-robotic coding of three-dimensional material composition, *Nat. Commun.*, **4**(3124), doi: 10.1038/ncomms4124.

119. Mehrkens A., Saxer F., Güven S., Hoffmann W., Müller A. M., Jakob M., Weber F. E., Martin I., Scherberich A. (2012) Intraoperative engineering of osteogenic grafts combining freshly harvested, human adipose-derived cells and physiological doses of bone morphogenetic protein-2, *Eur Cell Mater*, **24**, 308–319, PMID: 23018419.

Chapter 14

# *In vitro* and *in vivo* Testing

**Milena Fini, Nicolò Nicoli Aldini, and Paola Torricelli**
*Laboratory of Preclinical and Surgical Studies, Laboratory of Biocompatibility, Innovative Technologies and Advanced Therapies, Rizzoli Orthopedic Institute Bologna, Italy*
milena.fini@ior.it

Appropriate preclinical investigations, defined as the laboratory or animal studies performed to provide evidence that a new treatment is safe and effective enough to be administered in humans, are required to assess the scaffold properties. However, preclinical studies are only one step in the most complex and articulated route to validate materials proposed for clinical use; indeed the planning of a research project is based on the contributions of many skills, and the reliability of the results is the direct effect of the appropriate development of the experimental design. The ability of a material to perform a specific function cannot be evaluated only *in vitro* or *in vivo*. The *in vitro* tests do not replace the *in vivo* evaluations, and by only adopting both procedures the limits of each single method can be reduced and the benefits increased. Current worldwide regulations

*Scaffolds for Tissue Engineering: Biological Design, Materials, and Fabrication*
Edited by Claudio Migliaresi and Antonella Motta

ISBN 978-981-4463-20-1 (Hardcover), 978-981-4463-21-8 (eBook)
www.panstanford.com

are now mainly those of the U.S. Food and Drug Administration (FDA), the International Organization for Standardization (ISO) or the Japanese Ministry of Health and Welfare (JMHW). For the *in vivo* investigations, the accomplishment of strict ethical and legal rules is also required. In the European Union, the *Directive 86/609—Protection of Experimental Animals* was recently updated and replaced by the Directive 2010/63. Nowadays the evaluation of biocompatibility cannot ignore the development of very complex methods as well as the use of cells co-cultured and bioreactors that can better mimic some of the complexity of the whole organism. As the host variables such as the overall healthy status and the microenvironment at the implantation site will have a major effect on the biological behavior, preclinical models will also consider these aspects. In this perspective, the use of pathological tissue-derived cells (i.e., from osteoporotic or osteoarthritic bone and cartilage) and of *in vivo* animal models of diseases should be adopted and improved. Finally, the impact of sophisticated processing parameters to develop the proper chemistry and topography of scaffolds would require the use of advanced technologies.

## 14.1 Introduction

The general approaches in tissue engineering are to seed cells on scaffolds and proceed by *in vitro* culture of cells and then to the *in vivo* implantation (*in vitro* tissue engineering) or to implant bioactive scaffolds that are able to stimulate cell migration, adhesion, proliferation, protein synthesis and angiogenesis in the lesion site by directly interacting with the biological environment (*in situ* tissue engineering). In both cases, the scaffold may have specific chemical, biochemical, physical, topographical and mechanical properties depending on the biological function to be improved or replaced, and each of these properties influences the scaffold behavior and biocompatibility.

In the Consensus Conference on Biomaterials (Sorrento, Italy, 15–16 September 2005), the biocompatibility of tissue-engineered products was defined as follows: *The biocompatibility of a scaffold or matrix for a tissue-engineered product refers to the ability to perform as a substrate that will support the appropriate cellular activity, including the facilitation of molecular and mechanical signaling systems, in order to optimize tissue regeneration, without eliciting any undesirable effects in those cells, or inducing any undesirable local or systemic responses in the eventual host.* More recently, Williams redefined biocompatibility as follows: *Biocompatibility refers to the ability of a biomaterial to perform in a desired function with respect to a medical field, without eliciting any undesirable local or systemic effects in the recipient or beneficiary of that therapy, but generating the most appropriate beneficial cellular or tissue response in that specific situation, and optimizing the clinically relevant performance of that therapy.*[1]

The scaffold's biocompatibility is a function of the situations in which biomaterials are used; the reactions with the surrounding tissues, the degradation over time and, obviously, the material characteristics from which the scaffold is constructed;[1,2] and in the cases of scaffolds made of more than one material, biocompatibility data may be obtained on every component of the material used in the scaffold and, definitely, on the final one.

Several definitions of biomaterials exist, and with the development of tissue engineering and regenerative medicine, the definition has broadened as follows: *any material used in a medical field intended to interact with biological systems* in order to consider structures and combined devices to be included in the field of biomaterials.[3–5]

As reported by Binyamin *et al.*,[3] surface properties of a material are responsible for the interactions occurring at the interface with its environments, and for implanted biomaterials one of the key surface properties is biocompatibility.

### 14.1.1 Biocompatibility Evaluations: *in vitro* and *in vivo* Tests

Preclinical investigations are required to assess materials to investigate thrombogenic, toxic, immunogenicity, inflammatory response, carcinogenic, mutagenic or teratogen effects and can be defined as laboratory or animal studies that are performed to provide evidence that a new treatment is safe and effective enough to be administered in humans. However, biological preclinical studies are only one step in the most complex and articulated route to validate a material proposed for clinical use; indeed the planning of a research project is based on the contributions of many skills, and the reliability of the results is the direct effect of the appropriate development of the experimental design.

For the preclinical evaluation of materials and scaffolds, over the past decades a wide ranging and comprehensive set of *in vitro, in vivo* and, for some cases, *ex vivo* tests have been developed.

The European Centre for the Validation of Alternative Methods (ECVAM), the Interagency Coordinating Committee on the Validation of Alternative Methods (ICCVAM) and the Japanese Society for Alternative to Animal Experiments (JSAAE) represent some important International initiatives encouraging and developing alternative *in vitro* methods that need to be further extensively validated to achieve regulatory acceptance.[6]

By following the regulatory authorities nowadays, both *in vitro* and *in vivo* investigations are required because it is considered that the ability of a material to perform a specific function cannot be evaluated only *in vitro* or *in vivo.* In addition, *in vitro* tests till now do not exclude and replace the *in vivo* tests even if the development and evaluation of new or modified *in vitro* systems are strongly under study.[7] By only adopting both *in vitro* and *in vivo* procedures, the limits of each single method can be reduced and the benefits increased (Tables 14.1 and 14.2).

**Table 14.1** Main advantages and disadvantages of *in vitro* tests

| Benefits | Limits |
|---|---|
| High sensitivity | Specific sensitivity |
| Standard experimental conditions | Accumulation of metabolic products |
| Human cells and tissues | Incomplete information on biofunctionality |
| No ethical restraints | Lack of systemic factors |
| Fast and less expensive | Difficulty in long-term studies |
| Easy to perform and manage | |
| Reduction of animal use and suffering | |

A preclinical study that adopts only an *in vitro* or an *in vivo* methodology can in fact be considered to present limits for the subsequent safe use in humans and animals. Second, even though a study conducted only *in vitro* is acceptable as a first step, it becomes ethically unacceptable if performed only *in vivo* with no preliminary information on the biomaterial compatibility.

**Table 14.2** Main advantages and disadvantages of *in vivo* tests

| Benefits | Limits |
|---|---|
| Long-term investigations | Ethical restraints |
| Biofunctionality information | Compliance to legal requirements |
| Evaluation of the final prosthetic device | Appropriate facilities |
| Presence of systemic factors | Complex professional expertise |
| Biological environment | Prolonged time—high costs |
| | Difference between human and animals |

The limits of both *in vitro* and *in vivo* methods are derived from the complexity of tissues and organs, which cannot be experimentally reproduced:

- Heterogeneity of cells and extracellular matrix composition
- Influence of mechanical forces
- Modeling and remodeling
- Influence of many systemic and local factors
- Heterogeneity and influence of cells and proteins involved in the tissue healing and inflammatory phases following a surgical biomaterial implantation

*In vitro* cell cultures are increasingly employed to study biomaterial-tissue interactions and provide morphological, biochemical and molecular information regarding not only biocompatibility but also biofunctionality and bioactivity. The main advantages of *in vitro* tests are the easy of working, the cost-effectiveness, the large number of experiments that can be conducted on a wide range of biomaterials in a short time and the possibility of avoiding animal use and of using human cells and tissues. The use of tissues and cells from the prospective target organism greatly improves the transferability of *in vitro* test results. *In vitro* test sensitivity in discriminating between safe and toxic materials can be of great importance to reject inappropriate materials prior to the *in vivo* implantation studies.[8] On the other hand, the sensitivity of *in vitro* biological tests is sometimes reported to be too low to distinguish the effects of subtle changes in substratum surface chemistry.[9] It is also widely recognized that *in vitro* tests may be sometimes limited on account of their relative simplicity as compared to the complex interactions occurring in an organism in toto because of the lack of systemic and local factors due to the complexity of cell and tissue response.[8] The difficulty of obtaining information on biomaterial biofunctionality under proper physiological loads as well as on chronic and long-term biomaterial effects may bring about other important limits.

Interspecies differences are reported to be one of the main limits to animal experimentation. However, a sort of "interspecies difference" also affects the *in vitro* research results and, at least, impedes direct comparison. Immortalized cell lines are used, as well as primary cells from several species and anatomical locations. Differences in the behavior of primary cells and in their expression of specific markers have been found between the various species, as well as between various different immortalized cell lines currently used.[10]

*In vivo* tests better permit long-term and dynamic investigations in the presence of systemic factors, a biological local environment, but an almost insurmountable limit, is the differences existing between human and animals in tissue healing and response to an implanted biomaterial. However, to date medical science obtains much information from the experimental *in vivo* work especially to help answer questions that do not find solutions in *in vitro* studies.[7,11]

In summary both *in vitro* and animal models provide important knowledge on biomaterial/devices biocompatibility that can eventually lead to the development of more effective clinical treatments of diseases in both humans and animals, thus acting as a bridge between *in vitro* studies and clinical trials.

## 14.2 Regulations and International Standard Organization Rules

Standardization and regulatory bodies in different countries have adopted tests to license materials for use as "biomaterials" and the selection and evaluation of materials and devices intended for use in humans requires a defined and structured program of assessment to establish biocompatibility and safety. Current regulations, whether in accordance with the U.S. Food and Drug Administration (FDA), the International Organization for Standardization (ISO) or the Japanese Ministry of Health and Welfare (JMHW), require that manufacturers conduct adequate safety testing of their finished devices through pre-clinical and clinical phases as part of the regulatory clearance process.

The FDA and EU regulatory systems have similar objectives, but their systems of operation are different, and the approval of one of them does not imply the approval by the other. The most evident difference between FDA and EU regulations regards the risk perception and its management. The FDA uses a quantitative method of risk assessment, reviewing specific data about product complaints. The EU risk assessment process employs a more qualitative method, based on the entire set of available data.

Regarding tissue engineering and regenerative medicine, a legal definition of tissue-engineered products has been defined in Regulation EC No 1394/2007 amending Directive 2001/83/EC and Regulation No 726/2004.[12–14] Products containing cells of different origin (human or animal) and additional substances as growth factors, biomolecules, chemical substances, matrices or scaffolds are considered as therapy medicinal products.

According to this regulation, when a product contains viable cells or tissue, the pharmacological, immunological or metabolic action of those cells or tissue shall be considered as the principal mode of action of the product. Therefore, biomaterials could be not the only actor in these fields, however, the biomaterial biocompatibility is still an essential requisite, and the new products should be subjected to the same regulatory rules as the others biomedical devices (Regulation EC No 1394/2007).

In addition to the requirements laid down in Article 6 of Regulation No 726/2004 ("Authorization and supervision of medicinal products for human use"), the application for the authorization of an advanced therapy medicinal product containing medical devices, biomaterials, scaffolds or matrices shall include a description of the physical characteristics and performance of the product and the description of the product design methods, in accordance with the Annex 1 to Directive 2001/83/EC. The application for a marketing authorization for a combined advanced therapy medicinal product shall include evidence of conformity with the essential requirement referred to in the above-mentioned Article 6.

The ISO, JMHW and FDA guidelines provide a general testing framework to aid manufacturers in the assessment of device biocompatibility.[15]

The ISO is a worldwide federation of national standards bodies, whose task is to draft the international rules for a variety of technical subjects and applications. Among the International Standards, the ISO 10993 refers to the fundamental principles governing the biological evaluation of medical devices.[16] The definition of categories of devices is mainly based on the nature and duration of the contact with the body, and the selection of appropriate tests is performed consequently. Well-trained and experienced professionals should apply ISO tests with interpretation and judgement. The factors relevant to the material, its expected applications and the current knowledge provided by scientific literature as previous clinical experience must be considered. In particular, the biocompatibility tests must be performed on materials and the final products, taking into account the type, duration and conditions of the exposure in the human body, the physical and chemical features of the product, the toxicological activity of the chemical elements or compounds and the presence of leachable materials.

With respect to the type of contact with the body, the devices are grouped into three main classes: surface devices, external communicating devices and implant devices. With respect to the duration of the contact with the body, three time periods are considered: limited contact (<24 hours), prolonged contact (24 hours–30 days) and permanent contact (>30 days). Finally, a material or device will be submitted to those tests that are related to the specific contact with the body environment;[6,17] the researcher will select the most appropriate among them in relation to the purpose of its investigations.

## 14.2.1 *In vitro* Tests

Test that can be performed as required by ISO rules are summarized in Table 14.3.

### 14.2.1.1 Cytotoxicity

Test methods to assess the *in vitro* cytotoxicity of medical devices, according to the ISO10993-5,[18] are designed to determine the biological response of cultured mammalian cells, incubated in direct contact with a sample or with extracts of the device, using appropriate biological parameters. A positive, which provides a reproducible cytotoxic response, and a negative, which does not produce a cytotoxic response, control materials have to be included in each assay. Reference material, as a gold standard in clinical use, is recommended for comparison.

**Table 14.3** Summary of *in vitro* tests and related examples of biological environment, experimental times and information acquired

| Test | Examples of models | Experimental times | Information acquired |
|---|---|---|---|
| Cytotoxicity | Mammalian cell lines | 24–72 hours | Cell morphology, viability and proliferation |
| Genotoxicity | Mammalian cells (L5178Y mouse lymphoma cells) or bacteria (mutated *Salmonella typhimurium* strains) | 48–144 hours | Mutated cells or reverted colonies counting |
| Hemocompatibility | Human blood | Within 4 hours from blood collection | Test for thrombosis, coagulation, platelets, hematology, complement system |

Two main types of cell culture assays have been developed: the direct-contact test and liquid extract test. In the direct-contact test, the material samples of the required standard measure are cultured with cells. If their physical state is not appropriate, an extract shall be prepared from the material sample. Established cell lines are preferred to primary cultures and negative and positive controls have to be used for the purpose to demonstrate background and appropriate test system response respectively. Cytotoxic effects can be qualitatively and quantitatively evaluated at least after 24 hours of incubation, by measuring cell proliferation, cell death and release of enzymes, and by examining cell morphology, vacuolization, detachment, lysis, and membrane integrity.

To assess cell proliferation and viability, MTT or WST1 tests are suggested. These tests are based on the cleavage of the tetrazolium salt to soluble formazan salt by mitochondrial activity of viable cells. For example, after the medium is discarded, 100 ml of WST1 solution and 900 ml of fresh medium (1:10 final dilution) are added to the cell monolayers, and the multi-well plates are incubated at 37°C for further 4 hours. At the end of incubation time, supernatants are quantified spectrophotometrically at 450 nm with a reference wavelength at 640 nm. The results are presented as optical density (OD) and transformed in cytotoxicity % by means of the following formula:

$$\frac{\text{OD experimental sample}}{\text{OD negative control}} \times 100$$

Enzyme release is usually measured by testing lactate dehydrogenase (LDH) release in the supernatant of cultures. LDH is a cytoplasmic enzyme present in all cells and rapidly released in cell culture supernatant when membrane is damaged. LDH can be measured by enzymatic kinetic assay in supernatant to assess membrane integrity.

Cell morphology could be tested with the use of Neutral red dye: living cells that appeared red to the light microscope observation actively incorporate this dye. Vacuolization, detachment and lysis of cells can be evaluated and measured, according to a score from grade 0 (absent reaction: no cell lysis, no proliferation decrease, presence of cytoplasmic granules) to grade 4 (severe reaction: destruction of cell layer, more than 50% of proliferation inhibition, more than 70% of lysated and rounded cells) (Fig. 14.1a and b).

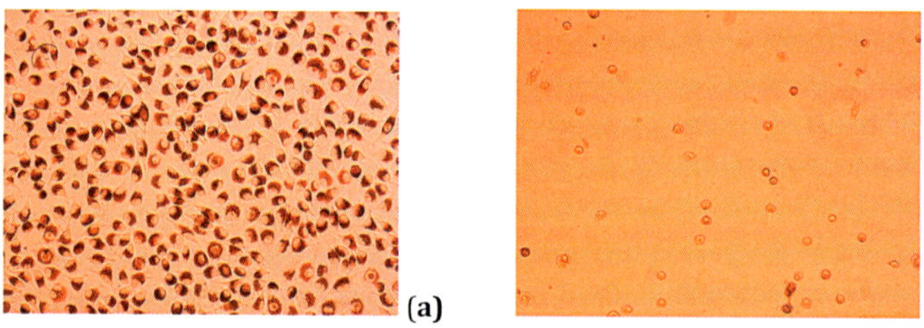

(a) (b)

**Figure 14.1** Neutral red staining. Murine fibroblasts L929 stained with neutral red. (a) Cells are well stained and display a normal morphology: fusiform shapes; and well-defined cell membrane margins can be observed. (b) Cell proliferation is markedly decreased and cells are not stained as they are dead, as also suggested by their rounded shape (4×).

#### 14.2.1.2 Genotoxicity

The possible mutagenic activity of materials that will have a prolonged body contact (>30 days) may be investigated (ISO 10993-3).[19] The test can be performed both *in vitro* and *in vivo*. The simplest and sensitive *in vitro* assays are those involving bacteria and cultured mammalian cells. Only if justified on scientific basis and if the *in vitro* test results indicate potential genotoxicity the tests *in vivo* can be performed (search of chromosomal aberration in rodents' bone marrow or peripheral blood cells).

#### 14.2.1.3 Hemocompatibility

Blood interaction is required for medical devices, which contact circulating, blood directly or indirectly (ISO 10993-4).[20] *In vitro* tests methods to assess hemocompatibility include different categories (thrombosis, coagulation, platelets, hematology and complement system), depending on the type of device. The international standard suggests a selection of the appropriate tests for the evaluation of different implant device category. Because of species differences in blood reactivity, human blood should be preferred.

### 14.2.2 *In vivo* Tests

For biomaterial validation, ISO rules accurately address researchers in the choice of animal model, test methodology, experimental times and result evaluation, but, in light of various ethical, legal and economic pressures, researchers have been searching for alternative methods for safety tests and there have been increasing efforts over the last few decades to reduce the number of animals needed for testing according to the three Rs concept[21] by addressing strategies leading to alternatives for reduction, refinement and replacement of laboratory animal use.[6] Recently some *in vivo* tests have been revised in order to perform alternative methods that do not require animals. For example, OECD (Organization for Economic Cooperation and Development) Guideline 439 approved in 2010 describes three validated models to assess skin irritation, based on *in vitro* reconstructed human epidermis that closely mimics the biochemical and physiological properties of skin. The test does not provide adequate information on skin corrosion, but covers cell damage that occurs during *in vivo* irritation. Other methods for skin sensitization and acute toxicity are moving toward validated methods and protocols.

To evaluate the *in vivo* behavior of the materials candidate to be employed for making medical devices, *in vivo* tests as detailed by the ISO 10993 in its different parts are reported in Table 14.4.

**Table 14.4** *In vivo* tests according to the ISO 10993

| Test | Model | Experimental Time | Evaluation |
|---|---|---|---|
| Irritation | Rabbit | 4 hours exposition, 24–48–72 hours evaluation | Skin reaction: erythema and oedema |
| Sensitization | Guinea pig | Topical and intradermal induction, and challenge phase 1 month | Skin reaction: erythema and oedema |
| Systemic toxicity | Mouse (nonrodents: if required and appropriate) | Acute <24 hours Subacute 24 hours–14 days, subchronic 14–28 days, chronic 6–12 months | Clinical signs, hematology, clinical chemistry, urinalysis, gross pathology, histopathology |
| Genotoxicity | Rodent | up to 14 days | Micronucleus test, metaphase analysis, DNA synthesis |
| Carcinogenicity | Rodent | Lifespan of the species | Toxicological signs and tumor development |
| Reproductive and developmental toxicity tests | Rat | 2 months | Alteration in the first generation |
| Hemocompatibility | Nonhuman primates, dog, pig | <24 hours, limited blood exposure devices >24 hours prolonged or permanent contact | Test for: thrombosis, coagulation, platelets, hematology, complement system |

| Test | Model | Experimental Time | Evaluation |
|---|---|---|---|
| Local effects after implantation | Mice, rat, hamster, rabbit (short-term tests) Rabbit, sheep, goat, pig (long-term testing) | Short-term: 1–4 weeks Long-term: 12–104 weeks | Macroscopically and histopathology response at the site of implant |

Animal models provide pivotal information in the route of validation of a biomaterial. It can be said that they act as a bridge between *in vitro* studies and clinical trials.[22]

However, to correctly perform *in vivo* studies, the knowledge and the compliance to the ethical and legal rules on animal experimentation are required. For this reason, it seems appropriate to include here a short presentation of the current regulations to be followed for managing laboratory animals and developing a research project involving their use. The use of laboratory animals for biomedical research is now a worldwide subject to precise rules, to ensure appropriate procedures and humane treatments.[23] In the United States, the *Animal Welfare Act* was approved in 1966, and more recently the *Public Health Service Policy on Human Care and Use of Laboratory Animals* and the *NIH Guide for the care and Use of Laboratory Animals* were enacted. All these documents contain the guidelines about the housing and management of the animals used for experiments.

In the European Union, the *Directive 86/609—Protection of Experimental Animals* indicated to the member states the guidelines to be applied in the regulation of animal research. This directive was recently updated and replaced by the Directive 2010/63 of 22 September 2010 on the protection of animals used for scientific purposes. In agreement to this directive, all the European countries will adapt their national rules.

Overall, these statements emphasize that strategies must be adopted to minimize the number of used animals and to avoid pain, suffering and stress and that animal tests cannot be performed before appropriate preliminary *in vitro* tests have been carried out with favorable results. Animal experiments must be carried out in authorized laboratories, with appropriate facilities for animal welfare. Cage size should be appropriate for small, medium and large animals. Surgical procedures must be performed in general anesthesia and in well-equipped operatory rooms, under the control and responsibility of well-trained scientists and veterinarians.

The prerequisites to be accomplished when animal tests should be performed are also pointed out in the part 2 of the ISO 10993 about the biological evaluation of medical devices entitled "Animal Welfare Requirements."[24] According to these guidelines, the tests are justified only when the expected results are not otherwise available and are essential for the material characterization, when no other scientifically validated method involving animals can be proposed and after the acquisition of data obtained by *in vitro* evaluation. The strategies to minimize pain, suffering, distress and lasting harm must be carefully considered, as well as the animal welfare, the anesthesia and the procedures of euthanasia. The research protocol must describe in detail the scientific aims of the research, the features of the material to be investigated, the rationale for the use of animals, stating clearly that alternative ways to *in vivo* experiments are not possible and must enclose a careful and complete study documentation. The protocol must also be approved by an Ethical Committee and submitted to the public authority, according to the regulations of each country.

Regarding the correct choice of the animal model, Einhorn emphasizes, "The appropriate use of animal models begins with careful considerations of the question asked."[25] During the *in vivo* validation of a material, the choice of more than one animal species, namely a small/medium and a large size animal, could therefore be required. The use of small animals could be considered appropriate in the early stage of testing, but the tissue healing characteristics of the large animals should better approximate those of humans during the last stages of research. In conclusion, the choice of an animal model must take into account regulatory requirements and assessments on ethics, availability of the animal, housing, easy of handling, costs, susceptibility to the disease and must be supported by an appropriate background and literature data.[26–29]

#### 14.2.2.1 Irritation and sensitization test

Irritation is the primary and direct effect of a substance when it is in contact with the skin, while sensitization reaction is an immune-mediated contact dermatitis.[30] *Irritation test* means the potential to produce skin irritation can be evaluated by applying the material on the shaved back in both the sides of the spine with a semi-occlusive dressing. The test can be performed in rabbits. In acute tests with a single exposure, the site of application must be controlled at 1, 24, 48 and 72 hours following the removal of the patches to detect the changes. For tests with repeated exposures, the appearance of the site will be recorded at 1, 24, 48 and 72 hours after the last exposure. The results are evaluated by applying a score on clinical parameters, as well as the erythema, eschar and the oedema formation. The intracutaneous reactivity assesses on the contrary the potential to produce irritation following intradermal injection of extracts of the material.

*Sensitization tests* are represented by the Magnusson–Kligman maximization test and the Buehler closed-patch test. Usually only one of these tests is sufficient. Among these methods, the preferred is the maximization test. A limit is that this test is performed by intradermal injections; therefore, if the material cannot be obtained in an injectable form, other methods must be used, as the closed patch test. The sensitizing capability of a substance is defined as the minimum concentration required to induce a predetermined level of sensitization.

#### 14.2.2.2 Systemic toxicity

These tests evaluate the possible effects of the absorption, distribution or metabolism of products that originate from a material or device, involving parts of the body or organs not in direct contact with them.[31] The tests will determine the systemic adverse effects that could be observed after the administration of a single dose or multiple doses of a test sample during a period of 24 hours (acute toxicity), or by repeated exposures for a prolonged time (subacute, subchronic and chronic systemic toxicity). *Pyrogenicity tests.* Pyrogens are substances that, when introduced in the organism, produce a febrile reaction. They can be present in a device and, therefore, after the implant they can cause a febrile response (material-mediated pyrogenicity). ISO 10993-11 recommends the evaluation of the pyrogenicity potential of extractable substances derived from implant materials and includes these tests in the category of systemic toxicity testing.[31]

### 14.2.2.3 Carcinogenicity and reproductive and development toxicity

These tests are considered in the same ISO 10993-3 standard.[18] *Carcinogenicity tests* explore the carcinogenetic potential of materials after a single or multiple exposures over a period of the total lifespan of the animal. They can require observation over a prolonged time span and can be planned to evaluate both carcinogenicity and chronic toxicity at the same time. They should be considered for those devices that will have prolonged contact (longer than 30 days) with tissues. *Reproductive and development toxicity tests* are planned to investigate the potential effects of the material/device on reproductive function, on the embryonic development and prenatal and early postnatal development. The tests should be performed only when the device has a potential impact on the reproductive ability, so they are indicated in particular for gynecologic devices.

### 14.2.2.4 Hemocompatibility

The interaction of materials with blood can have an impact both at cellular and plasmatic levels, that is, on coagulation, thrombosis, embolism and inflammation, and on the activation of complement and fibrinolytic systems. The cellular and humoral pathways are closely connected and reciprocally influenced. For this reason, *in vitro* studies allows a first screening, but definitive data are collected *in vivo* after the implant of the device into the vascular system for a scheduled period, that allows monitoring and systematic control of many hematologic parameters.[19]

### 14.2.2.5 Test for local effects after implantation in soft and hard tissues

The final destination of a biomaterial is a temporary or prolonged implantation in the organism. Therefore, the evaluation of the tissue reaction after the positioning of the device gives essential information about the expected clinical behavior. The part 6 of the ISO 10993 considers the test useful to detect the local effects after a material implant. The tests require that the sample is implanted into an appropriate anatomical site and that the implant is not subjected to mechanical or functional loading; the local effects are evaluated in comparison with the effects caused by control materials with well-assessed properties. It must be clear that the placement of a material or device in the tissues produces a host response that is dynamic and consequently changes with time.[32]

Surgical procedures used in animal experiments are like those used in humans. Consequently, the same iatrogenic potential related to the maneuvers of implant can be expected. An appropriate surgical technique is anyway a prerequisite for a correct *in vivo* evaluation of any device. Care must be taken to limit tissue damage, namely bleeding, excessive pressure and dead spaces because an inaccurate technique can produce the displacement of the implant. The size and shape of the implant as sheets or bulks must be appropriate for the site of implant and in agreement with the ISO 10093-6 rules (i.e., 10–12 mm of diameter and 0.3–1 mm in thickness; 1.5 mm of diameter and 5 mm of length). The implants can be positioned in the back subcutaneous tissue of adult mice, rats, guinea pigs, or rabbits. For short-term implants, the scheduled time is no more than 12 weeks. For long-term implants, the time is longer than 12 weeks. Several macroscopic and microscopic parameters are useful for the evaluation, including the presence of infection, fibrous tissue, degeneration, presence of phagocytic cells, necrosis, fatty infiltration, and foreign debris. Also the effects of the surgical procedure, and of the repair of the surgical wound, that affect the early periods after implantation will be considered.

In particular, macroscopic evidence of infection in the site of the implant as the presence of a fibrous reaction surrounding the material or necrotic tissues will be checked. The histological assessment is based on the following findings: (a) the extent of fibrosis, fibrous capsule and inflammation around the implant; (b) degeneration and changes in tissue morphology; (c) presence, number and distribution of the inflammation cells in relation with the distance from the material/tissue interface; (d) presence of necrosis; (e) material debris, fatty infiltration and granuloma; (f) for porous materials, quality and quantity of tissue ingrowth; and (g) for bone implants, study of the bone implant interface.

Bone implants are performed when the material is proposed mainly for orthopedic or dentistry applications. When the evaluation concern biomaterials intended for skeletal implants, some particular features of bone tissue must be keep in mind. Skeletal texture is in fact the combination of many components: bone, cartilage, connective, mesenchymal stromal tissue, vascular and hematopoietic. Due to the differences in architecture, vascularization and mechanical stress, the healing capacity of the site after the implant greatly affects both the osteointegration and the biodegradation of some biomaterials. The bone remodeling around an implant is a long-lasting process.

The sites of bone implants are usually the diaphysis (cortical bone), the epiphysis (cancellous bone) or both.[33] The samples are designed as screw or of cylindrical shape, and their size depends on the animal species in which the tests are performed, and of the skeletal segment in which the implant is placed (i.e., implants may have a diameter from 2 mm to 6 mm in length to 3 mm in diameter and 12 in length).

The contact between the tissue and the material at the bone/implant interface is matter of particular interest. Regarding this aspect, osteointegration is defined as the direct structural and functional connection between living bone and the surface of the implant.[34]

The histological evaluation, following the procedure used for the undecalcified or decalcified tissues, requires an accurate drawing of the samples. An adequate amount of tissue or bone surrounding the material must be withdrawn, to evaluate the quality of tissue around the implant (Fig. 14.2).

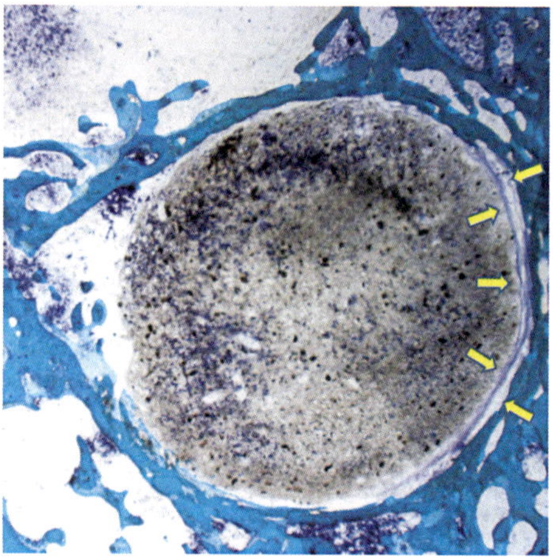

**Figure 14.2** A ceramic implant in trabecular bone. Evidence of a connective fibrous capsule between the bone and the material (arrows), Fast Green–acid fuchsin staining, 2.5×.

To evaluate tissue and/or inflammatory response, histomorphometric measures can be obtained with image analysis systems. A microscope composes these systems and a personal computer provided by appropriate software to perform interactive and automatic measurements, connected by a camera to capture images. Semiquantitative grading scores are used to analyse (a) the cellular infiltration (from 0 with a cellular infiltration of 0–50 cells) to 3 (with a cellular infiltration of >150 cells); (b) the presence of multinucleated giant cells (from 0 with no multinucleated giant cells to 3 presence of >5 cells); (c) the vascularity (from 0 with 0–1 vessels to 3 with >10 vessels); (d) the degree of connective tissue organization: from 0 (the original scaffold is intact) to 3 (original scaffold is disrupted with a dense and highly organized new host extracellular matrix).[35]

Considering bone implants, the rate of bone deposition and mineralization are evaluated by dynamic histomorphometry. For labeling the new bone deposition, fluorochromes (i.e., tetracycline, alizarin red, calcein blue, xylenol orange) are used, which are administered to the animal by injecting at scheduled times before the sacrifice. The fluorescence is evident on histological unstained sections at microscopic observation with appropriate filters and marks the new deposition of bone.[36]

Mechanical tests are also useful to evaluate the integration of the material with the bone and the bone quality. Implants in bone can be submitted to pull out and push out tests, and the whole bone to test of three point bending, torsion and compression.[37] Microhardness investigation, performed on the cutting surface of samples embedded in resin as for histology and carefully grounded and polished, evaluates the quality of bone surrounding the implant.[38]

Besides these investigations, more recently it is possible to perform computerized microtomography on the explanted samples; this is a nondestructive technique and the time required to capture the images is short; also, the same sample can therefore be used afterward for the histology and histomorphometry.

## 14.3 New Scenario and Further Biological Test Requirements

Artificial scaffolds are designed to guide cell proper behavior for tissue regeneration, and both material and biological requirements should be considered for each specific application. Recent technological advances in the material production methodology and the large progress in the development of biomaterials as scaffolds for tissue engineering with improved implantation regenerative performance were made efficient by the parallel establishment of adequate strategies for testing biocompatibility, biofunctionality, and bioactivity. Scaffolds, or artificial ECM, are designed to guide cell attachment, proliferation, and differentiation for tissue regeneration. Therefore, in a widest comprehension of biocompatibility, cell penetration, migration, and attachment onto the scaffold substrate and cell spreading, proliferation, and differentiation, and extracellular matrix synthesis, leading to neo-tissue formation, have to be preclinically evaluated both *in vitro* and *in vivo* and by taking more and more in consideration the final local implantation site and the host.

First, nowadays the evaluation of biocompatibility cannot ignore the development of very complex methods that also use co-cultured cells and bioreactors that can better mimic some of the complexity of the whole organism. While it is known that tissue healing involves the response of different kinds of resident or migratory cells, *in vitro* biocompatibility is quite often performed by a large battery of *in vitro* tests with different cell types other than in the contemporary presence of these, and despite abundant evidence that 2D and 3D models are superior to monolayer cultured cells, the majority of preclinical current research on biomaterial biocompatibility relies heavily on the last methods. Three-dimensional cultures are particularly useful for cells that require a tridimensional space for their differentiation maintenance: this technique uses, for example, micro-mass, 3D scaffold colonization, and cell seeding inside hydrogels.

Co-cultures are attractive also for *in vitro* biomaterial testing because they are one step closer to natural conditions and allow studying some aspects of the complex interactions among different cell types living in the same microenvironment, such as between bone-building and bone-resorbing cells, mesenchymal stem cell and differentiated cells, osteoblasts and chondrocytes, chondrocytes and ligamentocytes, and fibroblasts and macrophages.

The culture insertion system allows the diffusion of cell products through the interposed membrane, therefore, the effects of two different cell types crosstalk can be evaluated, even in presence of a material sample.

The development of 3D and multicellular *in vitro* models for biocompatibility testing will also reduce the number of animal experiments. From this point of view, even in the presence of a big cultural and economic support to the development of *in vitro* tests on biomaterial safety and biocompatibility, these may be improved and validated because there is still the lack of actual alternative methods. This is an important aspect to be supported because as reported by Lee (2012) and Höfer *et al.* (2004) in the topic developmental toxicology, under current guidelines of the European Union, the next 10–15 years the evaluation of chemicals would require the use of approximately 7.2 million laboratory animals.[39,40]

*In vitro* tissue cultures also reproducing tissue lesions and microwound *in vitro* may also be taken into consideration.[41–43]

Second, biocompatibility and biological biomaterial evaluations often forget that host variables are as important as material/scaffold variables and sometimes host variables such as the overall healthy status and the implantation site local microenvironment will have a major effect on the biological behavior.[44] Therefore, we need testing strategies that could take into consideration many other requirements from both the material and the host site (Table 14.3). Following ISO rules, sources of cells for cultures usually range from established cell lines to cells extracted from normal tissues and healthy animals. Investigations in a healthy cell or animals alone cannot be considered a final step in the preclinical evaluation of materials. In fact, the final product will be implanted usually in a patients of advanced age, with comorbidities, submitted to chronic drug treatments, or with uncorrect lifestyles, all affecting the local microenvironment of the implantation site. These conditions, together with the inflammatory microenvironment because of the pathology to be treated with the scaffold or because of the surgical intervention for its implantation, may affect the behavior and biocompatibility of the biomaterial and the final outcome in comparison with healthy standard preclinical conditions. In fact, when the scaffold and the engineered construct are implanted into the lesion site, the scenario completely differs from the strictly controlled *in vitro* culture conditions. The surgical implantation procedure triggers an inflammatory response.[45–47] Chemokynes, protease, growth factors, and pro-inflammatory cytokines are locally produced by cells and the effect of microenvironment on biomaterials and scaffolds may be marked especially in the first days after surgery. The potentially negative role of the inflammatory environment because of the lesion to be treated or surgery should be taken into account and, therefore, appropriate experimental models would be required to mimic local and systemic conditions of clinical relevance by changing the pH, oxygen tension, and for example, by adding the main cytokines, enzymes, and signal molecules involved in healing processes and present in the lesion site.

The use of pathological tissue–derived cells (i.e., from osteoporotic or osteoarthritic bone and cartilage) and of *in vivo* animal model of diseases should be improved and adopted. The use of *in vitro* tests with pathological bone-derived cell culture, such as osteoblasts from the osteoporotic bone,[48,49] in association with *in vivo* pathological models helps in the evaluation of the different responses of bone to candidate orthopedics materials by comparing healthy and pathological conditions.

| Material site | Biological site | Host site |
|---|---|---|
| Controlled degradation rate | Biocompatibility | Age |
| Surface properties | Non immunogenic, not cancerogenic | Comorbidities |
| Processability | No foreign body reactions | Drug assumption |
| Porosity | Absence of toxicity | Local pathological microenvironment |
| Mechanical properties | Degradation | Inflammation at the implantation site |
| Sterilization | Bioactivity | Lifestyles |

Finally, the impact of sophisticated processing parameters to develop the proper chemistry and topography of scaffolds would require the use of an advanced methodology for the *in vitro* and *in vivo* preclinical testing. Nanotechnology is considered as one of the key technologies of the 21st century but objects at nanoscale differ a lot from the bulk materials as far as biocompatibility and toxicology is concerned.[50]

Therefore, standardized, reliable, robust, reproducible, and intelligent testing strategies are urgently needed and should be validated to test the biocompatibility of new biomaterials and scaffolds.

## References


1. Williams, D. F. (2008). On the mechanism of biocompatibility, *Biomaterials*, **29**, 2941–2953.
2. Edwards, S. L., Mitchell, W., and Russel, S. J. (2004). Design of nonwoven scaffold structures for tissue engineering of the anterior cruciate ligament, *AUTEX Res. J.*, **4**(2), 86–94.
3. Binyamin, G., Shafi, B. M., and Mery, C. M. (2006). Biomaterials: A primer for surgeons, *Semin. Pediatric Surg.*, **15**, 276–283.
4. Williams, D. F. (1987). Definitions in biomaterials, in *Progress in Biomedical Engineering*, Vol. 4, New York: Elsevier.
5. Ratner, B. D., and Bryant, S. J. (2004). Biomaterials: Where we have been and where we are going, *Annu. Rev. Biomed. Eng.*, **6**, 41–75.
6. Ukelis, U., Kramer, P. J., Alejnuczak, K., and Mueller, S. (2008). Replacement of *in vivo* acute oral toxicity studies by *in vitro* cytotoxicity methods: opportunities, limits and regulatory status, *Regul. Toxicol. Pharm.*, **51**(1), 108–118.
7. Fini, M., and Giardino, R. (2003). *In vitro* and *in vivo* tests for the biological evaluation of candidate orthopedic materials: benefits and limits, *J. Appl. Biomater. Biomech.*, **1**, 155–163.
8. Belanger, M. C., and Marois, Y. (2001). Hemocompatibility, biocompatibility, inflammatory and *in vivo* studies of primary reference materials low density polyethylene and poly-dimethylsiloxane: A review, *J. Biomed. Mater. Res.*, **58**, 467–477.
9. Anselme, K. (2000). Osteoblast adhesion on biomaterials, *Biomaterials*, **21**, 667–681.
10. Torricelli, P., Fini, M., Giavaresi, G., Borsari, V., Carpi, A., Nicolini, A., and Giardino, R. (2003). Comparative interspecies investigation on osteoblast cultures: Data on cell viability and synthetic activity, *Biomed. Pharmacother.*, **57**(1), 57–62.
11. Pearce, A., Richards, R. G. G., Milz, S., Schneider, E., and Pearce, S. G. (2007). Animal models for implant biomaterial research in bone: A review, *Eur. Cells Mater.*, **13**, 1–10.
12. Regulation EC No 1394/2007 of the European Parliament and of the Council of 13 November 2007 on advanced therapy medicinal products and amending Directive 2001/83/EC and Regulation (EC) No 726/2004.
13. Directive 2001/83/EC of the European Parliament and of the Council of 6 November 2001 on the Community code relating to medicinal products for human use.

14. Regulation No 726/2004 of the European Parliament and of the Council of 31 March 2004 laying down Community procedures for the authorization and supervision of medicinal products for human and veterinary use establishing a European Medicines Agency.
15. http://www.wuxiapptec.com/biocompatibility
16. UNI EN ISO 10993-1 Biological evaluation of medical devices. Part 1: Evaluation and testing with a risk management process, 2010.
17. Gatti, A.M., and Knowles, J.C. (2002). Biocompatibility and biological tests, in *Integrated Biomaterial Science*, Barbucci, R. (Ed.), 793–813, New York: Kluwer Academic/Plenum Publishers.
18. UNI EN ISO 10993-5- Biological evaluation of medical devices. Part 5: Tests for *in vitro* cytotoxicity, 2009.
19. UNI EN ISO 10993-3 Biological evaluation of medical devices. Part 3: Tests for genotoxicity, carcinogenicity and reproductive toxicity, 2009.
20. UNI EN ISO 10993 Biological evaluation of medical devices. Part 4: Selection of tests for interactions with blood, 2002.
21. Russel, W., Burch, R. (1959). *The Principles of Humane Experimentation Technique*. London: Methuen.
22. Giardino, R., Fini, M., and Giavaresi, G. (1998). Experimental surgery in the research of artificial organs, *Int. J. Artif. Organs.*, **21**, 506–508.
23. Smith, A. C., Fasse, R. T., and Swindle, M. M. (1999). Ethics and regulations for the care and use of laboratory animals, in *Animal Models in Orthopaedic Research*, An, Y. H., and Friedman, R. J. (Eds.), 3–14, Oxford: Woodhead Publishing Ltd/CRC Press.
24. UNI EN ISO 10993 Biological evaluation of medical devices. Part 2: Animal welfare requirements, 2006.
25. Einhorn, T. A. (1999). Clinically applied models of bone regeneration in tissue engineering research, *Clin. Orthop. Relat. Res.*, **367** (Suppl.), S59–S67.
26. Torricelli, P., Fini, M., Giavaresi, G., and Giardino, R. (2004). *In vitro* models to test orthopedic biomaterials in view of their clinical application in osteoporotic bone, *Int. J. Artif. Organs.*, **27**(8), 658–663.
27. Muschler, G. F., Raut, V. P., Patterson, T. E., Wenke, J. C., and Hollinger, J. O. (2010). The design and use of animal models for translational research in bone tissue engineering and regenerative medicine, *Tissue Eng. Part B Rev.*, **16**(1), 123–145.
28. Pearce, A. I., Richards, R. G., Milz, S., Schneider, E., and Pearce, S.G. (2007). Animal models for implant biomaterial research in bone: A review, *Eur. Cell. Mater.*, **2**(13), 1–10.

29. O'Loughlin, P. F., Morr, S., Bogunovic, L., Kim, A. D., Park, B., and Lane, J. M. (2008). Selection and development of preclinical models in fracture-healing research, *Bone Joint Surg. Am.*, **90** (Suppl. 1), 79–84.

30. UNI EN ISO 10993 Biological evaluation of medical devices. Part 10: Tests for irritation and skin sensitization, 2010.

31. UNI EN ISO 10993 Biological evaluation of medical devices. Part 11: Tests for systemic toxicity, 2009.

32. UNI EN ISO 10993 Biological evaluation of medical devices. Part 6: Tests for local effects after implantation, 2009.

33. Giavarcsi, G., Fini, M., Chiesa, R., Giordano, C., Sandrini, E., Bianchi, A. E., Ceribelli, P., and Giardino, R. (2008). A novel multiphase anodic spark deposition coating for the improvement of orthopedic implant osseointegration: An experimental study in cortical bone of sheep, *J. Biomed. Mater. Res. A.*, **85**(4), 1022–1031.

34. Mavrogenis, A. F., Dimitriou, R., Parvizi, J., and Babis, G. C. Biology of implant osseointegration, *J. Musculoskelet. Neuronal. Interact.*, **9**(2), 61–71.

35. Valentin, J. E., Badylak, J. S., McCabe, G. P., and Badylak, S. F. (2006). Extracellular matrix bioscaffolds for orthopaedic applications. A comparative histologic study, *J. Bone Joint Surg. Am.*, **88**, 2673–2686.

36. Jee, W. S. (2005). The past, present, and future of bone morphometry: Its contribution to an improved understanding of bone biology, Review, *J. Bone Miner. Metab.*, **23** (Suppl.), 1–10.

37. Borsari, V., Fini, M., Giavaresi, G., Tschon, M., Chiesa, R., Chiusoli, L., Salito, A., Rimondini, L., and Giardino, R. (2009). Comparative *in vivo* evaluation of porous and dense duplex titanium and hydroxyapatite coating with high roughnesses in different implantation environments, *J. Biomed. Mater. Res. A.*, **89**(2), 550–560.

38. Bacchelli, B., Giavaresi, G., Franchi, M., Martini, D., De Pasquale, V., Trirè, A., Fini, M., Giardino, R., and Ruggeri, A. (2009). Influence of a zirconia sandblasting treated surface on peri-implant bone healing: An experimental study in sheep, *Acta Biomater.*, **5**(6), 2246–2257.

39. Lee, H. Y., Inselman, A. L., Kanungo, J., and Hansen, D. K. (2012). Alternative models in developmental toxicology. *Syst. Biol. Reprod. Med.*, **58**(1), 10–22.

40. Höfer, T., Gerner, I., Gundert-Remy, U., Liebsch, M., Schulte, A., Spielmann, H., Vogel, R., and Wettig, K. (2004). Animal testing and alternative approaches for the human health risk assessment under the proposed new European chemicals regulation, *Arch. Toxicol.* **78**(10), 549–564.

41. Maffulli, N., Even, S. W. B., Waterson, S. W., Reaper, J., and Barrass, V. (2000). Tenocytes from ruptured and tendinopathic Achilles tendons produce greater quantities of type III collagen than tenocytes from normal Achilles tendons, *Am. J. Sports Med.*, **28**, 499–505.

42. Hunter, C. J., and Levenston, M. E. (2004). Maturation and integration of tissue-engineered cartilages within an *in vitro* defect repair model, *Tissue Eng.*, **10**(5/6), 736–746.

43. De Vries-van Melle, M., Mandl, E. W., Kops, N., Koevoet, W. J. L. M., Verhaar, J., and van Osch, J. V. M. G. (2012) An osteochondral culture model to study mechanisms involved in articular cartilage repair, *Tissue Eng. Part C*, **18**(1), 45–53.

44. Williams D.F. (1998) General concepts of biocompatibility in *Handbook of Biomaterial Properties*, Black J. and Hastings G., (Eds.), 481–489, London: Chapman & Hall.

45. Francioli, S., Cavallo, C., Grigolo, B., Martin, I., and Barbero, A. (2011). Engineered cartilage maturation regulates cytokine production and interleukin-1b response, *Clin. Orthop. Relat. Res.*, **469**, 2773–2784.

46. Lima, F. G., Tan, A. R., Tai, T., Bian, L., Stoker, A. M., Athesian, G. A., *et al.* (2008). Differences in interleukin-1 response between engineered and native cartilage, *Tissue Eng. Part A*, **14**, 1721–1730.

47. Tanzil, G. S. (2009). Activation and dedifferentiation of chondrocytes: Implications in cartilage injury and repair, *Ann. Anat.*, **191**, 325–338.

48. Torricelli, P., Fini, M., Giavaresi, G., and Giardino, R. (2004). *In vitro* models to test orthopedic biomaterials in view of their clinical application in osteoporotic bone, *Int. J. Artif. Organs*, **27**(8), 658–663.

49. Torricelli, P., Fini, M., Giavaresi, G., Rimondini, L., Tschon, M., Rimondini, R., Carrassi, A., and Giardino, R. (2008). Chronic alcohol abuse and endosseous implants: Linkage of *in vitro* osteoblast dysfunction to titanium osseointegration rate, *Toxicology*, **243**(1/2), 138–144.

50. Hirsch, C., Roessien, M., Krug, H. F., and Wick, P. (2011). Nanomaterial cell interactions: Are current *in vitro* tests reliable? *Nanomedicine* (Lond.), **6**(5), 837–847.

Chapter 15

# Bioreactor Strategies for *in vitro* Conditioning of Engineered Tissues

**Aaron S. Goldstein**[a,b] **and Patrick Thayer**[b]
[a]*Department of Chemical Engineering, Virginia Tech, Blacksburg, VA 24061, USA*
[b]*School of Biomedical Engineering and Sciences, Virginia Tech, Blacksburg, VA 24061, USA*
thunserangst@gmail.com

In this chapter, we review a number of bioreactor strategies that have been used to develop neo-tissues *in vitro* under physiologically relevant stimuli. We begin with the application of perfusion, hydrodynamic shear stress, compression, and tension to the engineering of homogenous tissues for the regeneration of bone (Section 15.2), cartilage (Section 15.3), and tendon/ligament (Section 15.4). Then, we discuss cardiac tissue (Section 15.5) and the effects of mechanical strain, electrical fields, and perfusion on the establishment of synchronously contractile tissues. Next, we examine pulsatile flow for the development of vascular tissue and heart valves (Section 15.6). This section is followed by organ bath strategies for forming complex tissues (e.g., lung, liver, and kidney) by repopulating decellularized organs (Section 15.7). Finally, we end

*Scaffolds for Tissue Engineering: Biological Design, Materials, and Fabrication*
Edited by Claudio Migliaresi and Antonella Motta

ISBN 978-981-4463-20-1 (Hardcover), 978-981-4463-21-8 (eBook)
www.panstanford.com

with a description of efforts to incorporate nondestructive imaging modalities into bioreactors (Section 15.8) and conclude with a few general observations (Section 15.9).

## 15.1 Introduction

In its broadest definition, an engineered tissue consists of a biomaterial scaffold, pharmaceutical agents, and cells (Fig. 15.1a). However, such constructs have had only limited success to date *in vivo* integrating with host tissues, initiating vascularization and remodeling, and restoring the physiological functions of target tissues (e.g., synchronous contraction of cardiac tissue). Consequently, a broad array of *in vitro* bioreactor strategies have been developed with the two-fold purpose of inducing cells to assemble into functional tissues (Fig. 15.1b), and conditioning the cells to withstand the physiological mechanical and chemical stresses. Moreover, the bioreactor

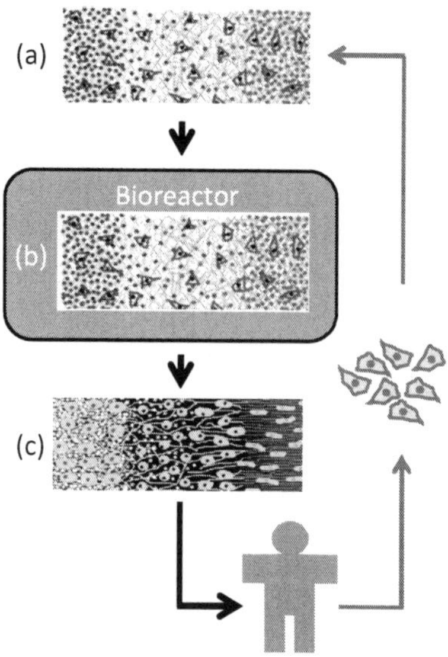


**Figure 15.1** Model for regenerative medicine. (a) Construct is prepared by combining a biomaterial scaffold, pharmaceutical agents, and the patient's cells. (b) Construct is cultured in a bioreactor under conditions that stimulate formation of a functional neo-tissue. (c) Neo-tissue marked by target cell phenotype(s) and mature ECM is suitable for clinical application.

conditions are frequently modeled after or designed to recapitulate the physiologic stresses of the target tissue.

The bioreactor strategies presented in this chapter are summarized in Table 15.1. Although this review is not comprehensive, we seek to emphasize three factors in bioreactor design. First, conditioning of neo-tissues under biologically relevant physical stimuli can aid the development of mature tissue phenotypes. Second, perfusion can be a valuable component of any bioreactor strategy, as it can facilitate the delivery of oxygen and nutrients to cells deep within neo-tissues. Third, nondestructive monitoring of the mechanical and physiological properties of the engineered tissue can provide meaningful data to assess tissue development.

**Table 15.1** Summary of stimuli used to condition engineered tissues

| Tissue | Stimulus | Ranges and references |
|---|---|---|
| Bone | Shear | 0.8–3 Pa [15] |
| Cartilage | Shear | 0 and 0.22 dynes/cm$^2$ [29]<br>0.24 dyne/cm$^2$ [27] |
| | Compression | 10% strain [34, 36] at 1 Hz [30, 34, 36] |
| | Oxygen tension | 5% $pO_2$ [38]<br>4, 10.5 and 21% $pO_2$ [39] |
| Tendon/ Ligament | Mechanical Stretch | 1.25 N at 0.017 Hz [44]<br>9% strain at 0.1 Hz [46]<br>10% strain at 1 Hz [47]<br>2.5–15% strain, 0.01–1 Hz [48] |
| | Torsion | 5% strain and 90% torsion [49] |
| Cardiac tissue | Electrical Stimulation | 5 V/cm at 1 Hz, 2 ms duration [51]<br>4 V/cm at 1 Hz ms duration [52]<br>0-8 V/cm at 1 Hz ms duration [54]<br>0-6 V/cm, 1–5 Hz [53]<br>74.4 mA/cm$^2$, 1 Hz [61] |
| | Mechanical Stretch | 10% strain, 2 Hz [56]<br>10% strain, 1–3 Hz [58] |
| | Perfusion | 100 ml/h, 120 bpm [59]<br>1.5 and 0.32 ml/min at 1 Hz [60] |
| Vascular tissue | Shear | 6 dyne/cm$^2$, 120/80 mmHg, 2 Hz [65]<br>40 ml/min at 1.5 Hz [67]<br>0 to 1.4 l/min, 120/80 mmHg, 70 bpm [69] |
| | Hoop stress | 10% circumferential Strain, 1 Hz [66] |

*(Continued)*

**Table 15.1** (*Continued*)

| Tissue | Stimulus | Ranges and references |
|---|---|---|
| Heart valve | Shear stress | 50–2000 ml/min [71] |
| | Hydrostatic pressure | 10–240 mmHg [71] |
| Lung | Stress | 180 breaths per minute [76, 77] |
| Liver/Kidney | Perfusion | General perfusion [79–82] |

## 15.2 Bone Tissue

We begin with *in vitro* bioreactors for the regeneration of mineralized bone tissue. Although bone is not crucial for patient survival, a competent skeleton enables normal daily activities and contributes positively to the patient's sense of independence and quality of life. Consequently, bone grafts are frequently used in spinal fusions, to increase bone mass around orthopedic and dental implants, and to repair defects resulting from disease, trauma, and congenital conditions.

*In vivo*, mineralized bone tissue is formed by osteoblasts, which are metabolically active cells derived from mesenchymal stem cells (MSCs). Drawing insights from bone biology, researchers have sought to form engineered bone tissue *in vitro* by incorporating MSCs within porous biomaterial scaffolds and culturing them under osteogenic conditions. However, early studies showed that a uniform cell distribution within physiologically sized porous scaffolds cannot be maintained under static culture conditions, as cell density declines substantially at depths greater than 1 mm from the outer surface of the constructs.[1] This phenomenon is illustrated in Fig. 15.2a, in which uniformly distributed cells (day 1, green) become nonhomogeneously distributed as proliferation occurs preferentially near the outer surfaces, and cell density decreases in the interior regions of the scaffold (blue and red curves). This phenomenon has been ascribed to the relatively slow oxygen and nutrient diffusion rates from the outer surface inward relative to the consumption rate by cells (Fig. 15.2b). From an engineering perspective, this system is marked by a large Damköhler number, $Da = kL^2/D$ (where $k$ represents the rate of consumption, $L$ is the dimension of the scaffold, and $D$ the diffusivity of the consumed species). This phenomenon is further compounded by the higher proliferation of cells at the outer surface, disproportionately increasing nutrient consumption and hindering inward diffusion.

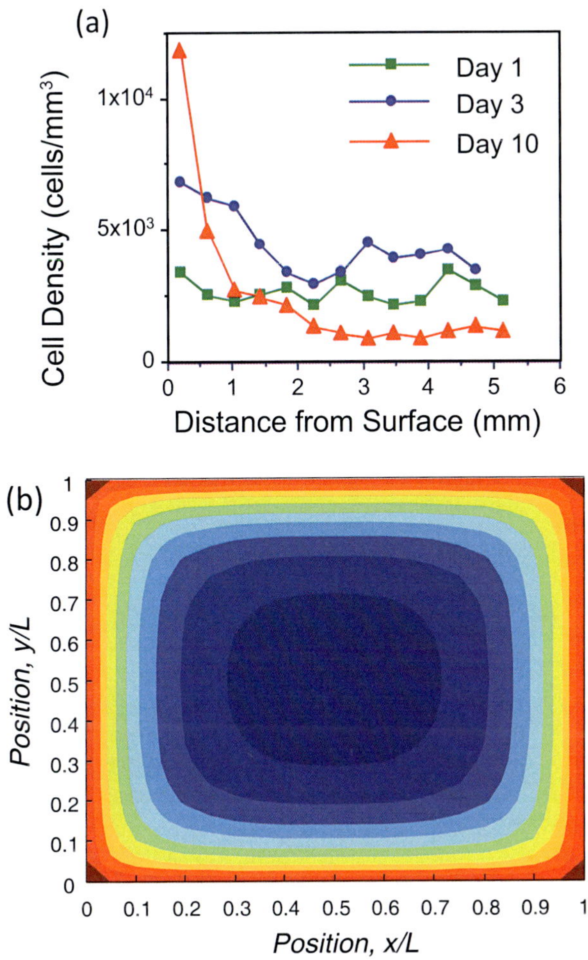


$$0 = \nabla^2 C_{O_2} - DaC_{O_2}, Da = 0.2$$

**Figure 15.2** (a) Densities within 6 mm thick foam scaffolds after 1, 3, and 10 days of static culture based on histological sections (Goldstein and Mikos, unpublished). (b) A simple 2D prediction for oxygen concentration, $CO_2$, within a scaffold assuming uniform availability at the edges, and first-order consumption kinetics within the scaffold.

### 15.2.1 Perfusion Bioreactors

Diffusional mass transport limitations can be overcome by convecting culture medium through the pores of three-dimensional scaffolds.[1–3] In particular, Goldstein *et al.* and Gomes *et al.* have shown that perfusion can achieve a fairly uniform cell density throughout scaffolds many millimeters in depth.[1–4] Moreover, perfusion can also enhance bone phenotypic markers, including collagen deposition and mineral accumulation,[1,3–9] making *in vitro* perfusion culture an attractive mechanism to induce bone tissue formation for clinical applications. Although the underlying mechanisms by which osteoblasts sense perfusion are not well understood, hydrodynamic shear stress, $\tau_w$ (exerted by the moving fluid), appears to stimulate cell surface mechanosensors (e.g., primary cilia,[10,11] stretch-activated channels[12,13]).

Two simple mathematical models to estimate the local shear stress can be derived from the steady-state Navier–Stokes Equation for an incompressible Newtonian fluid of viscosity $\mu$.

$$\tau_w = -\mu \left[ \frac{dv_z}{dr} \right]_{r=R} = \frac{4\mu Q}{\pi R^3} \tag{15.1}$$

$$\tau_w = -\mu \left[ \frac{dv_x}{dy} \right]_{y=h} = \frac{3\mu Q}{2wh^2} \tag{15.2}$$

Here, Eq. 15.1 predicts the wall shear stress, $\tau_w$, at the luminal surface of a tube of radius $R$, perfused with fluid at a volumetric flow rate of $Q$, and has been used to estimate shear stresses in perfused porous scaffolds.[1] In contrast, Eq. 15.2 predicts the wall shear stress for flow through a narrow slit of width $w$ and height $h$ ($w \gg h$) and has been used to estimate shear stress in model parallel-plate systems.[14] In this regard, precisely controlling the magnitudes of shear stresses is important because stress of 0.8–3 Pa has been estimated for fluid flow through the lacuna-canalicular spaces of cortical bone under normal physiological conditions.[15]

### 15.2.2 Mechanotransduction

Mechanotransduction is the cascade of signaling events by which mechanical forces affect gene expression. In osteoblastic cells, hydrodynamic shear stress has been shown to stimulate the expression of growth factors (e.g., bone morphogenetic protein-2, vascular endothelial growth factor),[10,14,16] through the activation of mitogen-activated protein (MAP) kinases ERK-1/2, and p38.[16] Furthermore, evidence exists that different perfusion conditions, such as flow rate and pulsatile flow profiles, result in varying biological responses. In particular, studies have shown that increasing local shear stresses can enhance cellular proliferation[10] and the accumulation of mineralized ECM.[8] Together, these findings indicate that *in vitro* perfusion bioreactor conditioning of MSCs may be a feasible approach to create a biologically active material for bone regeneration.[17,18]

### 15.2.3 Other Bioreactors

Other bioreactors have also been developed to facilitate formation of bone-like tissues *in vitro*. For example, a device that exerts both cyclic compression and perfusion has been described by David *et al.*[19] In addition, ultrasound[20,21] and mechanical stretch[22] have been explored as methods to stimulate osteoblastic differentiation.

## 15.3 Cartilage: Bioreactors for Hydrodynamic Shear Stress and Cyclic Compression

Cartilaginous tissue is found within a large number of anatomical sites (e.g., nose, ears, larynx/trachea, meniscus) in a broad variety of shapes and sizes. Consequently, a number of bioreactor strategies—designed to exert hydrodynamic shear stress, mechanical compression, or hydrostatic pressure—have been tailored for conditioning cartilage for specific anatomical applications. In general, the two-fold purpose of these bioreactors is to stimulate the deposition of a cartilage-specific ECM [rich in type II collagen and glycosaminoglycans (GAGs)] and to increase the mechanical strength of neo-tissue. In the following paragraphs, we focus on stirred-tank and cyclic compression bioreactors for the formation of engineered cartilage. We conclude this section with a mention of oxygen tension and its effect on cartilage formation.

### 15.3.1 Stirred-Tank Bioreactors

Stirred-tank bioreactors are attractive devices for developing complex-shaped pieces of cartilage, such as those needed for ear augmentation. The simplest bioreactors are spinner flasks (Fig. 15.3a,b), in which constructs are hung from wires and an impeller is used to circulate medium.[23,24] These devices have been shown to improve cell proliferation and collagen synthesis (relative to static culture),[23,24] and these positive effects have been attributed to increased convective mass transport. However, the formation of a thick, type I collagen containing fibrous capsule at the scaffold surface,[23,25] and loss of GAGs from the scaffold into the circulating medium[24] have been observed in spinner flasks. Consequently, efforts have been made to reduce the hydrodynamic shear stresses. One approach is a wavy wall spinner flask (Fig. 15.3c),[26] in which the curves act as baffles, dissipating circumferential momentum and enhancing axial flow. Computational models by Bilgen *et al.* suggest that the wavy wall results in 50% reduction in mean shear stress

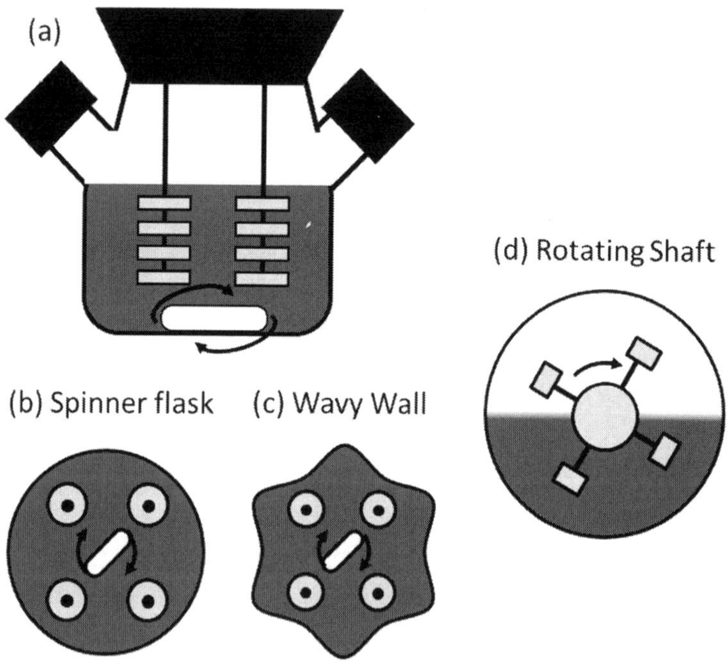


**Figure 15.3** Stirred bioreactors. (a) Side view of spinner-flask showing disc-shaped constructs on two wires. (b) Top view of a conventional spinner flask showing constructs on four wires. (c) Wavy-wall spinner-flask. Adapted from Ref. 27. (d) End view of a horizontal rotating shaft. Adapted from Ref. 29.

(from 0.49 dyne/cm$^2$ to 0.24 dyne/cm$^2$) at constant impeller rate (50 rpm),[27] while complementary experimental studies showed statistically significant increases in GAG, collagen, and cell density.[25,28] Furthermore, the authors predicted that the placement of the scaffolds in the lobes of the wavy wall spinner flask could further reduce the hydrodynamic shear stress. Nevertheless, the wavy wall spinner flask does not appear to mitigate accumulation of type I collagen on the scaffold surface.[25,26] A second approach to reduce shear stress is to affix scaffolds to a shaft that rotates slowly (e.g., 2–10 rpm) through a stationary fluid. In one particular study, Chen *et al.* oriented the shaft horizontally within a vessel half-filled with medium (Fig. 15.3d) to allow the scaffolds to be transported back and forth between gas and liquid phases, resulting in alternating periods of shear stress and no shear stress.[29] Using computational modeling of the fluid mechanics, the authors predicted that the surface shear stress of tissue discs steps between 0 and 0.22 dynes/cm$^2$ (Fig. 15.3a). Concurrently, cell culture studies indicated an increase in the collagen content of chondrocyte-seeded poly(lactic-co-glycolic) acid (PLGA) scaffolds in the rotating shaft bioreactor, relative to a conventional spinner flask.[29]

### 15.3.2 Cyclic Compression Bioreactors

Cyclic compression bioreactors that apply mechanical compression alone,[30,31] or in combination with shear stress[32] or perfusion,[33] have been developed to exert complex physiological stresses on constructs intended for repair of meniscus,[30,33] intervertebral disc,[31] and articular cartilage.[34] The simplest cyclic compression bioreactors involve the unconstrained compression of disc-shaped constructs between flat platens (Fig. 15.4),[35] but variation of the platen shape has permitted the compression of anatomically shaped constructs (e.g., meniscus).[30] Typical loading regimes for cyclic compression of engineered cartilage are a 10% strain[34,36] at 1 Hz,[30,34,36] applied intermittently over a period of 4 weeks.[34,36] Under these conditions, compression has been shown to increase GAG accumulation and the compressive modulus of neo-tissues.[30,34,36] Further, studies by both Mauck *et al.* in agarose[36] and Ballyns and Bonassar in alginate constructs[30] showed increases in collagen accumulation with cyclic compression. In contrast, Tran *et al.*[34]—who used a pellet culture approach to achieve neo-tissue discs (without a biomaterial scaffold)—observed lower collagen content with cyclic compression as compared with static culture, which they attributed to enzymatic degradation and/or convective loss of the protein. However, Mauck *et al.*[36] found a positive effect of mechanical stimulation only at lower chondrocyte densities; at higher cell densities, the levels of GAG and collagen were independent of mechanical loading. Finally, Chowdhury *et al.*,[37] who systematically varied duty cycle, showed that increasing the loading duration increased both GAG synthesis and neo-tissue modulus.

Despite the beneficial effects of cyclic compression, two limitations have been noted. First, Ballyns and Bonassar[30] found that both the GAG content and compressive modulus of their alginate-based neo-tissues decreased under cyclic loading between the 2- and 6-week timepoints. They speculate that this result may have been due to enzymatic degradation of GAGs, scaffold deterioration, and/or convective transport of GAGs from scaffolds. But it suggests that the beneficial effects of long-term mechanical conditioning are transient. Second, Mauck *et al.* reported a decline in neo-tissue thickness with cyclic compression[36] that could be attributed to creep (the process of plastic deformation over hours to weeks under mechanical forces). This can affect mechanical conditioning regimen, as simple compression bioreactors that are designed to apply a single prescribed cyclic displacement (e.g., Fig. 15.4) may exert a stress that changes over time. To a certain extent, this problem can be resolved using a bioreactor that is equipped with a feedback control system (i.e., an inline pressure transducer and a computer interface) that automatically adjusts the range of displacements to maintain a window of stresses. However, it does not resolve problems associated with creep.

### 15.3.3 Oxygen Tension

Oxygen tension can have a positive effect on cartilage formation and can be controlled using closed bioreactors. In a recent study, Sheehy *et al.* showed that expansion and subsequent differentiation of MSCs at 5% $pO_2$ (as compared with the typical 20% $pO_2$) can increase both GAG and collagen accumulation,[38] indicating a link between oxygen tension and chondrogenic development. Further, they showed in both pellet cultures and agarose hydrogels that the lower oxygen tension reduces synthesis of type I collagen and increases the synthesis of type II collagen. However, these results are in contrast with an earlier study by Malda *et al.*, who found no difference in the rates of proliferation, glucose consumption, and lactate production by bovine articular chondrocytes at 4%, 10.5%, and 21% $pO_2$.[39] Nevertheless, the control of oxygen tension in tandem with application of mechanical forces (e.g., shear stress, compression) can be readily achieved in closed bioreactors and may have beneficial effects on neo-tissue formation.

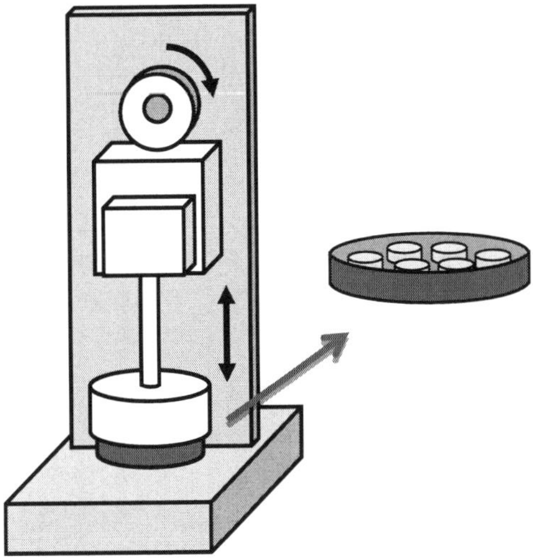

**Figure 15.4** Compression bioreactor. Adapted from Ref. 35.

## 15.4 Tendon and Ligament: Application of Cyclic Uniaxial Strain

Tendons and ligaments—which anchor muscles to bones and stabilize joints—consist of a sparse population of fibroblasts within a collagen-rich ECM (primarily types I and III, with lesser amounts of elastin, biglycan, decorin, fibronectin, and tenascin-C). On the basis of evidence that stresses associated with normal daily activities contribute to the maintenance of the ECM structure and tissue strength,[40] researchers have used cyclic mechanical forces to stimulate synthesis of tendon/ligament ECM proteins and increase the mechanical properties of neo-tissues.[41,42] In this section, we focus on bioreactors that recapitulate the cyclic tensile strain experienced *in vivo*,[43] discuss the effects of varying the strain regimen (e.g., frequency, strain magnitude, duty cycle), and describe the putative mechano-transductive signaling pathways.

### 15.4.1 Uniaxial Cyclic Strain

Uniaxial cyclic strain is the primary mode of *in vitro* mechanical stimulation of constructs for tendon/ligament applications, and typical conditions (4–10% strain at 0.01–1 Hz) result in increased expression of tendon/ligament proteins and neo-tissue stiffness. For example, Saber *et al.* demonstrated that 5 days of stimulation (1.25 N, at 0.017 Hz, with a 1 h on/1 h off duty cycle) could increase the ultimate tensile strength and elastic modulus of tenocyte-seeded decellularized rabbit tendons, relative to unstimulated tendons.[44] In a follow-up study, Woon *et al.* found that increasing or decreasing the load magnitude did not significantly affect mechanical strength after 5 days, but that increasing the duration of stimulation to 8 days further improved mechanical properties.[45] In a separate study, Androjna *et al.*—who compared static strain (ca. 2%) with cyclic strain (9% strain twice daily for 30 min at 0.1 Hz)—showed a greater increase in tissue stiffness with cyclic strain than with static strain over a 2-week period.[46] Lastly, Doroski *et al.* exposed MSCs—encapsulated in a three-dimensional PEG based hydrogel—to 10% strain at 1 Hz with a 3 h on/3 h off duty, and showed that expression of collagens I and III, and tenascin-C were increased at 21 days compared with statically cultured encapsulated MSC.[47] (Concurrently, they showed that osteogenic, chondrogenic, and adipogenic genes were not overexpressed, indicating that cyclic strain may be selective for differentiation into a tendon/ligament lineage.) Together, these data indicate that cyclic strain stimulates ECM deposition and that cell response may not be sensitive to strain magnitude and frequency.

### 15.4.2 Cell Signaling

Cell signaling in response to mechanical stimulation is not well understood, but the activation of ERK1/2 is thought to be involved in coupling mechanosensation to the induction of tendon/ligament genes. For example, Paxton *et al.*[48] showed that phosphorylation of ERK1/2 is independent of strain frequency and magnitude of cyclic strain (over the ranges of 0.1–1 Hz, and 2.5–15%, respectively), but dependent on duration of stimulation, with maximum activation after only 10 min. Consequently, the authors suggested that cells are refractory and require recovery periods between episodes of mechanical stimulation. Importantly, their results are consistent with general observations of long-term bioreactor studies: ECM protein deposition is improved when recovery periods are provided, but not significantly affected by strain magnitude and frequency.

We conclude this section with two final considerations of bioreactor design for tendon/ligament tissue engineering. First, cyclic stretch can be combined with other mechanical forces *in vitro*. For example, Sawguchi *et al.* examined concurrent application of 5% stretch and 90% torsion, and reported the enhanced expression of collagen types I and III, and fibromodulin relative to the individual stimuli.[49] Second, long-term cyclic strain on engineered tendon/ligament constructs may result in material creep. Similar to cartilage neo-tissues under cyclic compression (Section 15.3), the irreversible deformation of tendon/ligament neo-tissues under tensile strain can lead to a systematic decrease in mechanical stimulation over time. To address this problem, bioreactors have been designed with in-line force transducers that permit *in situ* measurement of stress/strain relationships.[50] Although they add to the cost, they also permit adjustment of displacements to ensure consistent loading regimens.

## 15.5 Cardiac Tissue

Myocardial infarction—resulting from occlusion of coronary arteries—has limited treatment options due to the inability of the myocardium to regenerate and a severe lack of donor tissue. Therefore, the goal of cardiac tissue engineering has been to develop neo-tissues that can contract in a synchronous manner and integrate with the host tissue to restore functionality. Cardiac tissue is phenotypically marked by aligned cardiomyocytes with well-developed parallel sarcomeres and high intercellular connectivity [e.g., connexin-43 (Cx43) gap junctions, adherens junctions, and desmosomes). In this section, we describe bioreactor strategies that employ electrical and mechanical stimuli to develop engineered cardiac tissue. In addition, we discuss bioreactors that incorporate perfusion to increase oxygen and nutrient availability.

### 15.5.1 Electrical Pulses

Electrical pulses have been used for more than a decade to condition cardiac cells within scaffolds for the purpose of creating contractile neo-tissues. Frequently, they are applied across a construct via a pair of carbon rods at strengths of 2–5 V/cm and frequencies of 1–5 Hz.[51–53] For example, Radisic *et al.*[51] stimulated neonatal rat ventricular myocytes in collagen sponges with 2 ms, 5 V/cm electrical pulses at 1 Hz for 5 days. This stimulation resulted in increased cell alignment and intercellular coupling. In addition, the stimulated cells developed more sarcomeric structures and contained more mitochondria than unstimulated cells. Moreover, the authors reported significant increases in gene expression of $\alpha$ and $\beta$-myosin heavy chain, Cx43, and integrin $\beta_1$ in the stimulated constructs. In a separate study, Lasher *et al.* electrically stimulated neonatal rat ventricular myocytes in fibrin gels at 150% of their excitation threshold to induce synchronous contractions.[52] Under electrical stimulation, the myocytes were more elongated than the non-stimulated myocytes, and similar in morphology to myocytes in age-matched myocardium. In addition, the Cx43 distribution in stimulated cells was similar to that of postnatal and adult myocytes. In another study, Tandon *et al.* systematically examined the effect of 2 ms pulses applied to cardiomyocyte-seeded collagen sponges at 1–5 Hz and 1–6 V/cm, and reported that 3 Hz, 2–3 V/cm was optimal for cardiomyocyte development.[53] In addition, the research team found that stimulation at or above 8 V/cm resulted in no contractile response.[54] Together, these results suggest that the magnitude of field strength is more important than its frequency for the development of engineered cardiac tissue.

### 15.5.2 Mechanical Stretch

Mechanical stretch has also been used by a number of researchers to achieve synchronously contractile cardiac neo-tissues. In a series of studies, Eshenhagen and colleagues formed contractile neo-tissues by suspending rat cardiomyocytes in Matrigel/collagen gels, allowing them to consolidate over several days (to achieve alignment and interconnectivity),[55] and then stretching them for 7 days at 10% strain at 2 Hz. This stimulation resulted in neo-tissues possessing a resting potential and resting tension[56] that could electrically integrate with cardiac tissue *in vivo*.[57] In a separate set of studies, Shimko and Claycomb subjected cardiac constructs to 10% cyclic mechanical strains at 1, 2, and 3 Hz.[58] After 3 days of stimulation, they found an upregulation of cardiac-specific genes for constructs stimulated at 3 Hz, but a downregulation of cardiac-specific genes for the 1 Hz group relative to nonstimulated constructs.

### 15.5.3 Perfusion Bioreactors

Perfusion bioreactors have also been used to develop cardiac neo-tissues. In particular, pulsatile flow mechanically stimulates cardiomyocytes, while convection improves the delivery of nutrients and oxygen. For example, Kofidis *et al.*[59] showed that pulsatile perfusion (100 ml/h, 2 Hz) increased cardiomyocyte viability throughout three-dimensional scaffolds compared with nonperfused scaffolds. In another study, Brown *et al.*[60] demonstrated that pulsatile flow (1 Hz) conditioning resulted in neo-tissues with a lower excitation threshold, higher capture rate, and higher contraction amplitude than statically cultured neo-tissues.

Perfusion loops have been integrated into bioreactors designed to apply electrical and mechanical stimulation. For example, Barash *et al.*[61] developed a perfusion bioreactor with integrated electrodes, by which an electrical field and perfusion could be applied simultaneously. Using computational modeling, they predicted optimal operating conditions and experimentally demonstrated that pulsatile stimulation (74.4 mA/cm$^2$, 2 ms, 1 Hz) and homogenous perfusion resulted in cell elongation, cell striation, and increased expression of Cx43. Separately, Lu *et al.*[62] developed and validated a bioreactor system that combined electrical stimulation, mechanical stretch, and perfusion. This bioreactor was able to induce rhythmic contractions in ventricular myocytes similar to those generated in an individual field stimulation chamber and to align the cells.

Together, these studies show that electrical stimulation, mechanical stretch, and perfusion, alone or together, can guide the organization of primary cardiomyocytes into functional neo-tissues. In addition, although the magnitude and frequency of electrical pulses is not particularly critical, the perfusion of medium is important for maintaining cell viability within large constructs.

## 15.6 Vascular Tissue and Heart Valves

Physiologically, both vascular tissues and heart valves are subject to hydrodynamic shear stresses and dynamic pressure gradients resulting from pulsatile blood flow. Consequently, bioreactors—to develop vascular and heart valve neo-tissues—are fundamentally similar, consisting of one or more constructs within a bath, a pump to circulate fluid, and a line conditioner to generate pulsatile wave-forms (Fig. 15.5a).

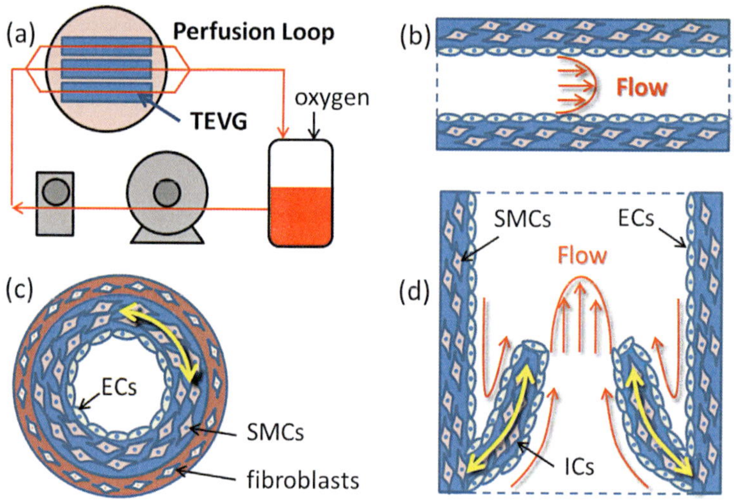


**Figure 15.5** Blood vessel bioreactor. (a) Diagram of a perfusion loop with three tissue engineered vascular grafts (TEVGs) in parallel within a culture bath, an oxygenated medium reservoir, peristaltic pump, and line conditioner (to produce pulsatile wave forms). Perfusion diagram adapted from Refs. 65 and 67. (b) Cross section of a blood vessel to illustrate the intimal (green), medial (blue), and adventitial (red) layers. (Yellow arrow indicates hoop stress.) (c) Cross section of blood vessel to indicate flow pattern (red). (d) Diagram of a heart valve indicating the bending stresses (yellow arrows, and the fluid flow patterns (red arrows).

### 15.6.1 Vascular Tissue Engineering

Vascular tissue engineering is motivated by the need for tissue grafts for patients suffering from atherosclerosis, aneurisms, and peripheral vascular disease. In particular, there is a significant need for engineered small diameter vascular grafts—such as those suitable for repair of the coronary arteries—because synthetics are not suitable for repair of blood vessels below 6 mm in diameter and autologous donor tissue is frequently in limited supply. Consequently, tissue engineering strategies have been developed with the goal of constructing vascular grafts that can withstand the stresses generated by circulating blood and recapitulate normal vascular functions.

Vascular tissues consist of intimal, medial, and adventitial layers (Fig. 15.5c). The intimal layer—composed of endothelial cells (ECs) on a basal lamina matrix—presents a non-thromobogenic surface to the circulating blood and responds to environmental conditions (e.g., shear stress, inflammation) by the release of nitric oxide and chemokines. The medial layer—consisting of circumferentially oriented vascular smooth muscle cells (SMCs) in an elastin and type I collagen matrix—provides hoop strength and allows blood vessels to dilate in response to nitric oxide. The adventitial layer consists of fibroblasts in collagen and is not discussed further.

Under normal physiological conditions, ventricular ejection of blood exerts a pulsatile hydrodynamic shear stress on the order of 10 dyne/cm$^2$ (but which can vary from 4 dyne/cm$^2$ to 80 dyne/cm$^2$ depending on anatomical location and phase of the cardiac cycle) on ECs.[63] At the same time, normal arterial blood pressure cycles between 120 and 80 mmHg (peak-to-trough waveforms) applying a cyclic hoop stress to the ECs.[64,65] Therefore, the purpose of vascular tissue bioreactors is to condition neo-tissues so that the intimal layer will not delaminate under hydrodynamic shear, and the medial layer will not mechanically fail under hoop stresses.

A common strategy for creating engineered vascular tissue involves developing a robust medial layer that is subsequently seeded with ECs.[66,67] Stegemann and Nerem formed a medial layer by casting suspensions of SMC in type I collagen into tubular molds, allowed them to contract around a mandrel for 2 days, and then subjected to cyclic hoop stress (10% strain, 1 Hz) for 4 days.[66] Compared with statically cultured tubes, cell number was greater in the stimulated samples, but the expression of smooth muscle actin was not different. Subsequently, the authors demonstrated that the expression of smooth muscle actin could be enhanced by the addition of transforming growth factor-$\beta$. Using the same approach, Seliktar *et al.*[68] formed collagen tubes containing a mixture of SMCs and fibroblasts (to mimic the medial and adventitial layers) and reported an increase in tube mechanical properties with 4 days of cyclic hoop stress, which was followed by a decrease in properties at 8 days (most likely due to degradation of the ECM). Using a similar strategy, Williams and Wick[67] seeded the lumen of tubular poly(glycolic acid) felts with SMCs and cultured them within a perfusion bioreactor for up to 25 days using a shear stress of 6 dyne/cm$^2$ (as estimated from Eq. 15.1). The researchers reported proliferation and circumferential alignment of SMCs, but very little accumulation of elastin. As a second step, they introduced ECs to the luminal surface of their tubular felts and reported evidence of a confluent EC monolayer after 2 days of perfusion culture. Iwasaki *et al.*[69] took this approach one step further, sequentially forming medial and adventitial layers by wrapping SMC-seeded and fibroblast-seeded poly(glycolic acid) sheets around a silicon tube. After the layers had consolidated, the silicone tube was removed and the lumen seeded with ECs. The resultant construct was subsequently cultured for 14 days, during which the flow rate and pressure were gently ramped to physiologic levels. These conditioned constructs developed similar mechanical elasticity and strength compared with native blood vessels and stained positively for elastin, collagen, and smooth muscle actin. Lastly, Jeong *et al.*[70] cultured SMCs on PLGA/collagen scaffolds for 14 days followed by the addition of ECs and reported that pulsatile stimulation resulted in an increased expression of $\alpha$-smooth muscle actin and smooth muscle myosin heavy chain by the SMCs, and von Willebrand factor by the ECs. Together, these studies demonstrate the feasibility of developing engineered vascular grafts with distinct layers, and the capacity of pulsatile perfusion to increase the mechanical properties of neo-tissue and to stimulate markers of the EC and SMC phenotypes.

### 15.6.2 Heart Valve

Heart valve disease stems from the abnormal functioning of one or more of the valves of the heart due to inflammation or calcification. The progressive changes in tissue mechanics can lead to altered blood flow including regurgitation or stenosis. At present, severe cases of valvular disease require the replacement of the diseased valve with a mechanical valve to restore basic function. However, flow profiles through mechanical valves are not physiologic, and the abnormal shear stresses can result in thrombosis and blood cell lysis. A tissue engineered heart valve replacement with a physiological structure could mitigate these effects. Therefore, in this section, we discuss work that has been undertaken using decellularize heart valves. The decellularization process removes the valvular ECs, SMCs, and interstitial cells (ICs) (Fig. 15.5d), but preserves the tissue architecture, ECM structure, and mechanical properties. The "scaffold" is then reseeded with ECs, which protect the leaflet interior from damage due to the flowing blood.

Bioreactors for heart valve engineering have focused on mimicking the complex mechanical loads that a heart valve is exposed to in its native environment *in vivo*, including the normal hemodynamic environment (Fig. 15.5d, red arrows) and flexural stresses (Fig. 15.5d, yellow arrows). For example, Hoerstrup *et al.*[71] designed a bioreactor that combined a perfusion flow system with a deflectable membrane, which could subject heart valve constructs to a pulsatile pressure, and showed that it could generate pulsatile flows from 50 ml/min to 2000 ml/min and pressures ranging from 10 mmHg to 240 mmHg at various stroke rates. A similar but simpler system was developed by Engelmayr *et al.*,[72] which could drive dynamic flexural stimulation of tissue-engineered heart valve materials and achieve native valvular deformations. Lichtenberg *et al.*[73] developed a physiologic circulation flow loop for the dynamic re-endothelialization of decellularized heart valves and showed that they could completely repopulate the tissue surface under physiological culture conditions. A similar approach was taken by Schenke-Layland *et al.* in which decellularized heart valves were seeded with myofibroblasts, cultured in a bioreactor under physiologic flow conditions for approximately 14 weeks, and finally seeded with ECs.[74] Immunohistochemical staining revealed EC markers on the surface of the leaflets and smooth muscle cells within. Together, these studies show the potential of decellularized heart valves as scaffolds for forming engineered neo-tissues with physiological mechanics and hemodynamics, and the potential for them to be recellularized within a pulsatile perfusion bioreactor.

## 15.7 Organ Bath Bioreactors for Hollow and Solid Organs

Although most techniques for engineering tissues utilize polymeric scaffolds that are fabricated using a bottom-up approach, organs with complex architectures (e.g., lung, kidney, liver) may be more easily engineered from the top-down utilizing decellularized whole-organ ECMs. This approach not only preserves the complex microstructures and high surface areas needed for mass transport and enzymatic processes but also maintains the macrostructures for the transport of blood and other fluids. Consequently, organ bath bioreactors—in which the tissue is bathed in media and fluid are perfused via existing ducts—may be an attractive approach. To date, few studies have described such organ baths, but in the following sections, lung, kidney, and liver are discussed.

### 15.7.1 Lung

Diseases related to lung account for approximately 400,000 deaths annually in the United States,[75] and treatment of diseased lungs is difficult, as lung tissue does not naturally regenerate when damaged, and sources of donor lung tissue are limited. Furthermore, lung transplantations have high failure rates, with half of patients suffering from chronic rejection symptoms and a 5-year survival rate of near 50%. Consequently, transplantation is considered a treatment of last resort. For these reasons, an engineered lung has been proposed as an alternative to traditional approaches; however, research in this area is scarce. Recently published studies have focused on the repopulation of decellularized lung tissue and its mechanical conditioning through simulated physiological breathing stresses and vascular perfusion. For example, Price *et al.*[76] described a bioreactor system for the cultivation of murine decellularized lung tissue reseeded with fetal alveolar type II cells. The seeded matrix was attached to a ventilation loop through the trachea and submerged in growth media, while external ventilation was applied through the airway to stimulate murine physiological breathing at 180 breaths per minute. After 7 days of culture, histological analysis revealed a uniform distribution of the cells throughout the matrix and the presence of alveolar phenotypic markers such as Pro-SP-C and cytokeratin 18. In another study, Petersen *et al.*[77]

used decellularized rat lungs that were reseeded with vascular and pulmonary ECs. The reseeded lungs were cultured in a bioreactor consisting of a vascular perfusion loop and a ventilation loop to simulate physiological perfusion and breathing. When implanted in rats, these reseeded lungs participated in successful gas exchange for a short period of time. In both cases, lungs were decellularized, reseeded, and subjected to physiological breathing loads. However, the biological activity of these cells was not observed beyond cellular distribution and expression of some relevant phenotypical markers.

### 15.7.2 Kidney and Liver

Kidney and liver are two of the most demanded organs for transplantation, according to the Organ Procurement and Transplantation Network, with waiting lists of 95,000 and 15,700 people in the United States, respectively, in 2013. Renal tissue contains a delicate microstructure of tortuous capillaries, known as a glomerulus, which surrounds each Bowman's capsule and provides extensive surface area for small molecule transport from the blood to the nephrons. Similar to kidney, the liver cleanses and detoxifies the blood through specialized cellular microstructures that make up the organ. Blood flowing from the hepatic artery and portal vein is branched into capillary beds lined with hepatocytes and Kupffer cells, which undertake the many roles of the liver (e.g., blood detoxification, breakdown of hormones and proteins). Thus, the microstructures of both the liver and kidney are vital for their functionality. However, neither can be effectively reproduced *ex vivo*, and consequently, the reseeding of decellularized kidneys and liver tissue has been pursued as a process to achieve functional organs.

Bioreactors for liver and kidney tissue development are currently in their infancy, and the current challenges facing researchers are decellularization of these organs (while preserving their architecture and ECM) and efficient engraftment of donor cells. For example, Sullivan *et al.*[78] showed that porcine kidneys—which are anatomically similar to human—could be decellularized in a manner that preserves the tubular microstructure and ECM. Separately, Ross *et al.*[79] used a perfusion device to decellularize rat kidneys and showed that they could be reseeded with murine embryonic stem cells through the artery and ureter. These donor cells subsequently localized to the glomeruli and arterioles, differentiated, and expressed phenotypic markers Pax-2 and Ksp-cadherin. Similarly, cell engraftment of decellularized liver tissue has been an ongoing challenge. For example, Soto-Gutierrez *et al.*[80] examined three different seeding techniques for introducing human fetal hepatocytes into decellularized rat livers, and reported that a multi-step approach was most effective at introducing hepatocytes and achieving production of albumin and urea and expression of cytochrome P450. Separately, Barakat *et al.*[81] recellularized porcine livers with a mixture of seeded human fetal hepatocytes and stellate cells. They showed that the engrafted cells could be maintained under perfusion culture and expressed hepatocyte markers such as lactate and urea production. A similar coengraftment approach was tested by Baptista *et al.*,[82] who introduced a mixture of ECs and human fetal liver cells, and showed that the cells adhered to their respective native locations within the liver and displayed physiologically relevant markers of endothelial, hepatic, and biliary epithelial phenotypes after perfusion culture.

## 15.8 *In situ* Imaging of Tissue Development

Most traditional assays for the evaluation of engineered tissue development are terminal, requiring the destruction or fixation of the construct to quantify the expression of genes and proteins or observe the organization of cells into tissue-like structures. Furthermore, the creation and destruction of multiple samples over an experiment is both expensive and time-intensive. Conversely, a noninvasive monitoring technique would allow real-time feedback of the biological development of a construct. Currently, only a few techniques permit *in situ* assessment of tissue development, including optical coherence tomography (OCT), confocal imaging, bioluminescence (BL), microcomputed tomography (μCT), and magnetic resonance imaging (MRI). Although noninvasive, these approaches share some limitations, including relatively low resolution, shallow penetration depth into large scaffolds, and the need to incorporate agents (e.g., fluorophores, contrast agents) into the engineered tissue or cells. In the following paragraphs, these five imaging modalities are discussed.

### 15.8.1 Optical Coherence Tomography

This technique can be used to form micrometer resolution, three-dimensional images through a fiberoptic cable from an optically scattering medium. OCT is essentially an optical ultrasound technique, recording the light reflections within a tissue to reconstruct cross-sectional images of the medium. It has several advantages such as real-time imaging, minimal sample preparation required, and no ionizing radiation. Although the resolution of the resulting images can be less than a micron, the depth of penetration is limited to approximately 2 mm. Nevertheless, OCT can be applied to thin tissues such as skin. In one particular study, Smith *et al.*[83] used OCT to visualize the development of the epidermis and dermis over time. They subsequently validated this technique by comparing their OCT cross sections with traditional histological sections.

### 15.8.2 Confocal Imaging

This technique utilizes a point source of light to excite a fluorescent marker and then records the resultant excitation from several spatial planes in the construct, localizing the signal within a three-dimensional medium. Microscopes that can utilize this imaging modality have been combined with bioreactors to culture constructs and apply mechanical stimulation, allowing real-time monitoring of cellular activity. One such study by Kluge *et al.*[84] visualized calcein-stained MSCs on silk scaffolds within a cyclic stretch bioreactor. The advantage of this system was that cell adhesion, morphology, and proliferation could be monitored in response to mechanical stimulation without removing the samples from the bioreactor. Limitations with this approach include the depth of field for confocal imaging and the need for fluorophores [either introduced into the bioreactor (e.g., calcein[84]) or into the cell genome (e.g., green fluorescent protein[85])].

### 15.8.3 Bioluminescence Imaging

This technique is a nondestructive imaging modality in which cells—modified with a reporter gene produce a protein (e.g., GFP) or enzyme (e.g., luciferase)—produce a fluorescent/luminescent signal that can be monitored over time. Several different bioreactor systems have been developed that permit real-time BL imaging of cells within a scaffold. One simple application of BL in tissue engineering was demonstrated by Logeart-Avramoglou *et al.*,[86] who seeded luciferase-tagged C3H10T1/2 cells within translucent AN69 scaffolds and opaque coral scaffolds. The seeded scaffolds were cultured for up to 21 days in a rotational bioreactor and removed at each time-point for bioluminescent imaging in a well-plate. The overall bioluminescent signal was recorded from both the translucent and opaque scaffolds and correlated to cell number, allowing the monitoring of cell proliferation over time. In another study, Napoli *et al.*[87] seeded luciferase-tagged MSCs within a hollow fiber membrane bioreactor and monitored cell density at various locations along the channel. These two studies showed how BL imaging could visualize cellular distribution in scaffolds on the macro-scale. An extension of the BL technique developed by Hofmann *et al.*,[85] who utilized a fiberoptic-based imaging system for the monitoring of individual cells on fibrous scaffolds. Specifically, GFP-expressing human ECs were monitored in a fibrous electrospun scaffold via a fiberoptic probe that could be rotated and translated within the scaffold and locally excite GFPs. The emission from the cells was captured and reconstructed to produce a two-dimensional image of cellular distribution, with which the investigators track cell migration and proliferation. One of the attractive features of BL imaging is that it can visualize the spatial and temporal distribution of specific cell phenotypes. By stable transfection of a fluorescent or luminescent reporter into the cell genome, it is possible to monitor cell differentiation *in situ* (Fig. 15.6). However, the caveat is that such transformed cells—although useful in research—may not be suitable for clinical application.

### 15.8.4 Microcomputed Tomography

This technique utilizes x-rays to generate two-dimensional cross sections of a three-dimensional object, which are reconstructed, recreating the three-dimensional object virtually. Clinically, this technology is utilized for the diagnosis and visualization of bony structures due to its excellent ability to differentiate between bone and soft tissues. Microcomputed tomography can generate images with resolutions of 10–100 μm, revealing fine details of bone microstructure. This technology has most commonly been utilized in tissue engineering applications for the monitoring of bone formation or mineral deposition on constructs. Hagenmüller *et al.*[88] utilized μCT to nondestructively monitor silk scaffolds seeded with MSCs to quantify the engineered bone volume density, surface-to-volume ratio, trabecular thickness, and mineral density. Bone-like tissue formation was observed to begin at day 10 and quantified via the μCT up to day 44. Porter *et al.*[89] utilized the same technique to monitor mineral growth on seeded PCL scaffolds nondestructively within a perfusion bioreactor. In both studies, contrast agents were not required to image mineral deposition, which is advantageous. However, other soft tissues cannot be visualized through this technique without the introduction of contrast agents, which could limit its application beyond bone.

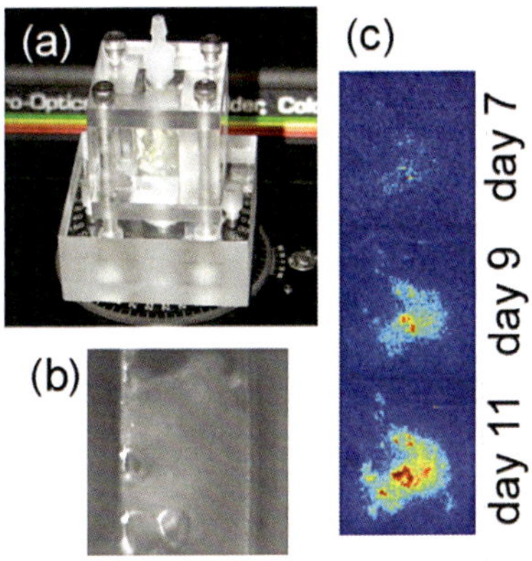


**Figure 15.6** BL imaging. (a) Perfusion chamber with four windows. (b) View of chitosan scaffold in chamber. (c) Set of three false-color images showing an increase in reporter gene expression over time.

## 15.8.5 Magnetic Resonance Imaging

This technique can be utilized for the visualization of super paramagnetic nanoparticle tagged cells within constructs. Terrovitis *et al.*[90] applied MRI to visualize paramagnetic nanoparticle labeled cells, appearing as darker regions of reduced signal intensity. However, as the labeled cells proliferated, the nanoparticles were diluted among the increasing cell population, weakening the magnetic signal. In another study, Ramaswany *et al.*[91] seeded unlabeled chondrocytes onto a hollow fiber perfusion bioreactor system, which were then labeled by perfusion of paramagnetic nanoparticles. The cellular distribution obtained from the MRI images was confirmed with histology. However, a limitation of this technique was that nanoparticles were only taken up by the cells directly exposed to the perfused nanoparticles, and regions away from the liquid layer could not be visualized. Together, these two approaches in labeling cells with paramagnetic nanoparticles demonstrate the inherent difficulties in visualizing cells with MRI.

## 15.9 Summary and Future Outlook of Bioreactors

In the preceding sections, we described bioreactor strategies that involved hydrodynamic shear, compression, tensile strain, electrical pulses, and dynamic pressure gradients. Although the specific applications and stimuli differ, cell responses to bioreactor stimuli share two similarities. First, cells appeared to be sensitive to the duty cycle of intermittent stimulation and responded more favorably (in terms of ECM deposition and function) if adequate periods of stimulation and recovery are provided (e.g., compression of cartilage[37]). In contrast, cells did not appear to be especially sensitive to the magnitude and frequency of the stimuli (e.g., cyclic stretch for tendon/ligament tissue,[48] electrical pulse for cardiac tissue[53]). Second, perfusion—either alone or in conjunction with other stimuli—improved cell viability, uniformity of cell distribution, and neo-tissue development (e.g., bone[1,4] and cardiac tissue[61]). The previous sections also highlight how bioreactors have become increasingly complex, and some can exert two stimuli (e.g., cyclic tension and torsion,[49] compression and perfusion[19]), or even three stimuli (e.g., electrical pulses, perfusion, and cyclic tension[62]). Concurrently, nondestructive *in situ* imaging modalities are increasingly being incorporated into the design of bioreactors (e.g., perfusion[89,91] and stretch[84]).

Future advancements in bioreactor design will likely focus on improving both batch-to-batch consistency and quality of neo-tissues. The former may be realized by the incorporation of sensors (to measure pH, oxygen tension, and mechanical properties) and feedback control systems. At the same time, the latter may be achieved through a number of advancements, including improved methods to image neo-tissues *in situ*, better diagnostic tools to characterize cell phenotype, and bioreactor strategies that more closely recapitulate the physiological micro-environment. In conclusion, bioreactors for tissue engineering are currently in their infancy, and tremendous opportunities exist for the development of better bioreactors to guide neo-tissue formation.

## References


1. Goldstein, A. S., Juarez, T. M., Helmke, C. D., Gustin, M. C., and Mikos, A. G. (2001) Effect of convection on osteoblastic cell growth and function in biodegradable polymer foam scaffolds, *Biomaterials*, **22**(11), 1279–1288.
2. Bancroft, G. N., Sikavitsas, V. I., and Mikos, A. G. (2003) Design of a flow perfusion bioreactor system for bone tissue-engineering applications, *Tissue Eng.*, **9**(3), 549–554.
3. van den Dolder, J., Bancroft, G. N., Sikavitsas, V. I., Spauwen, P. H., Jansen, J. A., and Mikos, A. G. (2003) Flow perfusion culture of marrow stromal osteoblasts in titanium fiber mesh, *J. Biomed. Mater. Res. A*, **64**(2), 235–241.
4. Gomes, M. E., Sikavitsas, V. I., Behravesh, E., Reis, R. L., and Mikos, A. G. (2003) Effect of flow perfusion on the osteogenic differentiation of bone marrow stromal cells cultured on starch-based three-dimensional scaffolds, *J. Biomed. Mater. Res. A*, **67**(1), 87–95.
5. Bancroft, G. N., Sikavitsas, V. I., van den Dolder, J., Sheffield, T. L., Ambrose, C. G., Jansen, J. A., and Mikos, A. G. (2002) Fluid flow increases mineralized matrix deposition in 3D perfusion culture of marrow stromal osteoblasts in a dose-dependent manner, *Proc. Natl. Acad. Sci. USA*, **99**(20), 12600–12605.
6. Glowacki, J., Mizuno, S., and Greenberger, J. S. (1998) Perfusion enhances functions of bone marrow stromal cells in three-dimensional culture, *Cell Transplant*, **7**(3), 319–326.
7. Mueller, S. M., Mizuno, S., Gerstenfeld, L. C., and Glowacki, J. (1999) Medium perfusion enhances osteogenesis by murine osteosarcoma cells in three-dimensional collagen sponges, *J. Bone Miner. Res.*, **14**(12), 2118–2126.
8. Sikavitsas, V. I., Bancroft, G. N., Holtorf, H. L., Jansen, J. A., and Mikos, A. G. (2003) Mineralized matrix deposition by marrow stromal osteoblasts

in 3D perfusion culture increases with increasing fluid shear forces, *Proc. Natl. Acad. Sci. USA,* **100**(25), 14683–14688.

9. Sikavitsas, V. I., Bancroft, G. N., Lemoine, J. J., Liebschner, M. A., Dauner, M., and Mikos, A. G. (2005) Flow perfusion enhances the calcified matrix deposition of marrow stromal cells in biodegradable nonwoven fiber mesh scaffolds, *Ann. Biomed. Eng.,* **33**(1), 63–70.
10. Hoey, D. A., Tormey, S., Ramcharan, S., O'Brien, F. J., and Jacobs, C. R. (2012) Primary cilia-mediated mechanotransduction in human mesenchymal stem cells, *Stem Cells,* **30**(11), 2561–2570.
11. Malone, A. M., Anderson, C. T., Tummala, P., Kwon, R. Y., Johnston, T. R., Stearns, T., and Jacobs, C. R. (2007) Primary cilia mediate mechanosensing in bone cells by a calcium-independent mechanism, *Proc. Natl. Acad. Sci. USA,* **104**(33), 13325–13330.
12. Wang, Y., McNamara, L. M., Schaffler, M. B., and Weinbaum, S. (2007) A model for the role of integrins in flow induced mechanotransduction in osteocytes, *Proc. Natl. Acad. Sci. USA,* **104**(40), 15941–15946.
13. Wu, D., Ganatos, P., Spray, D. C., and Weinbaum, S. (2011) On the electrophysiological response of bone cells using a Stokesian fluid stimulus probe for delivery of quantifiable localized picoNewton level forces, *J. Biomech.,* **44**(9), 1702–1708.
14. Sharp, L. A., Lee, Y. W., and Goldstein, A. S. (2009) Effect of low-frequency pulsatile flow on expression of osteoblastic genes by bone marrow stromal cells, *Ann. Biomed. Eng.,* **37**(3), 445–453.
15. Zeng, Y., Cowin, S. C., and Weinbaum, S. (1994) A fiber matrix model for fluid flow and streaming potentials in the canaliculi of an osteon, *Ann. Biomed. Eng.,* **22**(3), 280–292.
16. Kreke, M. R., Sharp, L. A., Lee, Y. W., and Goldstein, A. S. (2008) Effect of intermittent shear stress on mechanotransductive signaling and osteoblastic differentiation of bone marrow stromal cells, *Tissue Eng. Part A,* **14**(4), 529–537.
17. Datta, N., Pham, Q. P., Sharma, U., Sikavitsas, V. I., Jansen, J. A., and Mikos, A. G. (2006) *In vitro* generated extracellular matrix and fluid shear stress synergistically enhance 3D osteoblastic differentiation, *Proc. Natl. Acad. Sci. USA,* **103**(8), 2488–2493.
18. Pham, Q. P., Kasper, F. K., Scott Baggett, L., Raphael, R. M., Jansen, J. A., and Mikos, A. G. (2008) The influence of an *in vitro* generated bone-like extracellular matrix on osteoblastic gene expression of marrow stromal cells, *Biomaterials,* **29**(18), 2729–2739.

19. David, V., Guignandon, A., Martin, A., Malaval, L., Lafage-Proust, M. H., Rattner, A., Mann, V., Noble, B., Jones, D. B., and Vico, L. *Ex vivo* bone formation in bovine trabecular bone cultured in a dynamic 3D bioreactor is enhanced by compressive mechanical strain, *Tissue Eng. Part A*, **14**(1), 117–126.

20. Kang, K. S., Lee, S. J., Lee, H. S., Moon, W., and Cho, D. W. (2011) Effects of combined mechanical stimulation on the proliferation and differentiation of pre-osteoblasts, *Exp. Molec. Med.*, **43**(6), 367–373.

21. Li, L., Yang, Z., Zhang, H., Chen, W., Chen, M., and Zhu, Z. (2012) Low-intensity pulsed ultrasound regulates proliferation and differentiation of osteoblasts through osteocytes, *Biochem. Biophys. Res. Comm.*, **418**(2), 296–300.

22. Hanson, A. D., Marvel, S. W., Bernacki, S. H., Banes, A. J., van Aalst, J., and Loboa, E. G. (2009) Osteogenic effects of rest inserted and continuous cyclic tensile strain on hASC lines with disparate osteodifferentiation capabilities, *Ann. Biomed. Eng.*, **37**(5), 955–965.

23. Vunjak-Novakovic, G., Freed, L. E., Biron, R. J., and Langer, R. (1996) Effects of mixing on the composition and morphology of tissue-engineered cartilage, *AIChE. J.*, **42**(3), 850–860.

24. Gooch, K. J., Kwon, J. H., Blunk, T., Langer, R., Freed, L. E., and Vunjak-Novakovic, G. (2001) Effects of mixing intensity on tissue-engineered cartilage, *Biotechnol. Bioeng.*, **72**(4), 402–407.

25. Bueno, E. M., Bilgen, B., and Barabino, G. A. (2005) Wavy-walled bioreactor supports increased cell proliferation and matrix deposition in engineered cartilage constructs, *Tissue Eng.*, **11**(11–12), 1699–1709.

26. Bueno, E. M., Bilgen, B., and Barabino, G. A. (2009) Hydrodynamic parameters modulate biochemical, histological, and mechanical properties of engineered cartilage, *Tissue Eng. Part A*, **15**(4), 773–785.

27. Bilgen, B., and Barabino, G, A. Location of scaffolds in bioreactors modulates the hydrodynamic environment experienced by engineered tissues, *Biotechnol. Bioeng.*, **98**(1), 282–294.

28. Bilgen, B., Sucosky, P., Neitzel, G. P., and Barabino, G. A. Flow characterization of a wavy-walled bioreactor for cartilage tissue engineering, *Biotechnol. Bioeng.*,**95**(6), 1009–1022.

29. Chen, H. C., Lee, H. P., Sung, M. L., Liao, C. J., and Hu, Y.C. (2004) A novel rotating-shaft bioreactor for two-phase cultivation of tissue-engineered cartilage, *Biotechnol. Prog.*, **20**(6), 1802–1809.

30. Ballyns, J. J., and Bonassar, L. J. (2011) Dynamic compressive loading of image-guided tissue engineered meniscal constructs, *J. Biomech.,* **44**(3), 509–516.

31. Fernando, H. N., Czamanski, J., Yuan, T. Y., Gu, W., Salahadin, A., and Huang, C. Y. (2011) Mechanical loading affects the energy metabolism of intervertebral disc cells, *J. Orthop. Res.,* **29**(11), 1634–1641.

32. Martinez, H., Brackmann, C., Enejder, A., and Gatenholm, P. (2012) Mechanical stimulation of fibroblasts in micro-channeled bacterial cellulose scaffolds enhances production of oriented collagen fibers, *J. Biomed. Mater. Res. A,* **100**(4), 948–957.

33. Petri, M., Ufer, K., Toma, I., Becher, C., Liodakis, E., Brand, S., Haas, P., Liu, C., Richter, B., Haasper, C., von Lewinski, G., and Jagodzinski, M. (2012) Effects of perfusion and cyclic compression on *in vitro* tissue engineered meniscus implants, *Knee Surg. Sports Traumatol. Arthrosc.,* **20**(2), 223–231.

34. Tran, S. C., Cooley, A. J., and Elder, S. H. (2011) Effect of a mechanical stimulation bioreactor on tissue engineered, scaffold-free cartilage, *Biotechnol. Bioeng.,* **108**(6), 1421–1429.

35. Mauck, R. L., Soltz, M. A., Wang, C. C., Wong, D. D., Chao, P. H., Valhmu, W. B., Hung, C. T., and Ateshian, G. A. (2000) Functional tissue engineering of articular cartilage through dynamic loading of chondrocyte-seeded agarose gels, *J. Biomech. Eng.,* **122**(3), 252–260.

36. Mauck, R. L., Seyhan, S. L., Ateshian, G. A., and Hung, C. T. (2002) Influence of seeding density and dynamic deformational loading on the developing structure/function relationships of chondrocyte-seeded agarose hydrogels, *Ann. Biomed. Eng.,* **30**(8), 1046–1056.

37. Chowdhury, T. T., Bader, D. L., Shelton, J. C., and Lee, D. A. (2003) Temporal regulation of chondrocyte metabolism in agarose constructs subjected to dynamic compression, *Arch. Biochem. Biophys.,* **417**(1), 105–111.

38. Sheehy, E. J., Buckley, C. T., and Kelly, D. J. (2012) Oxygen tension regulates the osteogenic, chondrogenic and endochondral phenotype of bone marrow derived mesenchymal stem cells, *Biochem. Biophys. Res. Commun.,* **417**(1), 305–310.

39. Malda, J., van den Bring, P., Meeuwse, P., Grojec, M., Martens, D. E., Tramper, J., Riesle, J., and van Blitterswijk, C. A. (2004) Effect of oxygen tension on adult articular chondrocytes in microcarrier bioreactor culture, *Tissue Eng.,* **10**(7/8), 987–994.

40. Thampatty, B., and Wang, J. (2008) Mechanobiology of fibroblasts: mechanosensitive ion channels. In: Kamkin A, Kiseleva I, editors.: Springer Netherlands, 351–378.

41. Benhardt, H. A., and Cosgriff-Hernandez, E. M. (2009) The role of mechanical loading in ligament tissue engineering, *Tissue Eng. Part B,* **15**(4), 467–475.

42. Kuo, C., Marturano, J., and Tuan, R. (2010) Novel strategies in tendon and ligament tissue engineering: Advanced biomaterials and regeneration motifs, *Sports Med. Arthrosc. Rehabil. Ther. Technol.,* **2**(1), 1–14.

43. Wang, T., Gardner, B., Lin, Z., Rubenson, J., Wang, A., Kirk, T. B., Xu, J., Smith, D., Lloyd, D., and Zheng, M. H. Bioreactor design for tendon/ligament engineering, *Tissue Eng. Part B,* in press.

44. Saber, S., Zhang, A. Y., Ki, S. H., Lindsey, D. P., Smith, R. L., Riboh, J., Pham, H., and Chang, J. (2010) Flexor tendon tissue engineering: bioreactor cyclic strain increases construct strength, *Tissue Eng. Part A,* **16**(6), 2085–2090.

45. Woon, C. Y., Kraus, A., Raghavan, S. S., Pridgen, B. C., Megerle, K., Pham, H., and Chang, J. (2011) Three-dimensional-construct bioreactor conditioning in human tendon tissue engineering, *Tissue Eng. Part A,* **17**(19–20), 2561–2572.

46. Androjna, C., Spragg, R. K., and Derwin, K. A. (2007) Mechanical conditioning of cell-seeded small intestine submucosa: a potential tissue-engineering strategy for tendon repair, *Tissue Eng.,* **13**(2), 233–243.

47. Doroski, D. M., Levenston, M. E., and Temenoff, J. S. (2010) Cyclic tensile culture promotes fibroblastic differentiation of marrow stromal cells encapsulated in poly(ethylene glycol)-based hydrogels, *Tissue Eng. Part A,* **16**(11), 3457–3466.

48. Paxton, J. Z., Hagerty, P., Andrick, J. J., and Baar, K. Optimizing an intermittent stretch paradigm using ERK1/2 phosphorylation results in increased collagen synthesis in engineered ligaments, *Tissue Eng. Part A,* **18**(3–4), 277–284.

49. Sawaguchi, N., Majima, T., Funakoshi, T., Shimode, K., Harada, K., Minami, A., and Nishimura, S-I. (2010) Effect of cyclic three-dimensional strain on cell proliferation and collagen synthesis of fibroblast-seeded chitosan-hyaluronan hybrid polymer fiber, *J. Orthop. Sci.,* **15**(4), 569–577.

50. Kluge, J. A., Leisk, G. G., Cardwell, R. D., Fernandes, A. P., House, M., Ward, A., Dorfmann, A. L., and Kaplan, D. L. (2011) Bioreactor system

using noninvasive imaging and mechanical stretch for biomaterial screening, *Ann. Biomed. Eng.*, **39**(5), 1390–1402.

51. Radisic, M., Park, H., Shing, H., Consi, T., Schoen, F. J., Langer, R., Freed, L. E., and Vunjak-Novakovic, G. (2004) Functional assembly of engineered myocardium by electrical stimulation of cardiac myocytes cultured on scaffolds, *Proc. Natl. Acad. Sci. USA,* **101**(52), 18129–18134.
52. Lasher, R. A., Pahnke, A. Q., Johnson, J. M., Sachse, F. B., and Hitchcock, R. W. (2012) Electrical stimulation directs engineered cardiac tissue to an age-matched native phenotype, *J. Tissue Eng.*, **3**(1), 27.
53. Tandon, N., Marsano, A., Maidhof, R., Wan, L., Park, H., and Vunjak-Novakovic, G. (2011) Optimization of electrical stimulation parameters for cardiac tissue engineering, *J. Tissue Eng. Regen. Med.*, **5**(6), e115–e125.
54. Tandon, N., Marsano, A., Cannizzaro, C., Voldman, J., and Vunjak-Novakovic, G. (2008) Design of electrical stimulation bioreactors for cardiac tissue engineering, *Conf. Proc. IEEE Eng. Med. Biol. Soc.*, **7**(10), 4649983.
55. Zimmermann, W. H., Fink, C., Kralisch, D., Remmers, U., Weil, J., and Eschenhagen, T. (2000) Three-dimensional engineered heart tissue from neonatal rat cardiac myocytes, *Biotechnol. Bioeng.*, **68**(1), 106–114.
56. Zimmermann, W. H., Schneiderbanger, K., Schubert, P., Didie, M., Munzel, F., Heubach, J. F., Kostin, S., Neuhuber, W. L., and Eschenhagen, T. Tissue engineering of a differentiated cardiac muscle construct, *Circ. Res.*, **90**(2), 223–230.
57. Zimmermann, W. H., Melnychenko, I., Wasmeier, G., Didie, M., Naito, H., Nixdorff, U., Hess, A., Budinsky, L., Brune, K., Michaelis, B., Dhein, S., Schwoerer, A., Ehmke, H., and Eschenhagen, T. (2006) Engineered heart tissue grafts improve systolic and diastolic function in infarcted rat hearts, *Nature Med.*, **12**(4), 452–458.
58. Shimko, V. F., and Claycomb, W. C. (2008) Effect of mechanical loading on three-dimensional cultures of embryonic stem cell-derived cardiomyocytes, *Tissue Eng. Part A*, **14**(1), 49–58.
59. Kofidis, T., Lenz, A., Boublik, J., Akhyari, P., Wachsmann, B., Mueller-Stahl, K., Hofmann, M., and Haverich, A. (2003) Pulsatile perfusion and cardiomyocyte viability in a solid three-dimensional matrix, *Biomaterials*, **24**(27), 5009–5014.
60. Brown, M. A., Iyer, R. K., and Radisic, M. (2008) Pulsatile perfusion bioreactor for cardiac tissue engineering, *Biotechnol. Progress*, **24**(4), 907–920.

61. Barash, Y., Dvir, T., Tandeitnik, P., Ruvinov, E., Guterman, H., and Cohen, S. (2010) Electric field stimulation integrated into perfusion bioreactor for cardiac tissue engineering, *Tissue Eng. Part C,* **16**(6), 1417–1426.

62. Lu, L., Mende, M., Yang, X., Körber, H. F., Schnittler, H. J., Weinert, S., Heubach, J., Werner, C., and Ravens, U. (2013) Design and validation of a bioreactor for simulating the cardiac niche: a system incorporating cyclic stretch, electrical stimulation and constant perfusion, *Tissue Eng. Part A,* **19**(3–4), 403–414.

63. Nerem, R. M., Alexander, R. W., Chappell, D. C., Medford, R. M., Varner, S. E., and Taylor, W. R. (1998) The study of the influence of flow on vascular endothelial biology, *Am. J. Med. Sci.,* **316**(3), 169–175.

64. Brossollet, L. J. (1992) Mechanical issues in vascular grafting: a review, *Int. J. Artif. Organs,* **15**(10), 579–584.

65. Hahn, M. S., McHale, M. K., Wang, E., Schmedlen, R. H., and West, J. L. (2007) Physiologic pulsatile flow bioreactor conditioning of poly(ethylene glycol)-based tissue engineered vascular grafts, *Ann. Biomed. Eng.,* **35**(2), 190–200.

66. Stegemann, J. P., and Nerem, R. M. (2003) Phenotype modulation in vascular tissue engineering using biochemical and mechanical stimulation, *Ann. Biomed. Eng.,* **31**(4), 391–402.

67. Williams, C., and Wick, T. M. (2004) Perfusion bioreactor for small diameter tissue-engineered arteries, *Tissue Eng.,* **10**(5–6), 930–941.

68. Seliktar, D., Nerem, R. M., and Galis, Z. S. (2003) Mechanical strain-stimulated remodeling of tissue-engineered blood vessel constructs, *Tissue Eng.,* **9**(4), 657–666.

69. Iwasaki, K., Kojima, K., Kodama, S., Paz, A. C., Chambers, M., Umezu, M., and Vacanti, C. A. (2008) Bioengineered three-layered robust and elastic artery using hemodynamically-equivalent pulsatile bioreactor, *Circulation,* **118**(14 Suppl), S52–S57.

70. Jeong, S. I., Kim, S. Y., Cho, S. K., Chong, M. S., Kim, K. S., Kim, H., Lee, S. B., and Lee, Y. M. (2007) Tissue-engineered vascular grafts composed of marine collagen and PLGA fibers using pulsatile perfusion bioreactors, *Biomaterials,* **28**(6), 1115–1122.

71. Hoerstrup, S. P., Sodian, R., Sperling, J. S., Vacanti, J. P., and Mayer Jr, J. E. (2000) New pulsatile bioreactor for *in vitro* formation of tissue engineered heart valves, *Tissue Eng.,* **6**(1), 75–79.

72. Engelmayr, G. C., Hildebrand, D. K., Sutherland, F. W. H., Mayer, J. E., Sacks, M. S. (2003) A novel bioreactor for the dynamic flexural stimulation of tissue engineered heart valve biomaterials, *Biomaterials,* **24**(14), 2523–2532.

73. Lichtenberg, A., Tudorache, I., Cebotari, S., Ringes-Lichtenberg, S., Sturz, G., Hoeffler, K., Hurscheler, C., Brandes, G., Hilfiker, A., and Haverich, A. (2006) *In vitro* re-endothelialization of detergent decellularized heart valves under simulated physiological dynamic conditions, *Biomaterials,* **27**(23), 4221–4229.

74. Schenke-Layland, K., Opitz, F., Gross, M., Döring, C., Halbhuber, K. J., Schirrmeister, F., Wahlers, T., and Stock, U. A. (2003) Complete dynamic repopulation of decellularized heart valves by application of defined physical signals—an *in vitro* study, *Cardiovasc. Res.,* **60**(3), 497–509.

75. National Heart, Lung and Blood Insitute Fact Book, Fiscal Year 2011. Bethesda: National Institutes of Health; 2012.

76. Price, A. P., England, K. A., Matson, A. M., Blazar, B. R., and Panoskaltsis-Mortari, A. (2010) Development of a decellularized lung bioreactor system for bioengineering the lung: the matrix reloaded, *Tissue Eng. Part A,* **16**(8), 2581–2591.

77. Petersen, T. H., Calle, E. A., Zhao, L., Lee, E. J., Gui, L., Raredon, M. S. B., Gavrilov, K., Yi, T., Zhuang, Z. W., and Breuer, C. (2010) Tissue-engineered lungs for *in vivo* implantation, *Science,* **329**(5991), 538–541.

78. Sullivan, D. C., Mirmalek-Sani, S-H., Deegan, D. B., Baptista, P. M., Aboushwareb, T., Atala, A., and Yoo, J. J. (2012) Decellularization methods of porcine kidneys for whole organ engineering using a high-throughput system, *Biomaterials,* **33**(31), 7756–7764.

79. Ross, E. A., Williams, M. J., Hamazaki, T., Terada, N., Clapp, W. L., Adin, C., Ellison, G. W., Jorgensen, M., and Batich, C. D. (2009) Embryonic stem cells proliferate and differentiate when seeded into kidney scaffolds, *J. Am. Soc. Nephrol.,* **20**(11), 2338–2347.

80. Soto-Gutierrez, A., Zhang, L., Medberry, C., Fukumitsu, K., Faulk, D., Jiang, H., Reing, J., Gramignoli, R., Komori, J., and Ross, M. (2011) A whole-organ regenerative medicine approach for liver replacement, *Tissue Eng. Part C,* **17**(6), 677–686.

81. Barakat, O., Abbasi, S., Rodriguez, G., Rios, J., Wood, R. P., Ozaki, C., Holley, L. S., and Gauthier, P. K. (2012) Use of decellularized porcine liver for engineering humanized liver organ, *J. Surg. Res.,* **173**(1), e11–e25.

82. Baptista, P. M., Siddiqui, M. M., Lozier, G., Rodriguez, S. R., Atala, A., and Soker, S. (2011) The use of whole organ decellularization for the generation of a vascularized liver organoid, *Hepatology,* **53**(2), 604–617.

83. Smith, L. E., Bonesi, M., Smallwood, R., Matcher, S. J., and MacNeil, S. (2010) Using swept-source optical coherence tomography to monitor the formation of neo-epidermis in tissue-engineered skin, *J. Tissue Eng. Regen. Med.*, **4**(8), 652–658.

84. Kluge, J. A., Leisk, G. G., Cardwell, R. D., Fernandes, A. P., House, M., Ward, A., Dorfmann, A. L., and Kaplan, D. L. (2011) Bioreactor system using noninvasive imaging and mechanical stretch for biomaterial screening, *Ann. Biomed. Eng.*, **39**(5), 1390–1402.

85. Hofmann, M. C., Whited, B. M., Criswell, T., Rylander, M. N., Rylander, C. G., Soker, S., Wang, G., and Xu, Y. (2010) A fiber-optic-based imaging system for nondestructive assessment of cell-seeded tissue-engineered scaffolds, *Tissue Eng.Part C,* **18**(9), 677–687.

86. Logeart-Avramoglou, D., Oudina, K., Bourguignon, M., Delpierre, L., Nicola, M. A., Bensidhoum, M., Arnaud, E., and Petite, H. (2009) *In vitro* and *in vivo* bioluminescent quantification of viable stem cells in engineered constructs, *Tissue Eng. Part C,* **16**(3), 447–458.

87. De Napoli, I. E., Scaglione, S., Giannoni, P., Quarto, R., and Catapano, G. (2011) Mesenchymal stem cell culture in convection-enhanced hollow fibre membrane bioreactors for bone tissue engineering, *J. Membrane Sci.,* **379**(1), 341–352.

88. Hagenmüller, H., Hofmann, S., Kohler, T., Merkle, H. P., Kaplan, D. L., Vunjak-Novakovic, G., Müller, R., and Meinel, L. (2007) Non-invasive time-lapsed monitoring and quantification of engineered bone-like tissue, *Ann. Biomed. Eng.,* **35**(10), 1657–1667.

89. Porter, B. D., Lin, A. S. P., Peister, A., Hutmacher, D., and Guldberg, R. E. (2007) Noninvasive image analysis of 3D construct mineralization in a perfusion bioreactor, *Biomaterials,* **28**(15), 2525–2533.

90. Terrovitis, J. V., Bulte, J. W. M., Sarvananthan, S., Crowe, L. A., Sarathchandra, P., Batten, P., Sachlos, E., Chester, A. H., Czernuszka, J. T., and Firmin, D. N. (2006) Magnetic resonance imaging of ferumoxide-labeled mesenchymal stem cells seeded on collagen scaffolds-relevance to tissue engineering, *Tissue Eng.,* **12**(10), 2765–2775.

91. Ramaswamy, S., Greco, J. B., Uluer, M. C., Zhang, Z., Fishbein, K. W., and Spencer, R. G. (2009) Magnetic resonance imaging of chondrocytes labeled with superparamagnetic iron oxide nanoparticles in tissue-engineered cartilage, *Tissue Eng. Part A,* **15**(12), 3899–3910.

# Chapter 16

# Microvascular Imaging Methods for Tissue Engineering

**Benjamin W. Thimm, Sandra Hofmann, and Ralph Müller**
*Institute for Biomechanics, ETH Zürich, Zürich, Switzerland*
ram@ethz.ch

The formation of new blood vessels plays an important role not only in the progression of prevailing malignancies such as cancer or ischemia but also in physiological development, wound healing, and tissue engineering. Angiogenesis constitutes the formation of new blood capillaries from pre-existing blood vessels to provide tissues with blood, oxygen, and nutrients. It is a complex multistep process kept in a fine balance by pro-angiogenetic and anti-angiogenetic molecules. Due to the micro-scale of angiogenesis (capillary diameter of up to 10 μm), microvascular imaging is increasingly important in depicting and understanding vessel illnesses, vessel evolution, and cure. Several different vascular imaging technologies were developed over the course of the last century mostly for fast clinical cardiovascular examinations on the macrovascular level to the artery level. More recently, traditional technologies such as

*Scaffolds for Tissue Engineering: Biological Design, Materials, and Fabrication*
Edited by Claudio Migliaresi and Antonella Motta

ISBN 978-981-4463-20-1 (Hardcover), 978-981-4463-21-8 (eBook)
www.panstanford.com

computed tomography (CT), magnetic resonance imaging (MRI), positron emission tomography (PET), as well as newly developed imaging methods using quantum dots, fluorescently labeled proteins, or microbubbles have been highly refined in terms of resolution and practicability and could expand our view on the microvascular level *in vitro* and/or *in vivo*. Insight into the micrometric world of capillary structures has turned out to be very important for understanding the apithology and pathology of angiogenesis in more detail. This review focuses exclusively on microvascular imaging options in tissue engineering and summarizes the up-to-date *in vitro* and *in vivo* imaging methods on the microvascular level, with an emphasis on the options in two- and three-dimensional (2D/3D) imaging and the limitations of each technology in regard to spatial resolution and anatomic depiction.

## 16.1 Introduction

Angiogenesis, the development of new blood capillaries from existing blood vessels resulting in vascularization of a tissue, is a vital component of many normal physiological processes. The smallest vessel in the vascular system, the microvascular capillary, is located at the direct site of vessel-to-tissue oxygen exchange and exhibits diameters smaller than 10 μm (Fig. 16.1). Larger blood vessel types such as arterioles/venules (50 μm), arteries/veins (2 mm), and the largest human vessel, the aorta (3 cm), demonstrate the macrovascular system (Table 16.1). Both arteries and veins are composed of three distinct layers: the innermost, endothelium-covered tunica intima, the tunica media composed of smooth muscle cells, and the outermost, fibrous tunica adventitia.[1] With over 600 $m^2$ of endothelial surface area, capillaries provide an enormous surface for material and gas exchange[2] (Fig. 16.1; Table 16.1).

Endothelial cells form the main building unit of the capillary network, with pericytes and smooth muscle cells wrapped around arterioles in the precapillary and venules in the postcapillary network. Angiogenesis consists of a series of endothelial cell responses to angiogenic stimulation, including degradation of extracellular matrix, budding, proliferation, migration, tube formation, maturation, and maintenance of quiescent endothelium.[3] Under normal conditions, angiogenesis accounts for oxygenation of ischemic or newly formed tissues. However, angiogenesis is also an important factor in pathological conditions such as cancer, hypertension, rheumatoid arthritis, and diabetes, to name only a few.[4,5]

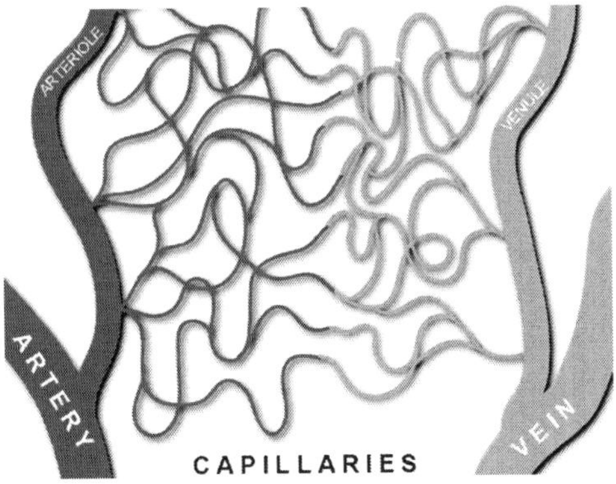


**Figure 16.1** Vessel network showing macrovascular (artery, arteriole, venule, and vein) and microvascular (capillaries) structures. The blood in the artery, arteriole, and their capillaries is oxygen-rich whereas that in the vein, venule, and their capillaries is oxygen-deficient. Graph not to scale.

**Table 16.1** Indication of human vessel sizes

| Vessel type | Average diameter | Amount in body | Length in body | Total length in body |
|---|---|---|---|---|
| Aorta | 10,000 μm | 1 | 400 mm | 40 cm |
| Artery | 3,000 μm | 40 | 200 mm | 800 cm |
| Artery arm | 1000 μm | 600 | 100 mm | 6,000 cm |
| Artery branch | 600 μm | 1,800 | 10 mm | 1,800 cm |
| Arteriole | 20 μm | 40,000,000 | 2 mm | 8,000,000 cm |
| Capillary | 8 μm | 1,200,000,000 | 1 mm | 120,000,000 cm |
| Venule | 30 μm | 80,000,000 | 2 mm | 16,000,000 cm |
| Vein branch | 1,500 μm | 1,800 | 10 mm | 1,800 cm |
| Vein arm | 2,400 μm | 600 | 100 mm | 6,000 cm |
| Vein | 6,000 μm | 40 | 200 mm | 800 cm |
| Cava | 1,500 μm | 1 | 400 mm | 40 cm |

*Source:* Rainer Flindt (2003) Biologie in Zahlen, 6th ed., Spektrum Akademischer Verlag, Germany; Konrad Kunsch and Steffen Kunsch (2007). Der Mensch in Zahlen, 2nd ed., Spektrum Akademischer Verlag, Germany.

Tissue engineering is an increasingly important part of modern biomedical therapy, which tries to replace damaged, injured, or missing body tissues or organs by growing donor tissue specific cells on 3D scaffolds for subsequent implantation. Here, vascularization is indispensable in supplying cells with oxygen and nutrients, transporting signaling proteins, and removing waste products. The maximum oxygen diffusion within the extravascular space is set at 100 μm.[6] The ability to create a functional microvascular system alongside tissue growth *in vitro* or *in vivo* to support the cells of the newly generated tissue with oxygen has been the limiting factor in producing tissues of clinically relevant sizes so far. However, engineered microvasculature is difficult to achieve and was unaccomplished until recent years for bone scaffolds *in vitro*.[7] Microvascular imaging presents an advantageous means of following desired scaffold vascularization *in vitro* and *in vivo*. The imaging of angiogenesis to understand/to interact with vascular development would render considerable progress in tissue engineering.

## 16.2 Imaging Methods for Tissue Engineering

The following paragraph summarizes all present literature reviews about imaging methods for tissue engineering in general. Two-dimensional and 3D imaging in tissue engineering play a vital role in characterizing and monitoring the dynamic interactions, as engineered tissues develop *in vitro*, *in situ*, or *in vivo*. This is necessary to understand the interdependence of processes that ultimately lead to tissue products.[8] Today, conventional histology combined with immunohistochemistry still presents the gold standard by which most tissue samples are evaluated.[9] Histology, however, is of destructive nature and requires the sacrifice of valuable 3D tissue replicas especially in monitoring studies, thus limiting the applicability of samples for *in vivo* use. Therefore, the need for sensitive, noninvasive, and nondestructive imaging approaches for assessing the quality of the engineered tissues and organs prior to surgical implantation is crucial in 2D and 3D tissue engineering.[10]

There has been substantial progress in tissue engineering in recent decades, including constant development of sophisticated imaging tools allowing qualitative and quantitative characterization of the biological integration of tissue engineered constructs *in vivo.*[11] *In vivo* imaging, in particular, may help to nondestructively monitor the dynamic and complex interactions of host tissue with newly implanted scaffold and its integration. Diverse imaging tools that are already used in prediction, production, and performance of tissue-engineered products can assist in creating functional tissues, as reviewed in detail by Pancrazio *et al.*[11] According to Pancrazio *et al.*, firstly, quantitative prediction by multi-scale computational modeling approaches exhibits an important role in the evolution of tissue engineering. The complexity in 3D tissue architecture, incorporating the cellular and molecular diversity, can be approached by computer models to enable rational design and implementation of engineered tissues.[12] Recent developments in computer-aided tissue engineering, that is, computer-aided design (CAD), computer-aided manufacturing (CAM), computer-aided anatomical modeling, 3D anatomic visualization and reconstruction (e.g., by CT or MRI), or fabrication techniques such as rapid prototyping (RP), injection molding, lithography, or sintering are explained and reviewed in detail by Sun *et al.* and Ballyns *et al.*[13–15] Secondly, production in tissue engineering reverts to a variety of fabrication technologies and bioreactors for which imaging plays a major part in examining extracellular matrix composition, inclusion of cells, scaffold architecture, porosity, surface chemistry, degradation dynamics, scaffold strength, shape, and compliance.[11] And thirdly, imaging plays another major role in evaluating performance of tissue-engineered products by repeated, noninvasive, fast, and accurate functional assessments of cultivated tissues at the cellular level.[11]

Optical imaging presents one of the most used methods in tissue engineering due to its noninvasive, cheap, and rather easy-to-use nature such as used in normal bright field microscopy or phase contrast microscopy.[16] Although high resolutions can be achieved with optical imaging, it is generally restricted to tissue surfaces due to the light scattering properties of typical biological tissues. To bypass this problem, near-infrared is preferably used in optical imaging due to the low tissue absorption of light at this wavelength.[11] Next to common optical light microscopy, other optical imaging options such as fluorescence microscopy, confocal laser scanning microscopy (CLSM), including multi-photon laser microscopy and second harmonic generation microscopy (SHG) constitute advancements in optical light microscopy, which result in high-resolution and depth-resolved images. These technologies represent the techniques of choice to image specific, fluorescently labeled cells, cell structures, or the orientation of extracellular matrix fibers allowing imaging of tissue structures within optically thick and opaque tissues *in vitro*, *in situ*, and *in vivo*.[8,17–23]

An advanced development in optical imaging presents optical coherence tomography (OCT). OCT allows real-time, nondestructive, and noninvasive monitoring of structural and functional properties of engineered tissue constructs during their development and growth *in situ* and *in vivo*, with a depth of tissue penetration of 2–3 mm. A focused beam of near-infrared light is applied to measure back-scattered/back-reflected light generating a 2D cross-sectional plane through the tissue.[8,10,11,16,24–26] Combined with Doppler velocimetry, OCT can follow fluid flow at the micron level,[27] which allows real-time imaging of rheological properties of tissue-engineered constructs.[11,28]

Optical coherence elastography (OCE) can further be used to optically image biomechanical tissue properties such as stiffness, softness, mechanical displacements, and strains of engineered tissues.[11,29,30]

Other optical imaging methods exist such as Fourier transform infrared (FTIR) spectroscopy and electron paramagnetic resonance, which may be used to analyze the performance of tissue-engineered constructs *in vitro*. FTIR, for instance, is able to characterize biochemical properties of tissues and analyze changes in molecular scaffold composition.[11,24] FTIR spectroscopy allows to use thick, unstained histological sections to obtain information on content and distribution of native and engineered molecular components.[31]

Infrared absorption, Raman scattering, Raman microspectroscopy, and surface plasmon resonance present further imaging modalities to yield spatially and temporally resolved chemical and physical information of tissue-engineered constructs; however, their usefulness for imaging in tissue engineering in general is rather small or not even obvious.[11]

Bioluminescent imaging (BLI) describes another noninvasive means of imaging in tissue engineering that allows capturing emitted light from grafted, luciferase transgenic cells. By putting luciferase expression under the control of tissue-specific regulatory elements, BLI can be used in noninvasively image processes such as cell tracking, cell differentiation, hypoxia, apoptosis, inflammation, signal transduction, and vasculogenesis. The procedure, however, requires to transgenically manipulate donor cells to image cells and cellular events.[32]

Scanning electron microscopy (SEM) and transmission electron microscopy (TEM) are increasingly being used in tissue engineering and in the evaluation of biomaterials to image scaffold topological properties, cell attachment, and cell morphology on tissue-engineered constructs providing high resolving power of ultrastructural feature in the nano domain.[16] Nevertheless and similar to histology, SEM and TEM present terminal technologies, as cells and tissues have to be fixed, processed, and gold sputtered for image retrieval.

Magnetic resonance imaging (MRI) has been assigned a growing role in the assessment of tissue-engineered constructs. MRI can be applied to monitor tissue structures, composition, and function to improve whole processes in tissue engineering *in vitro* and *in vivo*. It can assist in observing integration of an engineered tissue implant with the surrounding tissues and to check for early signs of immune rejection. Paramagnetic and superparamagnetic contrast agents help to quantify cell death, assess inflammation, and visualize cell trafficking and gene expression.[33] Other application fields of MRI present density imaging to examine electric fields in tissues, elastography to measure physical properties of tissues as well as spectroscopy for functional assessment of 3D tissues in bioreactors.[11] Although MRI will typically lack resolutions lower than 100 μm, the multitude of image contrasts that can be exploited by MRI and spectroscopy add valuable information in assessing tissue function *in vitro* and *in vivo*.[11]

Positron emission tomography (PET) and hybrid magnetic resonance–PET (MR-PET) present further imaging tools for tissue engineering, which may generate valuable information on inflammation, apoptosis, and gene expression in small animal imaging; however, the usage of exogenous molecular tags somehow limits their use for engineered products for clinical use in humans.[11]

Computed tomography (CT) and especially micro-computed tomography (microCT, or μCT) evolved in the last two decades into a key player for high-resolution, 3D, qualitative, and quantitative characterization of bone development, angiogenesis, and biomaterial morphology in tissue engineering by noninvasive and nondestructive means.[34,35] It has been shown that μCT can not only be used to localize ingrowth of bone-like tissue into biomaterials *in vitro* and *in vivo* but also to monitor scaffold degradation and to assess bone-to-scaffold interactions.[36,37] Upon application of a contrast agent into the vascular system, μCT imaging may be further used to study the vascular response to an implanted, engineered piece of tissue or biomaterial, thus characterizing angiogenesis and neovascularization in tissue engineering.[35]

The multitude of imaging modalities and their advantages to characterize engineered tissues *in vitro/in situ/in vivo* shows that imaging describes an indispensable means contributing significantly to the success in tissue engineering. Imaging in fact belongs to the most important elements in tissue engineering, as it adds to accelerated development of engineered tissues in research and production. Nondestructive, noninvasive, real-time, fast, accurate, low-cost, and automated imaging systems would present ideal attributes to survey cellular events in tissue engineering processes, which would be important for implant monitoring and quality control of tissue-engineered constructs.

## 16.3 Angiogenesis in Tissue-Engineered Constructs

Vascularization of engineered tissues and tissue substitution scaffolds is highly desired to keep newly grown tissue alive by nutrient and oxygen supply, for waste product removal, and to provide a livable environment for cells eventually recruited from the patient's own body. The artificial growth of microcapillary-like structures within or into tissue-engineered constructs from an early time point is indispensable in guaranteeing the survival of cells and a successful regeneration. To monitor the *status quo* in tissue/scaffold vascularization, microvascular imaging options are desirable to follow capillary formation qualitatively and quantitatively over time.

To date, *in vitro* vascularization of tissue-engineered body parts has not yet achieved the desired stage. Although many scientists have addressed this issue, cell survival within thick tissues is limited by insufficient oxygen diffusion, and thus, only thin tissue layers can be vascularized by endothelial cell application and reorganization *in vitro*. Therefore, vessel formation *in vitro* or accelerated vessel formation *in vivo* on biomaterials is currently being pursued. Cell growth enhancing bioreactors, endothelial cell-seeded constructs, or scaffolds with large pore sizes for eased ingrowth of cells or modified with growth factors have further been used to increase the vascularization success of scaffolds or tissues. The application of noninvasive and nondestructive imaging methods on the microscopic level would be helpful to follow endothelial cell spread and/or to monitor functional capillary formation. Both *in vitro* and *in vivo* monitoring would largely increase the value of tissue-engineered constructs for more successful implantations in future. Functional *in vitro* vessel formation by cell reorganization is regarded as particularly difficult to accomplish. For *in vivo* use, cell-seeded (and vascularized) or noncell-seeded scaffolds may be implanted into the defect site. Subsequently, peri-implant capillaries ideally connect the pre-existing vascular network within the construct or sprout into the plain scaffold. In both, *in vitro* and *in vivo*, a fully nondestructive and even noninvasive imaging approach would be useful to monitor vessel development and as a quality control tool.

## 16.4 Imaging of Angiogenesis

Because vascular imaging of scaffolds *in vitro*/*in vivo* has not been greatly pursued so far, most literature originates from cancer-derived research where vascularization is a central element of the disease. Cancer starts as a microscopic neoplasia turning into expansive tumor growth with intrinsic vessel development to provide intracancerous oxygen supply for cancer cell survival, ending in final metastatic spread. Thus, the interference with angiogenesis presents an important target for cancer therapy that has led to broad research aimed at understanding the mechanisms underlying the evolution of capillary networks. Despite their success, conventional 2D microscopy methods to visualize vascular structures, such as cytology and histopathology, suffer from significant limitations due to their need for biopsies, chemical tissue fixation, and observation of biological samples under nonphysiological conditions. In general, these methods cannot resolve the dynamics of cellular processes and, most importantly, show difficulties in generating quantitative data.[38] A few 2D histological stacks have been traditionally used to compute 3D vascular models of whole samples.

Concurrently, classical vascular imaging technologies such as MRI, PET, CT, and ultrasonography (US) have been further developed on the microscopic level. They are being incorporated as integral components of biomedical cancer research to observe cancer progression and to classify therapeutical progress due to their ability to increasingly monitor *in vivo* angiogenesis noninvasively.[5] Combining molecular cell biology with the latest high-tech noninvasive imaging methods, called 'molecular imaging', adds to the classical cross-sectional imaging techniques such as MRI, PET, CT, and US. Molecular imaging uses specific molecules as a contrast source compared with traditional imaging techniques based on physical tissue properties such as absorption and proton density. These marker molecules and imaging techniques have new potential for unraveling details such as the architecture of a microvascular network and their high sensitivity permits imaging at a higher resolution than before.[39] Apart from 3D anatomical information, another advantage of molecular imaging is its unique sensitivity for detecting functional information such as vessel permeability or vasoreactivity implemented by novel contrast agents. By using a contrast agent with a high affinity to special endothelial cell surface targets, molecular imaging facilitates the early detection of an angiogenetic disease before phenotypic changes become apparent.[40]

Both classical and modern biomedical imaging methods such as optical imaging (e.g., bioluminescence or fluorescence), radionuclide imaging [e.g., PET, single photon emission CT (SPECT)], CT, MRI, or US, all have their advantages and disadvantages especially on the spatial resolution and sensitivity level.[39,41] Due to enhancements over decades in the optical sector, resolution of a few millimeters down to the micrometer/nanometer range has improved. Imaging of macro-sized networks (i.e., arteries or veins up to large vessels such as the aorta) has been optimized and is performed on a daily basis, for instance in medical, cardiovascular diagnosis. However, vascular imaging of arterioles, the capillary bed, and venules *in vitro/in vivo* still needs to be refined to evaluate capillary-sized vessels in detail with the best possible resolution.

## 16.5 Major Problems of Vascular Imaging Modalities

Anatomically, vessel imaging aims to show details down to the smallest capillaries possible. Ideally, high image resolution of microvascular structures is necessary to depict not only blood-perfused capillaries but also to capture the event of nonperfused vessel sprouts during angiogenesis, known as "functional imaging." The general goals in imaging are (i) to obtain a maximum amount of signal, (ii) to localize the signal as accurately as possible, and (iii) to obtain high temporal and spatial resolution achieved with a minimum amount of contrast agent or molecular probe.[42]

In this regard, clinical PET and SPECT scanners offer favorable sensitivity for *in vivo* tomographic imaging with relatively precise localization of radiotracers. However, PET and SPECT technology suffer from a fundamental limitation of spatial resolution. As these approaches also do not provide true anatomical information, hybrid systems have been introduced including SPECT-CT and PET-CT. In this respect, CT is used to match a tracer signal with anatomical points of reference and to create attenuation maps to correct for nonuniform attenuation.[41]

CT and ultrasound render primarily anatomical conformation at limited spatial resolution, with CT also being limited in soft tissue resolution, making it necessary to administer contrast agents to delineate vascular structures.[42]

Clinical MRI offers superior penetration depth and good spatial resolution with the ability to delineate soft tissue without the need for contrast agents or ionizing radiation; however, it lacks MR-compatible molecular probes and shows relatively low sensitivity to the contrast agent with long scan times.[41] Molecular imaging is also limited due to poor sensitivity for low-level molecular activity of the relatively small proportion of endothelial cells within the area of interest.[5]

Optical imaging provides generally nonanatomical information. It can be divided into chemi-luminescence (bioluminescence) and photo-luminescence (fluorescence), which deliver planar images of subcutaneous vascular structures. Optical imaging shows disadvantages in very low penetration depth, as light scattering and absorption increase as a function of tissue depth.[41] Bioluminescence provides only low spatial resolution to properly image microvasculature *in vivo* so far. However, fluorescence imaging can be used with confocal light microscopy to yield micrometer spatial resolution.[2]

Each method has its strengths and weaknesses such as temporal and spatial resolution, penetration depth, sensitivity, and feasibility of quantitative measurements (Table 16.2). Although none of the modalities fulfills all the requirements for imaging microvascular structures, efforts have been made to optimize these imaging modalities, which will be summarized to the best of our knowledge in the following chapters (Table 16.2).

### 16.5.1 The Study of Microvasculature by Optical Light Microscopy

Tissue visualization with light is probably the most common imaging method in medical research.[45] Microscopic optical imaging methods are particularly useful for elucidating vessel structures. Optical light microscopic (OLM) devices use visible light and a set of glass lenses to magnify an object of interest. Optical microscopes have for the most part turned digitally using cooled coupled device (CCD) cameras for sample imaging, followed by analysis with computer software to process pictures. OLM methods are powerful tools for dissecting the cellular and molecular features of microvasculature. They offer high resolution (100 nm) and real-time functional readouts in animal models, such as in epi/transillumination and orthogonal polarization spectral (OPS) imaging.[46] However, a major limitation of optical microscopy techniques, which use linear (single-photon excitation) absorption processes for contrast generation, such as in fluorescence microscopy or CLSM, is the limited tissue penetration of light (100–300 μm) and light absorption by tissues.[47] Multi-photon laser scanning microscopy (MPLSM) demonstrates an advanced mode of CLSM resulting in better image quality and deeper light penetration (1 mm) into tissues. Similar deep penetration depths are also reached with near-infrared fluorescence (NIRF) microscopy exploiting the nonabsorbing properties of tissues with infrared light. Other means of labeling and imaging, such as quantum dots, fluorescent or bioluminescent proteins, have been invented to label specific cells of interest. Apart from 2D imaging of histological samples, a fundamental issue in optical light microscopy imaging is how to image 3D tissues for vascularization, which is generally

**Table 16.2** Characteristics of imaging modalities

| Modality | Electromagnetic radiation spectrum | Penetration depth | Spatial resolution | Advantages | Disadvantages |
|---|---|---|---|---|---|
| **Fluorescence** | Visible light or near-infrared | <1 cm | 2–3 mm | High sensitivity, detects fluorochromes in live and dead cells | Lack of anatomical information, low spatial resolution |
| **Bioluminescence** | Visible light or near-infrared | 1–2 cm | 3–5 mm | Highest sensitivity, quick, easy, low-cost, high-throughput | Very low penetration depth, low spatial resolution |
| **CT** | X-rays | No limit | 50–200 μm | Anatomical imaging | Limited soft tissue resolution, radiation, limited molecular applications |
| **μCT/SRμCT** | X-rays | No limit | 0.7–38 μm[43] | Anatomical imaging | Limited soft tissue resolution, radiation, limited molecular applications |
| **PET** | High-energy γ rays | No limit | 3–8 mm | High sensitivity | Low spatial resolution, radiation, PET cyclotron needed |
| **Small animal PET** | High-energy γ rays | No limit | 1 mm | High sensitivity | Low spatial resolution, radiation, PET cyclotron needed |

(*Continued*)

**Table 16.2** (*Continued*)

| Modality | Electromagnetic radiation spectrum | Penetration depth | Spatial resolution | Advantages | Disadvantages |
|---|---|---|---|---|---|
| **SPECT** | Lower-energy $\gamma$ rays | No limit | 1–2 mm | Many molecular probes available, can image multiple probes simultaneously | Low spatial resolution |
| **MRI** | Radiowaves | No limit | 25–100 μm | Highest spatial resolution, combines morphological and functional imaging | Low sensitivity |
| **μMRI** | Radiowaves | No limit | 4 μm[44] | Highest spatial resolution, combines morphological and functional imaging | Low sensitivity |
| **Ultrasound** | High-frequency sound | mm–cm | 50–500 μm | Real-time imaging, low cost | Limited spatial resolution, mostly morphological |

*Source:* Modified from Tarik F. Massoud and Sanjiv S. Gambhir. Molecular imaging in living subjects: seeing fundamental biological processes in a new light. *Genes & Development*, **17**, 545–580, 2003.

hampered again by light penetration depth and light scattering within tissues.[42]

#### 16.5.1.1 Epi/transillumination and orthogonal polarization spectral imaging

One of the simplest ways to image vascular networks *in vivo* is by 2D epi-illumination, implemented by superficial tissue illumination and microscopy of regions such as skin, lip, gingival tissue, or tongue.[48] Another way for *in vivo* imaging is transillumination of relatively thin tissues in animal models such as in the chronic dorsal window chamber, reviewed by Dewhirst *et al.*[49] In the dorsal window chamber, a transparent chamber is surgically introduced into the thin dorsal skin of a test animal to allow direct visualization of microcirculation by transillumination or fluorescence angiography.[45,50–52] A hole of approximately 10 mm in diameter is dissected into the epithelium of the window chamber, leaving a fascial plane with associated vasculature. For oncological studies, tumor cells or tumor tissues can be implanted onto the fascial plane, and the chamber sealed with glass cover slips. Tumor-induced vascularization may subsequently be monitored and quantified. For tissue-engineered vascularization studies, the same chamber model can be used to study the vascularization rate of subcutaneously implanted biomaterials (Fig. 16.2).[53–55]

In addition, other window models have been developed, for example, the mammary tumor window or the cranial tumor window model for the same monitoring purposes of angiogenesis.[56–58]

Due to the need for transillumination or injected contrast dyes for vessel contrast enhancement in tissues, microvascular imaging of human microcirculation has been limited to superficial surfaces such as nailfold or bulbar conjunctiva. Intravital imaging by transmission microscopy *in vivo* is unaccomplished for unapproachable sites or solid organs *per se*. To tackle this problem, OPS imaging was developed. Langer *et al.* yield clear images of microcirculation by light reflectance from solid organ surfaces using OPS imaging (Fig. 16.3).[59]

Here, image contrast is achieved without the use of dyes. OPS imaging relies on the absorbance of hemoglobin in red blood cells to create vessel contrast. Comparable to imaging by transilluminated intravital microscopy, OPS can visualize blood vessels of the peripheral organ microcirculation by applying a wavelength within the hemoglobin absorption spectrum (usually 548 nm). The resolving power of OPS is sufficient to resolve single red blood cells and thus single (functional) capillaries of 5 μm diameter.[60] This technique allows the real-time assessment of *in vivo* microvascular networks situated up to 3 mm depth at the site of interest.[61] OPS allows quantitative determination of parameters such as capillary density, microvessel morphology, and vascular dynamics.[60] *In vivo* assessment of microcirculation of solid organs by OPS has been reviewed in detail by Lupi *et al.*[61]

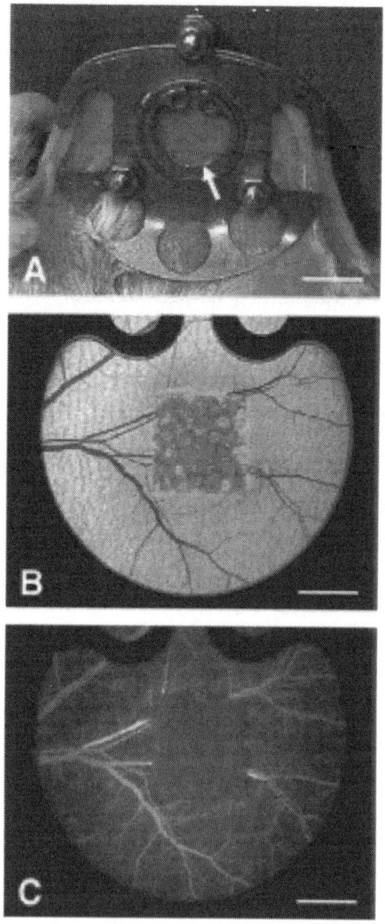


**Figure 16.2** Epi and transillumination of a mouse skinfold showing microcirculation. Reprinted from Ref. 55, Copyright (2009), with permission from Elsevier.

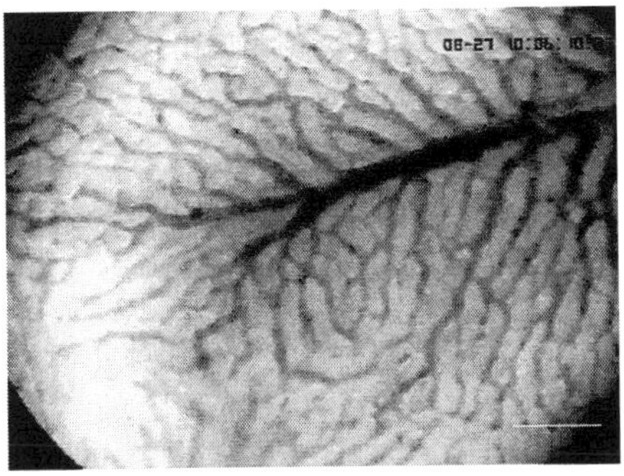

**Figure 16.3** Orthogonal polarization spectral (OPS) imaging of microcirculation by light reflectance from solid organ surfaces. Reprinted from Ref. 59, with permission of S. Karger AG, Basel.

### 16.5.1.2 Confocal laser scanning microscopy

Confocal laser scanning microscopy (CLSM) consists of a high spatial resolution fluorescence microscope that uses scanning laser light for excitation of fluorescent dyes, which can provide 3D spatial resolution pictures of 100 nm and allows reflected light imaging of the microvasculature to a tissue penetration depth of around 100 μm.[22,46] CLSM is mainly used for 2D histological tissue sections or fixed and fluorescently stained cell cultures. Fluorescent dyes, also known as fluorochromes, were developed in the past decade for fluorescent imaging and are omnipresent nowadays in the field of fluorescence microscopy, CLSM, MPLSM, and NIRF microscopy.

Fluorescence microscopy of tissues, cells, or subcellular structures is accomplished by labeling a targeting antibody with a fluorochrome and allowing the antibody to find its affinity antigen within the sample. Fluorochromes absorb energy of a specific excitation wavelength and successionally emit photons of a longer wavelength.[62,63] Various fluorochromes are available for targeting anatomical structures (e.g., cell surfaces, intracellular structures, or extracellular matrix components). Due to emission overlap, a limited number of different fluorescent reporters can be used simultaneously. However, the emission spectra/spectral overlap can be mathematically modeled to separate the contribution of each individual probe.[64] This makes optical light microscopy imaging a low-cost and simple technology, as optimized staining protocols are available and sample examination and validation can be implemented by high throughput screening in a fast mode.[5] Labeling different antigens with different fluorescently labeled antibodies additionally allows quantification of tissues, cells, and subcellular structures, reviewed by Wolf *et al.*[65]

In this respect, blood vessel detection is achieved by immunohistochemical staining of endothelial cell markers (most commonly the extracellular platelet endothelial adhesion molecule CD31 or the intracellular von Willebrand factor). However, the application of organic fluorochromes in traditional one-photon confocal microscopy poses several challenges for the visualization of vascular tissue, including autofluorescence of unlabeled tissue, fluorochrome emission crosstalk, and fluorochrome photobleaching by laser light.[46,66-68] CLSM vessel imaging by fluorescent capillary staining in 3D of fixed endothelial cells in coculture with osteoblastic cells was shown by Unger *et al. in vitro*[7,69] (Fig. 16.4).

*In vivo*, CLSM can be used, for example, in a vascular-specific reporter gene transgenic mouse model to track capillary networks. This model uses the natural occurring green fluorescent protein (GFP) from jellyfish *Aequorea victoria* that is a unique *in vivo* reporter to localize proteins or whole cells to which the fusion-tagged GFP is targeted.[70] GFP excitation with blue light is the only step needed for visualization and with an approximately 1000 times stronger signal intensity than bioluminescence, described later in this review, GFP is a potent candidate for *in vivo* imaging.[71] It can be used for living and fixed cells and no substrate is needed for its visualization.[42] GFP is very stable and comes in a multitude of colored variants such as red or yellow fluorescent protein allowing for multiple stainings.[72] For the transgenic mouse model, Roberts and Lin have used the endothelial reporter gene Tie2 to drive GFP expression in GFP-transgenic mice.[45] Tie2 belongs to the Tie family of receptor tyrosine kinases, which are constantly expressed predominantly on endothelial cells and are essential for vessel formation, control of microvessel sprouting, integrity, and maturation.[73] GFP under the control of the Tie2 promoter/enhancer element (Tie2-GFP) yields valuable vascular imaging in living animals.[74] Tie2-GFP modified mice were used to show bulbar conjunctiva capillaries and microcirculatory networks in skeletal muscle and ear dermis.[45] Others, such as Amoh *et al.*, have used nestin regulatory element driven GFP (ND-GFP) transgenic mice instead to visualize angiogenesis in skin and metastases.[75-79] Here, ND-GFP was highly expressed in proliferat-

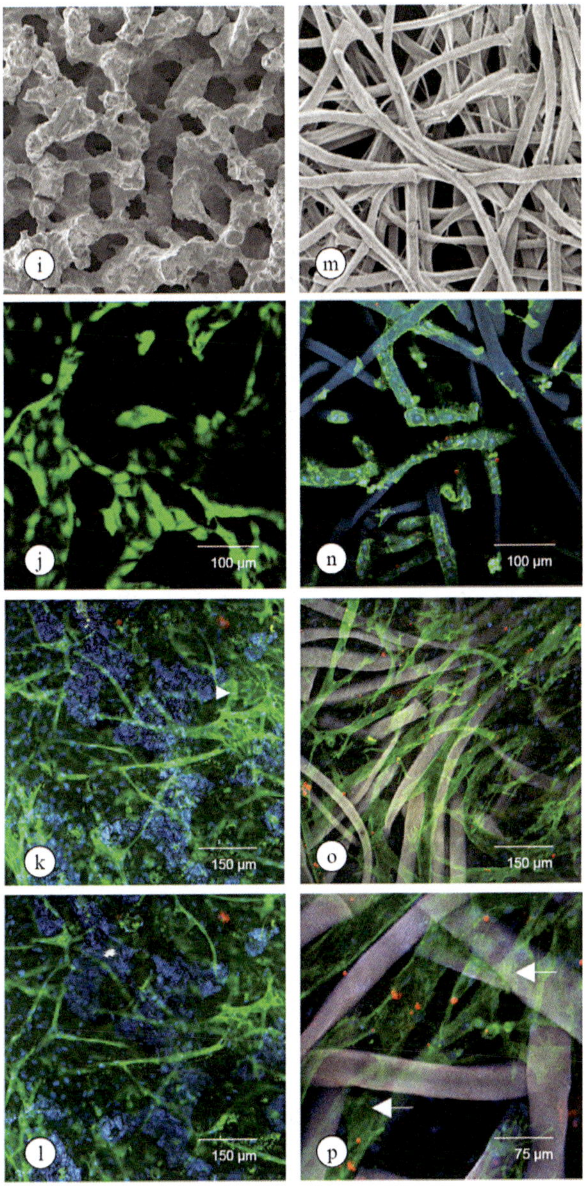


**Figure 16.4** Confocal laser scanning microscopy (CLSM) imaging of stained human dermal microvascular endothelial cells in coculture with primary human osteoblasts resulting in microcapillary-like structures on porous biomaterials. Reprinted from Ref. 7, Copyright (2007), with permission from Elsevier.

ing endothelial cells. In another experiment, the authors used dual-color fluorescence imaging to show tumor angiogenesis in ND-GFP transgenic mice. After orthotopic transplantation of a fibrosarcoma cancer expressing red fluorescent protein (RFP), nascent blood vessels in the tumor, expressing ND-GFP could be readily visualized and quantified.[76,77,79,80]

Cassot *et al.* who used an *in vitro* Indian ink-injected human brain for a new 3D computer-assisted microvascular cerebral network analysis showed another way of CLSM use for microvascular imaging on a detailed anatomical and topological resolution of a voxel size of 1.22 μm.[81,82] By developing a new computer-assisted skeletonization algorithm adapted to a very large dataset, Cassot *et al.* were able to generate a complete and automatic reconstruction of the vascular network to show the fractal nature of the microvascular network within a large area of the cortex (20 $mm^2$ × 300 μm). Contrary to former 2D image evaluation, this approach allows quantification of vessel properties, such as vessel volume, vessel diameter, and branching.

### 16.5.1.3 Multi-photon laser scanning microscopy

A technical advancement of CLSM is MPLSM, described in more detail by Piston or Williams *et al.*[22,83,84] Here, not one photon (as used for CLSM) but two or more photons that arrive simultaneously at the focal plane are needed to promote fluorochrome excitation. In contrast to CLSM, in which single photon absorption occurs within the entire excitation laser light cone, MPLSM causes minimal photodamage to the fluorescent marker due to the lack of out-of-focus excitation. This yields a prolonged fluorochrome lifetime, maintains resolution and contrast even in light scattering tissues, and allows an imaging penetration depth of up to 1 mm, which is 10 times better than traditional optical microscopy techniques.[46,47,64,84,85] MPLSM is particularly suitable for intravital microscopy to image fluorescent cells or other optical intravascular tracers [e.g., fluorescein isothiocyanate-labeled dextran (FITC-dextran) conjugates[86,87]], because tissue-penetrable, low scattering, near-infrared light sources (700–1000 nm) are used. Typical fields of application are the chronic-transparent skin window for anatomic cancer vessel imaging, *in situ* micro-lymphangiography, or *ex vivo* thick tissue preparations.[45,64]

#### 16.5.1.4 Near-infrared fluorescence microscopy

Analogous to CLSM/MPLSM imaging, near-infrared fluorescence (NIRF) imaging has emerged in the nineties as a powerful tool for studying the molecular basis of diseases in small animal studies.[42,88-90] NIRF imaging uses wavelengths larger than 700 nm, a range wherein absorbance and autofluorescence of water and biomolecules is minimal, thus allowing for efficient photon penetration through a tissue with minimal intratissue light scattering, yielding a good signal-to-noise ratio.[63,90] Cyanine dyes (e.g., Cy5.5 and Cy7) have been proven to act as convenient contrast agents, as they absorb and emit in the NIR range, provide a high quantum yield, are photostable, and are water soluble.[63] One example of use is anti-mouse CD31 antibodies covalently labeled with Cy5.5 fluorochrome, which showed an abundant staining of the endothelial vascular wall after *in vivo* intravenous injection.[91] The *in vivo* administration of anti-mouse CD31 antibodies had no toxic effect and showed high specificity to mouse endothelial cells, tested by *in vitro* mouse endothelial cell staining. Other researchers showed mouse ear vessel staining by qualitative image analysis after tumor necrosis factor-alpha (TNF-$\alpha$) treatment *in vivo* with subsequent injection of Cy5.5-coupled VCAM-1-antibodies (vascular cell adhesion molecule-1).[92,93] TNF-$\alpha$ induced an increased endothelial inflammatory reaction, resulting in an immediate expression of cell adhesion molecules, such as VCAM-1.

Hsu *et al.* and Wu *et al.* were able to stain single cancer cells for NIRF *in vitro*; however, *in vivo*, they showed only crudely resolved images of mouse angiogenetic areas by means of tumor integrin expression labeling using Cy5.5 or Cy7-linked cyclic Arg-Gly-Asp (RGD) peptides.[90,94,95] Funovics *et al.* followed a similar approach by coupling an E-Selectin-antibody to Cy5.5 fluorochromes for labeling and imaging endothelial cells *in vitro* and crudely imaging antibody binding to blood vessels in Lewis lung carcinoma *in vivo*.[96]

Alencar *et al.* who used the near-infrared vascular probe AngioSense 750 (VisEn Medical, USA) have shown higher resolved images (3.5 μm lateral resolution) of microvascular vessels *in vivo*. With confocal microscopy, they were able to image capillaries within mouse organs and to show actual blood flow within microvessels.[97,98] They used a biopsy needle-sized objective with a millimeter footprint to reach the intravital site of interest through a keyhole, minimal-invasive incision.[98,99]

#### 16.5.1.5 Quantum dot imaging

Apart from fluorochrome-coupled antibodies, nanocrystalline semiconductor materials, called quantum dots (QDots), have emerged as new agents for biological imaging over the past decade.[100] QDots have been covalently linked to biorecognition molecules such as peptides, antibodies, nucleic acids, or small-molecule ligands for application as fluorescent probes, reviewed by Zhou and Ghosh.[101] Compared with organic fluorochromes, QDots have unique optical properties such as size and composition-tunable fluorescence emission from visible to infrared wavelengths, large absorption coefficients across a wide spectral range, and very high levels of brightness and photostability. Their broad excitation profiles and narrow, symmetric emission spectra make QDots well suited to optical multiplexing, in which multiple colors and intensities can be combined to label different antigens simultaneously.[102] Concurrent with antibody-coupled fluorochromes (e.g., GFP or Alexa fluorochromes), QDot-antibody (QDot-Ab) bioconjugates can be synthesized to target common endothelial cell adhesion molecules. Jayagopal *et al.* specifically labeled vascular cells by using spectrally distinct QDot-Abs to label cell adhesion molecules such as PECAM-1 (CD31), intercellular adhesion molecule 1 (ICAM-1 or CD54), or VCAM-1 (CD106) on the retinal endothelium in a rat model of diabetes using *in vivo* fluorescence microscopy.[103] Ferrara *et al.* used multicolor profiling by QDot bioconjugates to image arterial walls, to quantify fluorescence cell wall antigen, and to visualize the vascular endothelium in 3D.[104] Åkerman *et al.* demonstrated selective targeting by peptide-coated QDots to the vasculature of tissues *in vivo* and to endothelial cells *in vitro*.[105–107] Smith *et al.* succeeded in using QDots to visualize, with a resolution of 2 μm, macro-sized blood vessels of the chick chorioallantoic membrane (CAM).[108] CAM is a popular model used for studying various research aspects such as general vessel development, surgical retinal research, or testing of biomaterials on its vascularization and biocompatibility.[108–111] The chick CAM, a part of the extraembryonic tissue, allows studying and interfering easily with vessel development under unsealed *in vivo* conditions. Upon intravital QDot injection, the authors showed that in comparison to FITC-dextran imaging probes, QDots could be kept in blood circulation over the course of 4 days without perfusion into the perivascular tissue, whereas FITC-dextrans used to leak from vessels into the allantoic sac after several hours. They could also show QDot endocytosis into each endothelial cell lining the blood vessel.[108] QDots were actively taken up without further cell permeabilization. For *in vivo* imaging, Larson *et al.* visualized QDots dynamically in living mice by mural tail vein injection of water-soluble QDots within a depth of nearly 100 μm through the skin and adipose tissue (Fig. 16.5).[109]

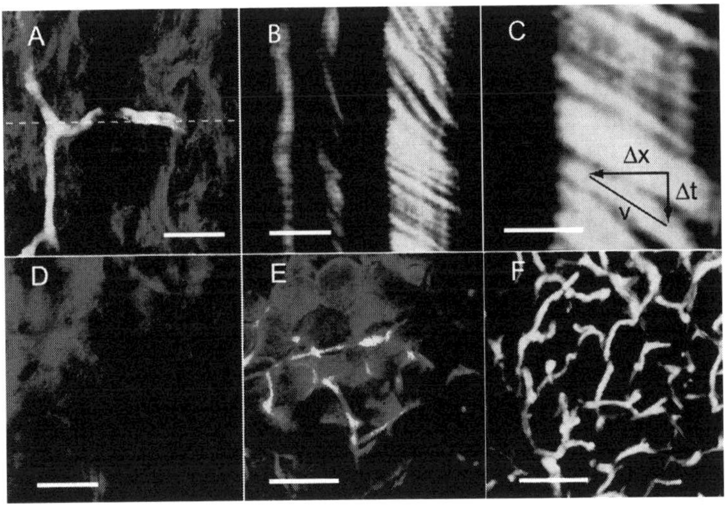


**Figure 16.5** Quantum Dot (QDot) imaging of microvasculature using fluorescence microscopy. From Ref. 109. Reprinted with permission from AAAS.

### 16.5.1.6 Bioluminescent imaging

*In vivo* BLI demonstrates another method of vascular imaging, which *per se* constitutes a relatively young research field developed within the last decade. Bioluminescence is a naturally occurring phenomenon and refers to the enzymatic generation of visible light, achieved by the bioluminescent reporter gene coding for luciferase from the North American firefly *Photinus pyralis*.[112] The enzymatic reaction releases light between 500 nm and 620 nm when luciferase is exposed to D-luciferin, ATP, magnesium, and oxygen. This light can cross tissues, including bone, and can be externally detected.[32] BLI encompasses noninvasive imaging of luciferase gene activity using CCD cameras in luciferase transgenic animals or in grafted, luciferase transfected cells, such as in a special reporter system to investigate cellular oxygen availability in tissue-engineered constructs both *in vitro* and *in vivo*.[113] The imaging procedure is technically simple and quantifiable by CCD photon counting. Analogous to the Tie2 reporter system for CLSM *in vivo* imaging, luciferase expression can be put under the control of tissue-specific regulatory elements, which allows BLI to be used in noninvasive imaging of processes such as vasculogenesis.[32]

In comparison to GFP-based imaging, luciferase-based imaging is much less powerful. No fine-resolved image can be formed with luciferase, as there are an insufficient number of photons emitted at a given time. The weak signal from luciferase requires photon counting, resulting in a pseudo-image *in vivo* rather than true light reflectance imaging, and thus greatly reduced resolution.[71] It has been reported that a minimum amount of 300 cells *in vitro* and 3000 cells *in vivo* are needed to be imaged with bioluminescence.[114] Furthermore, rapid clearance of the injected D-luciferin within an hour leads to an unstable signal that complicates comparison of data.[115] The main advantage of luciferase-based imaging is its photon self-emitting feature, which allows imaging from deeper regions within tissues, because GFP excitation is limited due to excitation light scatter and tissue attenuation. A detailed comparison of GFP-based and luciferase-based imaging is summarized by Hoffman.[71]

Nevertheless, direct BLI of vasculogenesis is feasible using endothelial-specific promoters driving luciferase expression to yield a crude image of vascularized areas only. For instance, Zhang *et al.* describe vascular endothelial growth factor receptor 2 (VEGFR2)-luciferase mice showing VEGFR2 activation during wound healing.[32,116] Others such as Dong *et al.* used stably transfected luciferase-expressing endothelial cells cultivated on scaffolds to track *in vivo* angiogenesis within those scaffolds or possible outgrowth of the cells.[117,118] Similar to *in vivo* experiments, cells in cell cultures expressing luciferase could not be resolved in high resolution.[119] To our knowledge, a microvascular anatomic depiction of capillaries using BLI has not been achieved so far.

### 16.5.2 The Study of Microvasculature by Radiation-Based Methods

Radiation-based methods such as PET and CT, developed in the early 1950s and 1970s, respectively, represent two frequently used methods for vessel imaging. Initially being coarse in resolution, research and development pushed resolution, especially in clinical CT scanners, toward increasingly smaller dimensions able to depict microvascular networks.

#### 16.5.2.1 Micro-computed tomography/synchrotron radiation micro-computed tomography

Over the last decade, the accelerated development in micro-CT (μCT) enabled a highly resolved replication of microvasculature *ex vivo*, allowing for quantitative assessment of vessel architecture in three dimensions.[120–123] *In vivo*, clinical CT has been used for tumor vessel imaging already; however, poor soft tissuc contrast presents a problem in clinical CT as well as *in vivo* μCT imaging to delineate structures of interest such as microvasculature.[42] Vessels with smaller diameters than 20 μm cannot be visualized without radio-opaque contrast agents such as iodine or barium sulfate, both used in *in vivo* X-ray angiography or *ex vivo* μCT imaging, respectively.[120,124] In this regard, Marxen *et al.* discussed three different contrast agents *in vivo* [Batson's™ No. 17 (Polysciences Inc., USA), Microfil® MV122 (Flow Tech Inc., USA), and Baritop®, respectively] to show kidney vascularization at a resolution of 20 μm.[125] Apart from this, other vessel contrast agents exist such as eXIA™160XL (Binitio, Canada), Fenestra VC (ART, Canada), Accupaque (GE Healthcare, Switzerland). eXIA™ represents a second-generation contrast agent (iodine-based contrast agents with a lipid formulation) that stays in blood over a longer time than traditional ones. Noniodinated contrast agents also exist containing gold or bismuth sulfide nanoparticles.[126,127] However, the use of iodinated CT contrast agents in patients with moderate-to-severe chronic kidney disease has been associated with adverse nephrotoxic side effects. In addition, other acute anaphylactoid and delayed adverse reactions are known.[128]

*Ex vivo*, several groups were able to visualize vasculature in isolated, tissue-fixed, and vasculature-stained rodent organs down to a voxel size of 5 μm by μCT analysis.[121,129,130] The main disadvantage of bench-top X-ray sources is their lack in sufficient flux in photons produced per unit-area per second for spatial and contrast resolution. This is why synchrotron radiation μCT (SRμCT) facilities based on radiation produced by the acceleration of electrons are more and more used as a source for tomographic acquisition of microvasculature. In this respect, SRμCT light was used for barium contrast-perfused rat brain tissue to image capillaries within several cubic millimeters by using pure absorption and edge enhancement at a voxel size of 1.4 μm.[123]

Heinzer *et al.* presented a novel powerful approach to hierarchically image vascular architecture (increasing magnification) with exceptional image quality and resolution at any depth of a brain vasculature network using vascular corrosion casting starting with μCT and ending with high-resolution SRμCT and SEM images (Fig. 16.6).[120]

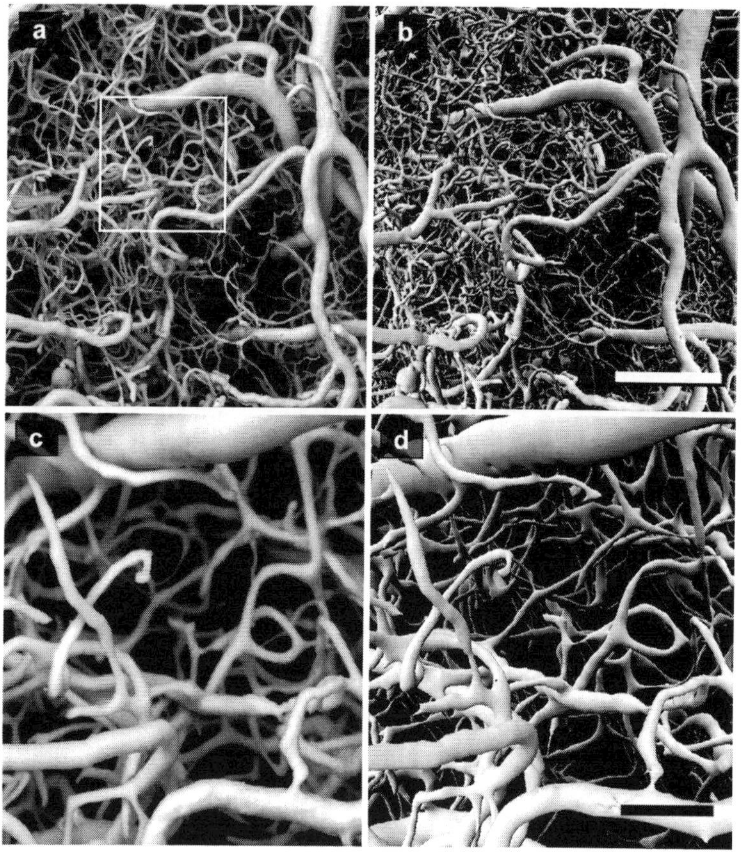


**Figure 16.6** Synchrotron radiation micro-computed tomography (SRμCT) imaging compared with SEM imaging of microvasculature. Reprinted from Ref. 120, Copyright (2006), with permission from Elsevier.

Their corrosion cast technology was based on a polymer resin (PU4ii, vasQtec, Switzerland) to fill capillaries via intracardial injection with subsequent soft tissue maceration and decalcification to yield a vascular skeleton of the mouse brain.[131–134] This approach allowed 3D visualization and quantification of large microvascular brain networks. Scans of entire mouse full brain vascular corrosion casts were performed at 16 μm resolution with desktop μCT scanners and 1.4 μm voxel size with SRμCT. This hierarchical imaging method demonstrated that desktop μCT can capture large extracerebral and intracerebral arteries and veins; smaller structures such as capillaries could only be resolved by synchrotron radiation light. Their comparative study showed that SRμCT capillary imaging could be very similar to SEM images in quality.[120]

Komlev *et al.* presented another interesting way how to visualize concurrently both mineralized and micro-vascularized soft tissues with no need for contrast agents by SRμCT pseudo-holotomography. In their approach, they combined the mapping of absorption with the phase shift to three-dimensional image soft tissue and vascular networks in tissue-engineered constructs using adequately chosen algorithms and threshold values (Fig. 16.7).[135]

#### 16.5.2.2 Positron emission tomography

Newer CT systems have combined CT with PET, enabling coacquisition of two-way images in the same object for further data interpretation. PET radiotracers demonstrate proton-rich atomic compounds, which are internalized by inhalation or injection and decay by emission of a positron. After unification of the positron with a close electron, two 511 keV photons are emitted in 180° opposite direction. The gamma ray photon registration at the PET cameras is coincidental and logged for time and location. Computing reconstruction algorithms can deduce spatiotemporal distribution of the PET radiotracer.[136] In comparison to other radiation-based methods, clinical PET scanner resolution (2 mm) is noticeably poorer compared with CT (200 μm) or MRI (100 μm). The $^{15}O$-oxygen isotope, with a half-life of 123 s, used in $C^{15}O_2$, $C^{15}O$, or $H_2{}^{15}O$, is predominately used to image cancer lesions and organs, with $H_2{}^{15}O$ currently being the most validated radiotracer system for PET.[136] Qualitative high-resolution PET findings, as in scanners for small animal PET imaging, start at a resolution of 1 mm scale, which may be enough for large arteries and veins, but excludes PET from being a successful application for microvascular imaging.[137]

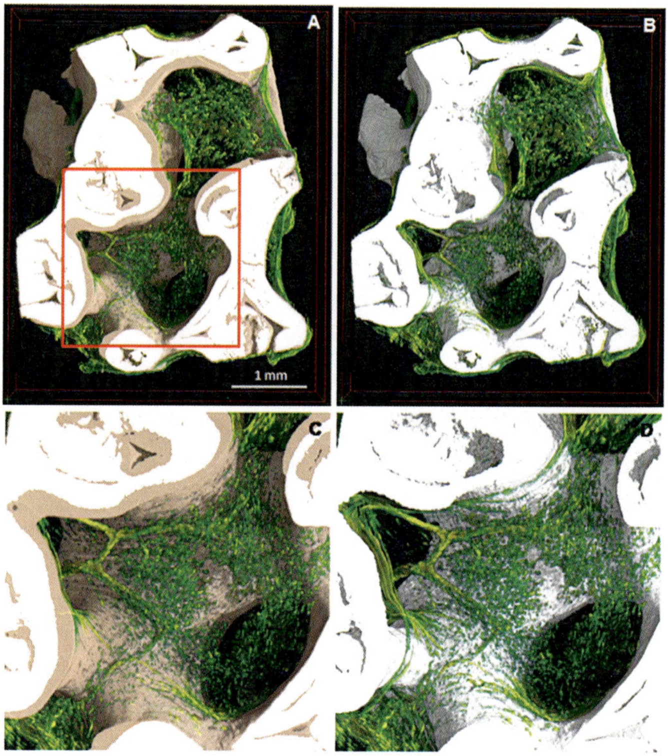


**Figure 16.7** Three-dimensional pseudo-holotomographic imaging of vessel ingrowth into a tissue-engineered construct. Reprinted from Ref. 135, with permission. The publisher for this copyrighted material is Mary Ann Liebert, Inc. publishers.

### 16.5.2.3 Magnetic resonance imaging

MRI constitutes another important field of research for vascular imaging, as its high temporal and spatial resolution of 100 μm are well suited for use in the assessment of angiogenesis.[128] Research in recent years has improved MRI setup and imaging processing methods to yield advances at the micron-sized field of MRI. MRI scanners set up a strong alternating magnetic field exciting the nuclear core of protons in the subject of investigation. MRI detectors, being exquisitely sensitive to soft tissue differences and abnormalities, generally omit a contrast agent for clinical imaging.[42] To increase vessel visibility, dynamic contrast-enhanced MRI (DCE-MRI) has regularly been used for the assessment of vasculature and tracking the entrance of diffusible contrast agents into tissue over time as a routine procedure. In comparison to CT measurements, MRI is considered less harmful due to the lack of ionizing radiation. It has been discovered that contrast agents such as gadolinium are valuable to image small vessel-like structures, as it generates a 3D layout of the vasculature and provides static architectural information.[40] However, side effects of gadolinium exist that may lead to nephrogenic systemic fibrosis for renally impaired patients.[128] It has been shown that vessels down to 100–200 μm in diameter can be resolved with clinical MR scanners. Smaller vessel structures require optimized magnetic resonance coils, longer scan times of more than 10 min, and contrast agents such as gadolinium-chelates covalently bound to macromolecular proteins (e.g., albumin), gadolinium-containing polymers/liposomes, or superparamagnetic iron oxide nanoparticles.[138] In practice, MRI has been used for gaining insight into tumor angiogenesis *in vivo*. Van Vliet *et al.* used a dorsal skin-fold window in a rodent model together with a 1 cm loop receiver antenna placed on the skin-fold window using a 3.0 Tesla clinical MRI for noninvasive quantification of tumor growth and vascularization over the course of several days.[139] With the application of the dorsal skin-fold window model, coacquisition of *in vivo* light microscopy pictures and MRI imaging was possible. Others used even higher magnetic fields (9.4 Tesla) to gain better signal-to-noise ratios to show intracortical artery and vein vasculature by means of an iron oxide contrast agent to yield *in vivo* MR images at resolutions smaller than 100 μm (Fig. 16.8).[140]

To our knowledge, a microvascular anatomic depiction of capillaries using MRI has not been shown so far; however, MRI has been used to monitor osteogenesis in tissue-engineered constructs *in vitro*.[141]

### 16.5.3 The Study of Microvasculature by Sound-Based Methods

#### 16.5.3.1 Ultrasonography

Ultrasound imaging has undergone revolutionary changes in the past decades that have improved its temporal and spatial resolution down to approximately 200 μm.[142] Pioneering work in medical ultrasound imaging was done by Karl Theodore Dussik in the late

1930s.[143] Traditionally, ultrasound has been utilized as a cheap and convenient method to image anatomical structures transdermally (e.g., organ echolocation or prenatal examination). Ultrasound imaging is based on differences in the density and compressibility of organs or macrovascular vessel structures (e.g., aorta, arteries) compared with the surrounding tissues resulting in an image contrast. Large differences in density and compressibility result in a strong echo. Although general ultrasound examinations do not need contrast agents *per se*, imaging down to the microvascular level does not happen without an ultrasound contrast agent (UCA) to depict vascularization smaller than approximately 200 μm. Blood vessels, in general, have poor backscatter characteristics than their surrounding tissues.

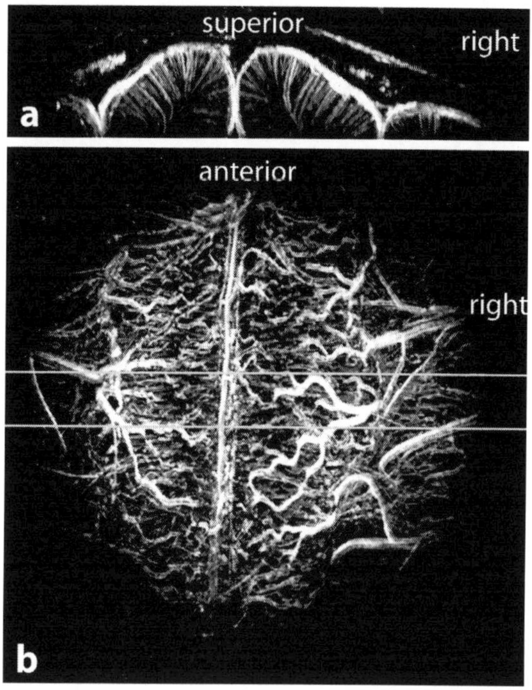


**Figure 16.8** Magnetic resonance imaging (MRI) of intracortical vessels using iron oxide contrast agent. Reprinted from Ref. 140, Copyright (2006), with permission from Elsevier.

In 1968, the cardiologists Gramiak and Shah coincidentally discovered that single air bubbles, accidentally introduced into the cardiac vessels system, were highly echogenic in ultrasound examinations, thus indirectly discovering a good UCA for vessel detection.[144] Today, nontoxic, FDA-approved, lipid, or protein-encapsulated microbubbles (mean diameter 1–2 μm) are used as UCA for vessel examinations being highly echogenic (with a 10,000 fold stronger impedance compared with blood and surrounding tissues) due to their ideal and efficient resonance oscillation under clinical ultrasound devices.[145–147] Typically, microbubbles are filled with air/nitrogen or heavy, inert gases such as octafluoropropane ($C_3F_8$), decafluorobutane ($C_4F_{10}$), or sulfur hexafluoride ($SF_6$), and $10^8$–$10^{10}$ microbubbles per human are applied per examination intravenously.[145,148] Microbubbles finally get internalized by macrophages or are exhaled when passing the capillary bed of the lungs.[149] Flow characteristics of microbubbles resemble that of red blood cells, which make microbubbles suitable for microvascular perfusion studies, shown by Kaufmann *et al.*[148] Microbubbles can be detected with a high degree of sensitivity compared with red blood cells, provide both anatomic and functional vessel information (such as tissue perfusion or vessel volume due to intermingling with red blood cells), and give additional data about intravascular vessel wall protein expression feasible by molecular UCA targeting.[145] Endothelial cell-specific molecules are targeted actively by binding microbubbles via antibodies to endothelial cell adhesion molecules on the intravascular vessel wall. On inflamed luminal endothelium, the passive binding of microbubbles to complement on leucocytes is possible as well.[145,150] In practice, Weller *et al.* who used microbubbles tagged with the tumor endothelial cell specific binding peptide sequence Arg-Arg-Lyc (RRL) or anti-ICAM-1 antibodies, respectively, actively imaged angiogenesis by targeting endothelial cells *in vitro* and endothelium of tumor microvascularity *in vivo*.[151–153] Others targeted VEGFR2 or used RGD motifs on microbubbles to target $\alpha_v\beta_3$ integrin receptors on leucocytes for vessel imaging.[145,150,154–156]

Spatial resolution of ultrasound imaging (50–500 μm) is comparable with MRI (25–200 μm) and CT (250–500 μm), whereas signal-to-noise ratio is similar to PET and SPECT.[145] Although capillary signals could be computer-processed to extract the unique echo of microbubbles, microvessels cannot clearly be resolved as discrete structures, as the spatial resolution of 50–500 μm is not high enough to resolve a single capillary.[157–159] A microvascular anatomic depiction of capillaries using ultrasound has not been shown so far.

## 16.6 Conclusion

Microvascular networks present an essential element both in the physiological development of the human body and in present-day pathologies such as cancer. Vascularization plays another important part in artificial tissue development and wound regeneration wherein microcapillary growth is indispensable. Medical expert groups stress the importance of understanding microcapillary evolution and more importantly to control cancer progression through the control of vessel growth for successful cancer therapies in future. This fact has led to the development and refinement of traditional imaging methods such as CT, MRI, or ultrasound, and the invention of novel methods such as fluorescent or bioluminescent methods, to image vasculature in a better resolution and/or noninvasively. The set of problems given with 3D reconstruction of histological 2D slices has further pushed the field in 3D imaging for faster and more precise results.

This review has focused on the knowledge on current methods in microvascular imaging, showing the possibilities and limitations on the micro scale. Light microscopy has progressed greatly in the last decade in the targeted and functional imaging of angiogenesis but is still hindered by the inaccessibility of sites embedded deeper in the body and light absorption at the scanning site. Radiation-based methods, especially μCT, have forged ahead in the *in vitro* resolution of microvessel networks of large explants as shown with synchrotron radiation based technology; however, resolution improvement and radiation reduction would be desirable for *in vivo* μCT blood capillary imaging. Sound-based capillary imaging has experienced success with the development of microbubbles for *in vivo* imaging. Although microbubbles have been reasonably optimized to achieve high-resolution standards, ultrasound of microvessels is still impaired by low-resolved signal detection and processing. Generally, further efforts need to be made to establish sensitivity, specificity, and the quantitative potential to resolve microvasculature structure and measure its functionality adequately. As microvascularization has already been extensively studied for cancerous diseases, a knowledge transfer of vessel imaging into the tissue engineering sector would be highly favorable to realize further progress. For *in vitro* and *in vivo* tissue engineering purposes, the discussed imaging modalities should be applied to enhance quality and to control vascularization progress, as tissue engineers endeavor to create artificial vasculature networks in a large-scale manner. Applied imaging technologies are increasingly refined and demonstrate impressive potential for diagnostic and prognostic purposes. They may strongly improve the pre-clinical assessment of anti-angiogenetic drugs and pave the way for even better clinical imaging results in future.

## Acknowledgments

The authors gratefully acknowledge funding from Angioscaff, EU Seventh Framework Program (FP7-NMP-2008-214402).

## References

1. Sherwood, L. (2001) Human physiology, Brooks/Cole, Belmont.
2. Jennings, D., Hatton, B. N., Guo, J., Galons, J. P., Trouard, T. P., Raghunand, N., Marshall, J. and Gillies, R. J. (2002) Early response of prostate carcinoma xenografts to docetaxel chemotherapy monitored with diffusion MRI, *Neoplasia*, **4**(3), 255–262.
3. Dufraine, J., Funahashi, Y. and Kitajewski, J. (2008) Notch signaling regulates tumor angiogenesis by diverse mechanisms, *Oncogene*, **27**(38), 5132–5137.
4. Carmeliet, P. and Jain, R. K. (2000) Angiogenesis in cancer and other diseases, *Nature*, **407**(6801), 249–257.
5. Costouros, N. G., Diehn, F. E. and Libutti, S. K. (2002) Molecular imaging of tumor angiogenesis, *J Cell Biochem Suppl*, **39**, 72–78.
6. Vunjak-Novakovic, G. and Freed, L. E. (1998) Culture of organized cell communities, *Adv Drug Deliv Rev*, **33**(1–2), 15–30.
7. Unger, R. E., Sartoris, A., Peters, K., Motta, A., Migliaresi, C., Kunkel, M., Bulnheim, U., Rychly, J. and Kirkpatrick, C. J. (2007) Tissue-like self-assembly in cocultures of endothelial cells and osteoblasts and the formation of microcapillary-like structures on three-dimensional porous biomaterials, *Biomaterials*, **28**(27), 3965–3976.
8. Georgakoudi, I., Rice, W. L., Hronik-Tupaj, M. and Kaplan, D. L. (2008) Optical spectroscopy and imaging for the noninvasive evaluation of engineered tissues, *Tissue Eng Part B Rev*, **14**(4), 321–340.
9. Lammers, G., Verhaegen, P. D., Ulrich, M. M., Schalkwijk, J., Middelkoop, E., Weiland, D., Nillesen, S. T., Van Kuppevelt, T. H. and Daamen, W. F. (2011) An overview of methods for the *in vivo* evaluation of tissue-engineered skin constructs, *Tissue Eng Part B Rev*, **17**(1), 33–55.

10. Mason, C., Markusen, J. F., Town, M. A., Dunnill, P. and Wang, R. K. (2004) The potential of optical coherence tomography in the engineering of living tissue, *Phys Med Biol*, **49**(7), 1097–1115.
11. Pancrazio, J. J., Wang, F. and Kelley, C. A. (2007) Enabling tools for tissue engineering, *Biosens Bioelectron*, **22**(12), 2803–2811.
12. Semple, J. L., Woolridge, N. and Lumsden, C. J. (2005) *In vitro, in vivo, in silico*: computational systems in tissue engineering and regenerative medicine, *Tissue Eng*, **11**(3–4), 341–356.
13. Sun, W. and Lal, P. (2002) Recent development on computer aided tissue engineering—a review, *Comput Methods Programs Biomed*, **67**(2), 85–103.
14. Ballyns, J. J. and Bonassar, L. J. (2009) Image-guided tissue engineering, *J Cell Mol Med*, **13**(8A), 1428–1436.
15. Tsang, V. L. and Bhatia, S. N. (2004) Three-dimensional tissue fabrication, *Adv Drug Deliv Rev*, **56**(11), 1635–1647.
16. Smith, L. E., Smallwood, R. and Macneil, S. (2010) A comparison of imaging methodologies for 3D tissue engineering, *Microsc Res Tech*, **73**(12), 1123–1133.
17. König, K. and Riemann, I. (2003) High-resolution multiphoton tomography of human skin with subcellular spatial resolution and picosecond time resolution, *J Biomed Opt*, **8**(3), 432–439.
18. Schenke-Layland, K. (2008) Non-invasive multiphoton imaging of extracellular matrix structures, *J Biophotonics*, **1**(6), 451–462.
19. Schenke-Layland, K., Riemann, I., Damour, O., Stock, U. A. and König, K. (2006) Two-photon microscopes and *in vivo* multiphoton tomographs--powerful diagnostic tools for tissue engineering and drug delivery, *Adv Drug Deliv Rev*, **58**(7), 878–896.
20. Werkmeister, E., Dumas, D., de Isla, N., Marchal, L. and Stoltz, J. F. (2008) Interest of multimodal imaging in tissue engineering, *Biomed Mater Eng*, **18**(4-5), 329–333.
21. Lavis, L. D. (2011) Histochemistry: live and in color, *J Histochem Cytochem*, **59**(2), 139–145.
22. Williams, R. M., Zipfel, W. R. and Webb, W. W. (2001) Multiphoton microscopy in biological research, *Curr Opin Chem Biol*, **5**(5), 603–608.
23. König, K. (2008) Clinical multiphoton tomography, *J Biophotonics*, **1**(1), 13–23.

24. Liang, X., Graf, B. W. and Boppart, S. A. (2009) Imaging engineered tissues using structural and functional optical coherence tomography, *J Biophotonics*, **2**(11), 643–655.

25. Tan, W., Sendemir-Urkmez, A., Fahrner, L. J., Jamison, R., Leckband, D. and Boppart, S. A. (2004) Structural and functional optical imaging of three-dimensional engineered tissue development, *Tissue Eng*, **10**(11–12), 1747–1756.

26. Matcher, S. J. (2011) Practical aspects of OCT imaging in tissue engineering, *Methods Mol Biol*, **695**, 261–280.

27. Yang, V. X., Mao, Y. X., Munce, N., Standish, B., Kucharczyk, W., Marcon, N. E., Wilson, B. C. and Vitkin, I. A. (2005) Interstitial Doppler optical coherence tomography, *Opt Lett*, **30**(14), 1791–1793.

28. Mason, C., Markusen, J. F., Town, M. A., Dunnill, P. and Wang, R. K. (2004) Doppler optical coherence tomography for measuring flow in engineered tissue, *Biosens Bioelectron*, **20**(3), 414–423.

29. Khalil, A. S., Chan, R. C., Chau, A. H., Bouma, B. E. and Mofrad, M. R. (2005) Tissue elasticity estimation with optical coherence elastography: toward mechanical characterization of *in vivo* soft tissue, *Ann Biomed Eng*, **33**(11), 1631–1639.

30. Ko, H. J., Tan, W., Stack, R. and Boppart, S. A. (2006) Optical coherence elastography of engineered and developing tissue, *Tissue Eng*, **12**(1), 63–73.

31. Boskey, A. and Pleshko Camacho, N. (2007) FT-IR imaging of native and tissue-engineered bone and cartilage, *Biomaterials*, **28**(15), 2465–2478.

32. de Boer, J., van Blitterswijk, C. and Lowik, C. (2006) Bioluminescent imaging: emerging technology for non-invasive imaging of bone tissue engineering, *Biomaterials*, **27**(9), 1851–1858.

33. Xu, H., Othman, S. F. and Magin, R. L. (2008) Monitoring tissue engineering using magnetic resonance imaging, *J Biosci Bioeng*, **106**(6), 515–527.

34. Müller, R. (2009) Hierarchical microimaging of bone structure and function, *Nat Rev Rheumatol*, **5**(7), 373–381.

35. Young, S., Kretlow, J. D., Nguyen, C., Bashoura, A. G., Baggett, L. S., Jansen, J. A., Wong, M. and Mikos, A. G. (2008) Microcomputed tomography characterization of neovascularization in bone tissue engineering applications, *Tissue Eng Part B Rev*, **14**(3), 295–306.

36. van Lenthe, G. H., Hagenmüller, H., Bohner, M., Hollister, S. J., Meinel, L. and Müller, R. (2007) Nondestructive micro-computed tomography

for biological imaging and quantification of scaffold-bone interaction *in vivo*, *Biomaterials*, **28**(15), 2479–2490.

37. Cancedda, R., Cedola, A., Giuliani, A., Komlev, V., Lagomarsino, S., Mastrogiacomo, M., Peyrin, F. and Rustichelli, F. (2007) Bulk and interface investigations of scaffolds and tissue-engineered bones by X-ray microtomography and X-ray microdiffraction, *Biomaterials*, **28**(15), 2505–2524.
38. Phair, R. D. and Misteli, T. (2001) Kinetic modelling approaches to *in vivo* imaging, *Nat Rev Mol Cell Biol*, **2**(12), 898–907.
39. Wessels, J. T., Busse, A. C., Mahrt, J., Dullin, C., Grabbe, E. and Mueller, G. A. (2007) *In vivo* imaging in experimental preclinical tumor research—a review, *Cytometry A*, **71**(8), 542–549.
40. Neeman, M., Gilad, A. A., Dafni, H. and Cohen, B. (2007) Molecular imaging of angiogenesis, *J Magn Reson Imaging*, **25**(1), 1–12.
41. Dobrucki, L. W. and Sinusas, A. J. (2007) Imaging angiogenesis, *Curr Opin Biotechnol*, **18**(1), 90–96.
42. Massoud, T. F. and Gambhir, S. S. (2003) Molecular imaging in living subjects: seeing fundamental biological processes in a new light, *Genes Dev*, **17**(5), 545–580.
43. Schneider, P., Voide, R., Stampanoni, M. and Müller, R. (2009) Post-processing technique for improved assessment of hard tissues in the submicrometer domain using local synchrotron radiation-based computed tomography, *Biomed Tech (Berl)*, **54**(1), 48–54.
44. Cho, Z. H., Ahn, C. B., Juh, S. C., Lee, H. K., Jacobs, R. E., Lee, S., Yi, J. H. and Jo, J. M. (1988) Nuclear magnetic resonance microscopy with 4-microns resolution: theoretical study and experimental results, *Med Phys*, **15**(6), 815–824.
45. Roberts, R. L. and Lin, P. C. (2004) Structural and functional optical imaging of angiogenesis in animal models, *Methods Enzymol*, **386**, 105–122.
46. McDonald, D. M. and Choyke, P. L. (2003) Imaging of angiogenesis: from microscope to clinic, *Nat Med*, **9**(6), 713–725.
47. Helmchen, F. and Denk, W. (2005) Deep tissue two-photon microscopy, *Nat Methods*, **2**(12), 932–940.
48. Monari, J. J. M. (2000) Human Capillaroscopy by Light Emitting Diode Epi-illumination, *Microvascular Research*, **59**, 172–175.
49. Dewhirst, M. W., Klitzman, B., Braun, R. D., Brizel, D. M., Haroon, Z. A. and Secomb, T. W. (2000) Review of methods used to study oxygen transport at the microcirculatory level, *Int J Cancer*, **90**(5), 237–255.

50. Lin, P. C. (2003) Optical imaging and tumor angiogenesis, *J Cell Biochem*, **90**(3), 484–491.

51. Endrich, B., Asaishi, K., Gotz, A. and Messmer, K. (1980) Technical report--a new chamber technique for microvascular studies in unanesthetized hamsters, *Res Exp Med (Berl)*, **177**(2), 125–134.

52. Gaustad, J. V., Brurberg, K. G., Simonsen, T. G., Mollatt, C. S. and Rofstad, E. K. (2008) Tumor vascularity assessed by magnetic resonance imaging and intravital microscopy imaging, *Neoplasia*, **10**(4), 354–362.

53. Laschke, M. W., Harder, Y., Amon, M., Martin, I., Farhadi, J., Ring, A., Torio-Padron, N., Schramm, R., Rucker, M., Junker, D., Haufel, J. M., Carvalho, C., Heberer, M., Germann, G., Vollmar, B. and Menger, M. D. (2006) Angiogenesis in tissue engineering: breathing life into constructed tissue substitutes, *Tissue Eng*, **12**(8), 2093–2104.

54. Abshagen, K., Schrodi, I., Gerber, T. and Vollmar, B. (2009) *In vivo* analysis of biocompatibility and vascularization of the synthetic bone grafting substitute NanoBone, *J Biomed Mater Res A*, **91**(2), 557–566.

55. Laschke, M. W., Strohe, A., Scheuer, C., Eglin, D., Verrier, S., Alini, M., Pohlemann, T. and Menger, M. D. (2009) *In vivo* biocompatibility and vascularization of biodegradable porous polyurethane scaffolds for tissue engineering, *Acta Biomater*, **5**(6), 1991–2001.

56. Shan, S., Sorg, B. and Dewhirst, M. W. (2003) A novel rodent mammary window of orthotopic breast cancer for intravital microscopy, *Microvasc Res*, **65**(2), 109–117.

57. Foltz, R. M., McLendon, R. E., Friedman, H. S., Dodge, R. K., Bigner, D. D. and Dewhirst, M. W. (1995) A pial window model for the intracranial study of human glioma microvascular function, *Neurosurgery*, **36**(5), 976–984; discussion 984–975.

58. Yuan, F., Chen, Y., Dellian, M., Safabakhsh, N., Ferrara, N. and Jain, R. K. (1996) Time-dependent vascular regression and permeability changes in established human tumor xenografts induced by an anti-vascular endothelial growth factor/vascular permeability factor antibody, *Proc Natl Acad Sci U S A*, **93**(25), 14765–14770.

59. Langer, S., von Dobschuetz, E., Harris, A. G., Krombach, F., Messmer, K. and Messmer, K. e. (2000) Validation of the orthogonal polarization spectral imaging technique on solid organs. *Orthogonal Polarization Spectral Imaging Prog Appl Microcirc.*, **24**, 32–46.

60. Groner, W., Winkelman, J. W., Harris, A. G., Ince, C., Bouma, G. J., Messmer, K. and Nadeau, R. G. (1999) Orthogonal polarization spectral imaging: a new method for study of the microcirculation, *Nat Med*, **5**(10), 1209–1212.

61. Lupi, O., Semenovitch, I., Treu, C. and Bouskela, E. (2008) Orthogonal polarization technique in the assessment of human skin microcirculation, *Int J Dermatol*, **47**(5), 425–431.

62. McVeigh, E. R. (2006) Emerging imaging techniques, *Circ Res*, **98**(7), 879–886.

63. Bremer, C., Ntziachristos, V., Mahmood, U., Tung, C. H. and Weissleder, R. (2001) [Progress in optical imaging], *Radiologe*, **41**(2), 131–137.

64. Jain, R. K., Munn, L. L. and Fukumura, D. (2002) Dissecting tumour pathophysiology using intravital microscopy, *Nat Rev Cancer*, **2**(4), 266–276.

65. Wolf, D. E., Samarasekera, C. and Swedlow, J. R. (2007) Quantitative analysis of digital microscope images, *Methods Cell Biol*, **81**, 365–396.

66. Hansen-Smith, F. M. (2000) Capillary network patterning during angiogenesis, *Clin Exp Pharmacol Physiol*, **27**(10), 830–835.

67. Ocak, I., Baluk, P., Barrett, T., McDonald, D. M. and Choyke, P. (2007) The biologic basis of *in vivo* angiogenesis imaging, *Front Biosci*, **12**, 3601–3616.

68. Morikawa, S., Baluk, P., Kaidoh, T., Haskell, A., Jain, R. K. and McDonald, D. M. (2002) Abnormalities in pericytes on blood vessels and endothelial sprouts in tumors, *Am J Pathol*, **160**(3), 985–1000.

69. Kirkpatrick, C. J., Fuchs, S. and Unger, R. E. (2011) Co-culture systems for vascularization—learning from nature, *Adv Drug Deliv Rev*, **63**(4–5), 291–299.

70. Gerdes, H. H. and Kaether, C. (1996) Green fluorescent protein: applications in cell biology, *FEBS Lett*, **389**(1), 44–47.

71. Hoffman, R. M. (2008) Recent advances on *in vivo* imaging with fluorescent proteins, *Methods Cell Biol*, **85**, 485–495.

72. Shaner, N. C., Campbell, R. E., Steinbach, P. A., Giepmans, B. N., Palmer, A. E. and Tsien, R. Y. (2004) Improved monomeric red, orange and yellow fluorescent proteins derived from Discosoma *sp*. red fluorescent protein, *Nat Biotechnol*, **22**(12), 1567–1572.

73. Hughes, D. P., Marron, M. B. and Brindle, N. P. (2003) The antiinflammatory endothelial tyrosine kinase Tie2 interacts with a novel nuclear factor-kappaB inhibitor ABIN-2, *Circ Res*, **92**(6), 630–636.

74. Motoike, T., Loughna, S., Perens, E., Roman, B. L., Liao, W., Chau, T. C., Richardson, C. D., Kawate, T., Kuno, J., Weinstein, B. M., Stainier, D. Y. and Sato, T. N. (2000) Universal GFP reporter for the study of vascular development, *Genesis*, **28**(2), 75–81.

75. Amoh, Y., Li, L., Yang, M., Jiang, P., Moossa, A. R., Katsuoka, K. and Hoffman, R. M. (2005) Hair follicle-derived blood vessels vascularize tumors in skin and are inhibited by Doxorubicin, *Cancer Res*, **65**(6), 2337–2343.

76. Amoh, Y., Bouvet, M., Li, L., Tsuji, K., Moossa, A. R., Katsuoka, K. and Hoffman, R. M. (2006) Visualization of nascent tumor angiogenesis in lung and liver metastasis by differential dual-color fluorescence imaging in nestin-linked-GFP mice, *Clin Exp Metastasis*, **23**(7–8), 315–322.

77. Amoh, Y., Yang, M., Li, L., Reynoso, J., Bouvet, M., Moossa, A. R., Katsuoka, K. and Hoffman, R. M. (2005) Nestin-linked green fluorescent protein transgenic nude mouse for imaging human tumor angiogenesis, *Cancer Res*, **65**(12), 5352–5357.

78. Hoffman, R. M. and Yang, M. (2006) Color-coded fluorescence imaging of tumor-host interactions, *Nat Protoc*, **1**(2), 928–935.

79. Hayashi, K., Yamauchi, K., Yamamoto, N., Tsuchiya, H., Tomita, K., Amoh, Y., Hoffman, R. M. and Bouvet, M. (2007) Dual-color imaging of angiogenesis and its inhibition in bone and soft tissue sarcoma, *J Surg Res*, **140**(2), 165–170.

80. Yang, M., Li, L., Jiang, P., Moossa, A. R., Penman, S. and Hoffman, R. M. (2003) Dual-color fluorescence imaging distinguishes tumor cells from induced host angiogenic vessels and stromal cells, *Proc Natl Acad Sci U S A*, **100**(24), 14259–14262.

81. Cassot, F., Lauwers, F., Fouard, C., Prohaska, S. and Lauwers-Cances, V. (2006) A novel three-dimensional computer-assisted method for a quantitative study of microvascular networks of the human cerebral cortex, *Microcirculation*, **13**(1), 1–18.

82. Tata, D. A. and Anderson, B. J. (2002) A new method for the investigation of capillary structure, *J Neurosci Methods*, **113**(2), 199–206.

83. Piston, D. W. (1999) Imaging living cells and tissues by two-photon excitation microscopy, *Trends Cell Biol*, **9**(2), 66–69.

84. Centonze, V. E. and White, J. G. (1998) Multiphoton excitation provides optical sections from deeper within scattering specimens than confocal imaging, *Biophys J*, **75**(4), 2015–2024.

85. Theer, P., Hasan, M. T. and Denk, W. (2003) Two-photon imaging to a depth of 1000 micron in living brains by use of a Ti:Al2O3 regenerative amplifier, *Opt Lett*, **28**(12), 1022–1024.

86. Vajkoczy, P., Ullrich, A. and Menger, M. D. (2000) Intravital fluorescence videomicroscopy to study tumor angiogenesis and microcirculation, *Neoplasia*, **2**(1–2), 53–61.

87. Tozer, G. M., Ameer-Beg, S. M., Baker, J., Barber, P. R., Hill, S. A., Hodgkiss, R. J., Locke, R., Prise, V. E., Wilson, I. and Vojnovic, B. (2005) Intravital imaging of tumour vascular networks using multi-photon fluorescence microscopy, *Adv Drug Deliv Rev*, **57**(1), 135–152.

88. Sevick-Muraca, E. M., Houston, J. P. and Gurfinkel, M. (2002) Fluorescence-enhanced, near infrared diagnostic imaging with contrast agents, *Curr Opin Chem Biol*, **6**(5), 642–650.

89. Ntziachristos, V., Bremer, C. and Weissleder, R. (2003) Fluorescence imaging with near-infrared light: new technological advances that enable *in vivo* molecular imaging, *Eur Radiol*, **13**(1), 195–208.

90. Hsu, A. R., Hou, L. C., Veeravagu, A., Greve, J. M., Vogel, H., Tse, V. and Chen, X. (2006) *In vivo* near-infrared fluorescence imaging of integrin alphavbeta3 in an orthotopic glioblastoma model, *Mol Imaging Biol*, **8**(6), 315–323.

91. Bogdanov, A. A., Jr., Lin, C. P. and Kang, H. W. (2007) Optical imaging of the adoptive transfer of human endothelial cells in mice using anti-human CD31 monoclonal antibody, *Pharm Res*, **24**(6), 1186–1192.

92. Tsourkas, A., Shinde-Patil, V. R., Kelly, K. A., Patel, P., Wolley, A., Allport, J. R. and Weissleder, R. (2005) *In vivo* imaging of activated endothelium using an anti-VCAM-1 magnetooptical probe, *Bioconjug Chem*, **16**(3), 576–581.

93. Kelly, K. A., Allport, J. R., Tsourkas, A., Shinde-Patil, V. R., Josephson, L. and Weissleder, R. (2005) Detection of vascular adhesion molecule-1 expression using a novel multimodal nanoparticle, *Circ Res*, **96**(3), 327–336.

94. Wu, Y., Cai, W. and Chen, X. (2006) Near-infrared fluorescence imaging of tumor integrin alpha v beta 3 expression with Cy7-labeled RGD multimers, *Mol Imaging Biol*, **8**(4), 226–236.

95. Cheng, Z., Wu, Y., Xiong, Z., Gambhir, S. S. and Chen, X. (2005) Near-infrared fluorescent RGD peptides for optical imaging of integrin alphavbeta3 expression in living mice, *Bioconjug Chem*, **16**(6), 1433–1441.

96. Funovics, M., Montet, X., Reynolds, F., Weissleder, R. and Josephson, L. (2005) Nanoparticles for the optical imaging of tumor E-selectin, *Neoplasia*, **7**(10), 904–911.

97. Alencar, H., Mahmood, U., Kawano, Y., Hirata, T. and Weissleder, R. (2005) Novel multiwavelength microscopic scanner for mouse imaging, *Neoplasia*, **7**(11), 977–983.

98. Montet, X., Figueiredo, J. L., Alencar, H., Ntziachristos, V., Mahmood, U. and Weissleder, R. (2007) Tomographic fluorescence imaging of tumor vascular volume in mice, *Radiology*, **242**(3), 751–758.

99. Lin, K. Y., Maricevich, M., Bardeesy, N., Weissleder, R. and Mahmood, U. (2008) *In vivo* quantitative microvasculature phenotype imaging of healthy and malignant tissues using a fiber-optic confocal laser microprobe, *Transl Oncol*, **1**(2), 84–94.

100. Jaiswal, J. K. and Simon, S. M. (2004) Potentials and pitfalls of fluorescent quantum dots for biological imaging, *Trends Cell Biol*, **14**(9), 497–504.

101. Zhou, M. and Ghosh, I. (2007) Quantum dots and peptides: a bright future together, *Biopolymers*, **88**(3), 325–339.

102. Gao, X., Cui, Y., Levenson, R. M., Chung, L. W. and Nie, S. (2004) *In vivo* cancer targeting and imaging with semiconductor quantum dots, *Nat Biotechnol*, **22**(8), 969–976.

103. Jayagopal, A., Russ, P. K. and Haselton, F. R. (2007) Surface engineering of quantum dots for *in vivo* vascular imaging, *Bioconjug Chem*, **18**(5), 1424–1433.

104. Ferrara, D. E., Weiss, D., Carnell, P. H., Vito, R. P., Vega, D., Gao, X., Nie, S. and Taylor, W. R. (2006) Quantitative 3D fluorescence technique for the analysis of en face preparations of arterial walls using quantum dot nanocrystals and two-photon excitation laser scanning microscopy, *Am J Physiol Regul Integr Comp Physiol*, **290**(1), R114–R123.

105. Akerman, M. E., Chan, W. C., Laakkonen, P., Bhatia, S. N. and Ruoslahti, E. (2002) Nanocrystal targeting *in vivo*, *Proc Natl Acad Sci USA* **99**(20), 12617–12621.

106. Cai, W., Shin, D. W., Chen, K., Gheysens, O., Cao, Q., Wang, S. X., Gambhir, S. S. and Chen, X. (2006) Peptide-labeled near-infrared quantum dots for imaging tumor vasculature in living subjects, *Nano Lett*, **6**(4), 669–676.

107. Cai, W., Chen, K., Li, Z. B., Gambhir, S. S. and Chen, X. (2007) Dual-function probe for PET and near-infrared fluorescence imaging of tumor vasculature, *J Nucl Med*, **48**(11), 1862–1870.

108. Smith, J. D., Fisher, G. W., Waggoner, A. S. and Campbell, P. G. (2007) The use of quantum dots for analysis of chick CAM vasculature, *Microvasc Res*, **73**(2), 75–83.

109. Larson, D. R., Zipfel, W. R., Williams, R. M., Clark, S. W., Bruchez, M. P., Wise, F. W. and Webb, W. W. (2003) Water-soluble quantum dots for multiphoton fluorescence imaging *in vivo*, *Science*, **300**(5624), 1434–1436.

110. Leng, T., Miller, J. M., Bilbao, K. V., Palanker, D. V., Huie, P. and Blumenkranz, M. S. (2004) The chick chorioallantoic membrane as a model tissue for surgical retinal research and simulation, *Retina*, **24**(3), 427–434.

111. Valdes, T. I., Kreutzer, D. and Moussy, F. (2002) The chick chorioallantoic membrane as a novel *in vivo* model for the testing of biomaterials, *J Biomed Mater Res*, **62**(2), 273–282.

112. Sato, A., Klaunberg, B. and Tolwani, R. (2004) *In vivo* bioluminescence imaging, *Comp Med*, **54**(6), 631–634.

113. Liu, J., Barradas, A., Fernandes, H., Janssen, F., Papenburg, B., Stamatialis, D., Martens, A. C., van Blitterswijk, C. A. and De Boer, J. (2009) *In vitro* and *in vivo* bioluminescence imaging of hypoxia in tissue engineered grafts, *Tissue Eng Part C Methods*, **16**(3):479–485.

114. Hoffman, R. (2002) Green fluorescent protein imaging of tumour growth, metastasis, and angiogenesis in mouse models, *Lancet Oncol*, **3**(9), 546–556.

115. Burgos, J. S., Rosol, M., Moats, R. A., Khankaldyyan, V., Kohn, D. B., Nelson, M. D., Jr. and Laug, W. E. (2003) Time course of bioluminescent signal in orthotopic and heterotopic brain tumors in nude mice, *Biotechniques*, **34**(6), 1184–1188.

116. Zhang, N., Fang, Z., Contag, P. R., Purchio, A. F. and West, D. B. (2004) Tracking angiogenesis induced by skin wounding and contact hypersensitivity using a Vegfr2-luciferase transgenic mouse, *Blood*, **103**(2), 617–626.

117. Dong, Z., Neiva, K. G., Jin, T., Zhang, Z., Hall, D. E., Mooney, D. J., Polverini, P. J. and Nor, J. E. (2007) Quantification of human angiogenesis in immunodeficient mice using a photon counting-based method, *Biotechniques*, **43**(1), 73–77.

118. Sanz, L., Santos-Valle, P., Alonso-Camino, V., Salas, C., Serrano, A., Vicario, J. L., Cuesta, A. M., Compte, M., Sanchez-Martin, D. and Alvarez-Vallina, L. (2008) Long-term *in vivo* imaging of human angiogenesis: critical role of bone marrow-derived mesenchymal stem cells for the generation of durable blood vessels, *Microvasc Res*, **75**(3), 308–314.

119. Roman, I., Vilalta, M., Rodriguez, J., Matthies, A. M., Srouji, S., Livne, E., Hubbell, J. A., Rubio, N. and Blanco, J. (2007) Analysis of progenitor cell-scaffold combinations by *in vivo* non-invasive photonic imaging, *Biomaterials*, **28**(17), 2718–2728.

120. Heinzer, S., Krucker, T., Stampanoni, M., Abela, R., Meyer, E. P., Schuler, A., Schneider, P. and Müller, R. (2006) Hierarchical microimaging for multiscale analysis of large vascular networks, *Neuroimage*, **32**(2), 626–636.

121. Jorgensen, S. M., Demirkaya, O. and Ritman, E. L. (1998) Three-dimensional imaging of vasculature and parenchyma in intact rodent organs with X-ray micro-CT, *Am J Physiol*, **275**(3 Pt 2), H1103–H1114.

122. Bentley, M. D., Ortiz, M. C., Ritman, E. L. and Romero, J. C. (2002) The use of microcomputed tomography to study microvasculature in small rodents, *Am J Physiol Regul Integr Comp Physiol*, **282**(5), R1267–R1279.

123. Plouraboue, F., Cloetens, P., Fonta, C., Steyer, A., Lauwers, F. and Marc-Vergnes, J. P. (2004) X-ray high-resolution vascular network imaging, *J Microsc*, **215**(Pt 2), 139–148.

124. Kiessling, F., Greschus, S., Lichy, M. P., Bock, M., Fink, C., Vosseler, S., Moll, J., Mueller, M. M., Fusenig, N. E., Traupe, H. and Semmler, W. (2004) Volumetric computed tomography (VCT): a new technology for noninvasive, high-resolution monitoring of tumor angiogenesis, *Nat Med*, **10**(10), 1133–1138.

125. Marxen, M., Thornton, M. M., Chiarot, C. B., Klement, G., Koprivnikar, J., Sled, J. G. and Henkelman, R. M. (2004) MicroCT scanner performance and considerations for vascular specimen imaging, *Med Phys*, **31**(2), 305–313.

126. Rabin, O., Manuel Perez, J., Grimm, J., Wojtkiewicz, G. and Weissleder, R. (2006) An X-ray computed tomography imaging agent based on long-circulating bismuth sulphide nanoparticles, *Nat Mater*, **5**(2), 118–122.

127. Hainfeld, J. F., Slatkin, D. N., Focella, T. M. and Smilowitz, H. M. (2006) Gold nanoparticles: a new X-ray contrast agent, *Br J Radiol*, **79**(939), 248–253.

128. Hasebroock, K. M. and Serkova, N. J. (2009) Toxicity of MRI and CT contrast agents, *Expert Opin Drug Metab Toxicol*, **5**(4), 403–416.

129. Duvall, C. L., Taylor, W. R., Weiss, D., Wojtowicz, A. M. and Guldberg, R. E. (2007) Impaired angiogenesis, early callus formation, and late stage remodeling in fracture healing of osteopontin-deficient mice, *J Bone Miner Res*, **22**(2), 286–297.

130. Toyota, E., Fujimoto, K., Ogasawara, Y., Kajita, T., Shigeto, F., Matsumoto, T., Goto, M. and Kajiya, F. (2002) Dynamic changes in three-dimensional architecture and vascular volume of transmural coronary microvasculature between diastolic- and systolic-arrested rat hearts, *Circulation*, **105**(5), 621–626.

131. Beckmann, N., Schuler, A., Mueggler, T., Meyer, E. P., Wiederhold, K. H., Staufenbiel, M. and Krucker, T. (2003) Age-dependent cerebrovascular

abnormalities and blood flow disturbances in APP23 mice modeling Alzheimer's disease, *J Neurosci*, **23**(24), 8453–8459.

132. Krucker, T., Lang, A. and Meyer, E. P. (2006) New polyurethane-based material for vascular corrosion casting with improved physical and imaging characteristics, *Microsc Res Tech*, **69**(2), 138–147.

133. Reina-De La Torre, F., Rodriguez-Baeza, A. and Sahuquillo-Barris, J. (1998) Morphological characteristics and distribution pattern of the arterial vessels in human cerebral cortex: a scanning electron microscope study, *Anat Rec*, **251**(1), 87–96.

134. Minnich, B., Bartel, H. and Lametschwandtner, A. (2002) How a highly complex three-dimensional network of blood vessels regresses: the gill blood vascular system of tadpoles of Xenopus during metamorphosis. A SEM study on microvascular corrosion casts, *Microvasc Res*, **64**(3), 425–437.

135. Komlev, V. S., Mastrogiacomo, M., Peyrin, F., Cancedda, R. and Rustichelli, F. (2009) X-ray synchrotron radiation pseudo-holotomography as a new imaging technique to investigate angio- and microvasculogenesis with no usage of contrast agents, *Tissue Eng Part C Methods*, **15**(3), 425–430.

136. Laking, G. R. and Price, P. M. (2003) Positron emission tomographic imaging of angiogenesis and vascular function, *Br J Radiol*, **76** Spec No 1, S50–S59.

137. Chatziioannou, A. F. (2002) Molecular imaging of small animals with dedicated PET tomographs, *Eur J Nucl Med Mol Imaging*, **29**(1), 98–114.

138. Kiessling, F., Jugold, M., Woenne, E. C. and Brix, G. (2007) Non-invasive assessment of vessel morphology and function in tumors by magnetic resonance imaging, *Eur Radiol*, **17**(8), 2136–2148.

139. van Vliet, M., van Dijke, C. F., Wielopolski, P. A., ten Hagen, T. L., Veenland, J. F., Preda, A., Loeve, A. J., Eggermont, A. M. and Krestin, G. P. (2005) MR angiography of tumor-related vasculature: from the clinic to the micro-environment, *Radiographics*, **25** Suppl 1, S85–S97; discussion S97–S88.

140. Bolan, P. J., Yacoub, E., Garwood, M., Ugurbil, K. and Harel, N. (2006) *In vivo* micro-MRI of intracortical neurovasculature, *Neuroimage*, **32**(1), 62–69.

141. Xu, H., Othman, S. F., Hong, L., Peptan, I. A. and Magin, R. L. (2006) Magnetic resonance microscopy for monitoring osteogenesis in tissue-engineered construct *in vitro*, *Phys Med Biol*, **51**(3), 719–732.

142. Elliott, M. R. and Thrush, A. J. (1996) Measurement of resolution in intravascular ultrasound images, *Physiol Meas*, **17**(4), 259–265.

143. Shampo, M. A. and Kyle, R. A. (1995) Karl Theodore Dussik---pioneer in ultrasound, *Mayo Clin Proc*, **70**(12), 1136.

144. Gramiak, R. and Shah, P. M. (1968) Echocardiography of the aortic root, *Radiology*, **3**, 356–366.

145. Dayton, P. A. and Rychak, J. J. (2007) Molecular ultrasound imaging using microbubble contrast agents, *Front Biosci*, **12**, 5124–5142.

146. Bushberg, J. T., Seibert, J. A., Leidholdt, E. M. and Boone, J. M. (2002) *The Essential Physics of Medical Imaging*, Lippincott Williams & Wilkins, Philadelphia.

147. Kaufmann, B. A. and Lindner, J. R. (2007) Molecular imaging with targeted contrast ultrasound, *Curr Opin Biotechnol*, **18**(1), 11–16.

148. Kaufmann, B. A., Wei, K. and Lindner, J. R. (2007) Contrast echocardiography, *Curr Probl Cardiol*, **32**(2), 51-96.

149. Unger, E. C., Porter, T., Culp, W., Labell, R., Matsunaga, T. and Zutshi, R. (2004) Therapeutic applications of lipid-coated microbubbles, *Adv Drug Deliv Rev*, **56**(9), 1291–1314.

150. Lindner, J. R. (2002) Evolving applications for contrast ultrasound, *Am J Cardiol*, **90**(10A), 72J–80J.

151. Weller, G. E., Wong, M. K., Modzelewski, R. A., Lu, E., Klibanov, A. L., Wagner, W. R. and Villanueva, F. S. (2005) Ultrasonic imaging of tumor angiogenesis using contrast microbubbles targeted via the tumor-binding peptide arginine-arginine-leucine, *Cancer Res*, **65**(2), 533–539.

152. Weller, G. E., Villanueva, F. S., Klibanov, A. L. and Wagner, W. R. (2002) Modulating targeted adhesion of an ultrasound contrast agent to dysfunctional endothelium, *Ann Biomed Eng*, **30**(8), 1012–1019.

153. Villanueva, F. S., Klibanov, A. and Wagner, W. R. (2002) Microbubble-endothelial cell interactions as a basis for assessing endothelial function, *Echocardiography*, **19**(5), 427–438.

154. Dayton, P. A., Pearson, D., Clark, J., Simon, S., Schumann, P. A., Zutshi, R., Matsunaga, T. O. and Ferrara, K. W. (2004) Ultrasonic analysis of peptide- and antibody-targeted microbubble contrast agents for molecular imaging of alphavbeta3-expressing cells, *Mol Imaging*, **3**(2), 125–134.

155. Korpanty, G., Carbon, J. G., Grayburn, P. A., Fleming, J. B. and Brekken, R. A. (2007) Monitoring response to anticancer therapy by targeting microbubbles to tumor vasculature, *Clin Cancer Res*, **13**(1), 323–330.

156. Brooks, P. C., Clark, R. A. and Cheresh, D. A. (1994) Requirement of vascular integrin alpha v beta 3 for angiogenesis, *Science*, **264**(5158), 569–571.

157. Stieger, S. M., Bloch, S. H., Foreman, O., Wisner, E. R., Ferrara, K. W. and Dayton, P. A. (2006) Ultrasound assessment of angiogenesis in a matrigel model in rats, *Ultrasound Med Biol*, **32**(5), 673–681.

158. Cosgrove, D. (2003) Angiogenesis imaging—ultrasound, *Br J Radiol*, **76 Spec No 1**, S43–S49.

159. Goh, V., Padhani, A. R. and Rasheed, S. (2007) Functional imaging of colorectal cancer angiogenesis, *Lancet Oncol*, **8**(3), 245–255.

## Chapter 17

# Tissue Engineering: Scope, Products, and Commercialization Strategies

**Arthur J. Coury**
*Coury Consulting Services, Boston, MA 02116, USA*
artjcoury@aol.com

This chapter provides information on the development of products that use tissue engineering for therapeutic purposes. It proceeds with rationale for the premise that tissue engineering is a broad segment of medical practice, which has enjoyed much success in its application as medical devices, combination products, and pharmacologic interventions as well as cell-based products. It embraces the fact that the nature of therapeutic medicine is shifting from replacement medicine to tissue engineering and that the forces impacting medical product development are changing. Therefore, successful product development strategies apply decades of research, development, manufacturing, regulatory, marketing, and other experiences while adapting to the ever-changing environment in effective ways.

*Scaffolds for Tissue Engineering: Biological Design, Materials, and Fabrication*
Edited by Claudio Migliaresi and Antonella Motta

ISBN 978-981-4463-20-1 (Hardcover), 978-981-4463-21-8 (eBook)
www.panstanford.com

## 17.1 Concepts and Scope of Tissue Engineering

The concept of tissue engineering has been interpreted in scope from rather narrow to quite broad. However, most of the definitions from the earliest days of tissue engineering reporting allow a broad range of therapeutic approaches. For example, an early definition by pioneers in the field, Vacanti and Langer is stated as: "Tissue engineering is an interdisciplinary field that applies the principles of engineering and the life sciences toward the development of biological substitutes that restore, maintain, or improve tissue function."[1] Williams' current definition focusing broadly on cell control, states: "Tissue engineering is the creation of new tissue for the therapeutic reconstruction of the human body, by the deliberate and controlled stimulation of selected target cells through a systematic combination of molecular and mechanical signals."[2] Since 2002, I have used the definition: "The generation, regeneration, augmentation or limitation of the structure and function of living tissues by the application of scientific and engineering principles."[3] In its simplest manifestation, tissue engineering is exerting systematic control of the body's cells matrices and fluids. Most importantly, none of these definitions limit the products to the use of exogenously manipulated cells in scaffolds or even such cells alone. Accordingly, this chapter incorporates into the field of tissue engineering the functions listed in Table 17.1 as an extensive, if incomplete list.

Products that effect tissue engineering functions can be categorized as: "Engineered tissue products" and "Tissue engineering products." The former category includes products that are formed outside the body, by processes that employ cells to produce enhanced cell numbers, cells with their intrinsic matrices, matrices produced by and isolated from cells, cells on scaffolds, or combinations of such. They also may include modified allogeneic or xenogeneic tissue. These "tissues" are then applied to the body or are used for diagnostic or testing purposes. Examples of engineered tissue products used clinically include autologous cultured cells for knee cartilage or dermal regeneration in culture or on a scaffold.[4] Other engineered tissue products used in clinical trials include autologous cell-based ureter segments and bladder constructs.[5] Tissue engineering products constitute a wide variety of medical products that affect the tissue engineering functions

**Table 17.1** Tissue engineering functions

| |
|---|
| Organ/tissue generation/regeneration |
| Guided tissue regeneration |
| Structural reinforcement via ingrowth |
| Tissue augmentation/space filling |
| Healing modulation |
| Hypertrophy inhibition |
| Fouling prevention |
| Tissue attachment/sealing |
| Surgical adhesion prevention |
| Enhanced mitosis/matrix deposition |
| Organ function modulation |
| Expression of bio-actives |

(Table 17.1) when applied to the body and include engineered tissue products (Table 17.2). An appropriate example of a non-cellular tissue engineering product is a modified hyaluronan film used for surgical abdominal adhesion prevention, which induces healing by guided tissue regeneration of mesothelium.[6] In this context, wound treatments ranging from bandages to occlusive dressings to surgical sutures and staples are tissue engineering products because they direct healing in favorable ways that would not have occurred in their absence. Taken together, an estimated 20–30% of the worldwide non-ethical drug medical products, which are worth $250 billion, are tissue engineering products. By this (albeit controversial) interpretation, the tissue engineering products industries have been very successful for several decades and continue to grow in favorable proportion to other medical products (e.g., non-tissue engineering replacement, adjunctive or delivery products, such as those listed in Table 17.3. Some products, which may or may not be tissue-engineering products depending on their function and tissue responses produced, are listed in Table 17.4.

**Table 17.2** Tissue engineering products

- Cell therapy (skin, bone, cartilage, etc.)*
- Gene therapy
- Dermal regeneration*
- Wound dressings
- Drug eluting stents
- Drug eluting pacemaker leads
- Sutures, staples
- Allogeneic grafts (soft, hard)*
- Natural, synthetic ingrowth scaffolds*
- Nerve regeneration sheaths*
- Adhesion prevention devices
- Tissue sealants/adhesives
- Periodontal, other guided tissue regeneration barriers
- Ophthalmic viscoelastics
- Soft tissue ingrowth-inducing devices (e.g., hernia mesh)
- Cementless hips, knees
- Bone fillers (alveolar, transplants, etc.)*
- Skin, bone expansion devices
- Dental braces
- Electronic bone growth stimulators
- Skin expanders
- Hydrocephalus shunts
- Spinal fusion products via bone generation (e.g., DBM, BMP products)
- Various resorbable implants*
- Combination products (device, drug biologics)*
- Drug eluting stents
- Drug eluting pacing leads
- Various drug-device combinations
- Porous ingrowth structural implants

*Some configurations may constitute engineered tissue.

**Table 17.3** Non–tissue engineering medical products

Replacement parts

    Smooth surfaced prosthetic hips, knees, shoulders

    Orthopedic plates, screws

Cranio-maxillofacial bone replacement

    Dental implants

| | |
|---|---|
| Dental crowns, fillings | |
| | Intraocular lens |
| | Auditory implants, hearing aids |
| | Vascular grafts |
| | Circulatory assist devices |
| | Heart valves |
| | "Bare metal" vascular stents |
| | Heart pacemakers |
| | Heart defibrillators |
| | Organ transplants |
| Adjunctive devices | |
| | Neurological stimulators |
| | Manual surgical instruments |
| | Robotic surgery instruments |
| | Endoscopic surgery instruments |
| All diagnostic devices | |

**Table 17.4** "Gray area" devices (tissue engineering or not, depending on application)

| | |
|---|---|
| Delivery devices | |
| | Indwelling catheters |
| | Drug pumps |
| | Percutaneous access devices |
| | Transcutaneous catheter access ports |
| | Osmotic pumps |
| | Sustained release composites |

## 17.2 Tissue Engineering Product Development

Medical products companies develop a wide variety of tissue engineering products according to general procedures used for regulated medical products, which involve designated stages of progression. Some regulations are common to all products of similar classes, although individualized approaches may be employed by the different companies to meet such requirements. Progression from early research to development stages requires management signoff at each stage and regulatory oversight commences when progress to development stages occurs.[8] The regulatory requirements associated with quality system, design control, and meeting various standards (e.g., within the international standard ISO 13485)[9] add most of the costs to the development program. A typical staged development process for tissue engineering products designated as medical devices is shown in Fig. 17.1.

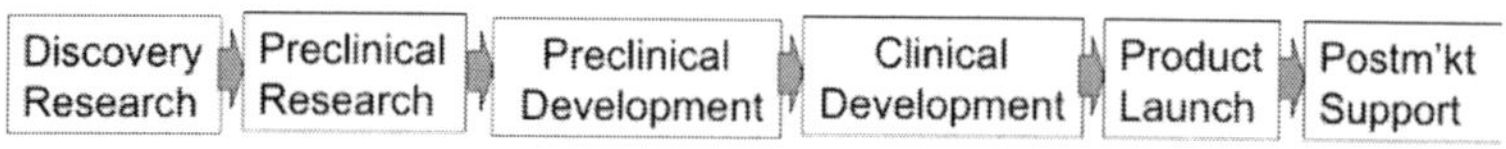


**Figure 17.1** Medical product development stages.

With tissue engineering products, as with medical products in general, certain criteria must be met to achieve a commercially successful product. Table 17.5 lists major variables that should be evaluated for potential success before a promising concept proceeds to the product development stage. Projected failure to achieve any of the goals associated with the listed variables would place the desired outcome at serious risk.

**Table 17.5** Medical product development: Justification criteria

| |
|---|
| Mission, vision, culture |
| Product mix strategy |
| Fiscal "health" (finances, stock movements, etc.) |
| Maturity: Start-up mature |
| Potential adoption curve |
| Market potential |
| Reimbursement potential |
| Intellectual property, internal, or external |
| Competition |
| Project champion |
| Staff capabilities: Lab, clinical, manufacturing (scalability), sales, marketing, etc. |
| Availability of staff for project, staff mix |
| Availability of materials, products from outside |
| Physical plant: Equipment, space, etc. |
| Project costs |
| Timing considerations: Handoffs, funding, regulatory, clinical, marketing, etc. |

Publication strategy
Business climate: Economic, societal, regulatory, liability, image, morale, etc.
Exit or survival strategy

Another process for deciding whether proceeding to product development is justified is to compare the attributes of the product in consideration to one that has already been successfully or unsuccessfully introduced to the market. The commercial product can be similar or dissimilar to the one being evaluated for development, but should provide insights into its potential for commercial success. Table 17.6 lists comparative variables that are relevant for product comparison. The exercise should help to guide a decision to proceed based on specifications such as minimum market size, adoption rate, etc., that are pre-set or determined from the comparison.

**Table 17.6** Strategic product comparisons

| **Comparing two products** | |
|---|---|
| Market | # Patients |
| Product Sophistication | High-Low |
| Development Costs | $$$-$ |
| Regulatory Pathway | PMA-501K |
| Implementation | Elective-Imperative |
| Utilization | % of Market |
| Unit Price | $ |
| Market Potential | $ |
| New Product Intro | Timing, Adoption |
| Implications | Product Improvements, Pipeline, etc. |

In Fig. 17.2, the two shaded areas below the product adoption curve represent the acceptance of a non-successful and a successful product. For the successful product, the "hockey stick" shape of the curve at early market times indicates slow acceptance of the product at introduction, followed by faster adoption at later stages with proof of its value verified. After revenues peak, competition or other factors cause a reduction in revenues until obsolescence causes the product to be withdrawn. The unsuccessful product never survived slow adoption and was withdrawn at an early stage.

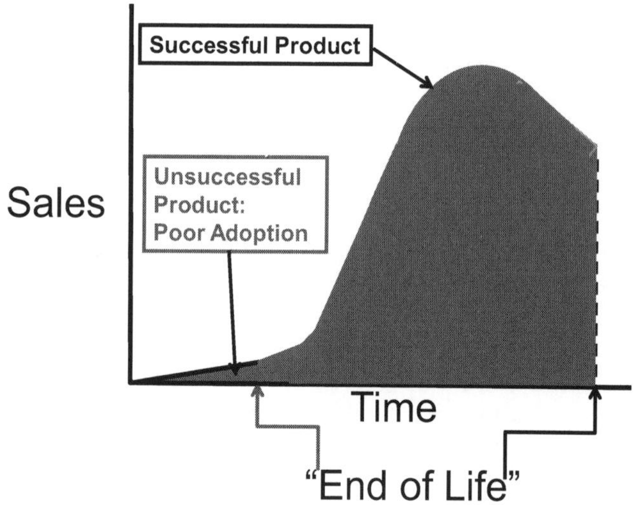


**Figure 17.2** Product adoption curve.

### 17.2.1 All-Important Resourcing during Product Development

The tissue engineering researcher may be adequately resourced with funds, personnel, and infrastructure to develop a concept to the stage of "proof of principle." The project may be still in the "discovery research" or "preclinical research" stage (Fig. 17.1). Thereafter, the costs of the project generally ramp up and achievement of further resourcing may become more difficult. When consideration of the product development justification criteria produces the signal to proceed, development of a strategy to further resource the project is critical. In a common scenario, "bridge funding" is necessary to develop the product to the stage where a capable entity (e.g., venture capitalist or for-profit corporation) can fully resource the product to market release. Failure to meet this requirement can stop the development program in its tracks, and this situation is sometimes referred to as "the valley of death" as illustrated by Fig. 17.3.

An example is provided in which it is assumed that a combination product based on a previously approved biologic and matrix composition will be developed for tissue engineering at a total cost of $200 million. This cost is intermediate between the recently estimated development cost of a new chemical entity ($4–11 billion)[10] and a medical device ($50–100 million).[11,12] The academic or corporate researcher had $2–3 million to advance the project to "proof of principle" during discovery research and must advance the project to the transition to preclinical research, whereupon a corporate entity will acquire the program and develop it to market and beyond.

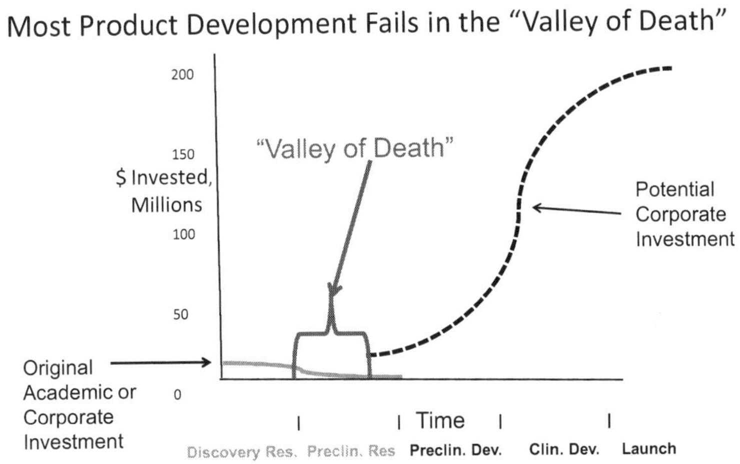


**Figure 17.3** "Valley of death" for tissue engineering products.

Overcoming this obstacle and achieving bridge funding from one or several sources, such as government, corporate, venture, angel, non-profit, or even personal sources, inevitably requires a "champion" who must exert entrepreneurial skills to influence the funding sources, which must be available during and after the "bridging" stage as shown in Fig. 17.4.

The mindset of those advancing the program should be of a "survivalist" nature, meaning that there is an appreciation that maximum progress must be shown with the limited resources at hand to assure continuing funding through the bridging stage and thereafter. Figure 17.5 illustrates two rates of progress and the value of urgency and efficiency in project execution. With the "fast" rate of progress, the research team reached the preclinical research milestone in a shorter time, potentially with less expenditure of initial resources (based largely on personnel costs). Depletion of funds did not reach a critical stage before assumption of costs by the corporate entity. Residual funds could be used to further accelerate the project or for other purposes. The product reached the market in a shorter time, although, in this example, overall costs were the same for the fast and slow progress scenarios. A crucial consideration is the requirement that product development conforms to a quality system. In actuality, costs may have been reduced by efficient and urgent progress at all stages of development.

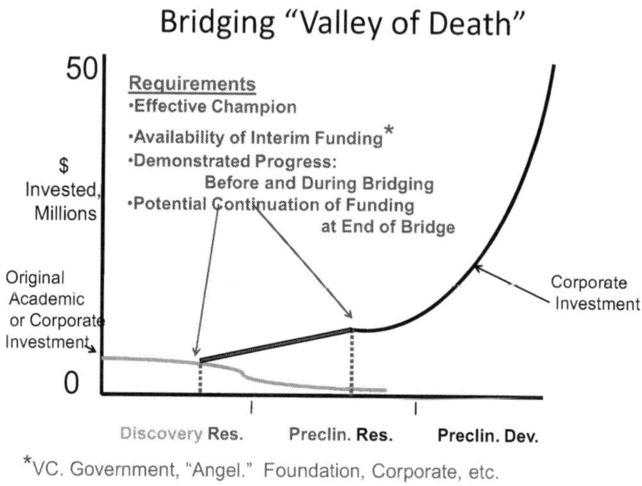


**Figure 17.4** Requirements for bridging the "valley of death".

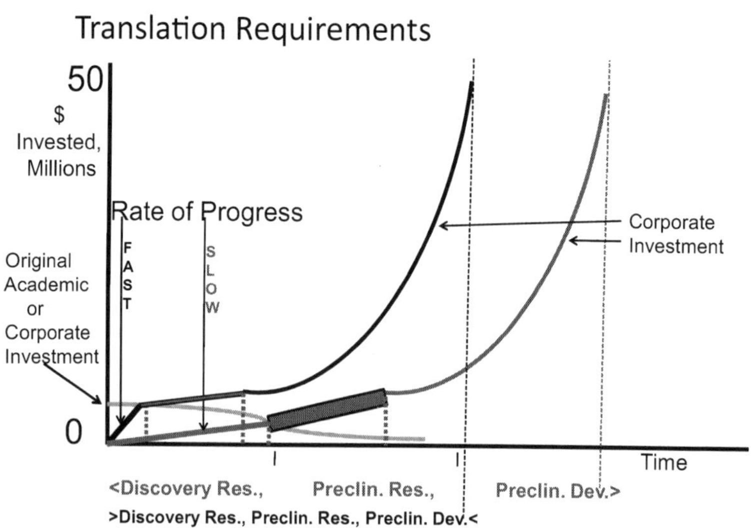


**Figure 17.5** Bridging the valley of death: Slow and fast progress rates.

## 17.3 Valuation of Product for Technology Transfer

Medical products are often developed to an early stage, such as "proof of concept," and sometimes patent applications are filed by academic investigators who, with the help of the university development office, seek to out-license the concept to a corporation. There is often a disparity between the academic and corporate sides opinion of the value of the license. Both sides generally agree that the further the development toward product, the more valuable the license. But often the academics believe the concept is farther along the development pathway than it actually is. Moreover, the concept may only provide a portion of the value of the final product. If royalties are involved, there may be several or many licenses associated with the product. The net income of any medical product corporation generally amounts to 20% or less annually, and any royalty comes right off the top. Moreover, development costs of the concept and associated risks assumed by the corporation usually involve many multiples of the academic investment. Many promising concepts never come to fruition because of the impasse in valuation between the university and the corporation. Table 17.7 lists some considerations in arriving to a realistic valuation of an academic license to a corporation.

**Table 17.7** Valuation of academic licenses

| **Comparing Academic vs. Corporate Investment** |
|---|
| • University vs. Corporate Expenditures |
| • Funding Sources, University vs. Company |
| • Contribution of University vs. Company to Intellectual Property |
| • Stage of Development at Time of Licensure |
| • Totality of Licenses for Product |
| • Projected Revenues and Profits at Time of Licensure |
| • Degree of Risk Assumption by Company |
| • Profitability of Product when Marketed |

## 17.4 Concluding Remarks

In this chapter, tissue engineering has been interpreted broadly in light of such allowances by most definitions of the concept. Products developed by tissue engineering, therefore, have been very successful, particularly over the last several decades, in generating new products and revenues. The trend is toward supplanting biomaterial hardware-based products that replace body parts and functions with technologies that restore health by exerting favorable control over the body's tissues (cells, matrices, and fluids).

Development of tissue engineering products as well as replacement medical products is often challenging and requires heroic efforts and a "survivalist," product-focused mentality to advance the products to the market. This is especially true if a new tissue engineering concept is being advanced toward a product in an academic laboratory or by a start-up company. Confidence in realistic chances for a successful product can be gained if the crucial medical product justification criteria are considered *a priori* and judged achievable. An effective, articulate "champion" is almost always required to advance the project technically, develop bridge funding, recruit strategic partners, and oversee the transitions under his or her control.

Human, economic, and equipment resourcing is essential in differing amounts throughout the product development process and is secured with more or less difficulty at different stages. Funding to bridge the "valley of death" between "proof of concept" and the product development stages may be especially difficult to obtain and may draw on all of the skills of the champion and efforts of the investigators. The farther and faster the development process goes, the earlier the entity supplying the final push to market will be recruited, and the higher the potential value to the licensor. That said, academic staff, investigators, and licensors are often unaware of the tremendous added costs of development driven by technical, regulatory, and market imperatives, and should assure that demands for compensation by the licensee are reasonable. A final observation: No matter how great the achievements of the research and development staff along the development pathway, majority of products never make it to market, do not become profitable, or change direction. However, I have found that no knowledge, experience, or reputation gained along the way is lost and can be exploited for future enterprises.

## References

1. Langer, R., Vacanti, J., *et al.* (1993) *Tissue Eng. Sci.*, 14 May, 920–926.
2. Williams, D. F. (2009) On the nature of biomaterials, *Biomaterials*, **30**, 5897–5909.
3. Hunziker, E., Spector, M., Libera, J., Gertzman, A., Woo, S. L., Ratcliffe, A., Lysaght, M., Coury, A., Kaplan, D., and Vunjak-Novakovic, G. (2006) Translation from research to applications, *Tissue Eng.*, **12**(12), 3341–3364.
4. Carticel® and Epicel® cartilage and skin regeneration products developed by Genzyme Corporation (now Sanofi Aventis). Carticel® Autologous Cultured Chondrocytes, © 2012 Genzyme Corporation, http://www.carticel.com/; Epicel® Cultured Epidermal Autografts: http://www.epicel.com/hcp/about/Epicel.aspx; "MACI®" Matrix-Induced Autologous Chondrocyte Implant© 2012 Genzyme Europe BV. http://www.maci.com/patient/maci-implant.aspx
5. Parson, A. (2006) Scientist at work: Anthony Atala—A tissue engineer sows cells and grows organs." Article first appeared in *The New York Times*, 7 November. Available at http://www.annparson.com/atala.htm
6. Seprafilm®, Adhesion Barrier developed by Genzyme Corporation (now Sanofi Aventis).
7. Markets http://www.espicom.com/Prodcat2.nsf/Product_ID_Lookup/00001500?OpenDocument *Epsicom Business Intelligence*, 12 May 12, 2012.
8. FDA Design Control Guidance for Medical Device Manufacturers: FDA 21 CFR 820.30 and Sub-clause 4.4 of ISO 9001 http://www.fda.gov/downloads/MedicalDevices/DeviceRegulationandGuidance/GuidanceDocuments/ucm070642.pdf
9. ISO 13485:2003 RSS Medical devices, Quality management systems—Requirements for regulatory purposes.
10. "The truly staggering cost of inventing new drugs," *Forbes* 2 October 2012 http://www.forbes.com/sites/matthewherper/2012/02/10/the-truly-staggering-cost-of-inventing-new-drugs/
11. Caplan, A., Baib, D., *et al.* (2004) *Circulation*, **109**, 3068–3072. doi: 10.1161/01.CIR.0000134695.65733.64
12. Estimate by Coury, A., 2012; based on summation of costs of medical device development stages.

# Index